Ingestion of Lead from Spent Ammunition:

Implications for Wildlife and Humans

Proceedings of the Conference
Ingestion of Spent Lead Ammunition:
Implications for Wildlife and Humans

12-15 May 2008, Boise State University,
Idaho, United States of America

Editors
Richard T. Watson
Mark Fuller, Mark Pokras
and Grainger Hunt

ISBN 0-9619839-5-7

Preferred Citation:
Watson, R. T., M. Fuller, M. Pokras, and W. G. Hunt (Eds.). Ingestion of Lead from Spent Ammunition: Implications for Wildlife and Humans. The Peregrine Fund, Boise, Idaho, USA.

Published by The Peregrine Fund,
5668 West Flying Hawk Lane, Boise, Idaho, USA.
www.peregrinefund.org

To view photos and figures in color, see
http://www.peregrinefund.org/Lead_conference/2008PbConf_Proceedings.htm

Cover Images

Front Cover, inset: A biologist radiographs a hunter-killed deer to count lead fragments and measure their dispersion from the bullet path. *Kathy Sullivan/Arizona Game and Fish Department.*

Background: Radiograph of bullet fragments in hunter-killed deer. *Oliver Krone/Leibniz-Institute for Zoo and Wildlife Research.*

Foreground: California Condors foraging on hunter-killed game are annually exposed to lead from spent ammunition. Fatal exposure levels occur often enough to prevent recovery of this critically endangered species.

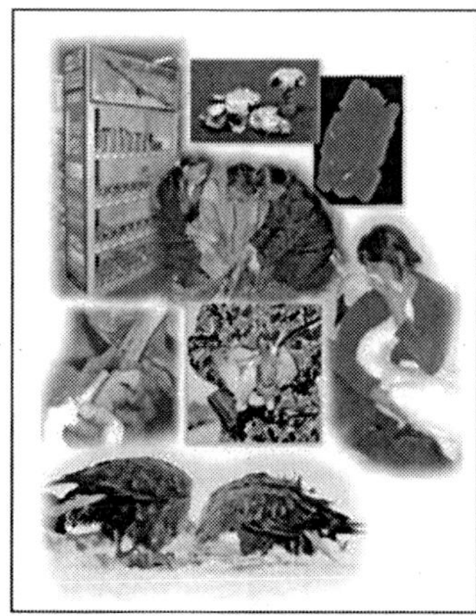

Back cover, clockwise from upper left:

A display features non-lead ammunition along with information for hunters at a sporting goods store in Arizona. *George Andrejko/Arizona Game and Fish Department.*

Lead bullets (left) are more likely to fragment and lose mass than solid copper bullets (right). *Chris Parish/The Peregrine Fund.*

Color encoded 3D reconstruction of CT scan data showing 20 packages of ground venison in red and metal fragments in white. The streak artifacts associated with the largest metal fragments in the dataset are a phenomenon common to larger metal objects in clinical CT imaging. *Edward Fogarty.*

A biologist massages the abdomen of a lead poisoned Trumpeter Swan to speed the passage of lead particles through the gastrointestinal tract to reduce lead absorption, while keeping the bird calm to reduce stress. *Raptor Education Group, Inc.*

White-tailed Sea Eagles feed on a carcass in Germany. *Oliver Krone/Leibniz-Institute for Zoo and Wildlife Research.*

Biologists collect a blood sample for lead analysis from a California Condor. *Arizona Game and Fish Department.*

Many Alaskans rely on gun-harvested wild game for a significant part of their total diet. *Susan Georgette/US Fish and Wildlife Service*

Upland game birds, including doves shown here, are commonly hunted with lead shot in the US. *Missouri Department of Conservation.*

Table of Contents

Preface
Richard T. Watson and W. Grainger Hunt 1

Introduction
Ian Newton 5

Review of Lead Uptake and Toxicosis in Humans and Wildlife

Understanding Lead Uptake and Effects across Species Lines: A Conservation Medicine Based Approach
Mark A. Pokras and Michelle R. Kneeland 7

History in Lead and Lead Poisoning in History
Jerome Nriagu 23

Health Effects of Low Dose Lead Exposure in Adults and Children, and Preventable Risk Posed by the Consumption of Game Meat Harvested with Lead Ammunition
Michael J. Kosnett 24

Biological and Societal Dimensions of Lead Poisoning in Birds in the USA
Milton Friend, J. Christian Franson, and William L. Anderson 34

Historical Perspective on the Hazards of Environmental Lead from Ammunition and Fishing Weights in Canada
Anton M. Scheuhammer 61

Technical Review of the Sources and Implications of Lead Ammunition and Fishing Tackle on Natural Resources
Barnett A. Rattner, J. Christian Franson, Steven R. Sheffield, Chris I. Goddard, Nancy J. Leonard, Douglas Stang, and Paul J. Wingate 68

Lead Poisoning in Wild Birds in Europe and the Regulations Adopted by Different Countries
Rafael Mateo 71

A Global Update of Lead Poisoning in Terrestrial Birds from Ammunition Sources
Deborah J. Pain, Ian J. Fisher, and Vernon G.Thomas 99

Gunshot Wounds: a Source of Lead in the Environment
Richard K. Stroud and W. Grainger Hunt 119

Human Exposure to Lead from Ammunition in the Circumpolar North
Lori A. Verbrugge, Sophie G. Wenzel, James E. Berner, and Angela C. Matz 126

Lead Exposure in Humans from Spent Ammunition

The Importance of Moose, Caribou, Deer and Small Game in the Diets of Alaskans
Kimberly Titus, Terry L. Haynes, and Thomas F. Paragi 137

Lead Bullet Fragments in Venison from Rifle-killed Deer: Potential for Human Dietary Exposure
W. Grainger Hunt, Richard T. Watson, J. Lindsay Oaks, Chris N. Parish, Kurt K. Burnham, Russell L. Tucker, James R. Belthoff, and Garret Hart 144

Qualitative and Quantitative Detection of Lead Bullet Fragments in Random Venison Packages Donated to the Community Action Food Centers of North Dakota, 2007
William E. Cornatzer, Edward F. Fogarty, and Eric W. Cornatzer 154

Distribution of Venison to Humanitarian Organizations in the USA and Canada
Dominique Avery and Richard T. Watson 157

Regulation of Lead-based Ammunition Around the World
Dominique Avery and Richard T. Watson 161

Hunters and Anglers at Risk of Lead Exposure in the United States
Richard T. Watson and Dominique Avery 169

Lead Exposure, Sources, and Toxicosis in Wildlife

Lead Isotopes Indicate Lead Shot Exposure in Alaska-breeding Waterfowl
Angela Matz and Paul Flint 174

Ingested Shot and Tissue Lead Concentrations in Mourning Doves
J. Christian Franson, Scott P. Hansen, and John H. Schulz 175

Acute Lead Toxicosis and Experimental Lead Pellet Ingestion in Mourning Doves
John H. Schulz, Xiaoming Gao, Joshua J. Millspaugh, and Alex J. Bermudez 187

Causes and Consequences of Ingested Lead Pellets in Chukars
R. Justin Bingham, Randy T. Larsen, John A. Bissonette, and Jerran T. Flinders 190

Lead Exposure in Wisconsin Birds
Sean M. Strom, Julie A. Langenberg, Nancy K. Businga, and Jasmine K. Batten 194

A Relationship Between Blood Lead Levels of Common Ravens and the Hunting Season in the Southern Yellowstone Ecosystem
Derek Craighead and Bryan Bedrosian 202

Lead Ingestion by Scavenging Mammalian Carnivores in the Yellowstone Ecosystem
Tom Rogers, Bryan Bedrosian, Derek Craighead, Howard Quigley, and Kerry Foresman 206

Potential Sources of Lead Exposure for Bald Eagles: a Retrospective Study
Patrick T. Redig, Donald R. Smith, and Luis Cruz-Martinez 208

Bald Eagle Lead Poisoning in Winter
Kay Neumann 210

Blood Lead Levels of Bald and Golden Eagles Sampled During and After Hunting Seasons in the Greater Yellowstone Ecosystem
Bryan Bedrosian and Derek Craighead 219

Blood-lead Levels of Fall Migrant Golden Eagles in West-central Montana
Robert Domenech and Heiko Langner 221

Survey of Lead Toxicosis in Free-ranging Raptors from Central Argentina
Miguel D. Saggese, Agustín Quaglia, Sergio A. Lambertucci, María S. Bo, José H. Sarasola, Roberto Pereyra- Lobos and Juan J. Maceda .223

Risk Assessment of Lead Poisoning in Raptors Caused by Recreational Shooting of Prairie Dogs
Robert M. Stephens, Aran S. Johnson, Regan E. Plumb, Kimberly Dickerson, Mark C. McKinstry, and Stanley H. Anderson 232

Lead in Griffon and Cinereous Vultures in Central Spain: Correlations Between Clinical Signs and Blood Lead Levels
Julia Rodriguez-Ramos, Valeria Gutierrez, Ursula Höfle, Rafael Mateo, Lidia Monsalve, Elena Crespo, and Juan Manuel Blanco 235

Long-term Effects of Lead Poisoning on Bone Mineralization in Egyptian Vulture *Neophron percnopterus*
Laura Gangoso, Pedro Álvarez-Lloret, Alejandro Rodríguez-Navarro, Rafael Mateo, Fernando Hiraldo, and José Antonio Donázar 237

Blood-lead Concentrations in California Condors Released at Pinnacles National Monument, California
James R. Petterson, Kelly J. Sorenson, Court VanTassell, Joe Burnett, Scott Scherbinski, Alacia Welch, and Sayre Flannagan 238

Blood Chemistry Values of California Condors Exposed to Lead
Molly Church, Karen Rosenthal, Donald R. Smith, Kathryn Parmentier, Ken Aron, and Dale Hoag 239

Effectiveness of Action to Reduce Exposure of Free-ranging California Condors in Arizona and Utah to Lead from Spent Ammunition
Rhys E. Green, W. Grainger Hunt, Christopher N. Parish, and Ian Newton 240

Bullet Fragments in Deer Remains: Implications for Lead Exposure in Scavengers
Grainger Hunt, William Burnham, Chris Parish, Kurt Burnham, Brian Mutch, and J. Lindsay Oaks 254

Lead Exposure Among a Reintroduced Population of California Condors in Northern Arizona and Southern Utah
Christopher N. Parish, W. Grainger Hunt, Edward Feltes, Ron Sieg, and Kathy Orr 259

Evidence for the Source of Lead Contamination within the California Condor
John Chesley, Peter Reinthal, Chris Parish, Kathy Sullivan, and Ron Sieg 265

Lead Intoxication Kinetics in Condors from California
Michael Fry, Kelly Sorenson, Jesse Grantham, Joseph Burnett, Joseph Brandt, and Michaela Koenig 266

Feather Pb Isotopes Reflect Exposure History and ALAD Inhibition Shows Sub-clinical Toxicity in California Condors
Kathryn Parmentier, Roberto Gwiazda, Joseph Burnett, Kelly Sorenson, Scott Scherbinski, Court VanTassell, Alacia Welch, Michaela Koenig, Joseph Brandt, James Petterson, Jesse Grantham, Robert Risebrough, and Donald Smith 267

Use of Machine Learning Algorithms to Predict the Incidence of Lead Exposure in Golden Eagles
Erica H. Craig, Tim H. Craig, Falk Huettmann, and Mark R. Fuller 269

Lead Shot Poisoning in Swans: Sources of Pellets within Whatcom County, WA, USA, and Sumas Prairie, BC, Canada
Michael C. Smith, Michael A. Davison, Cindy M. Schexnider, Laurie Wilson, Jennifer Bohannon, James M. Grassley, Donald K. Kraege, W. Sean Boyd, Barry D. Smith, Martha Jordan, and Christian Grue 274

Lead Poisoning of Trumpeter Swans in the Pacific Northwest: Can Recovered Shot Pellets Help to Elucidate the Source?
Laurie K. Wilson, Garry Grigg, Randy Forsyth, Monika Tolksdorf, Victoria Bowes, Michael Smith, and Anton Scheuhammer 278

Lead Objects Ingested by Common Loons in New England
Mark A. Pokras, Michelle R. Kneeland, Andrew Major, Rose Miconi, and Robert H. Poppenga 283

Difference Between Blood Lead Level Detection Techniques: Analysis Within and Among Three Techniques and Four Avian Species
Bryan Bedrosian, Chris N. Parish, and Derek Craighead 287

Remediation of Lead Exposure from Spent Ammunition

Lead Poisoning in White-tailed Sea Eagles: Causes and Approaches to Solutions in Germany
Oliver Krone, Norbert Kenntner, Anna Trinogga, Mirjam Nadjafzadeh, Friederike Scholz, Justine Sulawa, Katrin Totschek, Petra Schuck-Wersig, and Roland Zieschank 289

Lead Poisoning of Steller's Sea Eagle (*Haliaeetus pelagicus*) and White-tailed Eagle (*Haliaeetus albicilla*) Caused by the Ingestion of Lead Bullets and Slugs, in Hokkaido, Japan
Keisuke Saito 302

Success in Developing Lead-free, Expanding-nose Centerfire Bullets
Vic Oltrogge 310

Small Game Hunter Attitudes Toward Nontoxic Shot, and Crippling Rates with Nontoxic Shot
John H. Schulz, Ronald A. Reitz, Steven L. Sheriff, Joshua J. Millspaugh, and Paul I. Padding 316

Impacts of Lead Ammunition on Wildlife, the Environment, and Human Health—a Literature Review and Implications for Minnesota
Molly A. Tranel and Richard O. Kimmel 318

Policy Considerations for a Mourning Dove Nontoxic Shot Regulation
John H. Schulz, Joshua J. Millspaugh, and Larry D. Vangilder 338

Voluntary Lead Reduction Efforts Within the Northern Arizona Range of the California Condor
Ron Sieg, Kathy A. Sullivan, and Chris N. Parish 341

Taking the Lead on Lead: Tejon Ranch's Experience Switching to Non-lead Ammunition
Holly J. Hill 350

The Policy and Legislative Dimensions of Nontoxic Shot and Bullet Use in North America
Vernon G. Thomas 351

Commentaries on Research Needs and Remediation of Lead Exposure from Spent Ammunition in Wildlife and Humans

Commentary
John Freemuth 363

Commentary
Milton Friend 365

Commentary
Michael Kosnett 367

Commentary
Deborah Pain 369

Commentary
Mark Pokras 372

Commentary
Anton M. Scheuhammer 375

Commentary
Vernon G. Thomas 377

Commentary
Lori Verbrugge 379

Conference Summary

Summary of the Main Findings and Conclusions of the Conference "Ingestion of Spent Lead Ammunition: Implications for Wildlife and Humans"
Ian Newton 381

PREFACE

LEAD IS A POISONOUS METAL present in a variety of commercial products, as a pollutant from industrial activities, and as an environmental contaminant in many urban and rural habitats throughout the world. When ingested or inhaled, the body "mistakes" lead for calcium and other beneficial metals, and thus transports lead into nerve cells and other vital tissues. Consequences for wildlife include neural degeneration, modification of kidney structure and bone, inhibition of blood formation and nerve transmission, and numerous other harmful manifestations (Eisler 1988). Death can occur acutely, or the individual may become emaciated as a result of digestive paralysis (Locke and Thomas 1996). Clinical symptoms associated with blood lead concentrations exceeding one microgram per gram (one part per million) may include depression, lethargy, vomiting, diarrhea, nonregenerative anemia, anorexia, blindness, and seizures (Locke and Thomas 1996, Kramer and Redig 1997).

Mankind has long known about lead as an agent of sickness and death, but medical studies published in the last five to ten years are showing pernicious effects at unexpectedly low levels of exposure. There are permanent adverse impacts upon cognitive function and growth in children with histories of blood levels averaging less than one-tenth of one microgram per gram—a level formerly considered benign (Canfield et al. 2003, Hauser et al. 2008). Other newly-discovered manifestations of lead exposure include, for example, impaired motor function, decreased brain volume (Cecil et al. 2008), behavioral dysfunction, including criminality (Wright et al. 2008), and mortality from cancer and cardiovascular disease (Menke et al. 2006).

Scientific evidence of the effects of lead on human health has brought forth large scale restrictions on its use in the United States, including the prohibition of lead in many gasolines and paints. Responses on behalf of wildlife have been less forthcoming, but consumption of lead shot by ducks and geese and secondary poisoning of Bald Eagles contributed to the 1991 ban on lead shot for waterfowl hunting in the United States (United States Department of the Interior 1986). Other countries have instituted similar measures. Evidence of lead exposure in Arctic subsistence hunters continuing to use lead shot (Dewailly et al. 2001, Johansen et al. 2004) suggests that the ban on behalf of waterfowl and eagles has benefited humans as well.

Lead ammunition is still used in North America for purposes other than waterfowl harvest, and the extent to which lead is secondarily ingested by wildlife and humans has been the subject of some recent investigations. Mourning Doves, for example, confuse shotgun pellets for grit and grain around hunted stock ponds and accordingly die in large numbers (Schulz et al. 2002). Harmata and Restani (1995) found lead in the blood of 97% of 37 Bald Eagles and 85% of 86 Golden Eagles captured as spring migrants in Montana during 1985–1993; they implicated lead bullet fragments in ground-squirrel carcasses as one source. Pattee et al. (1990) reported that among 162 free-ranging Golden Eagles captured during 1985–86 in southern California, 36% had been exposed to lead. Six of nine dead eagles in Japan died of lead poisoning, and five had lead bullet fragments in their stomachs (Iwata et al. 2000). Lead ingestion was a principal cause of recorded death in wild California Condors during the 1980s when the population was brought into captivity (Wiemeyer et al. 1988).

Field studies by The Peregrine Fund from 2000 to the present show that ingestion of lead rifle bullet fragments and shotgun pellets from animal remains is likely the only significant obstacle to the establishment of the California Condor in the wilds of Arizona and Utah. Evidence includes (1) high rates of lead exposure and required treatment, (2) the presence of lead fragments or shot in radiographs of condors and their food, and (3) temporal and spatial associations of condors with the remains of gun-killed animals (Parish et al. 2007, Hunt et al. 2006, 2007, Cade 2007). During the 2006 hunting season,

90% of 57 free-ranging condors showed evidence of lead exposure, and four died of it, including a proven breeder almost 12 years old. This represents an 11% mortality rate for birds five years old or older, a meaningful consideration given the high sensitivity of condor populations to mortality within the older age categories. The additional unknown proportion of condors that would have died without treatment renders doubtful the survival of the species in the wild without continuing, intensive management.

Close monitoring of condors by radio tracking and blood testing, together with ancillary studies of lead prevalence in gun-killed deer and other animals, have produced new insights regarding the pervasive nature of lead contamination in scavenger food webs. One must now consider, on a global scale, the scope of sickness, death, and demographic impact inflicted upon a myriad of species by a contaminant now easily substituted with less toxic alternatives. The invention of highly efficacious non-lead bullets and pellets during recent decades parallels the discovery of lead's widespread impact on wildlife and coincides with additional studies documenting lead's effects on humans at even very low levels. It is evident that conditions now favor large scale mitigation.

An important step in understanding this problem is the gathering of relevant knowledge and scientific progress on these important topics. Nowhere would such an assembly of facts and interpretation be more useful than in the proceedings of a conference of world experts. Accordingly, The Peregrine Fund and organizing partners, Boise State University, US Geological Survey, and Tufts Center for Conservation Medicine, brought such a group together in the conference "**Ingestion of Spent Lead Ammunition: Implications for Wildlife and Humans**" held in Boise, Idaho, 12–15 May 2008.

The conference was attended by over 150 experts from around the world in the fields of human health, wildlife health and conservation, policy, and shooting sports. The resulting proceedings include 25 reviewed papers, 18 reviewed extended abstracts, five program abstracts, and nine expert commentaries including the conference summary by Professor Ian Newton.

Ingestion of Lead from Spent Ammunition: Implications for Wildlife and Humans, the proceedings of the conference, has been a collaborative effort. We thank our co-editors, Mark Fuller of the USGS and BSU Raptor Research Center, and Mark Pokras of Tufts University Veterinary School, Center for Conservation Medicine, for their critical partnership as peer editors of the proceedings and colleagues on the Scientific Program Committee of the conference. Their expert advice, experience in the field, and contacts over the year of work leading up to the conference were crucial for bringing together a unique diversity of professionals from fields that normally never meet. The depth, breadth, and diversity of conference participants were keys to its success.

We deliberated the choice between publishing papers in a recognized, abstracted scientific journal versus a book that may not receive the same level of readership and exposure.We chose to publish the proceedings as a book because we recognize the great value of the diversity of contributions from conference participants, and the likelihood that all would not meet the relevance criteria for inclusion in any one journal. The value of these proceedings resides in the mix of contributions that ranges from invited reviews to empirical studies, and a few anecdotal accounts. In format, contributions range from full papers to short notes, extended abstracts (with Tables, Figures, and citations), and program abstracts. Some extended abstracts have either been published elsewhere in full, or are slated for future publication with additional data. Three papers were reprinted with permission from the Wildlife Society Bulletin and PLoS ONE. The editorial team reviewed and edited all contributions for scientific and factual content. Papers beyond the technical and scientific expertise of the editorial team were sent out for external review. To expedite access to the important information in these papers, individual contributions were published separately online as soon as they were finalized, using the system of DOI numbers to permanently reference the early, online publications with the printed publication.

Readers of the vast literature on lead as an environmental contaminant encounter a confusing array of expressions pertaining to units of its concentration in blood and other tissues. Authors were asked to standardize expressions of blood concentration as µg/dL (micrograms per deciliter) and lead-in-tissue concentrations as µg/g (micrograms per gram). Where other units are used, readers can convert to these units by noting that a mass-to-volume blood lead level of 1 µg/dL is equivalent to 0.01 mg/l (milligrams per liter) or 0.048 µmol/l (micromoles per liter) or 10 ng/ml (nanograms per milliliter). For other tissues, 1 µg/g equals 1 ppm (part per million) or 1 mg/kg (milligram per kilogram). Help with other conversion issues can be obtained from http://www.onlineconversion.com/.

We thank the Conference Organizing Committee (Dominique Avery, Linda Behrman, Andrea Berkley, Kathy Bledsoe, Joell Brown, Patricia Burnham, Tom Cade, Bill Heinrich, J. Peter Jenny, Amy Siedenstrang, Rick Watson, Susan Whaley) and many volunteers and assistants for capably organizing the logistics of the conference. We especially thank David Dolton of the US Fish and Wildlife Service who took it upon himself to transcribe the comments of the expert panel, which we have included as contributions to the proceedings. We thank Amy and Mary Siedenstrang who typeset the text, tables, and graphics of the proceedings, and Terry Hunt who proofread the entire publication. We also express our sincere appreciation to the conference co-organizers, Boise State University, US Geological Survey, and Tufts Center for Conservation Medicine, and co-sponsors, the Charles Engelhard Foundation, National Park Service, Turner Foundation, and the National Aviary. Finally, we thank The Peregrine Fund, and in particular, Mr. and Mrs. Russell R. Wasendorf, Sr., for supporting the production of *Ingestion of Lead from Spent Ammunition: Implications for Wildlife and Humans*, the proceedings of the conference.

Ingestion of Lead from Spent Ammunition: Implications for Wildlife and Humans also would not have been possible without the enthusiastic participation of each and every contributor to the conference and proceedings. In the words of eminent scientist, Professor Ian Newton, "The meeting stands out among the many I have attended in the 45 years of my professional career as being one of a few really memorable and important ones in terms of both content and gravity of the problem discussed. I believe it could turn out, in the fullness of time, to have been a landmark conference, but this of course will be dependent on follow-up action from a range of people, including the public health delegates."

It is our hope that the findings and commentary brought forth during the conference and appearing in these proceedings will stimulate appropriate scientific and societal effort with respect to wildlife and human health.

Richard T. Watson, Ph.D.
Grainger Hunt, Ph.D.

The Peregrine Fund
Boise, Idaho
January 2009

LITERATURE CITED

CADE, T. J. 2007. Exposure of California Condors to lead from spent ammunition. Journal of Wildlife Management 71:2125–2133.

CANFIELD, R. L., C. R. HENDERSON, JR., D. A. CORY-SLECHTA, C. COX, T. A. JUSKO, AND B. P. LANPHEAR. 2003. Intellectual impairment in children with blood lead concentrations below 10 µg per deciliter. New England Journal of Medicine 348:1517–1526.

CECIL, K. M., C. J. BRUBAKER, C. M. ADLER, K. N. DIETRICH, M. ALTAYE, J. C. EGELHOFF, S. WESSEL, I. ELANGOVAN, R. HORNUNG, K. JARVIS, AND B. P. LANPHEAR. 2008. Decreased brain volume in adults with childhood lead exposure. PLoS Medicine 5:741–750.

DEWAILLY, E., P. AYOTT, S. BRUNEAU, G. LEBEL, P. LEVALLOIS, AND J. P. WEBER. 2001. Exposure of the Inuit population of Nunavik (Arctic Quebec) to lead and mercury. Archives of Environmental Health 56, 350–357.

EISLER, R. 1988. Lead hazards to fish, wildlife, and invertebrates: a synoptic review. US Fish and Wildlife Service, Biological Report 85(1.14).

HARMATA, A. R., AND M. RESTANI. 1995. Environmental contaminants and cholinesterase in blood of vernal migrant Bald and Golden Eagles in Montana. Intermountain Journal of Sciences 1:1–15.

HAUSER, R., O. SERGEYEV, S. KORRICK, M. M. LEE, B. REVICH, E. GITIN, J. S. BURNS, AND P. L. WILLIAMS. 2008. Association of blood lead levels with onset of puberty in Russian boys. Environmental Health Perspectives 116:976–980.

HUNT, W. G., W. BURNHAM, C. N. PARISH, K. BURNHAM, B. MUTCH, AND J. L. OAKS. 2006. Bullet fragments in deer remains: implications for lead exposure in scavengers. Wildlife Society Bulletin 34:168–171.

HUNT, W. G., C. N. PARISH, S. C. FARRY, T. G. LORD, AND R. SIEG. 2007. Movements of introduced condors in Arizona in relation to lead exposure. Pages 79–96 *in* A. Mee, L. S. Hall, and J. Grantham (Eds.). California Condors in the 21st Century. Series in Ornithology, no. 2. American Ornithologists' Union, Washington, DC, and Nuttall Ornithological Club, Cambridge, Massachusetts, USA.

IWATA, H., M. WATANABE, E. Y. KIM, R. GOTOH, G. YASUNAGA, S. TANABE, Y. MASUDA, AND S. FUJITA. 2000. Contamination by chlorinated hydrocarbons and lead in Steller's Sea Eagle and White-tailed Sea Eagle from Hokkaido, Japan. Pages 91–106 *in* M. Ueta and M. J. McGrady (Eds.). First Symposium on Steller's and White-tailed Sea Eagles in East Asia. Wild Bird Society of Japan, Tokyo, Japan.

JOHANSEN, P., H. S. PEDERSEN, G. ASMUND, AND F. RIGET. 2006. Lead shot from hunting as a source of lead in human blood. Environmental Pollution 142:93–97.

KRAMER, J. L., AND P. T. REDIG. 1997. Sixteen years of lead poisoning in eagles, 1980–95: An epizootiologic view. Journal of Raptor Research 31:327–332.

LOCKE, L. N., AND N. J. THOMAS. 1996. Lead poisoning of waterfowl and raptors. Pages 108–117 *in* A. Fairbrother, L. N. Locke, and G. L. Huff (Eds.). Noninfectious diseases of wildlife, second edition. Iowa State University Press, Ames, Iowa, USA.

MENKE, A., P. MUNTNER, V. BATUMAN, E. K. SILBERGELD, AND E. GUALLAR. 2006. Blood lead below 0.48 μmol/L (10μg/dL) and mortality among US adults. Circulation 114:1388–1394.

PARISH, C. N., W. R. HEINRICH, AND W. G. HUNT. 2007. Lead exposure, diagnosis, and treatment in California Condors released in Arizona. Pages 97–108 *in* A. Mee, L. S. Hall, and J. Grantham (Eds.). California Condors in the 21st Century. Series in Ornithology, no. 2. American Ornithologists' Union, Washington, DC, and Nuttall Ornithological Club, Cambridge, Massachusetts, USA.

PATTEE, O. H., P. H. BLOOM, J. M. SCOTT, AND M. R. SMITH. 1990. Lead hazards within the range of the California Condor. Condor 92:931–937.

SCHULZ, J. H., J. J. MILLSPAUGH, B. E. WASHBURN, G. R. WESTER, J. T. LANIGAN, III, AND J. C. FRANSON. 2002. Spent-shot availability and ingestion on areas managed for Mourning Doves. Wildlife Society Bulletin 30: 112–120.

UNITED STATES DEPARTMENT OF THE INTERIOR (USDI). 1986. Final supplemental environmental impact statement: use of lead shot for hunting migratory birds in the United States. US Fish and wildlife Service FES 86–16. Washington, DC, USA.

WIEMEYER, S. N., J. M. SCOTT, M. P. ANDERSON, P. H. BLOOM AND C. J. STAFFORD. 1988. Environmental contaminants in California Condors. Journal of Wildlife Management 52:238–247.

WRIGHT, J. P., K. N. DIETRICH, M. D. RIS, R. W. HORNUNG, S. D. WESSEL, B. P. LANPHEAR, M. HO, AND M. N. RAE. 2008. Association of prenatal and childhood blood lead concentrations with criminal arrests in early adulthood. PLoS Medicine 5:732–740.

INTRODUCTION

IAN NEWTON

Centre for Ecology and Hydrology, Monks Wood, Abbots Ripton, Huntington, PE28 2LS, England. E-mail: ine@ceh.ac.uk

Transcribed from oral opening address, 13 May 2008.

NEWTON, I. 2009. Introduction. *In* R.T. Watson, M. Fuller, M. Pokras, and W.G. Hunt (Eds.). Ingestion of Lead from Spent Ammunition: Implications for Wildlife and Humans. The Peregrine Fund, Boise, Idaho, USA. DOI 10.4080/ilsa.2009.0091

Key words: Ammunition, health, humans, hunting, ingestion, lead, regulation, wildlife.

WE ARE HERE FOR THE NEXT THREE DAYS to discuss findings on the impacts of lead exposure on people and wildlife. Many of the findings that will be presented here are new – the result of research over the last couple of years. To my knowledge, this is the first time that medics and wildlife biologists have been brought together in this way to share their knowledge, and hopefully come up with recommendations on what needs to be done: not only in future research, but in influencing societal views and government policy regarding the use of lead.

For much of my scientific career, I have worked on the impacts of pesticides and pollutants on wildlife. So the kinds of problems to be discussed at our conference are not entirely new to me—although I must say that I have been involved only very marginally with lead, and can claim no particular expertise in this field.

Lead is such a useful metal to us in so many different ways. It is not surprising it has a long history of use by human societies around the world, extending back for more than 3,000 years. To say that many of these uses have created serious health problems is an under-statement, and some of lead's effects on people are now well known. Depending on exposure, the effects range from relatively mild depression and lethargy, through to seizures, blindness, mental impairment and even death. These findings have led in the last 150 years to many of the uses of lead being outlawed in the interests of human health. The banning of its use in many paints and gasoline has occurred within our own recent lifetimes.

Most of these health problems result from the body mistaking lead for calcium and other essential metals, and thus incorporating lead into nervous and other vital tissues. Blood levels in adult people of only 1 ppm have been linked with many debilitating symptoms, and recent research has suggested that blood levels averaging as low as 0.1 ppm in children can be associated with measurable mental impairment. Clearly, the effects of low level lead exposure in humans are not trivial.

And nor are they in wildlife. For in birds and mammals, we are concerned not merely with the effects on individuals, but also on populations. Through the mortality it can impart, lead can be a key factor limiting population sizes and causing serious numerical declines. At least two major effects have been described. One involves mainly seed-eating birds, which deliberately ingest grit to assist in the digestion of seeds and other plant material. Some such species, from doves to swans, also ingest spent shotgun pellets along with grit, and in consequence have suffered huge mortalities. In many countries, the use of lead shot over wetlands has now been banned, but its use in the wider environment still persists.

The second major impact occurs in meat-eating animals, including humans, condors and eagles, which ingest lead bullet fragments and pellets from the carcasses of shot game animals. This is a problem that we are only now getting to grips with, and on which urgent action is required. This source of lead could be affecting people on a geographical scale and to a degree not previously appreciated. Fortunately, alternatives to lead ammunition exist, if only the will or the legal pressure was there to take them up.

The aim of this conference is to bring together experts from the medical and wildlife fields to pull together some of this information, and discuss ways of addressing the problem. It is becoming increasingly clear that we have on our hands a major environmental problem, but one which could in theory easily be solved, if only the will was there.

Perhaps at this early stage I could acknowledge the efforts of my colleagues in The Peregrine Fund and other members of the organizing committee in getting this conference off the ground. Their preparatory work has been exemplary and professional.

The rest of today will provide us with background information on lead and its uptake by people and animals. In particular, we will hear from Dr. Michael Kosnett about the damaging effects of very low levels of lead on children. We will also hear about society's response in different countries to the harmful effects of lead by means of both voluntary and legislative procedures.

Tomorrow is concerned mainly with studies of wildlife exposure to spent ammunition, and much new research will be presented, especially on problems resulting from game hunting. The third day is concerned with remediation of the problem from ammunition sources, and will include some interesting results on game consumption by people from Alaska. We end the conference in the afternoon with an expert panel session that will help identify gaps in our knowledge, and hopefully make recommendations for solutions to the problem.

I hope you all enjoy the conference, and your time in Boise.

Biography.—**Professor Ian Newton, D. Phil., D.Sc., FRS, OBE.** Senior Ornithologist (Ret.) Natural Environment Research Council, UK, Chairman of the Board, The Peregrine Fund, Chairman of the Council, Royal Society for the Protection of Birds in the United Kingdom. Ian received his D.Phil. and D.Sc. degrees from Oxford University. He has studied a wide range of bird species, but may be best known for his work on raptors, and his landmark book *Population Ecology of Raptors* first published in 1979. His 27-year study of a Sparrowhawk population nesting in southern Scotland resulted in what many consider to be the most detailed and longest-running study of any population of birds of prey. He is author of more than 300 papers and several books, including *The Sparrowhawk* (1986), *Population Limitation in Birds* (1998), *The Speciation and Biogeography of Birds* (2003), and *The Ecology of Bird Migration* (2007).

UNDERSTANDING LEAD UPTAKE AND EFFECTS ACROSS SPECIES LINES: A CONSERVATION MEDICINE BASED APPROACH

MARK A. POKRAS AND MICHELLE R. KNEELAND

Center for Conservation Medicine, Tufts University Cummings School of Veterinary Medicine, 200 Westboro Rd., North Grafton, MA 01536, USA.
E-mail: mark.pokras@tufts.edu

ABSTRACT.—Conservation medicine examines the linkages among the health of people, animals and the environment. Few issues illustrate this approach better than an examination of lead (Pb) toxicity. We briefly review the current state of knowledge on the toxicity of lead and its effects on wildlife, humans, and domestic animals.

Lead is cheap and there is a long tradition of its use. But the toxic effects of Pb have also been recognized for centuries. As a result, western societies have greatly reduced many traditional uses of Pb, including many paints, gasoline and solders because of threats to the health of humans and the environment. Legislation in several countries has eliminated the use of lead shot for hunting waterfowl. Despite these advances, a great many Pb products continue to be readily available. Conservationists recognize that hunting, angling and shooting sports deposit thousands of tons of Pb into the environment each year.

Because of our concerns for human health and over 100 years of focused research, we know the most about lead poisoning in people. Even today, our knowledge of the long-term sublethal effects of Pb on human health continues to grow dramatically. Our knowledge about lead poisoning in domestic animals is significantly less. For wild animals, our understanding of lead poisoning is roughly where our knowledge about humans was in the mid-1800s when Tanquerel Des Planches made his famous medical observations (Tanquerel Des Planches 1850).

From an evolutionary perspective, physiological processes affected by lead are well conserved. Thus, scientists are able to use rodents and fish to understand how lead works in people. Similarly, those of us interested in safeguarding wildlife health should consider humans as excellent models for lead's chronic and sublethal effects.

Given what we are learning about the many toxic effects of this heavy metal, there is every reason to switch to non-toxic alternatives. To accomplish this, a broad, cross-species ecological vision is important. All interest groups must work together to find safe alternatives, to develop new educational and policy initiatives, to eliminate most current uses of Pb, and to clean up existing problems. *Received 25 August 2008, accepted 21 November 2008.*

POKRAS, M. A., AND M. R. KNEELAND. 2009. Understanding lead uptake and effects across species lines: A conservation medicine approach. *In* R. T. Watson, M. Fuller, M. Pokras, and W. G. Hunt (Eds.). Ingestion of Lead from Spent Ammunition: Implications for Wildlife and Humans. The Peregrine Fund, Boise, Idaho, USA. DOI 10.4080/ilsa.2009.0101

Key words: Lead, conservation medicine, sentinel, public health, stakeholders.

IN 1987 A BIOLOGIST from New Hampshire brought us a dead Common Loon (*Gavia immer*). As wildlife veterinarians interested in conservation issues, we agreed to run a few tests and perform a necropsy on the loon. Examining the cadaver, the bird was found to be in good condition with perfect breeding plumage and no indications of parasites or disease. The only interesting piece of evidence was a radiographic image revealing a metal object within the bird's gizzard. During the necropsy, we were surprised to discover it was actually a lead (Pb) fishing sinker. We were even more intrigued when analysis of a liver sample concluded that the loon had died of lead poisoning. Inspired by this case, we have spent the past 20 years working on the issue of lead poisoning, examining well over a thousand dead loons and testing many other species for lead toxicosis. Many conservation and health professionals now are beginning to realize the extent to which accumulation of this toxic material has been underestimated in terms of its impact on human and animal health.

This paper will focus on a general review of a number of issues including:

- What is lead?
- How is lead handled inside animals' bodies?
- What are some of the short-term and chronic effects of lead?
- Are humans significantly different from other vertebrate animals in the ways in which our bodies handle lead?
- The importance of interdisciplinary, multispecies approaches for understanding the magnitude and threats posed by environmental lead.
- Do we really need more science to prove that lead is toxic?

WHAT DO WE KNOW ABOUT LEAD?

Lead is an element and a metal (atomic number 82). It is soft, has a low melting point (327.5 °C), a high density (11.34 g/cm^3) and is found naturally in a variety of minerals including galena, cerussite and anglesite. Unlike many natural elements, lead is not known to be required by any living organism. So what does lead poisoning do, and how do we know if an animal has lead poisoning?

Lead poisoning can have rapid, acute effects or chronic, long-term effects in people and other animals. Because of societal concerns for human health and over 100 years of focused research, we know the most about lead poisoning in people. Our knowledge about lead poisoning in domestic animals is significantly less. For wild animals, our understanding of lead poisoning is roughly where our knowledge about humans was in 1848 when Tanquerel Des Planches made his famous medical observations. We know a great deal about the more obvious cases of death and debility caused by lead, but extremely little about the more subtle, chronic, sublethal effects.

Acute and subacute effects are typically caused by relatively large doses of lead over a short period—often days to months. These effects can be dramatic and include sudden death, severe abdominal cramps, anemia, ataxia, strange headaches, and behavioral changes, such as irritability and appetite loss. Of course these signs are fairly non-specific and can be caused by many things besides lead. Chronic effects are most often the result of smaller amounts of lead being taken in over longer times – months to years. These effects can be quite subtle and nonspecific, but include all body systems. A brief list of effects documented in people includes such effects as lowered sex drive, decreased fertility (in males and females), miscarriages and premature births, learning problems, hypertension, cardiovascular disease, increased aggression and kidney problems.

We know many of the ways in which lead kills wildlife over a few weeks or months, but almost nothing of the chronic, low level effects that probably harm a great many more animals and upset ecosystem functions over time. From the perspective of conservation and ecology, this is quite frustrating. But the good news is that a great many of the physiological mechanisms by which lead acts upon bodily processes are well conserved among vertebrate species. Many references cite the fact that whether we are talking about people, condors or fish, the body handles lead in much the same way as calcium. Calcium is a crucial element for living things, being used in a wide variety of metabolic activities, signaling pathways and structural compounds. This has meant that understanding how lead (Pb) works in laboratory rodents and fish has helped us understand the chemistry and physiology of lead in people. Conversely, it also means that much of what we've learned about chronic, sub-lethal effects in people can reasonably be extrapolated to non-human animals. We often think of animals as sentinels for human health; but here it is we, the humans, who are the white mice. Understanding how lead works in *Homo sapiens* can play a significant role in protecting other species and environmental processes.

In all kinds of adult animals, most lead is absorbed through the digestive and respiratory systems. Under some circumstances, primarily occupational exposures, certain forms of lead (usually the more lipophilic ones) can also be absorbed through the skin. The key is that however it enters the body, the most important step is the absorption of lead into the bloodstream. Nearly all lead vapors getting into the lungs cross into the blood quickly and easily. Inhaled small particulate matter is coughed up and swallowed. Lead entering the digestive system is acted upon by stomach acids and made into soluble salts that can be absorbed by the intestine. In adult male people about 10–15% of ingested lead is usually absorbed, the rest leaves the body in feces. But in young children, up to 50% of ingested lead is absorbed. In all age groups, absorption can be increased by conditions that stress the body including pregnancy, injury and disease.

In adult people, lead remains in the blood stream for roughly 2 weeks. During that time some of the lead is excreted from the body, but much of it begins to be deposited in the soft tissues of the body, including the liver and kidneys. Residence time in these soft tissues depends on a great many variables, but is usually on the order of a few weeks to several months. The endpoint for most of the lead in the body is the skeleton. As lead leaves most of the body's other tissues some is excreted, but much is bound into the structure of bone (again, following calcium). The half-life of bone lead is on the order of decades. This means that the skeleton serves as a source pool from which low levels of lead are mobilized back into the bloodstream as the bones remodel throughout life.

In the body, lead does many things (Needleman 1991, Casas & Sordo 2006, Ahameda and Siddiqui 2007, Diertert et al. 2007). It can bind important enzymes (primarily at their sulfhydryl groups) and inactivate them. Lead (Pb) can also displace biologically important metals, such as calcium, zinc and magnesium, interfering with a variety of the body's chemical reactions (Tables 1, 2, and 3). Lead toxicity affects all organ systems, but the most profound effects are seen in the nervous, digestive, and circulatory systems. Every time nerves transmit messages around the body, calcium is required. Thus lead can interfere with functions dependent on nerve conduction such as learning, blood pressure, reaction time and muscle contraction. Contractions of smooth muscle (peristalsis) are required to move food through the esophagus, stomach and intestines. Lead can upset this ordered contraction leading to a great deal of stomach and abdominal pain, long referred to as "lead colic" in people. In the blood stream, lead interferes with the functions of hemoglobin, limiting the amount of oxygen that is carried to organs. Lead also interrupts the formation of new red blood cells in the bone marrow, leading to anemia. Effects in the skeletal and reproductive systems can cause problems such as stunted growth and infertility (in both genders). In situations where the body needs to use bone calcium stores, like growth, fracture healing, egg shell formation (in reptiles and birds), dietary imbalances, pregnancy and lactation (in mammals), and or bone loss due to aging or osteoporosis, lead is released from bones and can cause chronic, low level poisoning.

As we have learned more and more about the sublethal effects of lead in people over the last 40 or 50 years, government agencies have acted to lower the levels of blood lead that have been considered "safe." In 1968 a level of 80 μg/dL (micrograms per deciliter) in blood was considered the level of concern for children. But as we have learned more and more about the significant effects that even low levels of lead can have, these numbers have gradually been lowered. In the 1980s the limit was set at 50 μg/dL. That was changed to 30 μg/dL in 1995, then to the current 10 μg/dL. Recent studies have shown that even levels below 2 μg/dL can cause significant, long-term effects in children (Lanphear

Table 1. Mechanisms of Pb toxicity (Needleman 1991, Casas and Sordo 2006, Ahameda and Siddiqui 2007).

- Substitutes for and competes with Ca^{++}
- Disrupts Ca^{++} homeostasis
- Binds with sulfhydryl groups
- Stimulates release of Ca^{++} from mitochondria
- Damages mitochondria and mitochondrial membranes
- Substitutes for Zn in zinc mediated processes
- Increases oxidative stress
- Inhibits anti-oxidative enzymes
- Alters lipid metabolism

Table 2. Possible mechanisms by which Pb induces neurologic effects (Needleman 1991, Casas and Sordo 2006, Ahameda and Siddiqui 2007).

- Increase in affinity for Ca^{++} binding sites
- Disrupts Ca^{++} metabolism
- Substitutes for Ca^{++} in Ca/Na ATP pump
- Blocks uptake of Ca^{++} into mitochondria and endoplasmic reticula
- Interference with neural cell adhesion
- Impairment of cell to cell connections
- Alters some neurotransmitter function
- Activates protein kinase C
- Alters Ca^{++} mediated apoptosis

Table 3. Results of lead toxicity (Needleman 1991, Casas and Sordo 2006, Ahameda and Siddiqui 2007).

- Abnormal myelin formation
- Altered neurotransmitter density
- Altered neurotransmitter release
- Increase in lipid peroxidation
- Impaired heme biosynthesis leads to anemia
- Decreased cellular energy metabolism
- Altered apoptosis

2000, Canfield et al. 2003). In some areas of the USA blood lead testing is required for children entering kindergarten. This is not a very expensive test and only requires a couple of drops of blood, and yet it is still not generally required nationwide.

Detecting lead poisoning in people or other animals is done in a variety of ways. Radiographs (x-rays) can detect metal densities and their locations in the body, but cannot tell specifically if the metal is Pb. For that, we would have to remove and test the metallic object or look at lead levels in either blood or body tissues like liver, kidney or bone. In living animals, we usually look at blood samples. We can analyze blood either for the metal itself, or we can analyze blood for the levels of some enzymes which are affected by lead, typically ALAD or zinc protoporphyrin. If the person or animal is deceased, investigators typically send samples of liver or kidney to toxicologic laboratories to measure Pb levels. As will be reviewed in other presentations, determination of lead isotopes can give us important information about the origin of lead found in people and other species.

If we think about the mining, manufacturing and recycling of lead products in the USA, it is apparent that state and federal regulatory agencies try to minimize the amounts of lead that are released into the environment. State and federal permits are required for any sort of industrial lead release. But in recreational sporting goods, we have a whole class of products used for the shooting sports, hunting and fishing which, when used as intended, end up in the environment. It is difficult to come up with exact figures for how much lead is released through these activities. According to estimates from the US Geological Survey, roughly 10% of all the lead produced in the USA or imported goes for such sporting purposes. This amounts to approximately 6–10 thousand tons of lead being released into the environment annually in the USA.

Given that we have known about lead's toxicity for a very long time, you might reasonably ask why we are still using lead for so many purposes. Several possibilities come to mind. One is simply traditional practice. We have known about and used lead for a very long time, in fact the Latin name for lead, *plumbum,* is incorporated into quite a few modern terms including "plumber" (from the people who originally worked with lead pipes), "plumb bob" (for a lead weight on the end of a string) or "plumb stupid" to comment on the behavior of people who have had too much contact with lead. In following the early settlement patterns of the USA, it is interesting to note how many towns have names like Leadville or Galena. One reason we have used lead for so long is that it has a low melting point (621.43 °F). This means that no elaborate equipment is needed to extract the metal from rock. In fact it was probably just by noticing that a pure silvery/grey metal was left on the ground after campfires that early humans first figured out how to smelt metals, thus leading us from the stone age into the age of metals. The low melting point also means that one can melt lead and cast it into a variety of forms using reasonably inexpensive equipment. This can encourage many people to melt lead at home for making things like stained glass, fishing sinkers or lead soldiers, without using proper safeguards. Lead is quite dense, about 11 times heavier than an equal volume of water. This density is one of the characteristics that contributes to its usefulness for bullets, weights of various sorts and x-ray shielding. The fact that lead is soft, easy to work and resistant to many forms of corrosion also makes it attractive for a wide variety of uses. Last, lead has historically been fairly inexpensive; until recently only a few hundred dollars per ton. That price seems to be increasing rapidly due to new demands for lead to use in storage batteries. In addition, prices are rising because we have nearly exhausted many of the easy to obtain sources of ore and new mines are more expensive. Finally, in developed countries, new regulations to protect the health of miners, manufacturing workers, recyclers and the environment are increasing all the costs associated with this metal. Such new regulations certainly have beneficial effects for workers in the USA and other first world nations. But in this age of globalization, the rising costs of mining and production have driven many international corporations to move their activities to developing countries. In such locations there is frequently little environmental regulation, few protections for workers and little enforcement. Rather than solve the environmental and health problems inherent in making and using lead products, some of our industries have simply exported them.

Despite its utility, lead is toxic. The US Centers for Disease Control and Prevention states on its lead poisoning website: "Any combination of GI complaints, neurologic dysfunction and anemia should prompt a search for heavy metals toxicity." The federal agency charged with protecting workers' health, the Occupational Safety and Health Administration (OSHA), recommends on their website (http://www.osha.gov/SLTC/lead/) that all of us should:

- Avoid purchasing or using products known to contain lead
- Avoid inhaling dusts or fumes of lead or lead-containing compounds
- Avoid consuming food or beverages or putting items in the mouth in areas where lead-based compounds or lead-based materials are in use
- Wash hands with soap and water after handling lead.

Although we are still struggling to understand the full scope of the issue, lead poisoning is not a new problem. The toxic effects of lead have been reported for over 2000 years in both humans and animals (Nriagu 1983). Grinnell (1894) published the first USA report of lead poisoning in ducks caused by ingestion of spent shot more than 100 years ago. One of the most comprehensive clinical descriptions of human lead poisoning was undertaken by Tanquerel des Planches in 1848. Dr. Alice Hamilton, in originating the field of occupational health, performed extensive work on the social costs of industrial lead poisoning in the early 1900s (Sicherman 2003). Yet it took more than a half-century and the 1962 publication of Rachel Carson's momentous Silent Spring to refocus attention on the links between chemicals in the environment and the health of people and animals. That attention helped to catalyze a new generation of scientists to focus on lead's toxic legacy. In the public health arena, the work of Clair Patterson, Herbert Needleman, Ellen Silbergeld, Philip Landrigan and many others finally brought about the elimination of lead from most house paints and gasoline. While in the wildlife world, the efforts of a great many biologists (some of whom attended this meeting) brought about regulatory changes for hunting waterfowl, forcing the replacement of toxic lead shot with safer alternatives (Anderson and Havera

1989). Yet even today Hu et al. (2007) can state, "In the world of environmental health and environmental medicine, lead exposure remains one of the most important problems in terms of prevalence of exposure and public health impact." We have to ask ourselves why more progress to eliminate this persistent health threat has not been made.

A major impediment to progress is the disciplinary separation that has long existed among groups investigating issues related to lead poisoning. While there are multiple organizations currently working on the problem, most efforts are narrowly focused on one particular aspect of lead poisoning. We need a conservation medicine-based approach to the lead poisoning problem that overcomes barriers between the disciplines of human and animal health, barriers within the field of animal health, and barriers between researchers and the general public in order to finally eliminate this persistent health threat.

BRIDGING THE HUMAN/ANIMAL DIVIDE

While the concept of using animals as sentinels of human health is not new (Winter 2001), wildlife professionals seldom realize the wealth of information that can be gained by taking the opposite approach and using humans as sentinels for animal and environmental health. A representative medium size mammal, *Homo sapiens* is by far the best understood and most widely studied species on the planet, so why not utilize this abundance of data to help us understand our non-human counterparts? In fact, it is from the human literature that we get some of the best measures for sublethal effects of lead toxicity. Low level lead exposure has been associated with a wide range of conditions in humans, including cognitive deficiencies in children, renal impairment, hypertension, cataracts, and reproductive problems such as miscarriage, stillbirth, and decreased fertility in men and women (Patrick 2006).

Deficits in cognitive and academic skills have been reported in children with blood lead concentrations lower than 5 μg/dL (Lanphear et al. 2000). Another study found that a net increase of 1 μg/dL in the lifetime average blood lead level was correlated with a loss of 0.46 IQ points (Canfield et al. 2003). In light of these findings in humans, the cognitive effects of sublethal lead poisoning are beginning to be studied in wildlife. In Herring Gull (*Larus argentatus*) chicks, for example, effects on locomotion, food begging, feeding, treadmill learning, thermoregulation, and individual recognition were observed in nestlings dosed with lead to produce feather lead concentrations equivalent to those found in wild gulls (Burger and Gochfeld 2005). Several studies have found an association between sub-clinical lead toxicosis and delinquent, antisocial, and aggressive behaviors in humans (Sciarillo 1992, Needleman et al. 1996, Nevin 2000, Dietrich et al. 2001). Similarly, the development of aggressive behaviors has been documented in domestic dogs and cats with elevated blood lead levels, as well as rodents and songbirds (Koh 1985, Burright et al. 1989, Hahn et al. 1991, Delville 1999, Janssens et al. 2003, Li et al. 2003). Pattee and Pain (2003) document an increasing use of lead worldwide and state that "lead concentrations in many living organisms may be approaching thresholds of toxicity for the adverse effects of lead." Environmental lead exposure at low levels could very well be contributing to wildlife mortality by hindering the complex mental processes and social behaviors required for reproductive success and survival.

Abdominal pain and peripheral neuropathy are two symptoms of lead poisoning that have been described in human literature for centuries (Tanquerel des Planches 1850). Commonly referred to as "painter's colic" and "wrist drop" respectively, these conditions are not specific to humans. Abdominal pain is recognized as a clinical sign common to nearly all lead poisoned animals (Osweiler 1996). Sileo and Fefer (1987) observed "droop wing," the avian equivalent of human wrist drop, in Laysan Albatross (*Diomedea immutabilis*) fledglings that had ingested lead paint chips from abandoned buildings. Platt et al. (1999) document a similar wing droop in a Turkey Vulture (*Cathartes aura*), and this same symptom has been seen in other avian species.

Veterinarians and wildlife professionals are just beginning to investigate the potential effects of sublethal lead levels in animals, and the human lead poisoning literature serves as a wonderful resource to guide future research. Similarly, physicians and public health officials must also be willing to shift their anthropocentric focus in order to fully identify lead exposure risks to humans. For example, there have been numerous case reports in which a child was tested and found to have elevated blood lead levels through a pet dog first being diagnosed with lead poisoning. Thomas et al. (1976) reported that a blood lead concentration of diagnostic significance in a family dog resulted in a six-fold increase in the probability of finding a child in the same family with similarly elevated blood lead levels. There is also a possible connection between lead in the tissues of waterfowl or game animals and human health. Sportsmen and their families may be exposed to high lead concentrations from shot residue in the meat of hunted waterfowl. Johansen et al. (2003) found that hunters who reported regularly eating birds hunted with lead shot had significantly higher mean blood lead levels than hunters who reported not eating hunted birds, 128 μg/L and 15 μg/L respectively. Even when bullet fragments are not present, secondary lead ingestion in waterfowl hunters can also occur through consuming the livers of chronically lead poisoned birds (Guitart et al. 2002).

OVERCOMING SPECIES ISOLATION

In addition to the separation that exists between the realms of human and animal health, lead poisoning research also tends to be compartmentalized within specific taxonomic groups. The research on lead toxicosis in wildlife concentrates predominantly on avian species. Most of the current literature falls within discrete categories such as raptors, loons, or waterfowl, and discusses one specific route of exposure. For example, literature concerning lead poisoning in raptor species such as Bald Eagles (*Haliaeetus leucocephalus*) and California Condors (*Gymnogyps californianus*) focuses primarily on ingestion of lead gunshot embedded in prey or scavenged carcasses (Janssen et al. 1986, Mateo et al. 2001). Studies of lead-related mortality in the Common Loon (*Gavia immer*) have identified ingestion of fishing sinkers as the prime route of lead exposure (Pokras and Chafel 1992, Sidor et al. 2003). In a wide variety of waterfowl species, toxicosis resulting from accidental intake of spent lead shot has been reported in the literature for over 100 years (Pain 1992).

Increased collaboration among researchers with expertise in different taxa is needed in order to advance our knowledge of lead in the environment. Approaching lead poisoning as just a waterfowl problem or just a raptor problem impedes progress toward effective policy changes. For instance, the impact of lead fragments remaining in carcasses and gut piles on condor populations has been thoroughly investigated, but we know almost nothing about the effects on other avian and mammalian scavengers. While it is crucial to comprehend the pathology within a particular species, we must also gain a better understanding of lead effects at the level of ecosystems. We know lead can accumulate in organisms such as invertebrates and plants, but we still have much to learn about how this influences the rest of the food chain (Pattee and Pain 2003). Deciphering the intricate web of environmental lead sources and exposure routes will allow us to implement better strategies to reduce the occurrence of lead poisoning in all species.

The Gray Squirrel (*Sciurus carolinensis*) may be a prime illustration of our need for a more comprehensive understanding of lead exposure sources in wildlife. Recently, one of these rodents was brought to the Tufts Wildlife Clinic and found to have a markedly elevated blood lead level of over 65μg/dL. Anecdotal stories of squirrels chewing on lead chimney flashing have been reported by homeowners for years, and recently a New Hampshire Fish and Game biologist confirmed that she regularly receives calls about "problem" squirrels that continually gnaw on chimney flashing. Analogous to children eating flecks of lead-based paint, this may be evidence of pica in squirrels. It may also be one explanation for the lead-poisoned predators like Red-tailed Hawks and Barred Owls that are periodically submitted to Tufts' Wildlife Clinic. Because millions of homes nationwide have lead flashing around chimneys, doors, and other openings, this appears to represents an overlooked source of plumbism in wildlife.

The literature on lead poisoning in non-human species is both broad and deep. Virtually all vertebrate taxa have been documented as experiencing lead poisoning (Table 4) although adults of some species appear to be more resistant than others. Lead toxicosis is well documented in mammals, from marine

Table 4. Species in which lead poisoning has been well documented.

people, other primates
many songbirds
loons
woodpeckers
hawks, eagles (Bald, Golden, others)
herons, flamingos, pelicans
vultures (including condors)
gulls
waterfowl (many species)
turkeys, quail, grouse, bob-white, pheasant
cranes, rails
reptiles (snapping turtle, crocodile, iguana)
parrots (many species)
squirrels, rabbits
woodcock, snipe
horses and cattle
Mourning Doves, pigeons
sheep and pigs
bats (micro and macrochiropterans)
dogs and cats
fish (many species)
rats and mice

species (Shlosberg et al. 1997, Zabka et al. 2006) to cattle and horses (Palacios et al. 2002, Sharpe and Livesey 2006) to bats (Skerratt et al. 1998, Bennett et al. 2003, Walker et al. 2007) and rodents. Useful reviews are provided by Priester and Hayes (1972), Humphreys (1991) and Mautino (1997).

A great many avian species have been shown to experience lead poisoning, in the wild and in captivity. In addition to waterfowl, raptors, loons, songbirds, and psittacines, many upland species, including gamebirds, are regularly poisoned by lead (Artman and Martin 1975, Kennedy et al. 1977, Locke and Friend 1992, Platt et al. 1999, Morner and Petersson 1999, Burger and Gochfeld 2000, Lewis et al. 2001, Mateo et al. 2001, Vyas et al. 2001, Scheuhammer 2003, Rodrigue 2005, Fisher et al. 2006).

Reptiles and amphibians also are affected by lead. Although the adults of some reptile species seem relatively resistant, early developmental stages do appear sensitive, especially among those amphibian species that have aquatic eggs and larvae that are in contact with the sediments (Barrett 1947, Kober and Cooper 1976, Overmann and Kracjicek 1995, Stansley and Roscoe 1996, Stansley et al. 1997, Borkowski 1997, Burger 1998, Rice et al. 1999, Vogiatzis and Loumbourdis 1999, Rice et al. 2002, Rosenberg et al. 2003, Arrieta et al. 2004, Mouchet et al. 2007).

The situation seems to be similar in fish. Adult fish of some species appear to be relatively insensitive to acute toxicity, but their eggs and larvae can show dramatic effects at low levels of exposure, sometimes resulting in population level effects and ecosystem alteration (Carpenter 1924a, b, Dilling et al. 1926, Jones 1964, Srivastava and Mishra 1979, Birge et al. 1979, Johansson-Sjöbeck and Larsson 1979, Newsome and Piron 1982, Hodson et al. 1984, Coughlan et al. 1986, Dallinger et al. 1987, Tewari et al. 1987, Eisler 1988, Tulasi et al. 1989, Sorensen 1991, Weber et al. 1997, Kasthuri and Chandran 1997, Chaurasia et al. 1996, Chaurasia and Kar 1999, Shafiq-ur-Rehman 2003, Martinez et al. 2004, Shah 2006). It seems clear that as more studies explore the sub-lethal effects of lead exposure in non-human species, there will be increased emphasis on integrating our thinking so that threats to human health are understood in the context of an over all environmental well-being.

Although it is not central to the present discussion, it is worth noting that there is abundant literature on lead in invertebrates. Some of this regards direct toxicity, but there is also literature on the ability of some invertebrates to accumulate lead (and other heavy metals) and to cause indirect toxicity to vertebrates that eat them (Grosell et al. 2006, Ma 1982, 1987, 1989, Scheuhammer 2003).

Similarly there is evidence that plants can also be affected by lead; either experiencing toxicity or as bioaccumulators (Malanchuk and Gruendling 1973, Manninen and Tanskanen 1993, Xiong 1998, Terry and Bañuelos 2000). Much remains to be learned about the effects of these processes on animal lead accumulation and health.

THE LIMITATIONS OF REGULATIONS

In April of 2005, the US Consumer Product Safety Commission announced a nationwide recall of 1.5 million children's fishing rods because it was found that the paint on the rod exceeded the 0.06% limit for lead. Parents were instructed to discontinue the use of the product immediately. At the same time, an online retailer specializing in children's fishing gear was selling a product called "The Ultimate Fishing Kit for Kids." The kit comprised a plastic tackle box packed with 78 pure lead fishing sinkers. This and hundreds of other lead-stocked fishing kits designed for children are still widely available and have never been subject to a recall.

This example highlights the disjointed nature of current efforts to reduce lead exposure. Because many agencies in the USA regulate the various aspects of lead use—ranging from the Department of Labor for mining safety, to the Environmental Protection Agency for environmental pollution, to the Fish and Wildlife Service for the hunting of migratory birds—initiatives to limit lead's toxic effects have been myopic, lacking multidisciplinary perspective. In the USA, the banning of lead-based paint for residential use in 1978, the phasing-out of leaded fuel for on-road vehicles between 1973 and 1995, and most recently the mass recall of imported toys containing lead each represents an independent effort mobilized by separate groups. These measures have led to vast improvements in ecological health and helped protect human lives in much of the developed world. Annest et al. (1983) reported a 37% decrease in average blood lead levels in the USA between 1976 and 1980, associated with a reduction in the lead content of gasoline during this period. But this and other lead products are still widely available in developing countries. Lead is still a major component in some industrial paints, and leaded fuel continues to be sold for off-road uses such as aircraft, automobile racing, farm equipment, and marine engines. Even more alarming is the multiplicity of common household products that still contain lead – everything from curtain weights, solder, and batteries to imported ceramics, candy, and hair dyes.

Fortunately there is increased pressure to reduce lead in the USA and abroad. The European Union is engaged in efforts to eliminate several uses of lead. Groups in India, China, Australia, Nigeria and other countries are focused on eliminating leaded paints and gasoline and reducing human and animal exposure. Agencies concerned with the globalization of world economies and the internationalization of recycling (e.g., electronics) are increasingly taking steps to reduce toxic exports to developing countries in which protection for worker health may be lacking (Grossman 2006). Even in developed countries advocates for environmental justice have pointed out that exposure to lead and other toxic material falls disproportionately on many of the most vulnerable in our communities, often non-white populations with little education living in poorer neighborhoods (Bullard 1994). Clearly all stakeholders must be included in discussions of conservation and environmental health.

Although the toxicity of lead has been widely understood and reported for hundreds of years, progress to improve regulatory measures has been excruciatingly slow. Lead is cheap and easy to work with, and consequently the serious health risks associated with this metal are often overshadowed by its economic value. The continued prevalence of lead poisoning in both humans and animals is a signal that current policies are inadequate. A unified approach focused on interdisciplinary collaboration between specialists in human health, animal health, and ecological health is needed if we hope to make further progress in developing more protective legislation.

Like the classic story from the Far East of the five blind men and the elephant, each group of stakeholders perceives the lead (Pb) issue differently. Some will say, "What lead problem?" Others will say, "There may be a problem, but more research is needed." But for those of us who are convinced that it is necessary NOW to significantly decrease the quantities of this toxic metal that humans put into the environment, a number of approaches should be considered. These might include:

1. dramatically improving the marketing of non-toxic products,
2. exploring new business models, including the possibility of imposing taxes on Pb items (and perhaps investing those tax dollars in conservation), and/or providing tax and pricing incentives for non-toxic items,
3. improving educational efforts, especially those directed at sportsmen and their families,
4. developing new legislative approaches
5. encouraging technological innovation to find new non-toxic alternatives.

CONNECTING SCIENCE AND SPORT

The use of lead for hunting and fishing sports represents a particularly challenging situation. In the past 20 years, Vermont, New York, Maine, Massachusetts, New Hampshire, Great Britain, Denmark, and Canada have all passed legislation restricting the use of certain types of lead fishing gear. In 1991, lead shot was banned for use in hunting waterfowl in the United States. Most recently, California passed a bill in October 2007 banning the use of lead ammunition in areas of California Condor habitat. Unfortunately, these legislative initiatives designed to protect wildlife are often met with resistance from industry and sportsmen. Much of this opposition to proposed lead bans results from a lack of communication between scientists and the public. For example, many sportsmen's groups such as the US Sportsmen's Alliance condemn attempts to prohibit lead as an infringement on their rights, rather than a means to protect the health of people and wildlife. There is a misconception that lead prohibition laws are introduced by groups who oppose shooting and fishing sports solely as a tactic to limit these activities. Inaccurate information continues to plague a complete and successful transition to non-toxic hunting and fishing gear, and therefore establishing an open dialogue between researchers, sportsmen, and policy makers is critical.

Strict legislation banning the use of lead hunting and fishing gear that does not provide for the interests of sportsmen would result in ardent protest, low compliance, and ultimately would fail to resolve the lead poisoning issue. To bring an end to the problem once and for all, scientists and health professionals must find ways to better collaborate with hunting and fishing groups. We need to approach the issue in a way that encourages people to take a proactive role in eliminating this environmental crisis. Appealing to the conservationist roots of hunters and anglers is one way to do this. Many sportsmen are either unaware of the ecological damage caused by the use of lead gear or are skeptical of claims that a seemingly insignificant bullet or fishing weight could lead to such damaging effects for wildlife. It is therefore essential that we ramp up our efforts to reach out to sportsmen and educate them about the scientific rationale for moving away from lead, and to do so in a way that does not condemn their practices or their sport. In addition, working with manufacturers to provide sportsmen with more nontoxic alternatives that offer the same performance and practicality as lead is an important piece of a successful lead phase out. This cooperative approach will allow sportsmen to play a positive role in efforts to limit the use of lead, and in the end is the key to a permanent solution.

BREAKING DOWN THE BARRIERS

The effects of lead poisoning are not confined to human health nor to any one species of animal. Thus, we will never successfully gain control of the problem unless we take an approach that is all-inclusive. We cannot continue to view the different aspects of plumbism in isolation from one another. Paint, gasoline, occupational exposure, toys, bullets, fishing gear, and all the other sources of lead are not separate issues but rather are components of the same fundamental problem.

Developing strategies to achieve better integration among conservation and health disciplines will broaden our scientific understanding of lead poisoning and accelerate progress toward solutions. Currently, studies on lead poisoning in people, wildlife, and domestic animals are all published separately in journals devoted to those specific fields. Establishing resources that include lead poisoning literature from all domains will promote a better flow of ideas and scientific knowledge between disciplines and allow researchers to see the interconnectedness of human and animal plumbism. Conferences and meetings that address multiple aspects of lead poisoning provide a prime opportunity for researchers and action groups to network with experts in different disciplines. The EPA's National Lead Poisoning Prevention Week in October concentrates primarily on childhood poisoning from lead paint, but would be an ideal opportunity to spread public awareness and increase communication about all of the other issues associated with lead poisoning. Bringing together a wide range of stakeholders to participate in the lead poisoning dialogue will allow us to find solutions that are scientifically accurate, environmentally sound, economically viable, and socially acceptable.

An increasing number of organizations are now realizing the value of a conservation medicine-based approach to the lead poisoning issue and are making efforts to break down disciplinary barriers. Several groups such as the EPA's Leadnet, The Lead Education and Abatement Design Group in Australia, and the Tufts Veterinary School's Lead and Health group have formed listserves and contact databases to facilitate communications among people—diverse fields such as environmentalists, sportsmen, veterinarians, wildlife professionals, lead industry representatives, citizen action groups, and environmental justice groups. The Midwest Fish and Wildlife Conference held in December 2007 included a session on lead poisoning that was attended by public health professionals, veterinarians, wildlife biologists, sportsmen, and lead fishing tackle manufacturers. The Peregrine Fund's 2008 Lead Ammunition Conference focused on implications for both human and animal health. Finally, a session has been proposed for the EcoHealth II conference in December 2008 that would bring together a wide range of professionals to focus on lead and its many health and environmental effects.

CONCLUSIONS

Despite the fact that lead poisoning is well understood, it still threatens the health of millions of people, domestic animals and wildlife worldwide. Barriers among disciplines have impeded the scientific understanding of lead toxicosis and slowed policy initiatives to protect the health of animals and people. A conservation medicine-based approach focusing on increasing collaboration among professionals working in different fields offers the best hope for understanding and eliminating this ancient problem.

ACKNOWLEDGEMENTS

We gratefully acknowledge Springer Publishers, and The International Association for Ecology and Health and their journal EcoHealth for allowing us to utilize portions of a previously published article (Pokras and Kneeland 2008) in the current paper. We would like to thank the many veterinary students and volunteers whose help over the years has made our studies possible. We thank Mr. R. Wood, Dr. J. McIntyre of Utica College (Syracuse University), Dr. P. Spitzer of the Center for Northern Studies, and Mr. A. Major and Ms. C. Perry of the US Fish and Wildlife Service for their invaluable advice and assistance. We thank Drs. M. Friend and J. C. Franson of the National Wildlife Health Research Center (USGS), Dr. R. Haebler of the Atlantic Ecology Division (US EPA) and Dr. E. Silbergeld for their wise advice and encouragement over the years. The Pathology Department of Tufts Cummings School of Veterinary Medicine (TCSVM) generously contributed their expertise to this project.

LITERATURE CITED

AHAMEDA, M., AND M. K. J. SIDDIQUI. 2007. Low level lead exposure and oxidative stress: current opinions. Clinica Chimica Acta 383:57–64.

ANDERSON, W. L., AND S. P. HAVERA. 1989. Lead poisoning in Illinois waterfowl (1977–1988) and the implementation of nontoxic shot regulations. Illinois Natural History Survey. Champaign, Illinois, USA.

ANNEST, J. L., I. L. PIRKLE, D. MAKUC, J. W. NEESE, D. D. BAYSE, AND M. G. KOVAR. 1983. Chronological trend in blood lead levels between 1976 and 1980. New England Journal of Medicine 308:1373–1377.

ARRIETA, M. A., L. BRUZZONE, C. APARTIN, C. E. ROSENBERG, N. E. FINK, AND A. SALIBIAN. 2004. Biosensors of inorganic lead exposure and effect in an adult amphibian. Archives of Environmental Contaminants and Toxicology 46:224–230.

ARTMANN, J. W., AND E. M. MARTIN. 1975. Incidence of ingested lead shot in Sora Rails. Journal of Wildlife Management 39:514–519.

BARRETT, W. C. 1947. The effects of lead salts on the hemo-poietic and histiocytic systems of the larval frog. American Journal of Anatomy 81:117–133.

BENNETT, F., S. LOEB, P. VAN DEN HURK, AND W. BOWERMAN. 2003. The distribution and contaminant exposure of Rafinesque's Big-eared Bats in South Carolina with an emphasis on bridge surveys. DE-AI09-00SR22188 Technical Report 03-14-R. [Online.] Available at http://www.osti.gov/bridge/purl.cover.jsp?purl=/835177-FJC7so/.

BIRGE, W. J., J. A. BLACK, AND A. G. WESTERMAN. 1979. Evaluation of aquatic pollutants using fish and amphibian eggs as bioassay organisms. Pages 108–118 *in* F. Peter, P. Timmins, and D. Perry (Eds.). Symposium on Pathobiology of Environmental Pollutants: Animal Models and Wildlife as Monitors. National Academy of Sciences, Washington, DC, USA.

BORKOWSKI, R. 1997. Lead poisoning and intestinal perforations in a Snapping Turtle (*Chelydra serpentina*) due to fishing gear ingestion. Journal of Zoo and Wildlife Medicine 28(1):109–13.

BULLARD, R. D. (ED.). 1994. Unequal Protection: Environmental Justice and Communities of Color. Sierra Club Books, San Francisco, California, USA.

BURGER, J., AND M. GOCHFELD. 2005. Effects of lead on learning in Herring Gulls: an avian wildlife model for neurobehavioral deficits. NeuroToxicology 26:615–624.

BURGER, J., AND M. GOCHFELD M. 2000. Effects of lead on birds (Laridae): a review of laboratory and field studies. Journal of Toxicology & Environmental Health (Part B): Critical Reviews 3:59–78.

BURGER, J. 1998. Effects of lead on behavior, growth and survival of hatchling Slider Turtles. Journal of Toxicology and Environmental Health (Part A). 55:495–502.

BURRIGHT, R. G., W. J. ENGELLONNES, AND P. J. DONOVICK, 1989. Postpartum aggression and plasma prolactin levels in mice exposed to lead. Physiological Behavior 46:889–893.

CANFIELD, R. L., C. R. HENDERSON, JR., D. A. CORY-SLECHTA, C. COX, T. A. JUSKO AND B. P. LANPHEAR. 2003. Intellectual impairment in children with blood lead concentrations below 10 µg per deciliter. New England Journal of Medicine 348:1517–1526.

CARPENTER, K. 1924a. A study of the fauna of rivers polluted by lead mining in the Aberystwyth district of Cardiganshire. Annals of Applied Biology 11:1–23.

CARPENTER, K. 1924b. On the biological factors involved in the destruction of river-fisheries by pollution due to lead mining. Annals of Applied Biology 12:1–13.

CASAS, J. S., AND J. SORDO. 2006. Lead: chemistry, analytical aspects, environmental impact and health effects. Elsevier Sciences, New York, USA.

CHAURASIA, S. S., AND A. KAR. 1999. An oxidative mechanism for the inhibition of iodothyronine 5'-monodeiodinase activity by lead nitrate in the fish, *Heteropneustes fossilis.* Water Air and Soil Pollution 111:417–423.

CHAURASIA, S. S., P. GUPTA, A. KAR, AND P. K. MAITI. 1996. Lead induced thyroid dysfunction and lipid peroxidation in the fish *Clarias batrachus* with special reference to hepatic type I-5'-monodeiodinase activity. Bulletin of Environmental Contamination and Toxicology 56:649–654.

COUGHLAN, D. J., S. P. GLOSS, AND J. KUBOTA. 1986. Acute and sub-chronic toxicity of lead to the early life stages of Smallmouth Bass (*Micropterus dolomieui*). Water, Air, & Soil Pollution 28:265–275.

DALLINGER, R., F. PROSI, H. SEGNER, AND H. BACK. 1987. Contaminated food and uptake of heavy metals by fish: a review and a proposal for further research. Oecologia 73:91–98

DELVILLE, Y. 1999. Exposure to lead during development alters aggressive behavior in Golden Hamsters. Neurotoxicology and Teratology 21:445–449

DIETERT, R. R., AND M. J. MCCABE, JR. 2007. Lead immunotoxicology. Pages 591-605 *in* R. Luebke, R. House, and I. Kimber (Eds.). Immunotoxicology and Immunopharmacology, 3rd ed. CRC Press, Boca Raton, Florida, USA.

DIETRICH, K. N., M. D. RIS, P. A. SUCCOP, U. G. BERGER, AND R. L., BORNSCHEIN. 2001. Early exposure to lead and juvenile delinquency. Neurotoxicologic Teratology 23:511–518.

DILLING, W. J., C. W. HEALEY AND W. C. SMITH. 1926. Experiments on the effects of lead on the growth of Plaice *(Pleuronectes platessa).* Annals of Applied Biology 8:168–176.

EISLER, R. 1988. Lead hazards to fish, wildlife and invertebrates: a synoptic review. Biological Report 85(1.14) Contaminant Hazard Report No. 14. US Fish & Wildlife Service. Patuxent Wildlife Research Center, Laurel, Maryland, USA.

FISHER, I. J., D. J. PAIN, AND V. G. THOMAS. 2006. A review of lead poisoning from ammunition sources in terrestrial birds. Biological Conservation 131:421–432.

GRINNELL, G. B. 1894. Lead poisoning. Forest and Stream 42:117–118.

GROSELL, M., R. M. GERDES, AND K. V. BRIX. 2006. Chronic toxicity of lead to three freshwater invertebrates—*Brachionus calyciflorus, Chrionomus tentans*, and *Lymnaea stagnalis*. Environmental Toxicology and Chemistry 25:97–104.

GROSSMAN, E. 2006. High Tech Trash: Digital Devices, Hidden Toxics, and Human Health. Shearwater Press, Washington, DC, USA.

GUITART, R., J. SERRATOSA, AND V. G. THOMAS. 2002. Lead-poisoned wildfowl in Spain: a significant threat for human consumers. International Journal of Environmental Health Research 12:301–309.

HAHN, M. E., R. G. BURRIGHT, AND P. J. DONOVICK. 1991. Lead effects on food competition and predatory aggression in Binghamton HET mice. Physiological Behavior 50:757–764.

HODSON, P. V., D. M. WHITTLE, P. T. S. WONG, U. BORGMANN, R. L. THOMAS, Y. K. CHAU, J. O. NRIAGU, AND D. J. HALLET. 1984. Lead contamination of the Great Lakes and its potential effects on aquatic biota. *In:* J. O. Nriagu and M. S. Simmons (Eds.). Toxic contaminants in the Great Lakes. John Wiley and Sons, Indianapolis, Indiana, USA.

HU, H., R. SHIH, S. ROTHENBERG, AND B. S. SCHWARTZ. 2007. The epidemiology of lead toxicity in adults: measuring dose and consideration of other methodologic issues. Environmental Health Perspectives 115:455–462.

HUMPHREYS, D. J. 1991. Effects of exposure to excessive quantities of lead on animals. British Veterinary Journal 147:18–30.

JANSSEN, D. L., J. E. OOSTERHUIS, J. L. ALLEN, M. P. ANDERSON, D. G. KELTS, AND S. N. WIEMEYER. 1986. Lead poisoning in free-ranging California Condors. Journal of the American Veterinary Medical Association 189:1115–1117.

JANSSENS, E., T. DAUWE, E. VAN DUYSE, J. BEERNAERT, R. PINXTEN, AND M. EENS. 2003. Effects of heavy metal exposure on aggressive behavior in a small territorial songbird. Archives of Environmental Contamination and Toxicology 45:121–127.

JOHANSEN, P., H. S. PEDERSEN, G. ASMUND, AND F. RIGET. 2003. Lead shot from hunting as a source of lead in human blood. Environmental Pollution 142:93–97.

JOHANSSON-SJÖBECK, M-J., AND A. LARSSON. 1979. Effects of inorganic lead on delta-aminolevulinic acid dehydratase activity and hematological variables in the Rainbow Trout, *Salmo gairdnerii.* Archives of Environmental Contamination and Toxicology 8:419–431.

JONES, J. R. E. 1964. Fish and River Pollution. Butterworths, London, UK.

KASTHURI, J., AND M. R. CHANDRAN. 1997. Sublethal effect of lead on feeding energetics, growth performance, biochemical composition and accumulation of the Estuarine Catfish, *Mystus gulio* (Hamilton). Journal of Environmental Biology 18:95–101.

KENNEDY, S., J. P. CRISLER, E. SMITH, AND M. BUSH. 1977. Lead poisoning in Sandhill Cranes. Journal of the American Veterinary Medical Association 171:955–958.

KOBER, T. E., AND G. T. P. COOPER. 1976. Lead competitively inhibits calcium-dependent synaptic transmission in the bullfrog sympathetic ganglia. Nature 262:704–705.

KOH, T. S. 1985. Diagnosis of lead poisoning in dogs. Australian Veterinary Journal 62(12):434.

LANPHEAR, B. P., K. DIETRICH, P. AUINGER, AND C. COX. 2000. Cognitive deficits associated with blood lead concentrations <10 μg/dL in US children and adolescents. Public Health Reports 115:521–529.

LEWIS, L. A., R. J. POPPENGA, W. R. DAVIDSON, J. R. FISCHER, AND K. A. MORGAN. 2001. Lead toxicosis and trace element levels in wild birds and mammals at a firearms training facility. Environmental Contamination and Toxicology 41:208–214.

LI, W., S. HAN, T. R. GREGG, F. W. KEMP, A. L. DAVIDOW, D. B. LOURIA, A. SIEGEL, AND J. D. BOGDEN. 2003. Lead exposure potentiates predatory attack behavior in the cat. Environmental Research 92:197–206.

LOCKE, L.N., AND M. FRIEND. 1992. Lead poisoning of avian species other than waterfowl. Pages 19–22 *in* D. J. Pain (Ed.). Lead Poisoning in Waterfowl. Proceedings of an IWRB Workshop, Brussels, Belgium, 13–15 June 1991. International Waterfowl and Wetlands Research Bureau Special Publication 16, Slimbridge, UK.

MA, W. C. 1982. The influence of soil properties and worm-related factors on the concentration of heavy metals in earthworms. Pedobiologia 24:109–119.

MA, W. C. 1987. Heavy metal accumulation in the Mole, *Talpa europea*, and earthworms as an indicator of metal bioavailability in terrestrial environments. Bulletin of Environmental Contamination and Toxicology 39:933–938.

MA, W. C. 1989. Effect of soil pollution with metallic lead pellets on lead bioaccumulation and organ/body weight alterations in small mammals. Archives of Environmental Contamination and Toxicology 18:617–622.

MALANCHUK, J. L., AND G. K. GRUENDLING. 1973. Toxicity of lead nitrate to algae. Water, Air and Soil Pollution 2:181–190.

MANNINEN, S., AND N. TANSKANEN. 1993. Transfer of lead from shotgun pellets to humus and three plant species in a Finnish shooting range. Archives of Environmental Contamination and Toxicology 24:410–414.

MARTINEZ, C. B., M. Y. NAGAE, C. ZAIA, AND D. A. M. ZAIA. 2004. Acute morphological and physiological effects of lead in the neotropical fish *Prochilodus lineatus*. Brazilian Journal of Biology 65:797–807.

MATEO, R., R. CADENAS, M. MANEZ, AND R. GUITART. 2001. Lead shot ingestion in two raptor species from Doñana, Spain. Ecotoxicology and Environmental Safety 48:6–10.

MAUTINO, M. 1997. Lead and zinc intoxication in zoological medicine: a review. Journal of Zoo and Wildlife Medicine 28:28–35.

MORNER, T., AND L. PETERSSON. 1999. Lead poisoning in woodpeckers in Sweden. Journal of Wildlife Diseases 35:763–765.

MOUCHET, F., S. CREN, C. CUNIENQ, E. DEYDIER, R GUILET, AND L. GAUTHIER. 2007. Assessment of lead ecotoxicity in water using the amphibian larvae (*Xenopus laevis*) and preliminary study of its immobilization in meat and bone meal combustion residues. BioMetals 20:113–127.

NEEDLEMAN, H. L. 1991. Human Lead Exposure. CRC Press. Boca Raton, Florida, USA.

NEEDLEMAN, H. L., J. A. RIESS, M. J. TOBIN, G. E. BIESECKER, AND J. B. GREENHOUSE. 1996. Bone lead levels and delinquent behavior. Journal of the American Medical Association 275:363–369.

NEVIN, R. 2000. How lead exposure relates to temporal changes in IQ, violent crime, and unwed pregnancy. Environmental Research 83:1–22.

NEWSOME, C. S., AND R. D. PIRON. 1982. Aetiology of skeletal deformities in the Zebra Danio fish (*Bruchydanio rerio*, Hamilton-Buchanan). Journal of Fish Biology 21:231–237.

NRIAGU, J. 1983. Lead and Lead Poisoning in Antiquity. Academic Press. New York, USA.

OSWEILER, G. D. 1996. Toxicology. Williams & Wilkins. Philadelphia, Pennsylvania, USA.

OVERMANN, S. R., AND J. J. KRACJICEK. 1995. Snapping Turtles (*Chelydra serpentina*) as biomonitors of lead contamination of the Big River in Missouri's Old Lead Belt. Environmental Contaminants and Toxicology 14:689–695.

PAIN, D. J. 1992. Lead poisoning of waterfowl: a review. Pages 7–13 *in* D. J. Pain (Ed.). Lead Poisoning in Waterfowl. Proceedings of an IWRB Workshop, Brussels, Belgium, 13–15 June 1991. International Waterfowl and Wetlands Research Bureau Special Publication 16, Slimbridge, UK.

PALACIOS, H., I. HIBARREN, M. J. OLALLA, AND V. CALA. 2002. Lead poisoning of horses in the vicinity of a battery recycling plant. Science of the Total Environment 290:81–89.

PATRICK, L. 2006. Lead toxicity, A review of the literature: Part I: exposure, evaluation, and treatment. Alternative Medicine Review 11:2–22.

PATTEE, O. H., AND D. J. PAIN. 2003. Lead in the Environment. Pages 373–408 *in* D. J. Hoffman, B. A. Rattner, G. A. Burton, Jr., and J. Cairns, Jr. (Eds). Handbook of Ecotoxicology 2nd ed. Lewis Publishers, New York, USA.

PLATT, S. R., K. E. HELMICK, J. GRAHAM, R. A. BENNETT, L. PHILLIPS, C. L. CHRISMAN, AND P. E. GINN. 1999. Peripheral neuropathy in a Turkey Vulture with lead toxicosis. Journal of the American Veterinary Medical Association 214:1218–1220.

POKRAS, M. A., AND R. M. CHAFEL. 1992. Lead toxicosis from ingested fishing sinkers in adult Common Loons (*Gavia immer*) in New England. Journal of Zoo and Wildlife Medicine 23:92–97.

POKRAS, M. A., AND M. R. KNEELAND. 2008. Lead poisoning: using transdisciplinary approaches to solve an ancient problem. EcoHealth. electronic DOI: 10.1007/s10393-008-0177-x

PRIESTER, W. A., AND H. M. HAYES. 1972. Lead poisoning in cattle, horses, cats, and dogs as reported by 11 colleges of veterinary medicine in the United States and Canada from July, 1968 through June, 1972. American Journal of Veterinary Research 35:567–569.

RICE, T. M., B. J. BLACKSTONE, W. L. NIXDORF, AND D. H. TAYLOR. 1999. Exposure to lead induces hypoxia-like responses in bullfrog larvae (*Rana Catesbeiana*). Environmental Toxicology and Chemistry 18:2283–2288.

RICE T. M., J. T. ORIS, AND D. H. TAYLOR. 2002. Effects of growth and changes in organ distribution of bullfrog larvae exposed to lead throughout metamorphosis. Bulletin of Environmental Contamination and Toxicology 68:8–17.

RODRIGUE, J., R. MCNICOLL, D. LECLAIR, AND J.-F. DUCHESNE. 2005. Lead concentrations in Ruffed Grouse, Rock Ptarmigan, and Willow Ptarmigan in Quebec. Archives of Environmental Contaminants and Toxicology 49:97–104.

ROSENBERG, C. E., N. E. FINK, M. A. ARRIETA, AND A. SALIBIAN. 2003. Effect of lead acetate on the in vitro engulfment and killing capability of toad (*Bufo arenarum*) neutrophils. Comparative Biochemistry and Physiology (Part C) 136:225–233.

SCHEUHAMMER, A. W. 2003. Elevated lead exposure in American Woodcock (*Scolopax minor*) in eastern Canada. Archives of Environmental Contamination and Toxicology 36:334–340.

SCIARILLO, W. G., G. ALEXANDER, AND K. P. FARRELL. 1992. Lead exposure and child behavior. American Journal of Public Health 82:1356–1360.

SHAFIQ-UR-REHMAN, S. 2003. Lead-exposed increase in movement behavior and brain lipid peroxidation in fish. Journal of Environmental Science and Health (Part A) 38:631–643.

SHAH, S. L. 2006. Hematological parameters in Tench (*Tinca tinca*) after short term exposure to lead. Journal of Applied Toxicology 26:223–228.

SHARPE, R. T., AND C. T. LIVESEY, 2006. Lead poisoning in cattle and its implications for food safety. Veterinary Record 159:71–74.

SHLOSBERG, A. M. BELLAICHE, S. REGEV, R. GAL, M. BRIZZI, V. HANJI, L. ZAIDEL, AND A. NYSKA. 1997. Lead toxicosis in a captive Bottlenose Dolphin (*Tursiops truncatus*) consequent to ingestion of air gun pellets. Journal of Wildlife Diseases 33:135–139.

SICHERMAN, B. 2003. Alice Hamilton: A Life in Letters. University of Illinois Press, Champaign, Illinois, USA.

SIDOR, I. F., M. A. POKRAS, A. R. MAJOR, R. H. POPPENGA, K. M. TAYLOR AND R. M. MICONI. 2003. Mortality of Common Loons in New England, 1987–2000. Journal of Wildlife Diseases 39:306–315.

SILEO, L., AND S. I. FEFER. 1987. Paint chip poisoning of Laysan Albatross at Midway Atoll. Journal of Wildlife Diseases 23:432–437.

SKERRATT, L. F., R. SPEARE, L. BERGER, AND H. WINSOR. 1998. Lyssaviral infection and lead poisoning in Black Flying Foxes from Queensland. Journal of Wildlife Diseases 34:355–361.

SORENSEN, E. M. B. 1991. Lead. Pages 95–118 *in* E. M. B. Sorenson. Metal Poisoning in Fish. CRC Press, Boca Raton, Florida, USA.

SRIVASTAVA, A. K., AND S. MISHRA. 1979. Blood dyscrasia in a teleost, *Colisa fasdatus,* after acute exposure to sublethal concentrations of lead. Journal of Fish Biology 14:199–203.

STANSLEY, W., AND D. E. ROSCOE. 1996. The uptake and effects of lead in small mammals and frogs in a trap and skeet range. Environmental Contamination and Toxicology 30: 220–226.

STANSLEY, W., M. A. KOSENAK, J. E. HUFFMAN, AND D. E. ROSCOE. 1997. Effects of lead-contaminated surface water from a trap and skeet range on frog hatching and development. Environmental Pollution 96:69–74.

TANQUEREL DES PLANCHES, L. 1850. Lead Diseases: A Treatise. Tappan, Whittemore & Mason, Boston, Massachusetts, USA.

TERRY, N., AND G. S. BAÑUELOS (EDS.). 2000. Phytoremediation of Contaminated Soil and Water. Lewis Publishers. Boca Raton, Florida, USA.

TEWARI, H., S. G. TEJENDRA, AND J. PANT. 1987. Impact of chronic lead poisoning on the hematological and biochemical profiles of a fish, *Barbus conchonius* (Ham). Bulletin of Environmental Contamination and Toxicology 38:748–752.

THOMAS, C. W., J. L. RISING, AND J. K. MOORE. 1976. Blood lead concentrations of children and dogs from 83 Illinois families. Journal of the American Veterinary Medical Association 169:1237–1240.

TULASI, S. J., P. U. M. REDDY, AND J. V. RAMANA RAO. 1989. Effects of lead on the spawning potential of the fresh water fish, *Anabas testudineus*. Bulletin of Environmental Contamination and Toxicology 43:858–863.

VYAS, N. B., J. W. SPANN, AND G. H. HEINZ. 2001. Lead shot toxicity to passerines. Environmental Pollution 111:135–138.

VOGIATZIS, A. K., AND N. S. LOUMBOURDIS. 1999. Exposure of *Rana ridibunda* to lead. I. study of lead accumulation in various tissues and hepatic aminolevulinic acid dehydratase activity. Journal of Applied Toxicology 19:25–29.

WALKER, L. A., V. R. SIMPSON, L. ROCKETT, C. L. WIENBURG, AND R. F. SHORE. 2007. Heavy metal contamination in bats in Britain. Environmental Pollution 148:483–490.

WEBER, D. N., W. M. DINGEL, J. J. PANOS, AND R. H. STEINPREIS. 1997. Alterations in neurobehavioral responses in fishes exposed to lead and lead-chelating agents. American Zoologist 37:354–362

WINTER, M. 2001. The brain on lead: animal models are helping researchers understand the effects of lead exposure in children. Human Ecology 29:13–16.

XIONG, Z.-T. 1998. Lead uptake and effects on seed germination and plant growth in a Pb hyperaccumulator *Brassica pekinensis* Rupr. Bulletin of Environmental Contaminants and Toxicology 60:285–291.

ZABKA, T. S., M. HAULENA, B. PUSCHNER, F. M. D. GULLAND, P. A. CONRAD, AND L. J. LOWENSTINE. 2006. Acute lead toxicosis in a Harbor Seal (*Phoca vitulina richardsi*) consequent to ingestion of a lead fishing sinker. Journal of Wildlife Diseases 42:651–657.

HISTORY IN LEAD AND LEAD POISONING IN HISTORY

JEROME NRIAGU

Department of Environmental Health Sciences, School of Public Health, University of Michigan, Ann Arbor, MI 48109, USA. E-mail: jnriagu@umich.edu

ABSTRACT.—Lead ores were probably the first to be refined to the metallic state, hence the production and use of lead have been intertwined with the history of human culture for a very long time. Although there was never a "Lead Age" (lead romanticism did not dominate any period of human history), lead and its compounds were nevertheless present in all the metal ages and have certainly played important roles in industrial, scientific and military progress as well as in trade, material comfort, human vanity and curing of diseases. The use of lead had reached such an impressive level by about 2,000 years before present (estimated at 550 grams per capita per year) that lead is oftentimes referred to as a "Roman metal." The massive demand for lead that began subsequently with Industrial Revolution continues unabated until modern times. Throughout human history, it is estimated that over 350 million tons of lead have been mined, used and ultimately discarded in the environment. More lead has been discharged into the environment than any other toxic metal, and the consequences of this continuing human experiment of fouling our nest with lead remain to be fully understood. My presentation will focus on three periods in human history when lead poisonings were particularly pandemic: (a) Ancient Time especially during the Roman Empire period when many aristocrats were heavily exposed to lead in their foods and drinks; (b) Middle Ages when the adulteration and contamination of foods and wines were again rampant and saturnine drugs became a mainstay of the pharmacopoeia; and (c) Modern Times when poor people are being disproportionately exposed to lead in their environment. I shall let the dead in the cemetery population of Isola Sacra (in the outskirts of Rome) made up of middle class traders and craftsmen speak about their exposures during the 2nd and 3rd centuries AD through the lead in their bones. I shall talk about the mad and impotent princes (and princesses) of Europe during the Middle Ages and their affliction with gout of saturnine origin. I shall conclude with a brief discussion of changing phases of lead poisoning associated with environmental risk transition and how exposure of African children to lead in their environment may be altering the pathogenesis of some communicable diseases.

NRIAGU, J. 2009. History in lead and lead poisoning in history. Abstract *in* R. T. Watson, M. Fuller, M. Pokras, and W. G. Hunt (Eds.). Ingestion of Lead from Spent Ammunition: Implications for Wildlife and Humans. The Peregrine Fund, Boise, Idaho, USA. DOI 10.4080/ilsa.2009.0102

Key words: Lead poisoning, Roman, Middle Ages, modern.

HEALTH EFFECTS OF LOW DOSE LEAD EXPOSURE IN ADULTS AND CHILDREN, AND PREVENTABLE RISK POSED BY THE CONSUMPTION OF GAME MEAT HARVESTED WITH LEAD AMMUNITION

MICHAEL J. KOSNETT

University of Colorado, Denver, c/o 1630 Welton, Suite 300, Denver, CO 80202, USA.
E-mail: Michael.Kosnett@ucdenver.edu

ABSTRACT.—Research findings have heightened public health concern regarding the hazards of low dose lead exposure to adults and children. In adults, studies have established the potential for hypertension, decrements in renal function, subtle decline in cognitive function, and adverse reproductive outcome at blood lead levels less than 25 micrograms per deciliter (µg/dL). The developing nervous system of the fetus and young child is particularly sensitive to the deleterious effects of lead, with adverse impacts on physical growth and neurocognitive development demonstrable at blood lead levels less than 10 µg/dL. No low dose threshold for these adverse developmental effects has been discerned. Epidemiological studies, and risk assessment modeling presented in this paper, indicate that regular consumption of game meat harvested with lead ammunition and contaminated with lead residues may cause relatively substantial increases in blood lead compared to background levels, particularly in children. Because lead-free ammunition is an available substitute, this risk is amenable to the public health strategy of primary prevention. *Received 2 December 2008, accepted 12 December 2008.*

KOSNETT, M. J. 2009. Health effects of low dose lead exposure in adults and children, and preventable risk posed by the consumption of game meat harvested with lead ammunition. *In* R. T. Watson, M. Fuller, M. Pokras, and W. G. Hunt (Eds.). Ingestion of Lead from Spent Ammunition: Implications for Wildlife and Humans. The Peregrine Fund, Boise, Idaho, USA. DOI 10.4080/ilsa.2009.0103

ALTHOUGH THE TOXICITY OF LEAD has been known for millennia, recognition and management of the adverse effects in adults and children have often posed a clinical and public health challenge. This arises from two main features. Even at moderate to high levels of exposure, many of the overt multisystemic effects of lead, such as headache, fatigue, myalgias, arthralgias, abdominal discomfort, constipation, anemia, peripheral neuropathy, and renal insufficiency, are nonspecific, and might be attributable to other relatively common acute and chronic diseases. At the lower levels of exposure that are prevalent today, the effects of lead may not only be nonspecific, but also subclinical or asymptomatic. Nevertheless, these effects, which may include hypertension, decrement in renal function, subtle decline in cognitive function, and adverse reproductive function in adults, and developmental delay in children, are of considerable public health concern.

This article presents a short overview of selected health impacts of lead exposure. In this context, the focus is on some of the adverse effects of the low to moderate levels of lead exposure that might possibly result from the ingestion of lead residues remaining in the flesh of game birds or mammals harvested with lead ammunition. A detailed discussion of the vast literature on the health effects of lead exposure is well beyond the scope of this article, but may be accessed in part from other recent sources (EPA 2006, ATSDR 2007).

Whole Blood Lead as a Common Metric of Lead Dose.—Studies of lead in humans have often, though not exclusively, related the clinical effects of lead to the level measured in whole blood. Whether obtained through venipuncture or a capillary pinprick, lead in blood remains the mainstay of human biomonitoring. As such, it is useful to begin with a brief discussion of the levels of lead in blood that have been encountered in the general population. Large population studies conducted in the United States in the late 1970s found that most of the general population had a blood lead concentration between 10 to 20 micrograms per deciliter (µg/dL). In the NHANES II study conducted by the US Centers for Disease Control and Prevention (CDC) between 1976 to 1980, the geometric mean blood lead level in children age 1 to 5 was 15.0 µg/dL (NCHS 1984). As shown in Figure 1, this level fell dramatically over the following decade, largely due to the phase out of leaded gasoline, as well as to declining encounters with lead in residential paint, canned food, and other sources (EPA 2006). In its Third National Report on Human Exposure to Environmental Chemicals, the CDC estimated that the geometric mean blood lead concentration of children aged 1 to 5 years was 1.70 µg/dL, while that of adults aged 20 years and older was 1.56 (CDC 2005a).

For several decades, the US CDC has issued guidance that identified levels of lead in the blood of young children that were of concern with respect to public health intervention. In 1970, that value was 40 µg/dL. It fell to 30 µg/dL in 1975, 25 µg/dL in 1985, and 10 µg/dL in 1991 (CDC 1991). In a statement on the topic in 2005, the CDC noted that adverse effects of lead on cognitive development, a key health endpoint of concern, extend to blood lead concentrations less than 10 µg/dL, and that there is no value that constitutes a threshold or no-effect level (CDC 2005b). With respect to lead exposure to adults in the workplace, the U.S. Occupational Safety and Health Administration (OSHA), established general industry standards for lead in the late 1970s. Under the OSHA general industry lead standard, which remains in effect to the current time, a worker requires removal from lead exposure if a single blood lead level exceeds 60 µg/dL, or if the average of the three most recent blood lead measurements exceeds 50 µg/dL (provided the last is greater than 40 µg/dL). Nevertheless, studies conducted in recent decades, some of which are

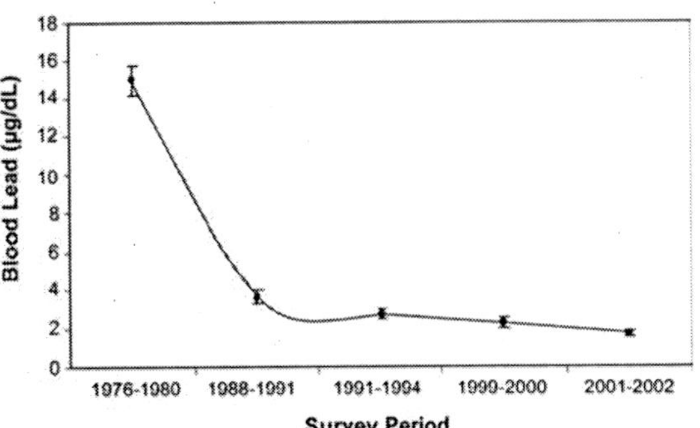

Figure 1. Blood lead concentration in U.S. Children. From: EPA, 2006 (Figure 4-3).

discussed below, demonstrate that the OSHA lead standards offer inadequate protection against the adverse effects of lead at low dose to adults (Kosnett et al. 2007).

Health Effects of Lead at Low Dose in Adults: Hypertension, Decrements in Renal Function, Cognitive Dysfunction.—Based on numerous recent studies, there is growing concern that the chronic impact of cumulative low dose lead exposure in adults may contribute to hypertension, decrements in renal function, and cognitive dysfunction. Evidence for the causal impact of lead on hypertension has emerged from multiple lines of investigation. From a mode of action standpoint, studies have identified impacts of lead on vascular smooth muscle (possibly mediated by interaction with intracellular calcium), and multi-organ oxidative stress (possibly effecting mediators such as nitrous oxide that influence vascular tone) (Chai and Webb 1988, Vaziri and Khan 2007). Laboratory studies have demonstrated that feeding lead to animals induces elevations in blood pressure. For example, a small but well designed study was conducted in six young female dogs and their matched litter mates (Fine et al. 1988). The animals were dosed with lead acetate (1 mg/kg/day x 5 months) or placebo. Blood pressure was measured regularly by Doppler in the foreleg without anesthesia or trauma by a blinded investigator. At 15 weeks, blood lead level of the exposed animals was 35.8 µg/dL versus 9.2 µg/dL in controls. The blood pressure of the exposed animals was consistently elevated compared to the controls. When the study was completed at 20 weeks, their mean blood pressure was 120 ± 2.1 mm Hg, compared to 108 ± 1.5 in controls.

Numerous human epidemiological studies have observed a robust relationship between blood lead and blood pressure. Findings obtained from the NHANES II survey are illustrative. NHANES II was a representative cross-sectional survey of the noninstitutionalized United States population examined between 1976 and 1980. Of the 20,322 persons examined, a blood lead sample was obtained in 9,933. In adults males aged 18 to 74, the geometric mean blood lead was 15.8 µg/dL (NCHS 1984). Blood lead was significantly associated with systolic and diastolic blood pressure, after controlling for age, body mass index, and demographic and nutritional factors (Schwartz 1988). Meta-analyses based on studies conducted on subjects with environmental and occupational lead exposure have found that the relationship between blood lead concentration and blood pressure can be described by a log-linear model. As blood lead level doubles (e.g. from 5 to 10 µg/dL), there is a corresponding increase in systolic blood pressure of either 1.0 mm Hg (Nawrot et al. 2002, see Figure 2), or 1.25 mmHg (Schwartz 1995). It should be realized that in studies such as these, the mean blood pressure increase (which from a clinical standpoint appears small) reflects observations in some individuals who may exhibit no pressor response, as well as those for whom the impact may be much higher. A mean blood pressure increase across the general population of only a few mmHg is of public health concern, since elevated blood pressure is a significant risk factor for cardiovascular, cerebrovascular and renovascular disease (Pirkle et al. 1985, EPA 2006). A recent 12-year follow-up study of subjects greater than 40 years of age enrolled in NHANES III (n = 9,757) observed that the subgroup with blood lead concentration ≥10 µg/dL (median 11.8) had a relative risk of cardiovascular mortality of 1.55 (95% C.I. 1.16–2.07) compared with subjects with blood lead <5 µg/dL (Schober et al. 2006).

Our understanding that the impact of lead on blood pressure is predominantly influenced by long-term, cumulative exposure has been derived in part from investigations that utilized noninvasive x-ray fluorescence measurement of lead in bone as a biomarker of exposure. Greater than 90 percent of the body lead burden is found in bone, where it resides with a half-life of years to decades. In a nested case-control study conducted in a subcohort of the Normative Aging Study, 146 hypertensive men were compared to 444 normotensive controls (Hu et al. 1996). The mean age of the study subjects was 66.6 ±7.2 years, and their mean blood lead concentration, 6.3 µg/dL, reflected background environmental lead exposure. Logistic regression analysis revealed three significant risk factors for hypertension: body mass index, family history of hypertension, and tibia bone lead concentration. From the lowest to the highest quintile of bone lead, the odds of being hypertensive increased by 50 % (O.R. = 1.5, 95% C.I. 1.1-1.8). The results of this and other studies conducted in middle aged to older adults suggest that long-term blood lead concentrations in the range of 10 to 25 µg/dL pose a significant risk of hypertension and cardiovascular disease (Navas–Acien et al. 2007).

Several studies conducted in general population samples have reported an association between blood lead concentration and biomarkers of renal function. For example, Staessen et al. (1992) examined the relationship between blood lead and creatinine clearance in 965 men and 1,016 women (age 20 to 88) recruited from a region with environmental cadmium exposure. The geometric mean blood lead concentration was approximately 10 µg/dL (range 1.7 - 72.5 µg/dL). There was a significant inverse correlation between age-adjusted creatinine clearance and blood lead, which persisted after excluding subjects with occupational lead exposure, or those with the highest tercile of blood lead (geometric mean 18.4 µg/dL). Some recent studies have suggested that the relationship between low level lead exposure and renal dysfunction may be accentuated in subjects with other risk factors for renal disease, such as hypertension or diabetes (Muntner et al. 2003, Tsaih et al. 2004).

Recent studies conducted in older adults have found cumulative lead exposure, as reflected by the concentration of lead in bone, to be a risk factor for poorer performance on some tests of cognitive function. The Baltimore Memory Study (Shih et al. 2006) examined 991 randomly selected, sociodemographically diverse community-dwelling adults aged 50 to 70 years. Blood lead (mean = 3.5 ±2.2

μg/dL) was not a predictor of neuropsychological performance. However, increasing tibia bone lead concentration was associated with deficits in visuoconstructive skill, such that an increase of 13 ppm bone lead yielded an impact equivalent to 4.8 years of aging. In a subcohort of the Normative Aging Study, 1089 older, mainly white men, (mean age 68.7 ±7.4 years) under repeat neuropsychological testing over an approximately 3.5 year interval (Weisskopf et al. 2007). Tibia bone lead concentration, but not blood lead concentration, was significantly associated with decreased visuospatial performance over time.

Adverse Reproductive Outcome in Women.—Concern over the adverse reproductive effects of low level lead exposure to women has emerged from studies of multiple endpoints. In a well-designed nested case control study conducted in Mexico City, 562 women seeking prenatal care were prospectively followed for the first 20 weeks of pregnancy (Borja-Aburto et al. 1999). The average blood lead at enrollment was 11 μg/dL. Using the quartile of women with blood lead <5 μg/dL as the referent group, the odds ratio for spontaneous abortion for the quartiles with blood lead of 5 - 9 μg/dL, 10 - 14 μg/dL, and ≥15 μg/dL were 2.3, 5.4, and 12.2 respectively (test for trend, P = 0.021). Overall, an increase in maternal blood lead of 5 μg/dL was associated with an odds ratio for spontaneous abortion of 1.8 (95% C.I. 1.1, 3.1). Other studies have associated maternal lead exposure with adverse effects on physical growth and neurocognitive development during infancy and childhood. Two long-term prospective studies examined the impact of low-level prenatal lead exposure on neurobehavioral development during childhood. In the Yugoslavia Prospective Lead Study, childhood IQ assessed at 3 to 7 years of age declined 1.8 points (95% C.I. 1.0, 2.6) for every doubling of prenatal blood lead, which was defined as the average of maternal blood lead at midpregnancy and delivery (mean, 10.2 ±14.4 μg/dL, n = 390) (Wasserman et al. 2000). Similarly, the Mexico City Prospective Lead Study found that IQ declined 2.7 points (95 C.I. 0.9, 4.4) for every doubling of third trimester maternal blood lead (geometric mean 7.8 μg/dL, n = 150) (Schnaas et al. 2006). In both studies, the relationship between prenatal blood lead and postnatal childhood IQ was characterized by a log-linear model, such that IQ decline was steepest at maternal blood lead levels less than 10 μg/dL.

Adverse Effects on Neurocognitive and Neurobehavioral Development in Children.—Much of the public health concern over low-level lead exposure in recent years has focused on adverse impacts to children. Children have heightened susceptibility to environmental lead exposure for several reasons. The developing nervous system of the fetus and young child is the human organ system most sensitive to the deleterious effects of lead. Compared to that of adults, the juvenile gastrointestinal tract absorbs a higher percentage of lead that is ingested. Normal mouthing behavior of young children results in greater intake of lead in environmental media such as soil or dust. Finally, in proportion to body size, children breathe more air, drink more liquid, and consume more food than adults.

The adverse effect of lead on children's intellectual function is well established by decades of extensive study. A recent analysis examined this association by pooling data on blood lead and intelligence quotient (IQ) on 1,333 children enrolled in seven prospective cohort studies from birth or infancy to age 5 to 10 years (Lanphear et al. 2005). The primary analysis examined full scale IQ as a function of concurrent blood lead level at 4 to 7 years of age, adjusting for HOME score (a measure of the child-rearing environment), birth weight, maternal IQ, and maternal education. The median concurrent blood lead level was 9.7 μg/dL (range 2.5 to 33.2). The relationship between blood lead and IQ was described by a log linear multiple regression model. This indicated a loss of 6.2 IQ points as blood lead increased from <1 to 10 μg/dL, and a loss of 3.0 IQ points as blood lead increased from 10 μg/dL to 30 μg/dL.

The impact of these decrements in IQ, which may be difficult to clinically discern in any one individual, is best appreciated in a broader societal context. Lead-related decrements in IQ are relatively uniform across a range of intelligence, and thus an overall downward shift in IQ in the general population not only increases the number of children with low test scores, but also decreases the number scoring in the gifted range. As illustrated in Figure 2, a five point IQ shift in the bell-shaped distribution of

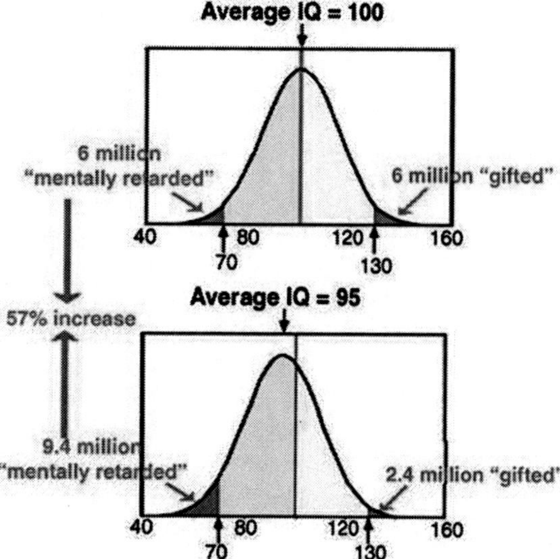

Figure 2. Changes in the number of children with IQ <70 and >130 per 100 million population based on a drop in the mean IQ of five points. Reprinted with permission, from Gilbert and Weiss, 2006.

IQ has its greatest impact at the tails of the distribution, increasing by approximately 57% the number of children with extremely low IQ scores (IQ <70), and decreasing by approximately 60% those with an IQ in the very superior range (IQ >130) (Gilbert and Weiss 2006). Given the capacity of individuals at the extremes of intelligence to have a disproportionate influence on societal function and resources, the overall cost of elevated lead exposure to society has been judged to be considerable.

The neurobehavioral effects of lead on child development, while less intensively studied than effects on cognition, may also have a significant societal impact. For example, some evidence suggests that blood lead levels in the mid-teens or higher may be associated with an increased risk of antisocial behavior and delinquency (Deitrich et al. 2001, EPA 2006).

Lead Ammunition and Primary and Secondary Prevention.—Public health action to reduce the risks associated with low level lead exposure may include elements of both primary and secondary prevention. With respect to adults, a recent recommendation has called for individuals to be removed from occupational lead exposure if a single blood lead concentration exceeds 30 µg/dL, or if two successive blood lead measurements over a 4 week interval are ≥20 µg/dL. It is further recommended that removal from lead exposure be considered to avoid long-term risk to health if exposure control measures over an extended period do not decrease blood lead concentrations to <10 µg/dL (Kosnett et al. 2007).

In its 2005 Statement on Preventing Lead Poisoning in Young Children, the US Centers for Disease Control and Prevention acknowledged that the effort to eliminate childhood lead poisoning would require a multitiered approach that included secondary prevention through case identification and management of elevated blood lead levels. However, because no threshold for the adverse effect of lead on neurodevelopment has been found, the CDC emphasized that "primary prevention must serve as the foundation of the effort" (CDC 2005b). It further noted that "efforts to eliminate lead exposures through primary prevention have the greatest potential for success....Ultimately, all nonessential uses of lead should be eliminated."

There is growing concern that the use of lead ammunition for the hunting of wild game, a nonessential use of lead, may increase the lead exposure of adults and children who consume the harvested meat. Case reports have described the occasional consumer of wild game who had markedly elevated blood lead concentrations and associated symptoms that were attributed to lead shotgun pellets retained in the appendix or ascending colon (Hillman 1967, Durlach et al. 1986, Gustavsson and Gerhardsson 2005). However, the more widespread public health issue is the risk of subclinical, low level lead exposure associated with the ingestion of lead-contaminated meat. Johansen and his colleagues (Johansen et al. 2004) measured the lead concentration in cooked whole breast tissue of seabirds from Greenland killed with lead shot. Visible lead pellets were identified by x-ray and removed by dissection prior to analysis. Breast tissue of the thick-billed murre (n =32) contained a mean lead concentration of 0.73 ±2.9 µg/g (wet weight), while that of the common eider (n = 25) contained 6.1 ±13 µg/g. By comparison, the lead content of breast meat in 25 common eiders that were accidentally drowned in fishing nets rather than being shot contained 0.14 ±0.13 μg/g. Another recent investigation measured the lead concentration in samples of raw muscle meat freshly harvested from red deer killed with lead bullets (Dobrowolska and Melosik 2008). Lead concentration in the muscle declined as a function of radial distance from the bullet pathway. However, at a radius of 15 cm (approximately 6 inches) from the bullet path, the muscle contained lead at a mean concentration of 8.5 μg/g wet weight (n = 10) above lead levels found in muscle far distant from the bullet pathway (mean = 0.16 μg/g). At a radius of 25 cm (approximately 10 inches) from the bullet pathway, the meat contained a mean lead concentration that was 1.16 μg/g above the values found in the far distant muscle.

A recent experimental study by Mateo and colleagues in Spain (Mateo et al. 2007) observed that cooking meat containing imbedded lead pellets of lead shot contributes to the transfer of lead contamination to other portions of the meat. In this investigation, 1, 2 or 4 pellets of pre-fired #6 lead shot were manually imbedded in the breast of non-shot farm raised quails obtained from a supermarket. The breasts (n = 3 per group) were subsequently cooked by boiling in a solution of water, sunflower oil, and spices. The lead pellets were then removed, and the breast meat was analyzed for lead. Compared to breast meat cooked without an imbedded pellet (mean lead concentration <0.01 μg/g wet weight, range <0.01 to 0.01), the breast cooked with 2 imbedded pellets contained 0.49 μg/g (range 0.10 to 1.19), and that with 4 pellets contained 1.64 μg/g (range 1.07 to 2.12). Substantially higher lead values were found when vinegar was added to the boiling water in a traditional pickling recipe.

A portion of game meat for an adult might weigh 141 g (approximately 5 ounces) (EPA 1997, Hogbin et al. 1999), and that for a 3 to 5 year old child might weigh 100 g (approximately 3.5 ounces). It can therefore be seen that a single serving of game meat containing 1 μg/g lead may result in ingestion of 141 μg lead in an adult and 100 μg lead in a child. These amounts are markedly elevated compared to the estimated daily dietary intake of 2 to 10 μg lead now considered prevalent in the American diet (EPA 2006).

Studies conducted in native peoples who regularly consume game birds harvested with lead ammunition have observed a relationship between blood lead concentration and bird meat consumption. In a study of adult male ethnic Greenlanders, mean blood lead was 1.5 μg/dL (n = 4) among control subjects consuming no bird meals, compared to 7.4 ±4.7 μg/dL (n = 31) among those consuming 5.1 to 15 bird meals per month, and 8.2 ±4.5 μg/dL among those consuming 15.1 to 30 bird meals per month (Johansen et al. 2006). In a study of native Cree adults residing in northern Ontario, Canada, the geometric mean blood lead concentration of adult males was approximately 6.3 μg/dL, compared to 2.1 μg/dL in a control group (n = 25) from the industrialized city of Hamilton, Ontario (Tsuji et al. 2008a). Isotopic lead ratio analysis of the blood lead samples, locally used lead ammunition, and lichen (an environmental biosensor) determined that ammunition was the main source of lead exposure in the native group (Tsuji et al. 2008b).

The extent to which human consumption of venison and breast meat from game birds such as mourning doves harvested with lead ammunition may contribute to lead exposure in the United States is a topic of increasing interest, sparked in part by the recent detection of lead fragments in ground venison submitted by hunters to food pantries in several Midwestern states (Bihrle 2008). The North Dakota Department of Health, in conjunction with the National Center for Environmental Health of the US CDC, recently conducted a survey of blood lead concentrations among a convenience sample of 740 individuals, 80.8% of whom reported a history of wild game consumption, predominantly venison (Iqbal et al. 2008). Almost all of the subjects were adults, with the exception of 7 subjects between the ages of 2 to 5 years (0.9%), and 12 subjects between the ages of 6 to 14 years (1.6%). The geometric mean blood lead concentration was 1.17 μg/dL (range 0.18 to 9.82 μg/dL), lower than the US population geometric mean of 1.56 μg/dL for adults 20 years of age and older (CDC 2005a). Eight participants (1.1%) had blood lead concentrations ≥ 5 μg/dL. In multivariate analysis that adjusted for age, sex, race, age of housing, and lead-related occupations and hobbies, individuals who reported consuming game meat had an increment in blood lead of 0.3 μg/dL (95% C.I. 0.157, 0.443). In like manner, individuals who had consumed game meat within the past month had a covariate-adjusted blood lead concentration that was 0.3 to 0.4 µg/dL higher than those who had last consumed it more than 6 months ago. Based upon the findings of this survey, the North Dakota Department of Health advised that pregnant women and children younger than 6 years of age should not eat venison harvested with lead bullets (NDDH 2008).

The magnitude of the health risk associated with consumption of game harvested with lead ammunition is likely to be influenced by multiple factors including, but not limited to the lead content of the ingested meat, the particle size and solubility of any ingested lead residues, the manner in which the meat is cooked or prepared, the frequency of consumption, and the age of the consumer. In a further effort to understand the potential impact, the LeadSpread model of the California Department of Toxic Substance Control (DTSC 2007) was used to estimate the median (50th percentile) and 95th percentile increment in blood lead concentration in adults and children consuming two or five portions of game meat per week containing *soluble* lead at a concentration of 1 µg/gram wet weight. This concentration was selected for modeling based on the findings of some analytical studies, summarized above, that suggest that a value of this magnitude might exist in servings of game meat harvested with lead ammunition, after intact pellets have been removed.

Table 1 presents estimates of the 50th percentile and 95th percentile blood lead concentrations of children and adults associated with consumption of either two or five game meals per week at two different levels of bioavailability. The relative bioavailability of lead residues present in cooked game meat harvested with lead ammunition has not been examined experimentally. However, a rough estimate of 0.2 was utilized after comparing the blood lead increment of rats fed a diet of 0.075% lead derived from small particles of metallic lead to that found in rats fed a diet of 0.02% lead derived from small particles of lead acetate (Barltrop and Meek, 1979). For purposes of discerning the upper bound of the influence of relative bioavailabilty of metallic lead, Table 1 also presents results obtained by setting relative bioavailability to 1.0.

The blood lead values in Table 1 represent the sum obtained by *adding* the estimated increment in blood lead attributed to the game meat consumption to the median (50th percentile) blood lead concentration for children or adults found in a recent U.S. population National Health and Nutrition Evaluation Survey (NHANES) (CDC 2005a). Although the NHANES general population estimates might have included some individuals who consume game meats, their impact on the general population median value can, for purposes of this example, be considered minor. The increment in blood lead attributed to game meat consumption in Table 1 can be found by subtracting 1.5 from the child values, and 1.6 from the adult values. It is notable that the median (50th percentile) increment calculated by the LeadSpread model for adults consuming two game meat meals per week containing 1 µg/g lead at a relative bioavailability of 0.2 is 0.3 µg/dL. This is the same increment in blood lead associated with game meat consumption in the North Dakota survey cited above (Iqbal 2008).

The main implication of the results yielded by the model is that regular consumption of game meat harvested with lead ammunition and contaminated with lead residues may cause relatively substantial increases in blood lead compared to background levels, particularly in children. Additional epidemiological investigations of potentially affected populations to further define the magnitude of the risk are warranted. Any such risk would be entirely avoidable by the use of the available alternatives to lead ammunition.

Table 1. Estimated blood lead distribution associated with regular consumption of game meat containing 1 ppm lead due to contamination from lead ammunition (background level plus game meat increment).[1]

Game meat meals per week[2]	Relative Bioavailability[3]	Estimated child blood lead level (µg/dL)[4]		Estimated adult blood lead level (µg/dL)[5]	
		50th percentile	95th percentile	50th percentile	95th Percentile
2	0.2	2.4	3.5	1.9	2.3
2	1.0	6.1	11.4	3.2	5.1
5	0.2	3.8	6.4	2.4	3.3
5	1.0	12.5	26.5	5.6	10.3

[1] Estimates derived from use of LeadSpread Version 7 (DTSC. 2007), assuming geometric standard distribution of 1.6; ingestion constant (µg/dL)/µg/day) of 0.16 for child (age 3 to 5 years); and ingestion constant for adult of 0.04.
[2] Child meal consists of 100 g of game meat with a lead concentration of 1 ppm. Adult meal consists of 141 g of game meat with a lead concentration of 1 ppm
[3] Relative to bioavailability of dietary lead acetate in rats (DTSC, 2007)
[4] Values shown represent blood lead increment attributed to game meat consumption *added* to 50th percentile blood lead reported in CDC Third National Report on Human Exposure to Environmental Chemicals, 2001 – 2002 (1.50 µg/dL for child 1-5 years of age) (CDC, 2005)
[5] Values shown represent blood lead increment attributed to game meat consumption *added* to 50th percentile blood lead reported in CDC Third National Report on Human Exposure to Environmental Chemicals, 2001 – 2002 (1.60 µg/dL for adults 20 years and older) (CDC, 2005)

ACKNOWLEDGEMENTS

The assistance of Estelle N. Shiroma, D. Env, and Ms Kun Zhao, of ENVIRON, Emeryville, CA with the LeadSpread blood lead modeling is gratefully acknowledged.

LITERATURE CITED

ATSDR. Agency for Toxic Substances and Disease Registry. 2007. Toxicological Profile for Lead. ATSDR, Atlanta, Georgia, USA.

BARLTROP, D., AND F. MEEK. 1979. Effect of particle size on lead absorption from the gut. Archives of Environmental Health 34:280 – 285.

BIHRLE, C. 2008. An evolving perspective on lead in venison. Pages 30-35 in North Dakota Outdoors Magazine. August-September 2008. North Dakota Game and Fish Department, Bismarck, North Dakota, USA.

BORJA-ABURTO, V. H., I. HERTZ-PICCIOTTO, M. R. LOPEZ, P. FARIAS, C. RIOS, AND J. BLANCO. 1999. Blood lead levels measured prospectively and risk of spontaneous abortion. American Journal of Epidemiology 150:590-597.

CDC. CENTERS FOR DISEASE CONTROL AND PREVENTION. 1991. Preventing Lead Poisoning in Young Children. CDC, Atlanta, Georgia, USA.

CDC. CENTERS FOR DISEASE CONTROL AND PREVENTION. 2005a. Third National Report on Human Exposure to Environmental Chemicals. NCEH Pub. No. 05-0570, Lead CAS No. 7439-92-1. CDC, Atlanta, Georgia, USA.

CDC. CENTERS FOR DISEASE CONTROL AND PREVENTION. 2005b. Preventing Lead Poisoning in Young Children. CDC, Atlanta, Georgia, USA.

CHAI, S. S., AND R. C. WEBB. 1988. Effects of lead on vascular reactivity. Environmental Health Perspectives 78:85-89.

DIETRICH, K. N., M. D. RIS, P. A. SUCCOP, O. G. BERGER, AND R. L. BORNSCHEIN. 2001. Early exposure to lead and juvenile delinquency. Neurotoxicology and Teratology 23:511-518.

DOBROWOLSKA, A., AND M. MELOSIK. 2008. Bullet-derived lead in tissues of the Wild Boar (*Sus scrofa*) and Red Deer (*Cervus elaphus*). European Journal of Wildlife Research 54:231-235.

DTSC. DEPARTMENT OF TOXIC SUBSTANCES CONTROL. 2007. LeadSpread 7. DTSC, Sacramento, California, USA.

DURLACH, V., F. LISOVOSKI, A. GROSS, G. OSTERMANN, M. LEUTENEGGER. 1986. Appendectomy in an unusual case of lead poisoning. Lancet 1(8482):687-688

EPA. ENVIRONMENTAL PROTECTION AGENCY. 1997. Exposure Factors Handbook. U.S. Environmental Protection Agency, National Center for Environmental Assessment, Washington, DC, USA.

EPA. ENVIRONMENTAL PROTECTION AGENCY. 2006. Air Quality Criteria for Lead (Final). U.S. Environmental Protection Agency, National Center for Environmental Assessment, Washington, DC, USA.

FINE, B. P., T. VETRANO, J. SKURNICK, AND A. TY. 1988. Blood pressure elevation in young dogs during low-level lead poisoning. Toxicology and Applied Pharmacology 93:388-393.

GILBERT, S. G., AND B. WEISS. 2006. A rationale for lowering the blood lead action level from 10 to 2 microg/dL. Neurotoxicology 27:693-701.

GUSTAVSSON, P., AND L. GERHARDSSON. 2005. Intoxication from an accidentally ingested lead shot retained in the gastrointestinal tract. Environmental Health Perspectives 113:491-493

HILMAN, F. E. 1967. A rare case of chronic lead poisoning: polyneuropathy traced to lead shot in the appendix. Industrial Medicine and Surgery 36:488-492

HOGBIN, M., A. SHAW, AND R. S. ANAND. 1999. Food portions and servings. How do they differ. Nutrition Insights, Volume 11, March 1999. USDA Center for Nutrition Policy and Promotion, Washington, DC, USA.

HU, H., A. ARO, M. PAYTON, S. KORRICK, D. SPARROW, WEISS S. T., AND A. ROTNITZKY. 1996. The relationship of bone and blood lead to hypertension. Journal of the American Medical Association 275:1171-1176.

IQBAL, S. 2008. Epi-Aid Trip Report:Assessment of human health risk from consumption of wild game meat with possible lead contamination among the residents of the State of North Dakota. National Center for Environmental Health, Centers for Disease Control and Prevention, Atlanta, Georgia, USA.

JOHANSEN, P., G. ASMUND, AND F. RIGET. 2004. High human exposure to lead through consumption of birds hunted with lead shot. Environmental Pollution 127:125-129.

JOHANSEN, P., H. S. PEDERSEN, G. ASMUND, AND F. RIGET. 2006. Lead shot from hunting as a source of lead in human blood. Environmental Pollution 142:93-97.

KOSNETT, M. J., R. P. WEDEEN, S. J. ROTHENBERG, K. L. HIPKINS, B. L. MATERNA, B. S. SCHWARTZ, H. HU, AND A. WOOLF. 2007. Recommendations for medical management of adult lead exposure. Environmental Health Perspectives 115:463-471.

LANPHEAR, B. P., R. HORNUNG, J. KHOURY, K. YOLTON , P. BAGHURST, D. BELLINGER, R. L. CANFIELD, K. N. DIETRICH, R. BORNSCHEIN, T. GREENE, S. J. ROTHENBERG, H. L. NEEDLEMAN, L. SCHNAAS, G. WASSERMAN, J. GRAZIANO, AND R. ROBERTS. 2005. Low-level lead exposure and children's intellectual function: An international pooled analysis. Environmental Health Perspectives 113:894-899.

MATEO, R., M. RODRIGUEZ-DE LA CRUZ, D. VIDAL, M. REGLERO, AND P. CAMARERO. 2007. Transfer of lead from shot pellets to game meat during cooking. Science of the Total Environment. 372:480-485.

MUNTNER P., S. VUPPUTURI, J. CORESH, AND V. BATUMAN. 2003. Blood lead and chronic kidney disease in the general United States population: Results from NHANES III. Kidney International 63:104-150.

NAVAS-ACIEN A., E. GUALLAR, E. K. SILBERGELD, AND S. J. ROTHENBERG. 2007. Lead exposure and cardiovascular disease – a systematic review. Environmental Health Perspectives 115:472-482.

NAWROT, T. S., L. THIJS, E. M. DEN HOND, H. A. ROELS, AND J. A. STAESSEN. 2002. An epidemiological re-appraisal of the association between blood pressure and blood lead: a meta-analysis. Journal of Human Hypertension 16:123-131.

NCHS. NATIONAL CENTER FOR HEALTH STATISTICS. 1984. Blood lead levels for persons ages 6 months to 74 years. United States, 1976-1980. Vital and Health Statistics. Series 11, No. 233. Pub. No. (PHS) 84-1683. National Center for Health Statistics, Washington, DC, USA.

NDDH. NORTH DAKOTA DEPARTMENT OF HEALTH. 2008. News release: State Health Department Announces Preliminary Findings in Blood Lead Level Study. November 5, 2008. North Dakota Department of Health, Bismarck, North Dakota, USA.

PIRKLE, J. L., J. SCHWARTZ., J. R. LANDIS., AND W. R. HARLAN. 1985. The relationship between blood lead levels and blood pressure and its cardiovascular risk implications. American Journal of Epidemiology 121:246-258.

SCHNAAS, L., S. J. ROTHENBERG, M. F. FLORES, S. MARTINEZ, C. HERNANDEZ, E. OSORIO, S. R. VELASCO, AND E. PERRONI. 2006. Reduced intellectual development in children with prenatal lead exposure. Environmental Health Perspectives 111:791-797.

SCHOBER, S. E., B. MIRAL, B. I. GRAUDBARD, D. J. BRODY, AND K. M. FLEGAL. 2006. Blood lead levels and death from all causes, cardiovascular disease, and cancer: Results from the NHANES III mortality study. Environmental Health Perspectives 114:1538- 1541.

SCHWARTZ, J. 1988. The relationship between blood lead and blood pressure in the NHANES II survey. Environmental Health Perspectives 78:15-22.

SCHWARTZ, J. 1995. Lead, blood pressure, and cardiovascular disease in men. Archives of Environmental Health 50:31-37.

SHIH, R. A., T. A. GLASS, K. BANDEEN-ROCHE, M. C. CARLSON, K. I. BOLLA, A. C. TODD, AND B. S. SCHWARTZ. 2006. Environmental lead exposure and cognitive function in community-dwelling older adults. Neurology 67:1556-1562.

STAESSEN, J. A., R. R. LAUWERYS, J. P. BUCHET, C. J. BULPITT, D. RONDIA, Y. VANRENTERGHEM, AND A. AMERY. 1992. Impairment of renal function with increasing blood lead concentrations in the general population. The Cadmibel Study Group. New England Journal of Medicine 327:151-156.

TSAIH. S. W., S. KORRICK, J. SCHWARTZ, C. AMARASIRIWARDENA, A. ARO, D. SPARROW, AND H. HU. 2004. Lead, diabetes, hypertension, and renal function: The normative aging study. Environmental Health Perspectives 112:1178-1182.

TSUJI, L. J. S., B. C. WAINMAN, I. D. MARTIN, J. P. WEBER, C. SUTHERLAND, E. N. LIBERDA, AND E. NIEBOER. 2008a. Elevated blood-lead levels in First Nation people of northern Ontario Canada: Policy implications. Bulletin of Environmental Contamination and Toxicology 80:14-18.

TSUJI, L.J.S., B. C. WAINMAN, I. D. MARTIN, C. SUTHERLAND, J. P. WEBER, P. DUMAS, AND E. NIEBOER. 2008b. The identification of lead ammunition as a source of lead exposure in First Nations: The use of lead isotope ratios. Science of the Total Environment 393:291-298.

VAZIRI, N. D., AND M. KHAN. 2007. Interplay of reactive oxygen species and nitric oxide in the pathogenesis of experimental lead-induced hypertension. Clinical and Experimental Pharmacology and Physiology 34:920-925.

WASSERMAN, G. A., X. LIU, D. POPOVAC, P. FACTOR-LITVAK, J. KLINE, C. WATERNAUX, N. LOIACONO, AND J. H. GRAZIANO. 2000. The Yugoslavia prospective lead study: contributions of prenatal and postnatal lead exposure to early intelligence. Neurotoxicology and Teratology 22:811-818.

WEISSKOPF, M. G., S. P. PROCTOR, R. O. WRIGHT, J. SCHWARTZ, A. SPIRO, D. SPARROW, H. NIE, AND H. HU. 2007. Cumulative lead exposure and cognitive performance among elderly men. Epidemiology 18:59-66.

BIOLOGICAL AND SOCIETAL DIMENSIONS OF LEAD POISONING IN BIRDS IN THE USA

MILTON FRIEND[1], J. CHRISTIAN FRANSON[1], AND WILLIAM L. ANDERSON[2]

[1]*US Geological Survey, National Wildlife Health Center, 6006 Schroeder Road, Madison, WI 53711, USA.*

[2]*Illinois Natural History Survey, 607 East Peabody Drive, Champaign, IL 61820, USA.*

ABSTRACT.—The ingestion of spent lead shot was known to cause mortality in wild waterfowl in the US a century before the implementation of nontoxic shot regulations began in 1972. The biological foundation for this transition was strongly supported by both field observations and structured scientific investigations. Despite the overwhelming evidence, various societal factors forestalled the full transition to nontoxic shot for waterfowl hunting until 1991. Now, nearly 20 years later, these same factors weigh heavily in current debates about nontoxic shot requirements for hunting other game birds, requiring nontoxic bullets for big game hunting in California Condor range and for restricting the use of small lead sinkers and jig heads for sport-fishing. As with waterfowl, a strong science-based foundation is requisite for further transitions to nontoxic ammunition and fishing weights. Our experiences have taught us that the societal aspects of this transition are as important as the biological components and must be adequately addressed before alternatives to toxic lead ammunition, fishing weights, and other materials will be accepted as an investment in wildlife conservation. *Received 16 May 2008, accepted 8 July 2008.*

FRIEND, M., J.C. FRANSON, AND W.L. ANDERSON. 2009. Biological and societal dimensions of lead poisoning in birds in the USA. *In* R. T. Watson, M. Fuller, M. Pokras, and W. G. Hunt (Eds.). Ingestion of Lead from Spent Ammunition: Implications for Wildlife and Humans. The Peregrine Fund, Boise, Idaho, USA. DOI 10.4080/ilsa.2009.0104

Key words: Lead poisoning, waterfowl, nontoxic shot, societal dimensions, litigation.

THE SCIENTIFIC ASPECTS of lead poisoning are but one component for consideration in addressing this disease in humans and animals alike. Here we address lead poisoning in waterfowl and key events associated with the transition from the use of lead shot to nontoxic shot for waterfowl hunting in the US. Historic documentation of lead poisoning in waterfowl is meshed with wildlife management factors and wildlife conservation transitions to provide issue context. We then consider the scientific foundation establishing lead poisoning as a mortality factor impacting waterfowl populations. Major issues that arose in the pursuit of a solution to address lead poisoning in waterfowl are then addressed. We conclude with an extended discussion that highlights key points for consideration by those engaged in attempts to further reduce lead exposures in wild birds.

Early Documentation.—Lead poisoning was first identified as a disease in wild birds in an 1842 scientific paper published in Berlin, Germany (von Fuchs 1842). The first published reports of this disease in the US appeared in the sporting and scientific literature of the late 1800s, and cited observations of lead poisoning of waterfowl appeared as early as the 1870s (Grinnell 1894, Hough 1894, Grinnell 1901). Additional sporadic reports can be

[1] The use of trade or product names in this report is for identification purposes only and does not constitute endorsement by the U.S. Government.

found in the literature during the first three decades of the 20th Century (Bowles 1908, McAtee 1908, Forbush 1912, Munro 1925, Van Tyne 1929, McGath 1931, Howard 1934, McLean 1939), clearly establishing lead poisoning as common in waterfowl and widely distributed geographically (Phillips and Lincoln 1930, Shillinger and Cottam 1937, Bellrose 1959, Mississippi Flyway Council Planning Committee 1965). Despite those early warnings, the first nontoxic shot requirements were not initiated until 1972, and the use of lead shot for waterfowl hunting was not completely banned in the US until the start of the 1991 waterfowl season (US Fish and Wildlife Service 1986a). The path leading to this endpoint traversed decades of observation, discovery, investigation, and controversy (Table 1), all of which are closely linked to the changing dynamics of the human interface with wildlife and that of wildlife conservation in general. Worldwide at least 20 countries had initiated some form of nontoxic shot requirements for waterfowl and/or for other hunting by the year 2000 (Figure 1). International efforts to "get the lead out" are being advanced by the African-Eurasion Migratory Waterbird Agreement and ongoing involvement of Wetlands International (formerly the International Waterfowl and Wetlands Research Bureau) (Beintema 2004).

Current arguments for retention of lead within the US for traditional uses in hunting, fishing, and shooting sports are similar to those of the past (Table 2). Thus, it is prudent for those seeking further reductions of lead poisoning in wildlife to be fully cognizant of the transition, conflicts, and factors that facilitated resolution of the lead poisoning issue in waterfowl. Application of this knowledge should expedite further transitions in the replacement of existing traditional lead uses in these sports so that past mistakes are not repeated. We highlight important benchmarks associated with the waterfowl lead poisoning issue and comment on important biological, social, economic, and political aspects of those benchmarks. In doing so, we identify motivating and inhibitory factors influencing the transition to nontoxic shot for hunting waterfowl.

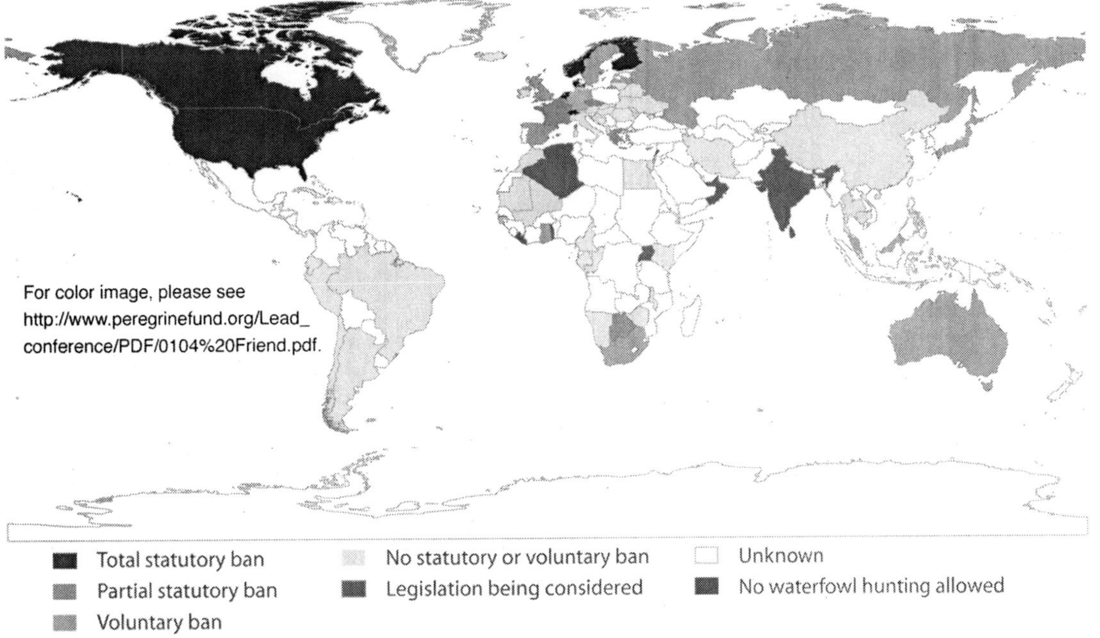

Figure 1. Countries reporting in 2000 to have various types of bans on the use of lead shot for waterbird hunting (Developed from Beintema 2004).

Table 1. Milestones in the transition to nontoxic shot use for waterfowl hunting in the US (see text for details).

Year	Discovery	Concern	Regression	First Actions
1874	Anecdotal mortality reports.			
1894	First documented mortality.			
1915	Numerous shot found in swan gizzards.			
1916	Numerous shot found in sediments near duck blinds.			
1919	First lead toxicity study in wild ducks.			
1930		Leading scientists report lead poisoning to be widespread.		
1936		Nontoxic shot development first pursued.		
1937	First broad-scale evaluation of shot ingestion by waterfowl.			
Early 1940s		Lead poisoning reported to be of great importance for ducks.		
1948		Olin Corporation initiates quest for nontoxic shot.		
Early 1950s		Expanded concerns and investigations		
Mid 1950s			Habitat conditions restore duck populations; interest in lead poisoning wanes; non-toxic shot stops development.	
1959	Bellrose report on lead as a waterfowl mortality factor.			
Early 1960s		Major waterfowl populations decline; interest in lead poisoning heightened.		
1965				First field test of nontoxic shot. Flyway Council urges development of non-toxic shot.
1972				First nontoxic shot use requirements.
1974–1976				FWS EIS proposing nationwide nontoxic shot use.
1976		First lawsuit opposing non-toxic shot regulations by FWS.	Government prevails.	
1978		Stevens Amendment prevents FWS from initiating or enforcing nontoxic shot requirements without State approval.		

Year	Discovery	Concern	Regression	First Actions
Early 1980s	NWHC documents numerous lead poisoning cases in Bald Eagles.	Lawsuits filed against state wildlife agencies and FWS nontoxic shot regulations.	Government prevails.	
Late 1980s		Lawsuit filed against FWS to prevent nontoxic shot use in California.		
1991			Nontoxic shot required nationwide for waterfowl hunting.	

[a] See Anderson 1992, Feierabend 1985.

Wildlife Management Factors.—Wildlife management within the US is often shaped by forces which are political, economic, and social, and is driven by the involvement of multiple segments of society (Heering 1986). The management of lead poisoning in waterfowl has involved all of these forces, and they will drive future efforts to manage lead poisoning in other wildlife. Therefore, it is useful to provide context for lead poisoning in birds over time, because current views of society towards wildlife and approaches towards wildlife conservation differ greatly from those of earlier decades. For example, wildlife conservation has broadened from a protective regulatory approach for limiting take as a means for species preservation to a focus on habitat management and other means to enhance and sustain wildlife populations (Leopold 1933). Further transition has redirected efforts from a species by species orientation to a biodiversity orientation in which ecosystems are the primary focus, even though species management continues. Also, today a far greater percentage of those involved in shaping the resolution of wildlife conservation issues are nonconsumptive resource-users, many of which seek different outcomes than their consumptive resource-user contemporaries (Sparrowe 1992).

Table 2. Common arguments by activity participants against nontoxic alternatives for lead uses in hunting, fishing and shooting sports.

Argument	Activity [a]				
	Waterfowl hunting	Other bird hunting	Other hunting	Shooting sports	Fishing
Magnitude of lead poisoning does not warrant ban (i.e., "the cure is worse than the disease").	X	X	X	X	X
Discrimination—alternatives not feasible for some uses (i.e., small shotshell gauges); gender and age group impacts.	X	X		X	X
Decreased achievement efficacy such as reduced effective range.	X	X			X
Increased secondary impacts (i.e., greater crippling loss).	X	X			
Equipment and personal safety hazards (i.e., ricochets, dental damage).	X	X	X	X	
Increased costs.	X	X	X	X	X
Make your own materials not available.	X				X
Loss of participants in activity if lead use is banned (i.e., lost revenues for conservation and local communities.	X	X	X		

[a] Ban on use of lead for waterfowl hunting completed in 1991 for US; few restrictions currently exist for other types of hunting. Shooting sports have not been subject to nontoxic ammunition requirements but have some restrictions on environments where spent ammunition can fall. Some requirements exist for nontoxic sinkers and jig heads for sport fishing and for some types of hunting in specific geographic and local areas.

Wildlife Conservation Transitions.—Although the origin of the US conservation movement can be traced back to late 19th century, major development did not occur until the 1920s and 1930s. Prior to that time, wildlife conservation in America was "...almost wholly a history of hunting controls" (Leopold 1933). During the 1920s and 1930s, American sportsmen were a major force in "...convincing the government to take a lead role in conserving and managing the nation's natural resources" (Heering 1986). At the federal level, the Bureau of Biological Survey (BBS) in the Department of Agriculture (USDA) was the organization addressing wildlife conservation issues. However, in 1939 the BBS was transferred to the Department of the Interior (DOI), where it was made part of the then US Fish and Wildlife Service (FWS) (Friend 1995). While part of the USDA, BBS scientists carried out landmark investigations of lead poisoning in waterfowl (Wetmore 1919, Shillinger and Cottam 1937).

Major foundational components that served the conservation and restoration of America's wildlife evolved during the pre-WWII era. The "Dirty Thirties" was a time of drought and the Dust Bowl, fiscal panic, and poverty that took a heavy toll on humans, wildlife habitat, and wildlife species that became food to sustain human life. However, the struggles of wildlife and society during those times provided stimulus for leading conservationists from the public sector and government to champion critical enactments and establish major programs that continue today. Thus, in 1935 the federal Duck Stamp and Fish and Wildlife Coordination Acts were passed, and the Cooperative Wildlife Research Unit Program was established. The National Wildlife Federation (NWF) and the first North American Wildlife and Natural Resources Conferences were initiated in 1936. The Federal Aid in Wildlife Restoration program, commonly referred to as the Pittman-Robertson (P-R) program, began in 1937 and is funded by an 11% manufacturer's excise tax on certain equipment used in hunting and by a 10% manufacturer's excise tax on handguns (Williamson 1987). The P-R program, and its later counterpart for fisheries (the D-J or Dingell-Johnson Program), were in essence sportsman-imposed taxes designed to enhance fish and wildlife resources and provide public areas for hunting, fishing, and shooting sports. The impetus for the P-R program followed WWI as the number of hunters greatly increased and impacts from the continuing diminishment of wildlife habitat due to other human needs greatly impacted opportunities for sport hunting (Kallman 1987). It is noteworthy that the first P-R project approved and funded was granted to the Utah Department of Fish and Game in 1938 to construct a 5-mile dike in the Weber River Delta of the Great Salt Lake to assist with combating avian botulism (Williamson 1987). Numerous lead poisoning investigations have been funded by the P-R program during its more than 60 years.

Following WWII, there was another major increase in the number of hunters, and interests grew in the new concepts of wildlife management championed by Aldo Leopold (Leopold 1933, McCabe 1987). By the 1950s, federal and state government agencies began to assume more and more responsibility for managing fish and game species (Heering 1986). A significant outcome of that transition was that wildlife professionals began calling for actions to address lead poisoning, partly because of greatly diminished waterfowl numbers in the early 1960s (Figure 2). At the same time, the hunting public was being disenfranchised from their leadership role in conservation. As noted by Heering (1986), "...by the 1970s, sportsmen's relationship to wildlife professionals had changed from one of 'co-worker' to 'customer.'" In retrospect, one can only wonder how the transition to nontoxic shot for waterfowl hunting might have differed if the relations between the principals involved had been more like "co-workers" rather than agency clients. A current question is what type of relations are now being forged to "get the lead out" of other sporting activities?

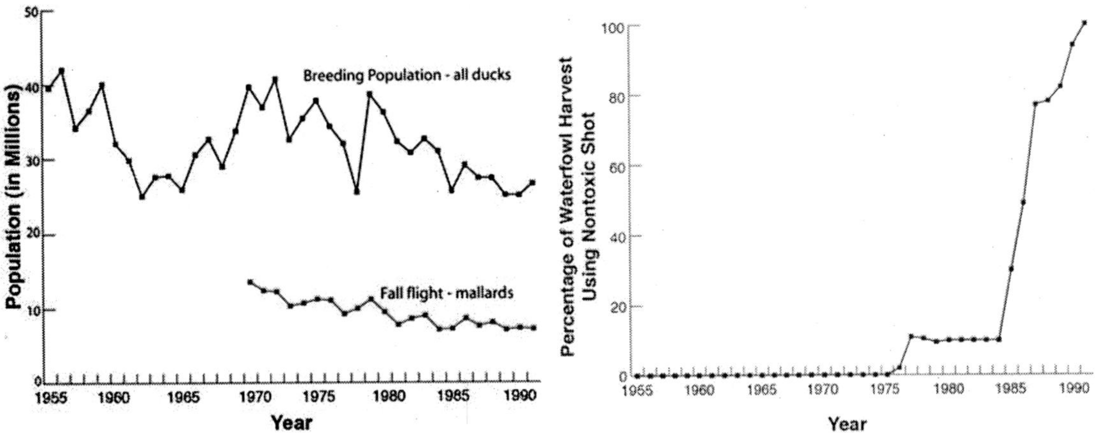

Figure 2a. North American Mallard Duck breeding population estimates, 1955–1991, and fall flight of Mallards, 1970–1991 (developed from records of the USGS Bird Banding Laboratory), and (**2b**) percentage of US waterfowl harvest using nontoxic shot during that same time period (developed from Anderson 1992).

METHODS

We utilize our personal involvement and experiences with lead poisoning in birds and as participants in the transition to nontoxic shot use (for waterfowl hunting in the US) as a foundation for our presentation of the issue. The scientific literature and other documents are used to support the evaluations provided. We begin by defining lead poisoning as a disease in birds and how exposure to lead is documented as the source for that disease.

Lead Poisoning in Birds.—We consider lead poisoning of birds to occur by primary, secondary, and environmental exposures (Figure 3). With environmental exposure, which is beyond the scope of this paper, it has been more difficult to document a cause and effect relationship for avian toxicity than for particulate lead ingestion. As a result, much work remains to be done in the area of environmental lead exposure as it relates to wild bird health. Further, some opposed to nontoxic shot have argued that environmental exposure rather than lead shot is a major source of lead residues found in waterfowl tissues (Winchester Group 1974, Sanderson and Bellrose 1986, Fisher et al. 2006).

Findings of particulate lead in the digestive tract of birds provide physical evidence of the ingestion of lead shot, as well as lead bullets, paint chips, solder, and other materials (Table 3). The presence of ingested lead shot has been an important factor

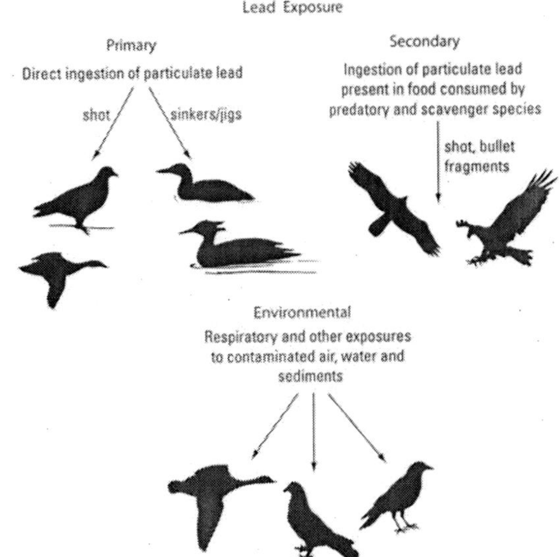

Figure 3. Primary routes for lead exposure in wild birds.

in assessing lead exposure rates in waterfowl because of the relationship between hunting pressure and shot deposition in waterfowl habitat (Bellrose 1959, US Fish and Wildlife Service 1976, 1986a). However, lead exposure and poisoning of waterfowl also has occurred in areas where few hunters are present, including remote areas of Alaska (Franson et al. 1995, Flint et al. 1997, Grand et al. 1998) and in Puerto Rico, the Virgin Islands, and similar areas (Morehouse 1992a). The initial DOI Environmental Impact Statements (EIS) to establish

Table 3. Relative frequency of ingested particulate lead and lead poisoning by various species groups.[a]

Species Group	Minimum no. of species affected	Lead Type: Spent Shot	Bullet Fragments	Sinkers	Paint chips[b]	Mine tailings[b]	Others[c]	Selected Citations
Waterfowl	19	● Frequent		Occasional		Common	○ Rare	Bellrose (1959), Anderson (1975), Blus et al. (1989), Franson et al. (1995), Beyer et al. (1998), Sileo et al. (2001), Franson et al. (2003), Degernes et al. (2006)
Coots and rails	6	Occasional						Jones (1939), Artmann and Martin (1975)
Shorebirds and gulls	4	Occasional						Kaiser and Fry (1980), Locke et al. (1991), NWHC files
Cranes	2	Occasional		○ Rare			Common	Windingstad et al. (1984), Franson and Hereford (1994)
Pelicans	1			Occasional				Franson et al. (2003)
Loons	2			● Frequent				Sidor et al. (2003), Wilson et al. (2004)
Other waterbirds	3	Occasional			Common			Sileo and Fefer (1987), Franson et al. (2003)
Raptors and scavengers	10	● Frequent	● Frequent					Reichel et al. (1984), Franson et al. (1996), Meretsky et al. (2000)
Gallinaceous birds	4	Occasional						Campbell (1950), Hunter and Rosen (1965), Lewis and Schweitzer (2000)
Doves	2	● Frequent						Locke and Bagley (1967), Castrale (1991), DeMent et al. (1987), Schulz et al. (2002)
Passerines	6	Occasional						Vyas et al. (2000), Lewis et al. (2001)

Frequency of Findings ● Frequent ◉ Common ⭘ Occasional ○ Rare

[a]See also Scheuhammer and Norris (1995). Scheuhammer et al. (2003), Fisher et al. (2006).
[b] Locally important as numerous birds may be poisoned from this source.
[c] Occasional to rare findings of ingested materials such as solder and lead fragments which have not been identified as a specific product.

nontoxic shot zones for waterfowl hunting were based on known lead poisoning losses and the amount of hunting pressure (US Fish and Wildlife Service 1974, 1976). However, that approach was abandoned because multiple variables, in addition to pellet deposition, influence shot ingestion (Bellrose 1959, Sanderson and Bellrose 1986, DeStefano et al. 1992). Thus, a documented 5% of gizzards with ingested lead shot became one of the criteria for recommending areas for nontoxic shot use (US Fish and Wildlife Service 1986a).

Investigations by the National Wildlife Health Center (NWHC) and others have demonstrated that rates of ingested lead shot in gizzards/stomachs do not adequately reflect lead exposure. Paired liver and gizzard analyses by the NWHC were part of the US Fish and Wildlife Service (FWS) lead monitoring program of the 1980s. Lead exposure rates based on elevated lead residues in livers (≥2.0 ppm, wet weight) were generally two-fold or more higher than rates based on ingested shot found in gizzards (US Fish and Wildlife Service 1986a). Similarly, DeStefano et al. (1991, 1995) found higher exposure rates based on elevated blood lead concentrations (≥0.18 ppm, wet weight) than on ingested shot in gizzards of Canada Geese (*Branta canadensis*) in nontoxic shot areas and those where lead was still used.

Lead residues in soft tissues have become generally accepted as criteria for evaluating lead exposure in waterfowl. The International Association of Fish and Wildlife Agencies (IAFWA) proposed blood and liver concentrations of 0.2 ppm and 2.0 ppm wet weight, respectively, when found in ≥ 5% of samples, as decision criteria for recommending conversions to nontoxic shot for waterfowl hunting (US Fish and Wildlife Service 1986a). Suggested interpretations of background, elevated, and toxic levels of lead in tissues of waterfowl and other avian species were proposed by Pain (1996) and Franson (1996), respectively, based on review of laboratory and field investigations.

Primary and secondary exposures of particulate lead are well-documented causes of mortality for a broad array of avian species (Table 3). Nevertheless, the mere presence or absence of lead fragments in the digestive tract of birds is insufficient to conclude that lead was, or was not, the cause of death. Such a conclusion is subjective unless supported by other appropriate findings (Locke and Thomas 1996, Friend 1999a). In the 1970s, inadequate documentation of lead poisoning resulted in mounting disagreements associated with factors influencing lead toxicity such as diet, weather, etc., causing the FWS to designate nontoxic shot zones primarily on the basis of state recommendations. Later, however, the FWS developed recommendations for uniform criteria for monitoring lead exposure and poisoning (US Fish and Wildlife Service 1986a).

A scientifically defensible diagnosis of lead poisoning in a bird carcass is based on a combination of postmortem findings and laboratory assays including lead residues in soft tissues (Table 4). Diagnosis of an epizootic is based on field observations and environmental conditions, and include signs in clinically ill birds, in addition to postmortem findings and tissue analyses (Sanderson and Bellrose 1986, Wobeser 1997, Friend 1999a). In waterfowl, highly visible evidence of lead poisoning is generally provided by clinical signs and gross pathology (Wobeser 1997, Friend 1999a) as described by early research studies involving the pathogenesis, toxicology, and other studies of lead intoxication in birds (Wetmore 1919, Coburn et al. 1951, Bates et al. 1968, Sileo et al. 1973, Clemens et al. 1975). Data of this breadth and quality became an important factor in resolving the waterfowl lead poisoning issue and were basic features of the FWS lead monitoring program conducted by the NWHC during the 1980s.

Mortality data supported by scientific investigations that correlated levels of exposure with population impacts were also needed to establish support for transition to nontoxic shot. Foundational studies by Wetmore (1919) and those that followed (Table 5) were invaluable in establishing the magnitude of losses due to lead poisoning (Bellrose 1959, Sanderson and Bellrose 1986). The robustness of this scientific foundation repeatedly overcame challenges to the toxicity of lead for birds and the magnitude of exposure occurring.

Table 4. Blood, liver, and bone concentrations associated with lead exposure in waterfowl.[a]

	Value Levels of Exposure		
Assay	**Background**	**Elevated**	**Toxic**
Blood (ppm wet weight)	<0.2	0.2–0.5	>1.0
Liver (ppm, wet weight)	<2.0	2.0–6.0	>6.0
Bone (ppm, dry weight)	<10	10–20	>20

[a]Adapted from Pain (1996).

Table 5. Important investigations that have provided a scientific foundation for evaluating the consequences of lead exposure in waterfowl populations.

Investigators (Year of Publication)	Primary Contributions
Wetmore (1919)	▪ Dosing experiments revealed that one No. 6 lead pellet might cause deaths of Mallards but six No. 6 pellets were always fatal. ▪ Results suggested population impacts might be inferred through frequency and amount of lead shot found in waterfowl gizzards.
Shillinger and Cottam (1937)	▪ Reported on the occurrence of shot in 8,366 gizzards of 14 species of ducks and concluded that the quantity of lead constituting a fatal dose was influenced by numerous factors. ▪ Study demonstrated that lead shot ingestion was widespread.
Jordan and Bellrose (1950)	▪ Reported that the nature of the diet rather than the dose of ingested lead was the more important variable determining lead toxicity. ▪ Found "game farm" Mallards to be less susceptible to lead poisoning than wild Mallards.
Bellrose (1951)	▪ Dosing and release of wild-trapped Mallards disclosed a progressive decrease in the rate of movement. ▪ Fluoroscoped wild-trapped Mallards had a progressive increase in ingested pellets from 3–4% prior to the hunting season to 12% by early December, thereby suggesting the importance of annual lead deposition on ingestion rates.
Coburn et al. (1951)	▪ Related the metabolism and deposition of lead in the tissues of ducks to the extent that clinical signs of intoxication could be predicted. ▪ Concluded that soft tissues from ducks could be collected as field samples for the determination of lead poisoning by chemical analysis.
Jordan and Bellrose (1951)	▪ Lead poisoning losses reported to be associated with waterfowl moving into heavily shot-over areas during late fall or winter. ▪ Reported that the majority (69.3%) of gizzards examined that contained shot had only 1 shot. ▪ Attributed differences in species ingestion rates of lead shot to variations in methods of feeding and types of habitat preferred.
Bellrose (1959)	▪ Classic report that placed lead poisoning in waterfowl in perspective. ▪ Documented the ecology of lead poisoning in waterfowl relative to the frequency and geographic distribution of epizootics, seasonality of occurrence, species affected, variations in shot ingestion among species and the effects of this disease on vulnerability to hunting, bird movements and year-of-banding mortality rate.
Longcore et al. (1974)	▪ Provided a basis for evaluating the significance of lead levels found in the tissues of waterfowl.
Clemens et al. (1975)	▪ Described the pathogenisis and associated pathology of lead poisoning in waterfowl. ▪ Found that rate of shot excretion by bird depends on shot size.
Sanderson and Bellrose (1986)	▪ Important comprehensive review of what was known about this disease, issued at a critical time in the transition to nontoxic shot use.
Rocke et al. (1997)	▪ Reported that the rates of lead exposure and lead poisoning mortality in sentinel Mallards maintained on previously hunted areas coincided with lead pellet density in sediments.

Adverse sublethal effects also have been reported in birds exposed to lead. Lead inhibits enzymes involved with heme synthesis, notably delta-aminolevulinic acid dehydratase (ALAD) and heme synthetase. Inhibition of heme synthetase allows protoporphyrin to accumulate in the blood. Thus, reduced ALAD activity and increased protoporphyrin concentrations in the blood have both been used as biomarkers for lead exposure (Finley et al. 1976, Roscoe et al. 1979). Kendall et al. (1996) reviewed sublethal effects of lead in non-waterfowl avian species and, although several studies reported no changes in reproductive parameters, reduced hatchability and testicular atrophy were reported in lead-dosed Mourning Doves (*Zenaida macroura*) and Ringed Turtle-doves (*Streptopelia risoria*), respectively (Veit et al. 1983, Buerger et al. 1986). In a study with Mallards (*Anas platyrhynchos*), no effects were noted on fertility, embryonic viability, or hatchability, but over a 2-year period, controls laid more eggs than lead-exposed birds (Elder 1954). Although the impact of lead exposure on the immune system of birds is poorly understood, studies with lead-exposed Mallards have demonstrated reduced antibody production and immunologic cell numbers (Trust et al. 1990, Rocke and Samuel 1991). Aspergillosis has been associated with lead exposure compromising the immune system of Canada Geese, thereby facilitating opportunistic infection by this fungal disease (Friend 1999b).

Although much of the information about lead poisoning in waterfowl is directly applicable for other species, extrapolation to other birds transforms specific findings to more general situations and consequently increases the risk of misuse and misinterpretation of data. A case in point is the Bald Eagle (*Haliaeetus leucocephalus*). Because documentation of lead poisoning in Bald Eagles did not occur until 1970 (Mulhern et al. 1970), and differences exist from waterfowl in food habits and in the digestive processing of food items, it became necessary to establish specific data for the Bald Eagle. A lead shot dosing study using non-releasable captive birds (Pattee et al. 1981) was conducted for this purpose as part of the assemblage of scientific findings used to assess lead poisoning in this then endangered species (US Fish and Wildlife Service 1986a).

RESULTS

The pursuit of nontoxic alternatives for lead shot was not closely associated timewise with initial recognition of lead poisoning as a cause of waterfowl mortality. Although lead poisoning of waterfowl was known to occur in the US as early as the 1870s, there was little motivation or urgency to act. Instead, from the late 1800s until the 1960s, leading conservationists of each decade drew attention to lead poisoning and noted the need to monitor the situation for possible future action (Grinnell 1901, McAtee 1908, Forbush 1912, Van Tyne 1929, Phillips and Lincoln 1930, Osmer 1940, Alder 1942, Day 1949, Bellrose 1951). As recently as 1959, noted waterfowl biologist, researcher, and lead poisoning investigator, Dr. Frank Bellrose, concluded his landmark scientific publications on lead poisoning with the following statement:

"At the present time, lead poisoning losses do not appear to be of sufficient magnitude to warrant such drastic regulations as, for example, prohibition of the use of lead shot in waterfowl hunting." (Bellrose 1959).

However, Bellrose went on to state that, "Should lead poisoning become a more serious menace to waterfowl populations, iron shot provides a possible means for overcoming it." That serious menace soon materialized. Now basic questions on this subject are: 1) What changed to create a need to act; 2) Why did it take nearly 20 years to complete the transition once there was involvement by the federal government to require nontoxic shot use; 3) Why was this effort totally focused on waterfowl hunting; 4) Was a nontoxic shot alternative already available at that time? The answer to the last question provides foundation for answers to the other three questions.

The Pursuit of Change.—There was early recognition by a number of conservationists that if lead poisoning was to be effectively addressed, it would be necessary to "get the lead out" (Phillips and Lincoln 1930, Alder 1942, Day 1959). In response to that need, the development of a lead-magnesium alloy shot pellet that would disintegrate in water, thus making spent shot unavailable to birds feeding in aquatic environments, was pursued (Dowdell and Green 1937, Green and Dowdell 1936). Those unsuccessful efforts of the mid-1930s were followed a decade later by major exploration for an alternative shot type. Olin Corporation assumed a leadership role in this effort, including a 1948 collaborative project involving its subsidiary, Western Cartridge (now Winchester), the Illinois Natural History Survey, and the University of Illinois School of

Table 6. Alternative shot types for lead tested during 1948–1949 collaborative project (developed from Jordan and Bellrose 1950).

Shot type	Comments
Lubaloy	▪ Thin copper coating over lead pellet Concept is delay of pellet erosion to provide time for pellet to be voided from the gizzard
Lead-tin-phosphorous	▪ Associated patent claims shot will be nontoxic if ingested ▪ Concept is phosphorus will act as neutralizing agent on lead dissolved in the gizzard
Lead-magnesium	▪ Magnesium used as substitute for arsenic and antimony components needed to provide hardness and spherical shape of lead shot ▪ Concept is magnesium will hydrolyze when shot is in water, causing irregular cracks across surface of the pellet and facilitate rapid disintegration of spent shot pellets
Lead-calcium	▪ Concept is that physiological advantage could result by introducing calcium along with lead since metabolism and storage of these elements follow similar pathways ▪ Thought that harmless storage in skeleton would take place following removal of harmful circulating lead

Veterinary Medicine (Jordan and Bellrose 1950). None of the alternatives evaluated by that project (Table 6) proved to be nontoxic. However, other research and development by Olin Corporation indicated that iron (steel) shot had potential as an alternative shot type (Bellrose 1959).

It also was recognized that there were considerations that needed to be overcome for further pursuit of iron shot. Specifically, "...the required manufacturing investment would be large, and this factor, coupled with uncertainty concerning customer acceptance, convinced Winchester-Western that manufacture of iron shot was not feasible unless drastic action was needed to save waterfowl from serious lead poisoning losses..." (Bellrose 1959). Further exploration of nontoxic shot began in the 1960s following major declines in North American waterfowl populations (Sparrowe 1992).

In 1964, the Mississippi Flyway Council Planning Committee formally recommended finding a nontoxic replacement for lead shot for waterfowl hunting. That recommendation was consistent with an opinion by many within the conservation community that lead poisoning was the easiest form of "wasted waterfowl" to address, and that by doing so continental waterfowl populations would benefit greatly (Mississippi Flyway Council Planning Committee 1965).

In the summer of 1964, the Olin Corporation offered the Mississippi Flyway Council the necessary materials and facilities for field testing iron and lead shot. Olin's Nilo Farms Shooting Preserve was used from December 1964 – January 1965 for a field test that yielded positive results for iron shot under standard shooting preserve conditions of pass shooting that averaged about 30 yards in distance (Mikula 1965). In 1965, staff of the FWS met with members of the Sporting Arms and Ammunition Manufacturer's Institute (SAAMI) to discuss the development of nontoxic shot. Research that followed at the Illinois Institute of Technology, the Patuxent Wildlife Research Center (PWRC) of the FWS, and elsewhere generally supported an evaluation by SAAMI that steel shot (soft iron) was the only viable substitute for lead shot for waterfowl hunting (SAAMI 1969).

Alternative shot types tested prior to FWS implementation of nontoxic shot requirements included various coatings on lead shot to prevent erosion and absorption of lead within the bird, alloys to reduce toxicity by reducing the amount of lead in the pellet, disintegrating shot to make the pellets/lead unavailable to birds, shot with antidotal components to offset the effects of lead, and substitute metals (Table 7). This testing included previously tested alternative shot types (Table 6) and a wide variety of other alternatives (US Fish and Wildlife Service 1986a). Although several of these shot types were found to be acceptable based on toxicology, industry criteria for production of acceptable shotshells were not satisfactory (Table 8). Tin was not toxic, but its density was too low for shotgun ballistics.

Table 7. Summary of alternative shot types tested prior to the initiation of nontoxic shot requirements for hunting waterfowl in selected areas of the US (developed, with modifications, from US Fish and Wildlife Service 1986 EIS on lead shot).

Shot type	Concepts	Types Tested
Shot coatings	Resistant enough to withstand acids in digestive tract and grinding action of gizzard, thereby facilitating expulsion by the bird. Overcome hardness and density issues associated with non-lead shot.	• Nickel on lead • Thin-nickel on lead • Lead on steel • Copper on lead • Plastic on lead • Zinc on iron • Molybdenum on iron • Teflon on steel
Alloys	Render the lead less toxic by reducing its content.	• Lead-tin-phosphorous • Lead-tin • Lead-iron
Disintegrating shot	Disintegrate in water to make shot unavailable to birds or disintegrate in digestive tract.	• Lead-magnesium • Lead-resin
Antidote components	Biochemical formation of a chelating ring to prevent lead absorption.	• Lead-calcium • Lead-tin-phosphorous • Others such as additives of EDTA and creatinine.
Substitute metals	Use a metal other than lead as the pellet core	• Copper • Zinc • Tin • Nickel • Steel • Iron • Uranium

Table 8. Industry criteria for acceptable shot characteristics (developed, with modifications, from US Fish and Wildlife 1986 EIS on lead shot).

Characteristic	Purpose
High density	Velocity and energy retention ($E = mv^2$) and weight effectiveness.
Reasonable cost	Readily available base material that is cost effective as shot is the most costly component of the shotshell.
Easily processed	Cost control issue for shot fabrication and facility/equipment requirements.
Relatively inert	Not reactive to other shotshell components and non-corrosive in shelf life.
Soft surface	Needed to prevent damage to gun barrel or chokes.
Nontoxic	Not poisonous to birds, must not contaminate the meat, and must be able to withstand corrosive acids of the digestive tract and grinding action of waterfowl gizzards to the extent that will be passed by the bird before any toxins are absorbed.

Limited testing with depleted uranium revealed no toxicity for Black Ducks (*A. rubripes*), but unanswered environmental fate questions also needed exploration. Neither steel nor nickel were found to be toxic, but had other issues that surfaced. Health impacts of copper were delayed but eventually manifested as significant weight losses from which captive birds could recover from but which might prove fatal for wild birds (Bellrose 1965, Irby et al. 1967, US Fish and Wildlife Service 1986a).

The Dilemma of Change.—Industry was faced with a dilemma. Steel shot (soft iron) was the only nontoxic shot option available in the 1960s, and the performance from steel shotshells being produced at that time was inferior to that from lead shotshells commonly used for waterfowl hunting. However, the major decline occurring in North American waterfowl populations (Figure 2) resulted in a focus on lead poisoning by the waterfowl management community as an issue requiring attention as a remedial action (Mississippi Flyway Council Planning Committee 1965). The magnitude of losses associated with this disease (Bellrose 1959) could no longer be tolerated, and the use of nontoxic shot alternatives was a logical means for significantly reducing those losses. However, major questions arose regarding the timing and the extent of government actions for the use of nontoxic shot.

By 1970, it was clear that nontoxic shot requirements were imminent, and that despite early discussions by some about lead poisoning in Mourning Doves (Shillinger and Cottam 1937) and other species, insufficient data existed to justify nontoxic shot use for hunting upland game birds (Jones 1939, Campbell 1950, Locke and Bagley 1967, Lewis and Legler 1968, McConnell 1968, Kendall 1980). Thus, the FWS limited their focus to waterfowl and associated wetland avian species hunted in waterfowl habitat (Morehouse 1992a).

Change and Controversy.—A paradox of the nontoxic shot controversy is that the Olin Corporation was a leader in early research and development efforts for alternative shot types, then subsequently opposed nontoxic shot requirements, and later once again became a leader in developing and producing high quality nontoxic shotshells. Substantial corporate investments were made in all of these situations. Concerns about lead shortages during the post-WWII era stimulated self-interest in pursuing shot alternatives. With similarity to the current larger-scale issue of alternative fuel to power transportation, those who first succeed in developing suitable alternatives would realize a market place advantage. Also, the 1948 alternative shot project enhanced the public image of the Olin Corporation as a contributor to wildlife conservation (Jordan and Bellrose 1951) and as a respondee to the highly visible dramatic lead poisoning die-offs of Mallards in nearby Grafton, Illinois, in 1947 and 1948.

The oppositional role of Olin Corporation when nontoxic shot requirements were being formulated by the FWS may have been driven by the changing dynamics of the marketplace. Profit margins were being challenged by increased cost for steel shot vs. lead shot. As noted by Coburn (1992), "we are not in either the lead or the steel business; we are in the ammunition business. The shot material is important to us only as it affects shotshell (cartridge) performance and cost." Steel and lead shot are produced by different methodologies, with the former requiring costly investments in manufacturing equipment for production.

Waterfowl hunting loads are not the major segment of the shotshell market. Thus, a total ban on lead shot use for any purpose may have been more acceptable across industry, even though the official Olin Industries position was that nontoxic shot use could only be justified on areas where lead poisoning of waterfowl was documented as a significant mortality factor (Coburn 1992). Previously, representatives of the ammunition industry had informally indicated that if a sufficiently competitive shotshell could be developed, "...ammunition companies would completely abandon the use of lead even for upland game shooting." Also noted was the need for legislation to provide a smooth transition over time and an opportunity to deplete existing lead stocks (Mississippi Flyway Council Planning Report 1965). It is noteworthy that a similar industry position of limiting nontoxic ammunition requirements for other species to well-documented lead poisoning problems was recently issued by Federal Cartridge Company, a leading proponent of the use of nontoxic shot for waterfowl hunting during the 1970s and 1980s: "...when regulating agencies seek to expand the use of lead-free ammunition to species other than waterfowl, the regulating agencies should do so only after they have fully gathered and analyzed thorough, scientifically based, and fully documented evidence that establishes a direct connection to the health and welfare of the species in question" (Federal Cartridge Company 2006). Similarly, the American Sportfishing Association has expressed concern "...about statewide restrictions on the sale and use of lead sinkers [to protect birds from lead poisoning] when research doesn't warrant such broad measures" (Tennyson 2002).

Transition to Nontoxic Shot Use.—Steel shot regulations for waterfowl hunting were initiated in 1972 with requirements to use nontoxic ammunition on seven NWRs selected to participate in field trials. Shotgun shells were provided by each of the major manufacturers: Winchester-Western, Remington Arms, and Federal Cartridge Company. The purposes for those trials was for the FWS to introduce hunters to the use of steel shot and to obtain data on steel shot performances and hunter reactions to its use (US Fish and Wildlife Service 1974). Author Milton Friend participated in the Monte Vista NWR trial in Colorado. The following year, special steel and lead shot comparisons, using unmarked shells, were held on selected public hunting areas. Observers accompanied hunters to dispense shells and record data. Results from these and other comparative studies were variable (Morehouse 1992b), but overall did not support the contention that the use of steel shot resulted in increased losses of waterfowl from crippling (Sanderson and Bellrose 1986a, US Fish and Wildlife Service 1986a). At the state level, Colorado and Oregon each required nontoxic shot on a single hunting area for the 1974 hunting season.

Despite these initiatives, little additional progress was realized in "getting the lead out" for waterfowl hunting until the 1985 hunting season. Until then, legislation was a major factor in suppressing nontoxic shot requirements. However, beginning in 1985, litigation opened the flood gates for the nontoxic shot use that followed. It is worth noting that all of the legal challenges to nontoxic shot regulations were instigated by organizations and/or groups of individuals affiliated with sport hunting. It is also of interest that, Sparrowe (1992) wrote, "Deposition of lead into the environment is still being used by major anti-hunting groups in the United States to argue against hunting. Removal of that argument is a big plus for retaining the social, economic, cultural, and recreational values of hunting." Sparrowe's comments from 16 years ago have even greater relevance today because of the continuing shift in the values of US society towards nonconsumptive wildlife uses.

Change and Litigation.—The first litigation challenge immediately followed issuance of the 1976 DOI Final Environmental Impact Statement (EIS) on the proposed use of nontoxic shot for waterfowl hunting (US Fish and Wildlife Service 1976). A previous proposal, drafted in 1974, called for phased-in flyway-wide bans of lead shot use beginning in 1976 for the Atlantic Flyway and 1977 for the Mississippi Flyway. Beginning in 1978, specifically designated areas in the Central and Pacific Flyways would implement nontoxic shot requirements for waterfowl hunting (US Fish and Wildlife Service 1974). The implementation time table was based on several factors, including needs by industry to deplete existing inventories of lead shot and time to produce adequate supplies of nontoxic shotshells to meet the needs of hunters across the US. Following public comment and review, the 1974 draft proposal was modified in 1976 to require nontoxic shot use for waterfowl hunting only in "problem areas" in each flyway and following the same time line for implementation as appeared in the 1974 draft (US Fish and Wildlife Service 1976).

The National Rifle Association (NRA), an organization whose leadership represents a large contingency of gun owners and sportsmen, challenged the proposal in court. They contended that steel shot (soft iron) posed human health and safety risks, increased waterfowl crippling, and damaged firearms, other property, and the environment. They also alleged that the EIS was inadequate. The court ruled in favor of the government, and an appeal by the NRA also failed (Anderson 1992).

In 1978, the "Stevens Amendment" to the Interior Department appropriations bill prohibited the FWS from using funds to implement or enforce nontoxic shot use for hunting waterfowl without approval of the states involved. That amendment was reenacted annually until 22 December 1987 when it was allowed to expire. During most of that time, the Stevens Amendment stifled implementation of nontoxic shot regulations. That effectiveness was a result of the individual states generally being more vulnerable to "political pressure" on this matter than the federal government.

Lawsuits were filed by groups of sportsmen against four states that initiated limited steel shot zones for waterfowl hunting following passage of the Stevens Amendment. Author Friend testified on behalf of the state in the South Dakota and Florida cases. Another NWHC staff member, Dr. Louis Locke, testified on behalf of the state of Texas in that lawsuit. The New York case was never tried, as the state was granted a motion for summary judgment. The states prevailed in all the judgments, including appeals filed in the South Dakota and Texas cases (Anderson 1992).

Illinois, Maryland, and Wyoming used the Stevens Amendment "states' rights" approach to pass legislation banning or restricting nontoxic shot use for waterfowl hunting in their states. Wisconsin was the only state to "swim upstream" while the Stevens Amendment was in place, passing 1985 legislation mandating statewide use of nontoxic shot for waterfowl hunting. Kansas followed with similar legislation five years later, after the Stevens Amendment had expired.

Lead poisoning of Bald Eagles eventually trumped the Stevens Amendment on the legislative/litigation battlefield. The first case of lead poisoning in a Bald Eagle was an incidental finding by the PWRC's environmental contaminant program (Mulhern et al. 1970). However, the 1975 establishment of the National Wildlife Health Center (NWHC) greatly expanded investigations of mortality in wildlife under FWS stewardship. The NWHC disease diagnostic database soon contained numerous records of Bald Eagle mortality due to lead poisoning (National Wildlife Health Center 1985). In response to what appeared to be a growing problem, the FWS proposed nontoxic shot regulations for waterfowl hunting in portions of eight states beginning in 1985.

Five of the states, acting under the auspices of the Stevens Amendment, did not consent to the regulations. As a result, the federal government announced that unless those states reversed their decisions, the FWS would not open the disputed areas for waterfowl hunting in 1986 as a means for protecting Bald Eagles from lead poisoning. At that point, the National Wildlife Federation (NWF) sued the federal government (Anderson and Havera 1989) to obtain a ban on lead shot for waterfowl hunting in the areas in question (Anderson 1992). The NWF prevailed because it was well-fortified with data obtained under the Freedom of Information Act and through other interactions with the NWHC and others (Feierabend and Myers 1984). The court ruled that the areas in question be closed to waterfowl hunting for the 1985 hunting season unless only nontoxic shot was used. That ruling increased the percentage of the US waterfowl harvest covered by nontoxic shot regulations three-fold (Anderson 1992) (Figure 4).

In 1986, the NWF again initiated legal action against DOI to prevent authorization of lead shot for waterfowl hunting throughout the continental United States beginning with the 1987 season. In response, the DOI unveiled a plan to phase out lead shot for waterfowl hunting over a period of several years, culminating in a nationwide ban in 1991. The court

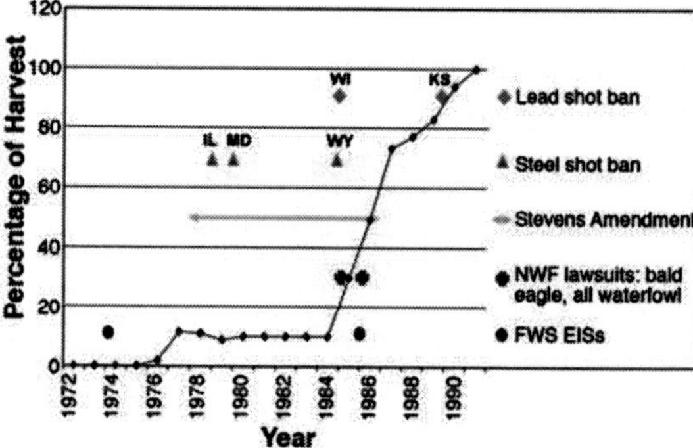

Figure 4. Effects of legislation on lead shot use for waterfowl hunting in the US as a function of the percentage of waterfowl harvest in nontoxic shot zones (Modified from Anderson 1992).

noted that the DOI had conceded on all aspects of the dispute except timing and dismissed the case for "want of ripeness." A countersuit and an appeal were unsuccessful (Anderson and Havera 1989). In response to the rulings of the court, the percentage of the waterfowl harvest in nontoxic zones increased to 49% in1986 as the FWS and some state wildlife agencies designed additional nontoxic shot areas (Figure 4).

The last lawsuit to challenge nontoxic shot regulations was initiated by the California Game and Fish Commission in the US District Court for the Eastern District of California in 1987. The NRA intervened on behalf of the plaintiff, and the NWF intervened on behalf of the defendant. Following the court's ruling

Table 9. Primary federal statutory authorities relevant to addressing lead poisoning in wildlife (developed from Anderson 1992).

Legislation		Relevance
Migratory Bird Treaty Act	1918	• Empowers federal government to *determine whether, to what extent, and by what means* hunting of migratory birds is allowed in the U. S.
Bald Eagle Protection Act	1940	• *Prohibits the take* of eagles without special authorization. • "Take" includes..."shoot at, poison, wound, kill..."
Bald and Golden Eagle Protection Act	1962	
Endangered Species Act	1976	• Bald Eagle listed as endangered in 43 of the conterminous states and as threatened in the other 5 states at the time of enactment. • Requires that listed species be conserved. • "Conserved" means *"...the use of all methods and procedures which are necessary to bring any endangered species..." to a point of recovery* consistent with delisting.
National Environmental Policy Act	1970	• Policy and responsibility for maintaining the quality of the environment and renewable resources. • Directs government to prevent environmental degradation "...or other undesirable and unintended consequences..." and to "...enhance the quality of renewable resources..."

in favor of nontoxic shot, plaintiffs filed a notice of appeal, but it was dismissed. The percentage of the waterfowl harvest in nontoxic shot zones increased to 73% in 1987 and to 100% in 1991 (Figure 4).

The arguments forwarded in the various lawsuits involved biological, socio-economic, and political issues – the last primarily consisting of challenges of agency authorities. In general, the issues involved had been identified in the 1965 Mississippi Flyway Planning Committee Progress Report as factors that needed to be addressed. In all cases, the courts ruled that the agencies whose actions were contested had fundamental authority and responsibilities under various statutes, legislation, and treaties to take those actions. Authority at the federal level relevant to addressing lead poisoning in migratory birds is vested in major legislation and international treaties (Table 9). In essence, in the California case, the Court found that, under the Migratory Bird Treaty Act and the Endangered Species Act, the FWS has almost carte blanche authority to take whatever steps are necessary to protect migratory birds and endangered species (US Fish and Wildlife Service 1988).

Biological challenges (Table 10) were answered by the sound science available to support problem identification and the need for action. However, science was helpful in addressing only a small number of the socio-economic issues (Table 11). These needs required continued attention throughout the transition process.

DISCUSSION

Little of what we have presented here reflects the bitterness that characterized much of the struggle to transition to the use of nontoxic shot for waterfowl hunting in the US. Nor does it reflect the heavy personal costs to those who championed the use of nontoxic shot, among them state and federal employees, outdoor columnists, members of the general public, academicians, researchers, and others. Although personal feelings on both sides were often emotionally charged, it would be folly to view this issue in terms of "we vs. them," for there was as much conflict within the professional wildlife conservation community as there was between agencies of that community and external parties. Similar conflicts existed within industry and elsewhere. For example, in contrast to Winchester, Federal Cartridge Company was aggressively pro-steel shot during the entire conflict period while Remington Arms remained rather neutral. In essence, both sides failed to adequately grasp the complexity of this issue, and in some instances, were so motivated for their causes that their actions ignored alternative

Table 10. Biological issues highlighted in court cases challenging nontoxic shot (steel) use for hunting waterfowl in the US.

Issue	Case(s)					
	NRA	South Dakota	New York	Texas	Florida	California
Poor/inadequate science			X	X		X
Population impacts from lead vs. benefits				X		
Increased crippling	X				X	
Steel shot boundary delineations		X	X	X		
Effects of diet			X			

Table 11. Socio-economic issues highlighted in court cases challenging nontoxic shot (steel) use for hunting waterfowl in the US.

	Case(s)					
Issue	NRA	South Dakota	New York	Texas	Florida	California
Human health and safety risks (ricochets and dental)	X				X	
Firearms damage, other property damages	X	X	X	X	X	
Reloading components not available				X		
Availability of steel shot shells				X		
Discrimination against females, children, aged				X		
Increased costs to hunters			X		X	
Economic losses from lead shot stocks				X		

considerations. The following statement, made in 1965, serves as an example of an overly simplified perspective of the challenges involved:

"Public relations experts are confident that paving the way for public acceptance of a new type of shotgun shell is not a difficult problem...a well-planned program should be ready to go as soon as the new product is announced. Preliminary conditioning of the public can even precede that." (Cox 1965).

To a large extent, "public education" needed to begin within the conservation agencies, because there were many employees who interfaced with hunters and other members of the public who knew too little about lead poisoning and/or were opposed to nontoxic shot use. Further, reaction drove public education efforts for too long and was a poor substitute for a progressive, well-rounded education program. The US Cooperative Lead Poisoning Control Information Program (CLPCIP) arose from this need (Bishop and Wagner 1992). However, that program was not initiated until 1982, well after the conflict over nontoxic shot use had become well entrenched. CLPCIP has since become the Cooperative Non-toxic Shot Education Program.

Informational and educational activities assumed many forms during the conflict over nontoxic shot. They also consumed large amounts of personnel and fiscal resources from agency, non-government organization (NGO), industry, and others. At times, prolonged skirmishes initiated by both sides took place via articles in sporting magazines and newspapers and via news releases and various public consumption publications. Anti-steel shot factions did verbal battle with pro-steel shot advocates over the interpretation of shooting trials and ballistics data. Unique statistical approaches and carefully worded evaluations were sometimes used to interpret data and support conclusions (Kozicky and Madson 1973, Lowry 1974, Bockstruck 1978, Coburn 1992, Carmichael 2002). As a result, waterfowl hunters and others were left awash in a sea of conflicting information and presentations.

The encouragement of hands-on involvement by the public were powerful tools for "perspective and attitude adjustments" regarding the lead poisoning issue. A case in point is the assistance of Wisconsin hunters in the clean-up of a major lead poisoning die-off of Canada Geese. Another example is the encouragement of hunters in conducting their own lead shot ingestion studies using gizzards from birds they personally harvested. Steel shot shooting clinics sponsored by the CLPCIP were invaluable educational forums for influencing hunters about steel shot and enhancing their shooting skills. A modified shooting clinic and education program organized by the FWS for the Congressional Sportsmen Coalition and held at Andrews Air Force Base in the Washington, DC area was attended by the Secretary of the Interior, the FWS Director, and others involved in managing the direction of FWS actions on nontoxic shot requirements.

Two graphic movies also served important educational roles. The first, on lead poisoning in waterfowl and Bald Eagles, utilized footage from lead poisoning field outbreaks and other visual materials to address commonly asked questions about this disease (US Fish and Wildlife Service 1986b). The second focused on bagging waterfowl with steel shot. Footage of waterfowl being taken under field conditions and follow-up laboratory measurements of those birds were used to assess steel shot performance (US Fish and Wildlife Service 1986c). The lead poisoning movie was widely viewed and has served as an educational tool for use by those in other countries pursuing nontoxic shot use for waterfowl hunting. The steel shot shooting movie was found to be unacceptable by some DOI administrators after a preliminary showing, and was not released for general use. However, an unofficial copy obtained by non-government sources was seen by numerous audiences. Eventually, a shorter version of this movie was released by the FWS for general use. Both movies were converted to video format to enhance distribution and use by external parties. A variety of industry and other public sector videos on shooting steel shot followed (Table 12).

Table 12. Examples of video presentations on lead poisoning in waterfowl and on the use of steel shot for waterfowl hunting.

Title	Year of Issue	Source	Running Time (minutes:seconds)
Steel Shot Facts for the Waterfowl Hunter	1986	Federal Cartridge Company Minneapolis, MN	11:18
World Champion's Guide to Hunting Waterfowl with Steel Shot	1988	W.C. Badorek and D. Beleha Klamath Falls, OR	45:00
The Duck Hunter Shooting and Shot	1987	Videolore Emeryville, CA	Not given
Field Testing Steel Shot	1986	FWS Washington, DC	30:00
Lead Poisoning in Waterfowl	1986	FWS Washington, DC	27:49

The FWS utilized a three person team of subject matter experts with established credentials in waterfowl/migratory bird management (Office of Migratory Bird Management), in lead poisoning/avian diseases (NWHC), and in shooting steel shot (contractor) to provide presentations and respond to questions at various public forums involving the nontoxic shot issue. The same individuals appeared together at nearly all of those events. That approach provided consistent commentary, technical breadth to respond to the broad array of questions received and a great deal of useful insight for guiding FWS management of this issue by listening to concerns and commentary by others across the country.

An organized "opposition team" often also presented at these forums and typically included at least one individual known to the local audience. There also was a private sector individual that attended a number of these forums and tape-recorded commentary, by at least the nontoxic shot advocates, as a basis for his "writings." Some of these forums were aggressively hostile towards presenters supporting nontoxic shot. Their commentary and response to questions were disrupted, and in some instances, their safety was threatened. Some forums were conducted in ways that facilitated comment from those opposing nontoxic shot, but suppressed commentary from those supporting nontoxic shot. Those situations were preplanned and orchestrated to prevent dissemination of information on non-toxic shot and open discussion.

In addition to the larger public forums/hearings, a great deal of time was devoted by NWHC personnel to speaking on lead poisoning at sportsman clubs and civic group meetings, and conducting agency workshops involving state and federal biologists, law enforcement, and management personnel. Information from ongoing NWHC disease investigations and the FWS lead monitoring program was incorporated in workshop presentations. Hard copy brochures, pamphlets, and other materials specifically developed for informational and educational purposes were provided as handouts for important adjuncts to presentations. Some handouts, such as the "Kansas Wildlife" magazine reprint Are We Wasting Our Waterfowl? (Kraft 1984) and a FWS brochure depicting the clinical signs and gross lesions of lead poisoning in waterfowl (US Fish and Wildlife Service 1986d) were made available for additional use and distribution by workshop participants that requested them.

A relevant question is, "What has been accomplished?" Clearly, the implementation of nontoxic shot requirements for hunting waterfowl has dramatically reduced lead shot ingestion by waterfowl and subsequent losses from lead poisoning. Anderson et al. (2000) found in their 16,651 samples from the Mississippi Flyway during 1996 and 1997 that gizzards of 44% to 71% of major duck species contained only nontoxic shot. These authors estimated that nontoxic shot reduced mortality from lead poisoning in Mississippi Flyway Mallards by 64% and extrapolated their data to a saving from lead poisoning of 1.4 million ducks nationwide in the 1997 fall continental flight of 90 million ducks (Anderson et al. 2000). Smaller scale post-nontoxic shot implementation evaluations also disclosed major reductions in lead exposure (DeStefano et al. 1995, Calle et al. 1982). Additionally, it appears that reduced exposure to lead shot has not been offset by increased crippling caused by the use of nontoxic shot. A recent review of historical waterfowl harvest data by Schulz et al. (2006) revealed that after an initial increase in reported crippling rates, current reported rates were lower than those of pre-nontoxic shot rates for both ducks and geese (see also: US Fish and Wildlife Service 1986a, Morehouse 1992b).

Once the FWS implemented a total ban on the use of lead shot for hunting waterfowl, industry responded by developing high quality nontoxic shotshells. As noted by Coburn (1992), Winchester was marketing a total of 59 different steel shot loads, including 10-, 12-, 16-, and 20-gauge sizes by 1992. This variety of shells reflects major advancement from the first steel load marketed in 1976, a 12-gauge, 2¾", 1¼ ounce shotshell. Hunter education programs and shooting clinics also have enhanced hunter performance in bagging waterfowl with nontoxic shells.

Not only are steel shot shells of today far superior in quality and performance than those first provided to hunters in 1972, but steel shot is no longer the only choice (Table 13). In addition to multiple types of nontoxic shotshells produced by an increased number of manufacturers, these shells are now available in a greater range of gauges and shell lengths. Industry has developed a 3½" chamber 12-gauge shotgun as a new consumer product. Hand-loading components and equipment for nontoxic shotshells are also marketed along with a wide array of educational materials for improving hunter performance. These transitions have involved social change for waterfowl hunters by replacing traditional methods with new ones. The variety and high quality of today's nontoxic shotshells are clearly products of the competitive marketplace responding to firm regulatory schedules for nontoxic shot implementation (Coburn 1992). Current nontoxic shotshells also represent a response by industry to the needs of both conservation and contemporary hunters.

Points to Ponder.—We close by suggesting that further resolution of lead poisoning issues will benefit from a three-legged stool approach (Figure 5a). The legs supporting the platform for that proverbial stool are problem identification, acceptable alternatives, and authority to act. Each leg is comprised of a different substance. Material failure in any of the legs will cause the platform to collapse because the tensile strength achieved by the

Table 13. Entry of various nontoxic shotshells into the United States market.

Year	Approximate percentage of waterfowl harvest within non-toxic shot zone[a]	Shell type entry by Federal Cartridge Company[b]
1972	<1	2 ¾" 12-gauge steel[c]
1977	12	3" 12-gauge steel
1980	10	3" 20-gauge and 3 ½" 10-gauge steel
1982	10	2 ¾" 20-gauge steel
1987	73	2 ¾" 16-gauge steel
1989	82	3 ½" 12-gauge steel
1991	100	(start of nationwide requirement for nontoxic shot use for waterfowl hunting)
1992–1995	100	Bismuth
1996	100	Tungston-iron alloy 2 ¾" and 3" 12-gauge)
1997	100	Tungston-polymer (3 ¾" and 3" 12-gauge)[d]
1998	100	Tungston-iron alloy (3 ½" 10- and 12-gauge)[e]
2003–2004	100	Tungsten-iron-nickel alloy[f]
		Other[g]

[a] Anderson 1992.
[b] Contributed by William F. Stevens.
[c] Federal, Remington and Winchester all introduced this shell type at the same time; Federal was the first company to offer all of the other loads listed except for bismuth shot.
[d] Discontinued after 3 or 4 years due to high production cost; marketed by Kent Cartridge Company under the trade name of Tungsten Matrix.
[e] Currently sold under the trade name of High Energy.
[f] Sold under the trade name of Heavyweight; most dense nontoxic shot pellet on the market.
[g] Remington, Winchester, and small independent companies have contributed other shell types since nationwide nontoxic shot requirements were implemented; for example, Hevi-shot loads have a tungsten base combined with iron, nickel, bronze, and perhaps something else; bismuth/tin alloy only nontoxic shot pelts in 28 and .410 gauges but Hevi-Shot Classic Doubles Shot (tungsten alloy) plans to provide 16, 28, and .410 gauge loads in the near future (Bourjaily 2008).

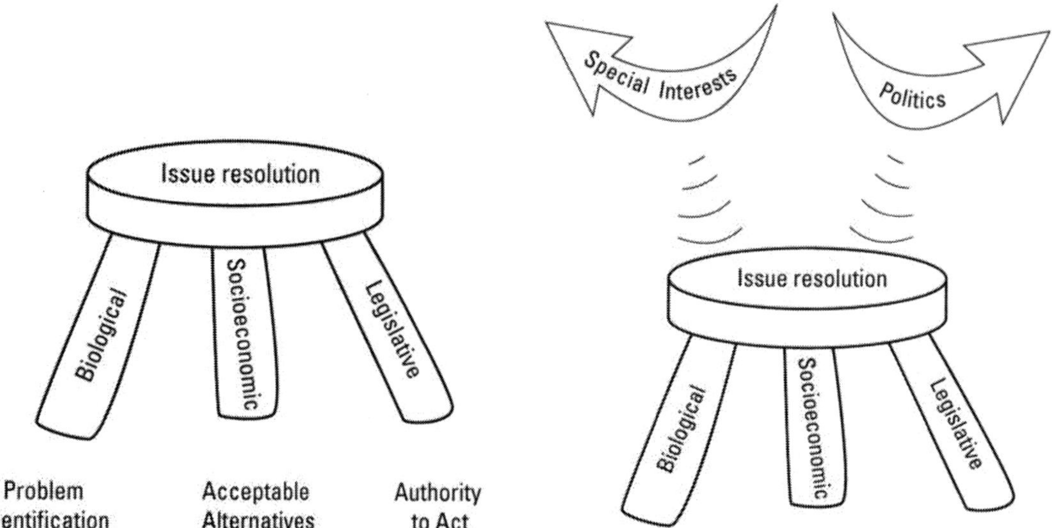

Figure 5. Three-legged stool approach for addressing lead poisoning in wildlife. (A) Each leg provides support for a different aspect of the issue. (B) The collective strength of the platform supported by the legs can withstand the pressures exerted upon it by external forces, thereby facilitating issue resolution.

combination of all three substances is required to withstand the pressures that will be exerted on the stool (Figure 5b).

Problem identification is the biological leg and provides foundation by soundly identifying the who, what, when, where, and why aspects of the problem (Table 14). The socio-economic leg provides support through the continued conduct of acceptable alternatives for traditional activities. Authority to act is the legislative leg, and it provides support through a willingness to act by discharging vested authorities and responsibilities.

The challenges in moving forward in reducing lead poisoning in avian species are great. Nevertheless, we believe that by applying the lessons learned in the waterfowl lead poisoning struggle, a sound platform for success can be built that will withstand forthcoming pressures. Many of the lessons learned are highlighted in the conclusions section of the "Lead Poisoning in Waterfowl" proceedings from the 1991 International Waterfowl Research Bureau workshop held in Brussels, Belgium (Pain 1992). Among the concluding comments are nine recommendations for implementing the solutions to lead poisoning (Moser 1992). We offer one of those for our concluding comments:

"It is essential to have an effective information, awareness and education programme prior to, and during, the implementation of a lead shot replacement programme. This should include definition of the problem, an explanation of the options considered for the solution, and hands-on demonstrations for hunters to see for themselves the efficacy of non-toxic shot..."

Table 14. Basic foundational considerations needing to be addressed in developing a biological justifications for the replacement of lead used in recreational sports.

Considerations	Dimensions
Who is affected?	Species impacts that are to be addressed.
What is the problem?	Magnitude of impacts relative to population or other costs that require remedial action.
When does it occur	Seasonality considerations that may guide approaches for addressing the problem.
Where does it occur?	Delineation of the problem geographically to guide remedial and preemptive actions.
Why does it occur?	Determination of the factors contributing to the undesirable outcome(s) to be addressed.

The concept cited is as valid for the replacement of lead in fishing tackle, shooting sports, and other sporting activities that deposit lead in the environment as it is for waterfowl hunting. A critical aspect governing the effectiveness of this concept is the involvement of all stakeholders in its development and implementation.

LITERATURE CITED

ALDER, F. E. W. 1942. The problem of lead poisoning in waterfowl. Wisconsin Conservation Bulletin 7:5–7.

ANDERSON, W. L. 1975. Lead poisoning in waterfowl at Rice Lake, Illinois. Journal of Wildlife Management 39:264–270.

ANDERSON, W. L. 1992. Legislation and lawsuits in the United States and their effects on nontoxic shot regulations. Pages 56–60 *in* D. J. Pain (Ed.). Lead Poisoning in Waterfowl. Proceedings of an IWRB Workshop, Brussels, Belgium, 13–15 June 1991. International Waterfowl and Wetlands Research Bureau Special Publication 16, Slimbridge, UK.

ANDERSON, W. L., AND S. P. HAVERA. 1989. Lead Poisoning in Illinois Waterfowl (1977–1988) and the Implementation of Nontoxic Shot Regulations. Illinois Natural History Survey Biological Notes 133, Urbana, Illinois, USA.

ANDERSON, W. L., S. P. HAVERA, AND B. W. ZERCHER. 2000. Ingestion of lead and nontoxic shotgun pellets by ducks in the Mississippi Flyway. Journal of Wildlife Management 64:848–857.

ARTMANN, J. W., AND E. M. MARTIN. 1975. Incidence of ingested lead shot in Sora Rails. Journal of Wildlife Management. 39:514–519.

BATES, F. Y., D. M. BARNES, AND J. M. HIGBEE. 1968. Lead toxicosis in Mallard ducks. Bulletin of the Wildlife Disease Association 4: 116–125.

BEINTEMA, N. 2004. Non-toxic Shot. United Nations Environment Program (UNEP)/African-Eurasian Waterbird Agreement (AEWA), Bonn, Germany.

BELLROSE, F. C. 1951. Effects of ingested lead shot upon waterfowl populations. North American Wildlife Conference Transactions 16:125–135.

BELLROSE, F. C. 1959. Lead poisoning as a mortality factor in waterfowl populations. Illinois National History Survey Bulletin 27:235–288.

BELLROSE, F. C. 1965. The toxicity of ingested copper pellets in wild Mallards. Pages 70–72 *in* Wasted Waterfowl. Mississippi Flyway Council Planning Committee Report.

BEYER, W. N., J. C. FRANSON, L. N. LOCKE, R. K. STROUD, AND L. SILEO. 1998. Retrospective study of the diagnostic criteria in a lead-poisoning survey of waterfowl. Archives of Environmental Contamination and Toxicology 35:506–512.

BISHOP, R. A. AND W. C. WAGNER II. 1992. The US cooperative lead poisoning control information program. Pages 42–45 *in* D. J. Pain (Ed.). Lead Poisoning in Waterfowl. Proceedings of an IWRB Workshop, Brussels, Belgium, 13–15 June 1991. International Waterfowl and Wetlands Research Bureau Special Publication 16, Slimbridge, UK.

BLUS, L. J., R. K. STROUD, B. REISWIG, AND T. MCENEANEY. 1989. Lead poisoning and other mortality factors in Trumpeter Swans. Environmental Toxicology and Chemistry 8:263–271.

BOCKSTRUCK, H. 1978. Steel shot: a ballistician's view. Ducks Unlimited 42:26–28, 88, 90.

BOURJAILY, P. 2008. New age birdshot. Field and Stream 113:26.

BOWLES, J. H. 1908. Lead poisoning in ducks. Auk 25:312–313.

BUERGER, T. T., R. E. MIRARCHI, AND M. E. LISANO. 1986. Effects of lead shot ingestion on captive Mourning Dove survivability and reproduction. Journal of Wildlife Management 50:1–8.

CALLE, P. P., D. F. KOWALCZYK, F. J. DEIN, AND F. E. HARTMAN. 1982. Effect of hunters' switch from lead to steel shot on potential for oral lead poisoning in ducks. Journal of the American Veterinary Medical Association 181:1299–1301.

CAMPBELL, H. 1950. Quail picking up lead shot. Journal of Wildlife Management 14:243–244.

CARMICHEL, J. 2002. After the ban. Outdoor Life 209:20–23.

CASTRALE, J. S. 1991. Spent shot ingestion by Mourning Doves in Indiana. Proceedings of the Indiana Academy of Science 100:197–202.

CLEMENS, E. T., L. KROOK, A. L. ARONSON, AND C. E. STEVENS. 1975. Pathogenesis of lead shot poisoning in the Mallard duck. Cornell Veterinarian 65:248–285.

COBURN, C. 1992. Lead poisoning in water fowl: the Winchester perspective. Pages 46–50 *in* D.

J. Pain (Ed.). Lead Poisoning in Waterfowl. Proceedings of an IWRB Workshop, Brussels, Belgium, 13–15 June 1991. International Waterfowl and Wetlands Research Bureau Special Publication 16, Slimbridge, UK.

COBURN, D. R., D. W. METZLER, AND R. TREICHLER. 1951. A study of absorption and retention of lead in wild waterfowl in relation to clinical evidence of lead poisoning. Journal of Wildlife Management 15: 186–192.

COX, N. A. 1965. Progress report on lead poisoning investigations. Pages 1–5 *in* Wasted Waterfowl. Mississippi Flyway Council Planning Committee.

DAY, A. M. 1959. North American Waterfowl. The Stackpole Company, Harrisburg, Pennsylvania, USA.

DEMENT, S H., J. J. CHISOLM, JR., M. A. ECKHOUS, AND J. D. STRANDBERG. 1987. Journal of Wildlife Diseases 23: 1987:273–278.

DESTEFANO, S., C. J. BRAND, D. H. RUSCH, D. L. FINLEY, AND M. M. GILLESPIE. 1991. Lead exposure in Canada geese of the eastern prairie population. Wildlife Society Bulletin 19:23–32.

DESTEFANO, S., C. J. BRAND, AND M. D. SAMUEL. 1995. Seasonal ingestion of toxic and nontoxic shot by Canada Geese. Wildlife Society Bulletin 23: 502–506.

DESTEFANO, S., C. J. BRAND, AND D. H. RUSCH. 1992. Prevalence of lead exposure among age and sex cohorts of Canada Geese. Canadian Journal of Zoology 70: 901–906.

DEGERNES, L., S. HEILMAN, M. TROGDON, M. JORDAN, M. DAVISON, D. KRAEGE, M. CORREA, AND P. COWEN. 2006. Epidemiologic investigation of lead poisoning in Trumpeter and Tundra Swans in Washington State, USA, 2000–2002. Journal of Wildlife Diseases 42: 345–358.

DOWDELL, R. L., AND R. G. GREEN. 1937. Lead-magnesium alloys for the prevention of lead poisoning in waterfowl. Mining and Metallurgy 18: 463–466.

ELDER, W. H. 1954. The effect of lead poisoning on the fertility and fecundity of domestic Mallard ducks. Journal of Wildlife Management 18: 315–323.

FEDERAL CARTRIDGE COMPANY. 2006. Federal Cartridge Company Position on the Implementation of New Non-lead Regulations. Statement released December 12, 2006. Anoka, Minnesota, USA.

FEIERABEND, J. S., AND O. MYERS. 1984. A National Summary of Lead Poisoning in Bald Eagles and Waterfowl. National Wildlife Federation, Washington, DC, USA.

FEIERABEND, J. S. 1985. Legal challenges to nontoxic (steel) shot regulations. Proceedings of the Annual Conference of the Southeastern Association of Fish and Wildlife Agencies 39: 452–458.

FINLEY, M. T., M. P. DIETER, AND L. N. LOCKE. 1976. δ-aminolevulinic acid dehydratase: inhibition in ducks dosed with lead shot. Environmental Research 12: 243–249.

FISHER, I. J., D. J. PAIN, AND V. G. THOMAS. 2006. A review of lead poisoning from ammunition sources in terrestrial birds. Biological Conservation 131: 421–432.

FLINT, P. L., M. R. PETERSEN, AND J. B. GRAND. 1997. Exposure of Spectacled Eiders and other diving ducks to lead in western Alaska. Canadian Journal of Zoology 75: 439–443.

FORBUSH, E. H. 1916. A History of the Game Birds. Pages 547–548 *in* Wild-Fowl and Shore Birds of Massachusetts and Adjacent States. Massachusetts State Board of Agriculture, Boston, Massachusetts, USA.

FRANSON, J. C., AND S. G. HEREFORD. 1994. Lead poisoning in a Mississippi sandhill crane. The Wilson Bulletin 106: 766–768.

FRANSON, J. C., N. J. THOMAS, M. R. SMITH, A. H. ROBBINS, S. NEWMAN, AND P. C. MCCARTIN. 1996. A retrospective study of postmortem findings in Red-tailed Hawks. Journal of Raptor Research 30: 7–14.

FRANSON, J. C. 1996. Interpretation of tissue lead residues in birds other than waterfowl. Pages 265–279 *in* W. N. Beyer, G. H. Heinz, and A. W. Redmon-Norwood (Eds.). Environmental Contaminants in Wildlife: Interpreting Tissue Concentrations. Lewis Publishers, Boca Raton, Florida, USA.

FRANSON, J. C., S. P. HANSEN, T. E. CREEKMORE, C. J. BRAND, D. C. EVERS, A. E. DUERR, AND S. DESTEFANO. 2003. Lead fishing weights and other fishing tackle in selected waterbirds. Waterbirds 26: 345–352.

FRANSON, J. C., M. R. PETERSEN, C. U. METEYER, AND M. R. SMITH. 1995. Lead poisoning of Spectacled Eiders (*Somateria fischeri*) and of a

Common Eider (*Somateria mollissima*) in Alaska. Journal of Wildlife Diseases 31:268–271.

FRIEND, M. 1995. Conservation landmarks: Bureau of Biological Survey and National Biological Service. Pages 7–9 *in* E. T. LaRoe, G. S. Farris, C. E. Puckett, P.D. Doran, and M. J. Mac (Eds.). Our Living Resources: A Report to the Nation on the Distribution, Abundance, and Health of US Plants, Animals, and Ecosystems. US Department of the Interior, National Biological Service, Washington, DC, USA.

FRIEND, M. 1999a. Lead. Pages 317–334 *in* M. Friend and J. C. Franson (Eds). Field Manual of Wildlife Diseases: General Field Procedures and Diseases of Birds. US Geological Survey, Biological Resources Division, Information and Technology Report 1999–0001.

FRIEND, M. 1999b. Aspergillosis. Pages 129–133 *in* M. Friend and J.C. Franson (Eds). Field Manual of Wildlife Diseases: General Field Procedures and Diseases of Birds. US Geological Survey, Biological Resources Division, Information and Technology Report 1999–0001.

VON FUCHS, C. J. 1842. Die schädlinchen Einflüsse der Bleibergewerke auf die Gesundheit der Haustiere, insbesondere des Rindviehes (The Detrimental Effect of Lead Mines on the Health of Animals, Especially those with Horns). Veit, Berlin, Germany.

GRAND, J. B., P. L. FLINT, M. R. PETERSEN, AND C. L. MORAN. 1998. Effect of lead poisoning on Spectacled Eider survival rates. Journal of Wildlife Management 62: 1103–1109.

GREEN, R. G., AND R. L. DOWDELL. 1936. The prevention of lead poisoning in waterfowl by the use of disintegrable lead shot. North American Wildlife Conference Proceedings 1: 486–489.

GRINNELL, G. B. 1894. Lead poisoning. Forest and Stream 42: 117–118.

GRINNELL, G. B. 1901. American Duck Shooting. Forest and Stream Publishing Company, New York, USA.

HEERING, S. G. 1985. 1985 Study of American Hunting Issues. Institute for Social Research, University of Michigan, Ann Arbor, Michigan, USA.

HOUGH, B. 1894. Lead-poisoned ducks. Forest and Stream 42: 117.

HOWARD, W. J. 1934. Lead poisoning in *Branta c. canadensis*. Auk 51: 513–514.

HUNTER, B. F., AND M. N. ROSEN. 1965. Occurrence of lead poisoning in a wild Pheasant (*Phasianus colchicus*). California Fish and Game 51: 207.

IRBY, H. D., L. N. LOCKE, AND G. E. BAGLEY. 1967. Relative toxicity of lead and selected substitute shot types to game farm Mallards. Journal of Wildlife Management 31: 253–257.

JONES, J. C. 1939. On the occurrence of lead shot in stomachs of North American Gruiformes. Journal of Wildlife Management 3: 353–357.

JORDAN, J. S., AND F. C. BELLROSE. 1950. Shot alloys and lead poisoning in waterfowl. Transactions of the North American Wildlife Conference 15: 155–168.

JORDAN, J. S., AND F. C. BELLROSE. 1951. Lead poisoning in wild waterfowl. Illinois Natural History Survey Biological Notes No. 26.

KAISER, G. W., K. FRY, AND J. G. IRELAND. 1980. Ingestion of lead shot by Dunlin. The Murrelet 61: 37.

KALLMAN, H. (ED.). 1987. Restoring America's Wildlife 1937–1987. United States Department of the Interior, Washington, DC, USA.

KENDALL, R. J. 1980. The toxicology of lead shot and environmental lead ingestion in avian species with emphasis on the biological significance in Mourning Dove populations. Ph.D. dissertation, Virginia Polytechnical Institute, Blacksburg, Virginia, USA.

KENDALL, R. J., T. E. LACHER, JR., C. BUNCK, B. DANIEL, C. DRIVER, C. E. GRUE, F. LEIGHTON, W. STANSLEY, P. G. WATANABE, AND M. WHITWORTH. 1996. An ecological risk assessment of lead shot exposure in non-waterfowl avian species: upland game birds and raptors. Environmental Toxicology and Chemistry 15: 4–20.

KOZICKY, E., AND J. MADSON. 1973. Nilo shotshell efficiency test on experimental Mallard ducks, 1972–73. Presentation at Annual Meeting of the International Association of Game, Fish and Conservation Commissioners. Conservation Department Winchester-Western, East Alton, Illinois, USA.

KRAFT, M. 1984. Lead Poisoning. Are we wasting our waterfowl? Kansas Wildlife 41: 13–20.

LEOPOLD, A. 1933. Game Management. Charles Scribner's Sons, New York, USA.

LEWIS, J. C., AND E. LEGLER, JR. 1968. Lead shot ingestion by Mourning Doves and incidence in soil. Journal of Wildlife Management 32: 476–782.

LEWIS, L. A., AND S. H. SCHWEITZER. 2000. Lead poisoning in a northern bobwhite in Georgia. Journal of Wildlife Diseases 36: 180–183.

LEWIS, L. A., R. J. POPPENGA, W. R. DAVIDSON, J. R. FISCHER, AND K. A. MORGAN. 2001. Lead toxicosis and trace element levels in wild birds and mammals at a firearms training facility. Archives of Environmental Contamination and Toxicology 41: 208–214.

LOCKE, L. N., AND G. E. BAGLEY. 1967. Lead poisoning in a sample of Maryland Mourning Doves. Journal of Wildlife Management 31: 515–518.

LOCKE, L. N., M. R. SMITH, R. M. WINDINGSTAD, AND S. J. MARTIN. 1991. Lead poisoning of a Marbled Godwit. Prairie Naturalist 23:21–24.

LOCKE, L.N., AND N. J. THOMAS. 1996. Lead poisoning of waterfowl and raptors. Pages 108–117 *in* A. Fairbrother, L. N. Locke, and G. L. Hoff (Eds.). Noninfectious Diseases of Wildlife, 2nd Ed. Iowa State University Press, Ames, Iowa, USA.

LONGCORE, J. R., L. N. LOCKE, G. E. BAGLEY, AND R. ANDREWS. 1974. Significance of Lead Residues in Mallard Tissues. Special Scientific Report—Wildlife Number 182. US Department of the Interior, Fish and Wildlife Service, Washington, DC, USA.

LOWRY, E. D. 1974. Shotshell efficiency: the real facts. Sports Afield 172: 148–153.

MCATEE, W. L. 1908. Lead poisoning in ducks. Auk 25: 472.

MCCABE, R. A. 1987. Aldo Leopold, the Professor. Rusty Rock Press, Madison, Wisconsin, USA.

MCCONNELL, C. A. 1968. Experimental lead poisoning of Bobwhite Quail and Mourning Doves. Pages 208–219 *in* Proceedings 21st Annual Conference of the Southeastern Association of Game and Fish Commissioners.

MCGATH, T. B. 1931. Lead poisoning in wild ducks. Proceedings of the Staff Meeting of the Mayo Clinic 6: 749–752.

MCLEAN, D. D. 1939. Shot poisoning of wild ducks. California Conservation 4: 9.

MERETSKY, V. J., N. F. R. SNYDER, S. R. BEISSINGER, D. A. CLENDENEN, AND J. W. WILEY. 2000. Demography of the California Condor: implications for re-establishment. Conservation Biology 14: 957–967.

MIKULA, E. J. 1965. Iron shot field study. Pages 61–65 *in* Wasted Waterfowl. Mississippi Flyway Council Planning Committee Report.

MISSISSIPPI FLYWAY COUNCIL PLANNING COMMITTEE. 1965. Wasted Waterfowl.

MOREHOUSE, K. A. 1992a. Lead poisoning of migratory birds: the US Fish and Wildlife Service position. Pages 51–55 *in* D. J. Pain (Ed.). Lead Poisoning in Waterfowl. Proceedings of an IWRB Workshop, Brussels, Belgium, 13–15 June 1991. International Waterfowl and Wetlands Research Bureau Special Publication 16, Slimbridge, UK.

MOREHOUSE, K. A. 1992b. Crippling loss and shottype: The United States experience. Pages 32–37 *in* D. J. Pain (Ed.). Lead Poisoning in Waterfowl. Proceedings of an IWRB Workshop, Brussels, Belgium, 13–15 June 1991. International Waterfowl and Wetlands Research Bureau Special Publication 16, Slimbridge, UK.

MOSER, M. 1992. Conclusions. Pages 95–97 *in* Lead Poisoning in Waterfowl (D. J. Pain, Ed.). Proceedings of an IWRB Workshop, Brussels, Belgium, 1991. IWRB Special Publication 16, Slimbridge, United Kingdom.

MULHERN, B. M., W. L. REICHEL, L. N. LOCKE, T. G. LAMONT, A. BELISLE, E. CROMARTIE, G. E. BAGLEY, AND R. M. PROUTY. 1970. Organochlorine residues and autopsy data from Bald Eagles 1966–68. Pesticide Monitoring Journal 4: 141–144.

MUNRO, J. A. 1925. Lead poisoning in Trumpeter Swans. Canadian Field Naturalist 39: 160–162.

NATIONAL WILDLIFE HEALTH CENTER. 1985. Bald Eagle Mortality from Lead Poisoning and Other Causes, 1963–1984. National Wildlife Health Center, Madison, Wisconsin, USA.

OSMER, T. 1940. Lead shot: its danger to waterfowl. Science Monthly 50:455–459.

PAIN, D. J. (ED.). 1992. Lead Poisoning in Waterfowl. *In* D. J. Pain (Ed.). Lead Poisoning in Waterfowl. Proceedings of an IWRB Workshop, Brussels, Belgium, 13–15 June 1991. International Waterfowl and Wetlands Research Bureau Special Publication 16, Slimbridge, UK.

PAIN, D. J. 1996. Lead in waterfowl. Pages 251–264 *in* W. N. Beyer, G. H. Heinz, and A. W.

Redom-Norwood (Eds.). Environmental Contaminants in Wildlife: Interpreting Tissue Concentrations. Lewis Publishers, Boca Raton, Florida, USA.

PATTEE, O. H., S. N. WIEMEYER, B. M. MULHERN, L. SILEO, AND J. W. CARPENTER. 1981. Experimental lead-shot poisoning in Bald Eagles. Journal of Wildlife Management 45: 806–810.

PHILLIPS, J. C., AND F. C. LINCOLN. 1930. American Waterfowl. Houghton Mifflin Company, Boston, Massachusetts, USA.

REICHEL, W. L., S. K. SCHMELING, E. CROMARTIE, T. E. KAISER, A. J. KRYNITSKY, T. G. LAMONT, B. M. MULHERN, R. M. PROUTY, C. J. STAFFORD, AND D. M. SWINEFORD. 1984. Pesticide, PCB, and lead residues and necropsy data for Bald Eagles from 32 states—1978–81. Environmental Monitoring and Assessment 4: 395–403.

ROCKE, T. E., C. J. BRAND, AND J. G. MENSIK. 1997. Site-specific lead exposure from lead pellet ingestion in sentinel Mallards. Journal of Wildlife Management 61: 228–234.

ROCKE, T. E., AND M. D. SAMUEL. 1991. Effects of lead shot ingestion on selected cells of the Mallard immune system. Journal of Wildlife Diseases 27: 1–9.

ROSCOE, D. E., S. W. NIELSEN, A. A. LAMOLA, AND D. ZUCKERMAN. 1979. A simple, quantitative test for erythrocytic protoporphyrin in lead-poisoned ducks. Journal of Wildlife Diseases 15:127–136.

SAAMI. 1969. Super-soft iron may solve waterfowl lead poisoning problems. September 19, 1969 News Release, Sporting Arms and Ammunition Manufacturers Institute, New York, USA.

SANDERSON, G. C., AND F. C. BELLROSE. 1986. A review of the problem of lead poisoning in waterfowl. Illinois Natural History Survey Special Publication 4. Illinois Natural History Survey, Champaign, Illinois, USA.

SCHEUHAMMER, A. M., AND S. L. NORRIS. 1995. A review of the environmental impacts of lead shotshell ammunition and lead fishing weights in Canada. Occasional Paper No. 88. Canadian Wildlife Service, Ottawa, Ontario, Canada.

SCHEUHAMMER, A. M., S. L. MONEY, D. A. KIRK, AND G. DONALDSON. 2003. Lead fishing sinkers and jigs in Canada: Review of their use patterns and toxic impacts on wildlife. Occasional Paper No. 108. Canadian Wildlife Service, Ottawa, Ontario, Canada.

SCHULZ, J. H., P. I. PADDING, AND J. J. MILLSPAUGH. 2006. Will Mourning Dove crippling rates increase with nontoxic shot regulations? Wildlife Society Bulletin 34: 861–865.

SCHULZ, J. H., J. J. MILLSPAUGH, B. E. WASHBURN, G. R. WESTER, J. T. LANIGAN III, AND J. C. FRANSON. 2002. Spent shot availability and ingestion on areas managed for Mourning Doves. Wildlife Society Bulletin 30: 112–120.

SHILLINGER, J. E., AND C. C. COTTAM. 1937. The importance of lead shot in waterfowl. Transactions of the North American Wildlife Conference 2: 398–403.

SIDOR, I. F., M. A. POKRAS, A. R. MAJOR, R. H. POPPENGA, K. M. TAYLOR, AND R. M. MICONI. 2003. Mortality of Common Loons in New England, 1987 to 2000. Journal of Wildlife Diseases 39: 306–315.

SILEO, L., R. N. JONES, AND R. C. HATCH. 1973. The effect of ingested lead shot on the electrocardiogram of Canada Geese. Avian Diseases 17: 308–313.

SILEO, L., AND S. I. FEFER. 1987. Paint chip poisoning of Laysan Albatross at Midway Atoll. Journal of Wildlife Diseases 23: 432–437.

SILEO, L., L. H. CREEKMORE, D. J. AUDET, M. R. SNYDER, C. U. METEYER, J. C. FRANSON, L.N. LOCKE, M.R. SMITH, AND D.L. FINLEY. 2001. Lead poisoning of waterfowl by contaminated sediment in the Coeur D'Alene River. Archives of Environmental Contamination and Toxicology 41: 364–368.

SPARROWE, R. D. 1992. Side issues: environmental, economic, and social. Pages 38–41 *in* D. J. Pain (Ed.). Lead Poisoning in Waterfowl. Proceedings of an IWRB Workshop, Brussels, Belgium, 13–15 June 1991. International Waterfowl and Wetlands Research Bureau Special Publication 16, Slimbridge, UK.

TENNYSON, J. 2002. American Sportfishing Association Encourages Sound and Consistent Lead Fishing Sinker Policy. Press Release, American Sportfishing Associates, Alexandria, Virginia. [Online.] Available at http://www.asafishing.org/asa/newsroom/newspr_121902e.html. (Accessed 05 May 2008).

TRUST, K. A., M. W. MILLER, J. K. RINGELMAN, AND I. M. ORME. 1990. Effects of ingested lead on antibody production in Mallards (*Anas platyrhynchos*). Journal of Wildlife Diseases 26: 316–322.

US FISH AND WILDLIFE SERVICE. 1974. Proposed Use of Steel Shot for Hunting Waterfowl in the United States. Draft Environment Statement DES 74–76. Department of the Interior, Washington, DC, USA.

US FISH AND WILDLIFE SERVICE. 1976. Proposed Use of Steel Shot for Hunting Waterfowl in the United States. Final Environmental Statement, Department of the Interior, Washington, DC, USA.

US FISH AND WILDLIFE SERVICE. 1986a. Final Supplemental Environmental Impact Statement, Use of Lead Shot for Hunting Migratory Birds in the United States. Department of the Interior, Washington, DC, USA.

US FISH AND WILDLIFE SERVICE. 1986b. Lead Poisoning in Waterfowl (video). Department of the Interior, Washington, DC, USA.

US FISH AND WILDLIFE SERVICE. 1986c. Field Testing Steel Shot (video). Department of the Interior, Washington, DC, USA.

US FISH AND WILDLIFE SERVICE. 1986d. Lead Poisoning in Waterfowl (brochure). Department of the Interior, Washington, DC, USA.

US FISH AND WILDLIFE SERVICE. 1988. Appendix 13, A synopsis of the nontoxic shot issue. Pages 317–319 *in* SEIS 88, Final Supplemental Environmental Impact Statement: Issuance of Annual Regulations Permitting the Sport Hunting of Migratory Birds. Department of the Interior, Washington, DC, USA.

VAN TYNE, J. 1929. Greater Scaup affected by lead poisoning. Auk 46: 103–104.

VEIT, H. P., R. J. KENDALL, AND P. F. SCANLON. 1983. The effect of lead shot ingestion on the testes of adult Ringed Turtle Doves (*Streptopelia risoria*). Avian Diseases 27: 442–452.

VYAS, N. B., J. W. SPANN, G. H. HEINZ, W. N. BEYER, J. A. JAQUETTE, AND J. M. MENGELKOCK. 2000. Lead poisoning of passerines at a trap and skeet range. Environmental Pollution 107: 159–166.

WETMORE, A. 1919. Lead Poisoning in Waterfowl. Agriculture Bulletin Number 793–19. Department of Agriculture, Washington, DC, USA.

WILLIAMSON, L. E. 1987. Evolution of a landmark law. Pages 1–17 *in* H. Kallman (Ed.). Restoring America's Wildlife 1937–1987: the First 50 Years of the Federal Aid in Wildlife Restoration (Pittman-Robertson) Act. Fish and Wildlife Service, United States Department of the Interior, Washington, DC, USA.

WINCHESTER GROUP. 1974. A summary of comments on Department of Interior draft environmental statement DES 74–76: the proposed use of steel shot for waterfowl hunting in the United States. Olin Corporation, New Haven, Connecticut, USA.

WINDINGSTAD, R. M., S. M. KERR, AND L. N. LOCKE. 1984. Lead poisoning of Sandhill Cranes (*G. canadensis*). Prairie Naturalist 16: 21–24.

WOBESER, G. A. 1997. Diseases of Wild Waterfowl, 2nd ed. Plenum Press, New York, USA.

HISTORICAL PERSPECTIVE ON THE HAZARDS OF ENVIRONMENTAL LEAD FROM AMMUNITION AND FISHING WEIGHTS IN CANADA

ANTON M. SCHEUHAMMER

Environment Canada, National Wildlife Research Centre, Carleton University, Ottawa, ON, K1A 0H3, Canada. E-mail: Tony.Scheuhammer@ec.gc.ca

ABSTRACT.—In Canada, the environmental hazards of lead ammunition first gained prominence with waterfowl managers during the 1960s after publication of the seminal paper by Bellrose (1959); however, concerns regarding possible increased crippling of waterfowl from use of early steel shot ammunition, and a belief that lead shot ingestion and poisoning was primarily a USA problem, delayed the regulation of lead shot in Canada for many years. During the 1980s, increased national and regional research by the Canadian Wildlife Service and partners established the frequency and extent of shot ingestion and elevated lead accumulation in waterfowl, and led to the creation of the first non-toxic shot zones in Canada in 1990 and 1991. Gradual development of a wider array of non-toxic shot products, and increasingly broad bans on the use of lead shot, culminated in a national regulation in 1999 prohibiting the use of lead shot for the purpose of hunting all migratory game birds anywhere in Canada (exempting upland species—American Woodcock [*Scolopax minor*], Mourning Doves [*Zenaida macroura*], and Rock Doves [*Columbia livia*]). After non-toxic shot regulations were established in Canada, the incidence of elevated lead exposure in hatch year (HY) ducks declined dramatically, testifying to the effectiveness of the regulations and a generally high compliance by hunters. Compared with waterfowl, ingestion and accumulation of lead in upland birds has received relatively little attention in Canada, and elsewhere. However, the frequency of elevated lead accumulation in some HY upland game birds (e.g., Hungarian Partridge [*Perdix perdix*]) in Canada can be comparable to that experienced by HY ducks prior to non-toxic shot regulations. In addition, lead poisoning is a significant cause of death in some upland-foraging raptors (e.g., Golden Eagles [*Aquila chrysaetos*]) that may feed on dead or wounded upland prey with embedded lead shot or bullet fragments. Embedded fragments of metallic lead from ammunition projectiles are also a source of dietary lead exposure in humans who consume hunted game animals. The environmental impacts of small lead fishing sinkers and jigs in Canada have also been assessed. A major conclusion from a review of available data was that ingestion of small lead sinkers or jigs accounts for about 20–30% of recorded mortality of breeding adult Common Loons in habitats that experience high recreational angling activity. A common thread in these issues is the risk of elevated lead exposure and toxicity from ingestion of small metallic lead items deposited into the environment, largely through recreational activities. These risks can be avoided by replacing the use of lead in these activities by less toxic materials. *Received 29 May 2008, accepted 28 June 2008.*

SCHEUHAMMER, A.M. 2009. Historical perspective on the hazards of environmental lead from ammunition and fishing weights in Canada. *In* R. T. Watson, M. Fuller, M. Pokras, and W. G. Hunt (Eds.). Ingestion of Lead from Spent Ammunition: Implications for Wildlife and Humans. The Peregrine Fund, Boise, Idaho, USA. DOI 10.4080/ilsa.2009.0105

Key words: Ammunition, bullet, Common Loon, fishing, lead, shot, sinker, waterfowl, wildlife.

LEAD IS A TOXIC METAL with no known biological function. Lead's low melting point, malleability, ease of processing, and low cost have resulted in its use in a wide range of applications for over a thousand years. However, the recognition of adverse effects of lead at increasingly lower levels of exposure in humans and other animals has led to the gradual elimination of lead from many of its once common uses. A major continuing use of lead is in the manufacture of projectiles for ammunition used in hunting and target shooting; and for terminal tackle (sinkers and jigs) used in recreational angling. Here, the hazards resulting from the release of metallic lead into the environment from these activities are discussed from a Canadian perspective.

Lead in Canadian Wetland (Waterfowl) Hunting.— In Canada, significant concern over possible lead poisoning of waterfowl from lead shot ingestion first surfaced in the late 1960s and early 1970s, following publication of the comprehensive USA study by Bellrose (1959) and subsequent regulations by the US government to establish non-toxic shot zones for waterfowl hunting. During this time, Environment Canada (the federal environment department) contemplated banning the use of lead shot; and the Canadian Wildlife Service (CWS), in collaboration with the National Research Council, conducted studies to investigate the manufacture and ballistic properties of various alternative, non-lead shot. However, little research on shot ingestion rates in waterfowl beyond a few local or regional studies was conducted in Canada during this period. This was partly due to a belief by many waterfowl managers that, because waterfowl did not overwinter in Canada, lead shot ingestion was probably not a significant problem in Canadian wetlands; and that the use of steel shot might cripple far more waterfowl than would be saved by prohibiting the use of lead.

Concerted research and monitoring efforts to study the nature and extent of lead shot ingestion nationally were initiated in Canada when the USA announced, in the late 1980s, its intent to completely ban the use of lead shot for waterfowl hunting by 1991. Once research had been completed that demonstrated significant rates of shot ingestion by waterfowl in several locations across the country (collated in Kennedy and Nadeau 1993), the CWS, using its regulatory authority under the Migratory Birds Convention Act, and with Provincial agreement, established the first Canadian non-toxic shot zones in British Columbia, Manitoba, and Ontario in 1989 and 1990. At that time, the CWS judged that there was insufficient evidence to justify a national ban on the use of lead shot for waterfowl hunting, and it developed a set of criteria for assessing whether local lead exposure in waterfowl was sufficiently severe to require non-toxic shot regulations. This framework (Wendt and Kennedy 1992) and its subsequent modifications are typically referred to as the "hot spot" approach to regulating the use of lead shot. The CWS criteria were accepted in 1990 by federal and provincial wildlife Ministers as an interim policy for managing the problems associated with the use of lead shot for waterfowl hunting. However, a national wing bone survey to determine the pattern of elevated lead exposure in hatch-year (HY) ducks in Canada reported a widespread geographic association between elevated bone lead concentrations and waterfowl hunting, rather than a few, local sites of high lead exposure (Scheuhammer and Dickson 1996). In addition, lead poisoning of Bald Eagles (*Haliaeetus leucocephalus*), which in the USA had been linked to feeding on dead or wounded waterfowl containing embedded lead shot (Pattee and Hennes 1983), was also documented in Canada (Elliott et al. 1992). Scheuhammer and Norris (1995, 1996) reviewed the environmental impacts of lead from shotshell ammunition and lead fishing weights in Canada. Studies such as these, combined with the development of a wider variety of non-toxic shot products by the ammunition industry, and decisions by some Canadian provinces to completely phase out the use of lead shot for waterfowl hunting within their jurisdictions, resulted in a policy evolution away from the "hot spot" management approach in favour of broader controls on the use of lead shot. Beginning in 1997, a national regulation came into effect prohibiting the use of lead shot for the purpose of hunting migratory game birds within 200 meters of a watercourse anywhere in Canada (exempting upland migratory species—American Woodcock [*Scolopax minor*], Mourning Doves [*Zenaida macroura*], and Band-tailed Pigeons [*Columba fasciata*]); and in 1999, this regulation was expanded to include dry land as well as wetland areas (although upland migratory species were still exempt).

The regulation prohibiting the use of lead for migratory bird hunting was expected to reduce the deposition of lead into the Canadian environment by approximately 1,000 metric tonnes per year. Stevenson et al. (2005) confirmed that bone lead concentrations in HY ducks in Canada declined by 50%–90% (depending on species and location) after non-toxic shot regulations were established. Declines in bone-lead concentration were consistent with the results of a large anonymous hunter survey, which indicated a high level of reported compliance (>80%) with the nontoxic shot regulation among Canadian waterfowl hunters (Stevenson et al. 2005). Conversely, American Woodcock, an important upland game species not affected by the nontoxic shot regulation, showed no decrease in mean bone-lead concentration in samples collected after the national regulation came into effect; and a majority (70%) of Canadian waterfowl hunters who also hunt upland game birds reported continued (legal) use of lead shot for upland game bird hunting (Stevenson et al. 2005). These findings are in agreement with those of USA studies that also reported a generally low rate of lead shot ingestion in waterfowl and a high compliance by hunters after non-toxic shot regulations were established (Anderson et al. 2000, Moore et al. 1998). However, lead shot ingestion and poisoning remains a significant problem in some aquatic environments; for example, for swans in southern British Columbia and northern Washington (Degernes et al. 2006, Wilson et al. 2004), in environments where the availability of "old" lead shot may still be high.

Lead in Canadian Upland Hunting.—The harvest of upland game birds in Canada was estimated to be roughly equal to that of waterfowl (Scheuhammer and Norris 1995), suggesting that shot deposition rates may be similar for each type of hunting. However, relatively little research has been undertaken to estimate rates of shot ingestion or elevated lead exposure in upland birds. In Canada, migratory upland species such as American Woodcock were exempted from the national regulation requiring nontoxic shot for migratory game bird hunting, based on a lack of data indicating lead shot ingestion or elevated lead accumulation in these species; and there are no provincial regulations prohibiting the use of lead for hunting non-migratory game species. However, Kendall et al. (1996) concluded that concern for lead shot impacts on upland game birds and raptors was warranted, and that the issue merited continued scrutiny to protect upland game bird and raptor resources. In this context, Scheuhammer et al. (1999) investigated the degree of lead exposure in American Woodcock in eastern Canada, reporting that HY woodcock had median bone-lead concentrations of 11 μg/g, almost twice as high as HY Mallard (*Anas platyrhyncos*) and American Black Ducks (*Anas rubripes*) prior to restrictions on the use of lead shot for waterfowl hunting. Similar findings of elevated lead were later reported in American Woodcock from Wisconsin (Strom et al. 2005). Although the finding of a high incidence of elevated lead exposure in young woodcock was surprising, the source of exposure was uncertain. In a follow-up study, Scheuhammer et al. (2003a) sampled American Woodcock, surface soil, and earthworms from a number of sites in eastern Canada, and reported that, although lead concentrations in soils were low in most sites—typical of uncontaminated rural soils—lead concentrations in earthworms could be many times that of soil, and might therefore be a significant source of elevated lead exposure in American Woodcock as this species feeds extensively on earthworms and other soil invertebrates (Keppie and Whiting 1994). However the stable lead isotopic signatures (Scheuhammer and Templeton 1998) for bones of American Woodcock with elevated lead accumulation were often significantly different from signatures for both worms and soils sampled from the same areas. The range of $^{206}Pb/^{207}Pb$ ratios in wing bones of American Woodcock with elevated lead exposure was not consistent with exposure to lead in earthworms, nor to environmental lead from past gasoline combustion or mining wastes related to Precambrian lead ores, but was consistent with ingestion of spent lead shotgun pellets sold in Canada (Scheuhammer et al. 2003a). Ingestion of lead shot used for upland game bird hunting is a likely source of high bone-lead accumulation for American Woodcock in eastern Canada.

Table 1. Percent of hatch year (HY) upland game birds and ducks from Saskatchewan, Canada, with elevated (>10 μg/g dry wt.) bone lead concentrations.

Species	Percent with >10 μg/g lead in bone	Year Sampled
Hungarian Partridge (*Perdix perdix*)	11 (19/169)	2001
Sharp-tailed Grouse (*Tympanuchus phasianellus*)	5 (4/79)	2001
Mallard (*Anas platyrhyncos*)	8 (17/214)	1989/90

In addition to American Woodcock, lead exposure has been studied in a few other upland game bird species in Canada. For example, 11% of HY Hungarian Partridge (*Perdix perdix*) from Saskatchewan had elevated (>10 μg/g) bone-lead concentrations (Table 1), higher than the percentage of dabbling ducks from the same province that had elevated bone-lead levels prior to the establishment of non-toxic shot regulations (~8%: Scheuhammer and Norris, 1995). Chukar (*Alectoris chukar*) and Ring-necked Pheasant (*Phasianus colchicus*) from a heavily hunted area in southern Ontario, Canada, had shot ingestion rates of 8% and 34%, respectively (Kreager et al. 2008). A lower ingestion rate (1.2%) was reported for Ruffed Grouse (*Bonasa umbellus*) in Quebec (Rodrigue et al. 2005). Studies such as these demonstrate that the ingestion of lead shot in upland game birds can be comparable to that documented for waterfowl prior to restrictions on the use of lead shot for waterfowl hunting.

In a study examining lead exposure in 184 individuals of 16 upland raptor species found dead across Canada, Clark and Scheuhammer (2003) reported that 3–4% of total mortality in the three most commonly encountered species (Red-tailed Hawk [*Buteo jamaicensis*], Great Horned Owl [*Bubo virginianus*], and Golden Eagles [*Aquila chrysaitos*]) was attributable to lead poisoning. Golden Eagles, which feed exclusively on terrestrial/upland prey, die of lead poisoning on the Canadian Prairies as frequently as Bald Eagles that typically feed on more aquatic prey (waterfowl, fish) (Wayland and Bollinger 1999), indicating that some upland environments can present a risk of lead exposure and poisoning for raptors. A major source of elevated lead exposure in upland predators and scavengers is lead fragments from ammunition used for upland hunting, and for varmint shooting (Cade 2007, Craighead and Bedrosian 2008, Fisher et al. 2006, Hunt et al. 2006, Knopper et al. 2006). Replacing lead shot and high velocity lead bullets used in upland shooting with non-toxic alternatives would eliminate the only significant source of high lead exposure and poisoning for large avian predators and scavengers.

Small fragments of metallic lead are often present in tissues of animals shot with lead ammunition, and can be detected using radiography. In Canada, approximately 11% of breast muscle samples from a large number of hunter-killed game birds had elevated (>2 μg/g dry wt.) lead concentrations, with some samples having >1,000 μg/g (Scheuhammer et al. 1998). All muscle samples were examined visually prior to analysis, and none contained detectable lead pellets; however, radiography confirmed the presence of numerous small (<1 mm diameter) metallic fragments (Scheuhammer et al. 1998). Similar results have been reported by others (Frank 1986, Hunt et al. 2006, Johansen et al. 2004). Small pieces of metallic lead from shotgun pellet and bullet fragmentation embedded in the flesh of game animals are potential sources of dietary lead exposure for predators, and for human consumers of wild game, especially in communities that rely on subsistence hunting and for whom hunter-killed wild game represents a major food source (Hanning et al. 2003, Levesque et al. 2003, Tsuji et al. 1999). This risk can be eliminated by the use of non-toxic materials for manufacturing bullets and shotgun pellets.

Lead in Recreational Angling in Canada.—As reviewed in Scheuhammer et al. (2003b), more than 5 million Canadians take part in recreational angling each year, spending over 50 million days fishing on open water. Recreational anglers contribute to environmental lead deposition through the loss of lead fishing sinkers and jigs. Lost or discarded fishing sinkers and jigs amounting to an estimated 500 tonnes of lead, and representing up to 14% of all non-recoverable lead releases in Canada, are deposited in the Canadian environment annually. Wildlife, primarily Common Loons (*Gavia immer*) and other water birds, ingest fishing sinkers and jigs during feeding, when they mistake the sinkers and jigs for food items or grit, or consume lost bait fish with the line and weight still attached. Lead sinkers and jigs weighing <50 g and smaller than 2 cm in any dimension are generally the size found to be ingested by wildlife. Ingestion of a single lead sinker or lead-headed jig, representing up to several grams of lead, is sufficient to expose a loon or other bird to a lethal dose of lead. Lead sinker and jig ingestion has been documented in 10 different wildlife species in Canada; and in 23 species of wildlife in the United States, including loons, swans, other waterfowl, cranes, pelicans, and cormorants. Lead sinker and jig ingestion is the only significant source of elevated lead exposure and lead toxicity for Common Loons, and the single most important cause of death reported for adult Common Loons in eastern Canada and the United States during the breeding season, frequently exceeding deaths associated with entanglement in fishing gear, trauma, disease, and other causes of mortality. In addition to lead poisoning of wildlife, there are also risks of lead exposure in humans during both the home manufacture and handling of lead sinkers and jigs.

There are numerous viable non-toxic materials for producing fishing sinkers and jigs, including tin, steel, bismuth, tungsten, rubber, ceramic, and clay. Tin, steel, and bismuth sinkers and bismuth jigs are the most common commercially available alternatives in Canada. Many of the available non-lead products are currently more expensive than lead; however, switching to these products is estimated to increase the average Canadian angler's total yearly expenses by only about $2.00.

Regulatory action has been taken by some nations to reduce the use of lead sinkers and jigs. In 1987, Britain banned the use of lead fishing sinkers weighing less than one ounce. The United States has banned the use of lead sinkers and jigs in some National Wildlife Refuges, and in Yellowstone National Park. New Hampshire, Maine, and New York have established statewide regulations prohibiting the use of lead sinkers. Environment Canada and Parks Canada prohibited the possession of lead fishing sinkers or lead jigs weighing less than 50 g by anglers fishing in National Wildlife Areas and National Parks under the *Canada Wildlife Act* and the *National Parks Act*, respectively, in 1997. Environment Canada is currently considering broader controls on the manufacture, import, and sale of small lead sinkers and jigs.

CONCLUDING REMARKS

Although lead has been mined, smelted, and used in western society for hundreds of years, over the last half century many of its once common uses have been phased out in favour of less toxic materials. Thus, lead in house paints, water pipes, solder, pottery glazes, and gasoline has been eliminated or greatly reduced. A major remaining source of continuing environmental deposition of lead is metallic lead used in ammunition for hunting and target shooting, and in sinkers and jigs for recreational angling. These uses of lead have both environmental and human health consequences. Given the availability and continued development of functional, affordable, non-toxic alternatives to lead, the development of policies aimed at phasing out the use of lead for these activities is highly probable. It is important in the crafting of such policies that all significant stakeholders be actively involved in a consultative approach, including federal and state/provincial environmental and human health agencies, ammunition and tackle industries, non-governmental environmental and wildlife organizations, and hunting and angling communities.

LITERATURE CITED

ANDERSON, W. L., S. P. HAVERA, AND B. W. ZERCHER. 2000. Ingestion of lead and nontoxic shotgun pellets by ducks in the Mississippi flyway. Journal of Wildlife Management 64:848–857.

BELLROSE, F. C. 1959. Lead poisoning as a mortality factor in waterfowl populations. Illinois Natural History Survey Bulletin 27:235–288.

CADE, T. J. 2007. Exposure of California Condors to lead from spent ammunition. Journal of Wildlife Management 71:2125–2133.

CLARK, A. J., AND A. M. SCHEUHAMMER. 2003. Lead poisoning in upland-foraging birds of prey in Canada. Ecotoxicology 12:23–30.

CRAIGHEAD, D., AND B. BEDROSIAN. 2008. Blood lead levels of Common Ravens with access to big-game offal. Journal of Wildlife Management 72:240–245.

DEGERNES, L., S. HEILMAN, M. TROGDON, J. JORDAN, J. DAVISON, D. KRAEGE, M. CORREA, AND P. COWEN. 2006. Epidemiological investigation of lead poisoning in Trumpeter and Tundra Swans in Washington State, USA, 2000–2002. Journal of Wildlife Diseases 42:345–358.

ELLIOTT, J. E., K. M. LANGELIER, A. M. SCHEUHAMMER, P. H. SINCLAIR, AND P. E. WHITEHEAD. 1992. Incidence of lead poisoning in Bald Eagles and lead shot in waterfowl gizzards from British Columbia, 1988–91. Canadian Wildlife Service Progress Note, no. 220, Ottawa, Canada.

FISHER, I. J., D. J. PAIN, AND V. G. THOMAS. 2006. A review of lead poisoning from ammunition sources in terrestrial birds. Biological Conservation 131:421–432.

FRANK, A. 1986. Lead fragments in tissues of wild birds: a cause of misleading results. Science of the Total Environment 54:275–281.

HANNING, R. M., R. SANDHU, A. MACMILLAN, L. MOSS, L. J. S. TSUJI, AND E. NIEBOER. 2003. Impact on blood Pb levels of maternal and early infant feeding practices of First Nation Cree in the Mushkegowuk Territory of northern Ontario, Canada. Journal of Environmental Monitoring 5:241–245.

HUNT, G. W., W. BURNHAM, G. N. PARISH, K. K. BURNHAM, B. MUTCH, AND J. L. OAKS. 2006. Bullet fragments in deer remains: Implications for lead exposure in avian scavengers. Wildlife Society Bulletin 34:167–170.

JOHANSEN, P., G. ASMUND, AND F. RIGET. 2004. High human exposure to lead through consumption of birds hunted with lead shot. Environmental Pollution 127:125–129.

KENDALL, R. J., T. E. LACHER, C. BUNCK, B. DANIEL, C. DRIVER, C. E. GRUE, F. LEIGHTON, W. STANSLEY, P. G. WATANABE, AND M. WHITWORTH. 1996. An ecological risk assessment of lead shot exposure in non-waterfowl avian species: Upland game birds and raptors. Environmental Toxicology and Chemistry 15:4–20.

KENNEDY, J. A., AND S. NADEAU. 1993. Lead shot contamination of waterfowl and their habitats in Canada. Canadian Wildlife Service Technical Report Series, no. 164, Canadian Wildlife Service, Ottawa, Canada.

KEPPIE, D. M., AND R. M. WHITING, JR. 1994. American Woodcock (*Scolopax minor*). *In* A. Poole and F. Gill (Eds.). The Birds of North America, no. 100. Academy of Natural Sciences, Philadelphia, and American Ornithologists' Union, Washington, DC, USA.

KNOPPER, L. D., P. MINEAU, A. M. SCHEUHAMMER, D. E. BOND, AND D. MCKINNON. 2006. Carcasses of shot Richardson's Ground Squirrels may pose lead hazards to scavenging hawks. Journal of Wildlife Management 70: 295–299.

KREAGER, N., B. C. WAINMAN, R. K. JAYASINGHE, AND L. J. S. TSUJI. 2008. Lead pellet ingestion and liver-lead concentrations in upland game birds from southern Ontario, Canada. Archives of Environmental Contamination and Toxicology 54:331–336.

LÉVESQUE, B., J. F. DUCHESNE, C. GARIÉPY, M. RHAINDS, P. DUMAS, A. M. SCHEUHAMMER, J. F. PROULX, S. DÉRY, G. MUCKLE, F. DALLAIRE, AND E. DEWAILLY. 2003. Monitoring of umbilical cord blood lead levels and sources assessment among the Inuit. Occupational and Environmental Medicine 60:693–695.

MOORE, J. L., W. L. HOHMAN, T. M. STARK, AND G. Q. WEISBRICH. 1998. Shot prevalences and diets of diving ducks five years after the ban on use of lead shotshells as Catahoula Lake, Louisiana. Journal of Wildlife Management 62:564–569.

PATTEE, O. H., AND S. K. HENNES. 1983. Bald Eagles and waterfowl: the lead shot connection. Transactions of the North American Wildlife and Natural Resources Conference 48:230–237.

RODRIGUE, J., R. MCNICOLL, D. LECLAIR, AND J.-F. DUCHESNE. 2005. Lead concentrations in Ruffed Grouse, Rock Ptarmigan, and Willow Ptarmigan in Quebec. Archives of Environmental Contamination and Toxicology 49:97–104.

SCHEUHAMMER, A. M., AND K. DICKSON. 1996. Patterns of environmental lead exposure in waterfowl in eastern Canada. Ambio 25:14–20.

SCHEUHAMMER, A. M., AND S. L. NORRIS. 1995. A review of the environmental impacts of lead shot shell ammunition and lead fishing weights in Canada. Canadian Wildlife Service Occasional Paper, no. 88, Environment Canada, Ottawa, Canada.

SCHEUHAMMER A. M., AND S. L. NORRIS. 1996. The ecotoxicology of lead shot and lead fishing weights. Ecotoxicology 5:279–295.

SCHEUHAMMER, A. M., AND D. M. TEMPLETON. 1998. Use of stable isotope ratios to distinguish sources of lead exposure in wild birds. Ecotoxicology 7:37–42.

SCHEUHAMMER, A. M., J. A. PERRAULT, E. ROUTHIER, B. M. BRAUNE, AND G. D. CAMPBELL. 1998. Elevated lead concentrations in edible portions of game birds harvested with lead shot. Environmental Pollution 102:251–257.

SCHEUHAMMER, A. M., C. A. ROGERS, AND D. BOND. 1999. Elevated lead exposure in American Woodcock (*Scolopax minor*) in eastern Canada. Archives of Environmental Contamination and Toxicology 36:334–340.

SCHEUHAMMER, A. M., D. E. BOND, N. M. BURGESS, AND J. RODRIGUE. 2003a. Lead and stable lead isotope ratios in soil, earthworms, and bones of American Woodcock (*Scolopax minor*) from eastern Canada. Environmental Toxicology and Chemistry 22:2585–2591.

SCHEUHAMMER, A. M., S. L. MONEY, D. A. KIRK, AND G. DONALDSON. 2003b. Lead fishing sinkers and jigs in Canada: Review of their use patterns and toxic impacts on wildlife. Canadian Wildlife Service Occasional Paper, no. 108, Environment Canada, Ottawa, Canada.

STEVENSON, A. L., A. M. SCHEUHAMMER, AND H. M. CHAN. 2005. Effects of lead shot regulations on lead accumulation in ducks in Canada. Archives of Environmental Contamination and Toxicology 48:405–413.

STROM, S. M., K. A. PATNODE, J. A. LANGENBERG, B. L. BODENSTEIN, AND A. M. SCHEUHAMMER. 2005. Lead contamination in American Woodcock (*Scolopax minor*) from Wisconsin. Archives of Environmental Contamination and Toxicology 49:396–402.

TSUJI, L. J. S., E. NIEBOER, J. D. KARAGATZIDES, R. M. HANNING, AND B. KATAPATUK. 1999. Lead shot contamination in edible portions of game birds and its dietary implications. Ecosystem Health 5:183–192.

WAYLAND M., AND T. BOLLINGER. 1999. Lead exposure and poisoning in Bald Eagles and Golden Eagles in the Canadian prairie provinces. Environmental Pollution 104:341–350.

WENDT, S., AND J. A. KENNEDY. 1992. Policy considerations regarding the use of lead shot for waterfowl hunting in Canada. Pages 61–67 *in* D.J. Pain (Ed.). Lead Poisoning in Waterfowl. Proceedings of an IWRB Workshop, Brussels, Belgium, 13–15 June 1991. International Waterfowl and Wetlands Research Bureau Special Publication 16, Slimbridge, UK.

WILSON, L. K., M. DAVISON, AND D. KRAEGE. 2004. Lead poisoning of Trumpeter and Tundra Swans by ingestion of lead shot in Whatcom County, Washington, USA, and Sumas Prairie, British Columbia, Canada. Bulletin of the Trumpeter Swan Society 32:11–13.

TECHNICAL REVIEW OF THE SOURCES AND IMPLICATIONS OF LEAD AMMUNITION AND FISHING TACKLE ON NATURAL RESOURCES

BARNETT A. RATTNER[1], J. CHRISTIAN FRANSON[2], STEVEN R. SHEFFIELD[3], CHRIS I. GODDARD[4], NANCY J. LEONARD[5], DOUGLAS STANG[6] AND PAUL J. WINGATE[7]

[1]*US Geological Survey, Patuxent Wildlife Research Center, Beltsville Laboratory c/o BARC-East, Building 308, 10300 Baltimore Avenue, Beltsville, MD 20705, USA.*
E-mail: Barnett_Rattner@usgs.gov

[2]*US Geological Survey, National Wildlife Health Center, 6006 Schroeder Road, Madison, WI 53711-6223, USA.*

[3]*Department of Natural Sciences, Bowie State University, 4000 Jericho Road, Bowie, Maryland 20715 and College of Natural Resources, Virginia Polytechnic Institute and State University, National Capital Region – Northern Virginia Center, 7054 Haycock Road, Falls Church, VA 22043, USA.*

[4]*Great Lakes Fishery Commission, 2100 Commonwealth Boulevard, Suite 100, Ann Arbor, MI 48105-1563, USA.*

[5]*Michigan State University, Center for Systems Integration and Sustainability, Department of Fisheries and Wildlife, 115 Manly Miles Building, 405 South Harrison, East Lansing, MI 48823-5243, USA.*

[6] *Bureau of Fisheries, New York State Department of Environmental Conservation, 625 Broadway, Albany, NY 12233-4753, USA.*

[7]*Minnesota Department of Natural Resources, 500 Lafayette Road, St. Paul, MN 55155-4020, USA.*

EXTENDED ABSTRACT.—A technical review of lead sources that originate from hunting, shooting sports, and fishing activities (Rattner et al. 2008) was undertaken to provide background information for the preparation of policy statements by the American Fisheries Society and The Wildlife Society. Lead is a naturally occurring metal in the environment. In biological systems, it is a nonessential metal with no functional or beneficial role at the molecular and cellular levels of organization. Its use in ammunition and fishing tackle dates back hundreds to thousands of years, respectively. Realization of the hazards of lead shot to waterfowl can be traced to the late 1870s, while the hazards of lead fishing sinkers to birds became well-recognized in the 1970s with lead poisoning of swans in Britain. By the 1980s, Britain and some jurisdictions within the United States and Canada began placing restrictions on the use of lead shot and fishing sinkers.

Large quantities of lead ammunition and fishing tackle are produced annually. Estimates of lost fishing tackle are much less than the quantity of spent ammunition at waterfowl hunting areas and target ranges. Nonetheless, lost fishing tackle poses a toxicological threat to some waterbird species. Lead from spent ammunition and lost fishing tackle is not readily released into aquatic and terrestrial systems under most environmental conditions. Lead artifacts can be relatively stable and intact for decades to centuries. Nevertheless, under some environmental conditions (e.g., soft acidic water, acidic soil), lead can weather and be mobilized from such artifacts, yielding free dissolved lead, precipitates, and chemical species that complex with inorganic and organic matter. Dissolved, complex species and particulate lead can be adsorbed onto or incorporated into the surface of plants. In soil and sediment, various forms of lead can become adsorbed, taken up by tissues, and entrained in the digestive tract of invertebrates. Lead that is released from artifacts can evoke a range of biochemical, physiological, and behavioral effects in some species and life stages of invertebrates, fish, amphibians, and terrestrial vertebrates and can exceed criteria for protecting some biota (e.g., water quality criteria for invertebrates). Lead in soil, adsorbed or incorporated into food items, and fragments emanating from shooting ranges can ultimately result in elevated tissue concentrations in birds and small mammals and cause hematological changes and pathological lesions. For anthropogenic activities such as mining and smelting, lead concentrated in sediments (lead in silt, fine particulates, and pore water) can also be lethal to aquatic invertebrates, fish and waterbirds.

There is evidence documenting ingestion of spent shot and bullets, lost fishing sinkers and tackle, and related fragments by reptiles, birds, and mammals. Ingestion of some of these elemental lead artifacts can be accompanied by a range of effects (molecular to behavioral) in individuals and potentially even population-level consequences in some species (e.g., waterfowl, eagles, condors). Fish can ingest sinkers, jigs, and hooks, but unlike higher vertebrates, fatality seems to be related to injury, blood loss, exposure to air and exhaustion rather than lead toxicosis. There are no data demonstrating the ingestion of spent shot or bullets by invertebrates, fish or amphibians. Numerous reports in the medical literature describe accidental or purposeful ingestion of lead fragments in humans. Lead shot and sinkers can be retained in the appendix and digestive tract, and in some instances (particularly in children) lead artifacts are surgically removed to minimize exposure and adverse effects. Lead poisoning related to spent ammunition and lost fishing tackle has been most studied in avian species, and at least two studies indicate that the ban on the use of lead shot for hunting waterfowl and coots in North America has been successful in reducing lead exposure in waterfowl. Nonetheless, other species including upland game (e.g., doves, quail) and scavengers (e.g., vultures, eagles) continue to be exposed, and in some instances populations (e.g., California Condor, *Gymnogyps californianus*) may be at risk. Accordingly, many states have instituted restrictions on the use of lead ammunition to minimize effects to upland game birds and scavengers. The hazard of ingested lead sinkers and fishing tackle is well documented in swans and loons, and restrictions on the sale and use of lead weights have been instituted in the United Kingdom, Canada, numerous other countries, and several states in the United States to minimize effects on these and other species. There are only limited data on lead ingestion at shooting ranges by terrestrial vertebrates, and reproductive rates and estimation of population parameters of wildlife at these sites have not been adequately investigated. The hazards of spent ammunition and lost fishing gear to fish populations are unknown, but suspected to be minimal.

There has been an extensive effort in the development, efficacy testing, and regulation of alternatives to lead shot for hunting waterfowl and coots. Environmentally safe alternatives have been approved and currently are available in North America and elsewhere. Environmentally safe (non-lead) alternatives for some other types of hunting (e.g., shot for some upland birds, bullets for large game) and for target shooting are more recent developments, but use of these alternatives is not widespread. Many substitutes for lead fishing tackle have entered the marketplace in recent years. Some, but not all lead-substitute metals in fishing tackle have been previously deemed safe if ingested by waterfowl and some other birds and mammals. Less is known about the potential hazard of these alternatives to lower vertebrates.

The overall understanding of the hazards of lead used in shot, bullets, and fishing tackle would benefit from research generating toxicological and environmental chemistry data, and monitoring and modeling of exposure and effects. Those of highest priority include:

(1) broad scale monitoring on the incidence of lead poisoning in wildlife in countries where the extent of the problem is poorly documented or unknown, (2) data on the prevalence of lead poisoning related to fishing tackle in reptiles and aquatic birds, (3) information on the weathering, dissolution and long-term fate of lead fragments, and bioavailability of lead, in various aquatic and terrestrial ecosystems, (4) the hazards of spent ammunition and mobilized lead to wildlife at or near shooting ranges, and (5) evaluation of the results of regulations restricting the use of lead ammunition and fishing tackle on exposure and health of biota in various ecosystems.

The American Fisheries Society and The Wildlife Society seek to prepare position statements on the continued use of lead in ammunition and fishing tackle. There are at least three position options. Namely, the introduction of lead into the environment from hunting, shooting sports and fishing activities (1) is adequately regulated and the toxicological consequences of ingestion of lead are currently considered acceptable, (2) could be restricted in locations where lead poses an unacceptable hazard to biota and their supporting habitat, or (3) could be phased out with a goal of complete elimination. The leadership of both the American Fisheries Society and The Wildlife Society could interact with various entities to disseminate information about hazards and toxic effects of lead ammunition and tackle, as well as the availability and ecological benefits of safe alternatives. *Received 11 June 2008, accepted 28 July 2008.*

RATTNER, B. A., J. C. FRANSON, S. R. SHEFFIELD, C. I. GODDARD, N. J. LEONARD, D. STANG, AND P. J. WINGATE. 2009. Technical review of the sources and implications of lead ammunition and fishing tackle on natural resources. Extended abstract *in* R. T. Watson, M. Fuller, M. Pokras, and W. G. Hunt (Eds.). Ingestion of Lead from Spent Ammunition: Implications for Wildlife and Humans. The Peregrine Fund, Boise, Idaho, USA. DOI 10.4080/ilsa.2009.0106

Key words: Ammunition, fishing, hunting, lead, shooting, sinkers, wildlife.

LITERATURE CITED

RATTNER, B. A., J. C. FRANSON, S. R. SHEFFIELD, C. I. GODDARD, N. J. LEONARD, D. STANG, AND P. J. WINGATE. 2008. Sources and Implications of Lead Ammunition and Fishing Tackle on Natural Resources. Technical Review 08-01, The American Fisheries Society, The Wildlife Society, Bethesda, Maryland, USA.

LEAD POISONING IN WILD BIRDS IN EUROPE AND THE REGULATIONS ADOPTED BY DIFFERENT COUNTRIES

RAFAEL MATEO

Instituto de Investigación en Recursos Cinegéticos IREC (CSIC, UCLM, JCCM), Ronda de Toledo, s/n, E-13071 Ciudad Real, Spain. E-mail: rafael.mateo@uclm.es

ABSTRACT.—Lead poisoning in birds was described in Europe at the end of the 19th century, but the first epidemiological studies were not done until the second half of the 20th century. So far, most work has focused on waterfowl and birds of prey, with very few reports evaluating the impact on upland game birds.

The density of lead shot in sediments of wetlands has been studied in several countries, with maximal densities observed in southern Europe where up to 399 shot/m^2 in the upper 30 cm of sediment has been reported. Similarly, the highest prevalence of lead shot ingestion has been found in waterfowl wintering in the dry Mediterranean region, where birds concentrate in a limited number of wetlands which have also been intensively hunted for decades. If we consider the Mallard (*Anas platyrhynchos*) as a bioindicator species, the prevalence of lead shot ingestion varies from 2-10% in the wetlands of northern Europe, up to 25-45% in the Mediterranean deltas in southern Europe. The species with the highest prevalence of lead shot ingestion are the Northern Pintail (*Anas acuta*) and the Common Pochard (*Aythya ferina*) with values around 60-70%. Lead poisoning has been identified as an important cause of death for the endangered White-headed Duck (*Oxyura leucocephala*) and for swans (*Cygnus sp.*).

The accumulation of lead shot in upland ecosystems was recently studied in intensively hunted areas and estates used for driven shooting of Red-legged Partridge (*Alectoris rufa*) where densities of 7.4 shot/m^2 in soil prevailed. Prevalence of lead shot ingestion in Red-legged Partridge varies between 1.4% in Britain and 3.9% in Spain. Cases of lead poisoning have also been described in Grey Partridge (*Perdix perdix*, 1.4% of birds found dead), Common Pheasant (*Phasianus colchicus*), and Wood Pigeon (*Columba palumbus*).

Lead poisoning has been described in 17 species of birds of prey in Europe, some of which have been near-threatened (NT) such as the White-tailed Eagle (*Haliaeetus albicilla*), or endangered (EN), specifically, the Spanish Imperial Eagle (*Aquila adalberti*). Some studies have been conducted to evaluate the exposure in populations of raptors based on tissue analysis or the presence of lead shot in regurgitated pellets. Significant rates of lead shot ingestion have been observed in the Spanish Imperial Eagle (11% of pellets with lead shot). Further, 26-40% of Marsh Harriers (*Circus aeruginosus*) had blood lead >30 µg/dL, 91% of Griffon Vultures (*Gyps fulvus*) had blood lead >20 µg/dL, and 28% of White-tailed Eagles found dead or moribund had liver lead >5 µg/g (wet weight).

The ban on lead shot for hunting in wetlands, and/or for the hunting of waterbirds, was adopted by Denmark in 1985, and some years later by Norway, the Netherlands, Finland and Sweden. Other European countries agreed to implement bans on the use of lead for shooting over wetlands by the year 2000 following the African Eurasian Migratory Waterbird Agreement (AEWA). However, only Denmark, Norway, and the Netherlands have extended the ban to all hunted species. *Received 22 May 2008, accepted 6 September 2008.*

MATEO, R. 2009. Lead poisoning in wild birds in Europe and the regulations adopted by different countries. *In* R. T. Watson, M. Fuller, M. Pokras, and W. G. Hunt (Eds.). Ingestion of Lead from Spent Ammunition: Implications for Wildlife and Humans. The Peregrine Fund, Boise, Idaho, USA. DOI 10.4080/ilsa.2009.0107

Key words: Conservation, game, heavy metal, lead, raptor, partridge, waterfowl.

BIRDS CAN BE LETHALLY EXPOSED to different sources of lead, such as contaminated sediments from mining and smelting activities (Beyer et al. 2000), leaded paint (Sileo and Fefer 1987) or lead fishing weights (Birkhead 1983), but the most common source associated with clinical lead poisoning is by far the ingestion of lead ammunition, mainly shot pellets, used for hunting as reviewed by several authors (Thomas 1982, Sanderson and Bellrose 1986, Locke and Friend 1992, Pain 1992, Sanderson 1992, Guitart et al. 1999, Fisher et al. 2006). There are two causes of ingestion of lead shot pellets in birds. The species with a more developed muscular stomach (gizzard) like waterfowl or upland game birds (pheasants, partridges or pigeons) usually feed on plant matter or animals with hard shells that make necessary the regular ingestion of sand or gravel (gastroliths or grit) to break and grind the food (Gionfrido and Best 1999). Lead shot pellets accumulated in hunted environments are ingested by these species, confused as grit or possibly, as seeds (Trost 1981, Pain 1990, Moore et al. 1998, Mateo and Guitart 2000). The second cause of lead shot ingestion is found in birds of prey, especially obligatory and occasional scavengers that feed on carcasses with embedded lead ammunition in their flesh (Kenntner et al. 2001, Fisher et al. 2006). Although in many cases, the lead shot or bullet fragments are regurgitated with undigested material such as hair and feathers in the form of pellets, the absorption of lead during food digestion can be enough to kill the raptors.

Lead poisoning in birds by the ingestion of lead ammunition was described in Europe at the end of the 19th century in Common Pheasants (*Phasianus colchicus*) in the UK (Calvert 1876). The first reports of lead poisoning in European waterfowl were from the 1960-70s in France (Hoffmann 1960, Hovette 1971, 1972), the UK (Olney 1960, Beer and Stanley 1965), Italy (del Bono 1970) and the Scandinavian countries (Erne and Borg 1969, Danell and Anderson 1975, Holt et al. 1978). The first cases in birds of prey were reported in the 1980s in falconry and wild birds (MacDonald et al. 1983, Lumeij et al. 1985).

The present review compiles the available data about lead poisoning in Europe due to the ingestion of lead ammunition and the regulatory actions adopted by the different countries to reduce its incidence.

LEAD SHOT INGESTION IN WATERBIRDS

Types of Hunting Activities and Lead Shot Densities in Wetlands.—Waterfowl hunting is done in Europe by several techniques, depending on the species and the wetland. Eurasian Coots (*Fulica atra*) are hunted by pushing them with boats to force birds to fly over the circle of hunters located in boats or on the shore of a lagoon. Ducks are mainly hunted from blinds located on the border of lagoons, marshes, or in rice fields. Hunting is generally restricted to daylight in lagoons, but in some places, like the Ebro Delta (Spain), hunting in rice fields is allowed during several nights around the full moon. Hunters usually bait with grain around their blinds in rice fields to attract birds several days before hunting. Similarly, wildfowling in the UK is done on flight ponds to which waterfowl return for food at dawn and dusk from their daytime resting places on estuaries or large water bodies. These flight ponds are repeatedly baited to attract birds and therefore accumulate high lead shot concentrations (Thomas 1982). In the past, in areas like the Cerro de los Ánsares dune in Doñana (Spain), Greylag Geese (*Anser anser*) were shot when they concentrated there to ingest coarse sand as grit for the gizzard (Mateo et al. 2000a). Also in Doñana, one of the largest wetlands in Europe, flocks of ducks were followed through marshlands on horseback and hunted with guns capable of shooting heavy loads of pellets. The horse ("cabresto") served both as a hide and a support for the gun (Chapman and Buck 1893). This type of hunting produced a more diffuse contamination of lead shot in the marshes than the techniques based on fixed blinds.

The density of lead shot in wetland sediments has been studied in several countries, with maximal densities observed in southern Europe where up to 399 shot/m^2 in the upper 30 cm of sediment has been reported in the Medina Lagoon in southern Spain (Mateo et al. 2007a). The lead shot densities in Europe have been >100 shot/m^2 in the upper 20 cm of wetland sediments from Denmark, France and Spain, and between 10 and 50 shot/m^2 in most of the wetlands from the UK (Table 1). The highest lead shot densities have been found around shooting ranges located in wetlands. Petersen and Meltofte (1979) found lead shot densities ranging from 44 to 2,045 shot/m^2 at four Danish shallow water localities with shooting ranges. Smit et al. (1988a) found 400 and 2,195 shot/m^2 at two clay pigeon grounds in the Netherlands. At Lough Neagh, Co. Antrim, in Ireland, 2,400 spent gunshot/m^2 in the upper 5 cm were found along 100 m of shore in front of a clay pigeon shooting site and on the lake bed up to 60 m from the shore (O'Halloran et al. 1988b). A similar scenario was found in the El Hondo Natural Park in Spain, where a shooting range was located in a temporary marshland, yielding a density of 1,432 gunshot/m^2 (Bonet et al. 2004). Although no data about lead shot densities exist for many European countries, the highest densities are clearly in the Mediterranean wetlands (Figure 1).

Table 1. Lead shot densities in European wetlands with waterfowl hunting.

Country	Area	Site	Depth (cm)	Year	shot/m^2
Ireland	Cork	Kilcolman W. R.	-	1985-86[a]	7
United Kingdom	Moray/Beauly F.	Longman Bay	15	1981-82[b]	nd
		Lentral Point			2.57
		Easter Lovat			nd
	Loch of Strathbeg	Starnakeppie			2.04
		Back Bar			10.29
		Savoch Burn mouth			7.18
		Savoch Farm			2.04
		Starnafin			3.11
	Caelaverock	The Merse			3.95
	Gayton Sands	Marsh End			nd
		Railing Flash			9.77
	Llyn Ystumllyn	The marsh			3.04
	Gloucestershire	Flight pond			30.00
		Saul Warth			5.45
		The Pill meadow			9.44
		The Pill mud			3.04
	Elmley	Shellfleet Creek			7.44
		Shellfleet Creek			4.88
		Brick fields			13.08
	Norfolk	Flight pond 1			26.80
		Flight pond 2			8.22
	Ouse Washes	The washes			16.00
Denmark	Western Jutland	Agger Fjord	20	1978[c]	14.10
		Thyborøn Fjord			0
		Harboøre Fjord			25.90
		Ringkøbing Fjord			35.70
		Ho Bugt			0
	Ringkøbing Fjord	Klægbanken			53.30
		Haurvig Grund			12.20
		Skjern Åś munding			65.80
		Tipperne øst			88.30

Country	Area	Site	Depth (cm)	Year	shot/m^2
		Tippersande			166.80
		Tipperne vest			183.70
		Nymindestrømmen			145.90
	Sjæelland, Køge	Ølsemagle Revle			70.00
The Netherlands	Overissjel	Ketelmeer	7	1979-84[d]	20.20
	Zuid-Holland	Beninger Slikken			18.60
					14.00
	Zuid-Holland	Dordtsche Biesboch			23.10
					43.50
Hungary	Six areas	Ce.	-	-[e]	0.60
		So.			10.41
		Cs.			5.71
		Ur.			0.07
		Vá.			2.58
		Al.			1.82
France	Camargue	Mejanes 1	15-20	1987[f]	6.40
		Mejanes 2			41.90
		North Vaccares 1			6.40
		North Vaccares 2			nd
		North Vaccares 3			25.00
		Fangouse 1			6.40
		Fangouse 2			26.40
		Cameroun			6.40
		Pebre			170.30
		Beluge			12.70
		Tortue			nd
		Paty			199.50
		Consecan 1			nd
		Consecan 2			nd
		La Saline			83.90
	L. de Grand Lieu	La Morne	5	1988[g]	80.00
		La Ségnaigerie 1			46.00
		La Ségnaigerie 1		1989	50.00
Spain	Ebro delta	Buda Island 1	20	1991[h]	28.20
		Buda Island 2		1992	54.50
		Canal Vell rice			6.00
		Buda Island 3		1993[i]	97.10
		Encanyissada			266.10
		Punta de la Banya			nd
		La Llanada			48.50
		L'Aufacada		1996[j]	82.70
		Migjorn			13.90
		Dapsa			66.50
	Tablas de Daimiel	Puesto del Rey		1993[k]	99.40
	Alb. de València	Sueca			287.50
	El Hondo	Embalse de Levante		1993[k,l]	163.00
		Charca Sur			123.60
	Cádiz-Sevilla	Medina 1	10	2002[m]	148.30
		Medina 2	30		398.90
		Salada del Puerto	10		58.90
		Chica del Puerto			12.10
		Jeli de Chiclana			21.60
		Zorrilla de Espera		2001	26.70
		Taraje de Sevilla		2002	8.50
	Guadalquivir M.	Salinas de Sanlúcar		2002	18.30

Country	Area	Site	Depth (cm)	Year	shot/m²
		Santa Olalla		2001	11.80
		Lucio de Marilópez		2002	nd
		Veta la Palma		2002	nd
		Brazo del Este		2001	24.60
		L. Caraviruelas	15	1993[k]	14.40
		Hato Blanco		1997	nd
		C. de los Ánsares	20	1997[n]	16.20
		L. Caballero	15	1997[o]	7.20

[a]O'Halloran et al. 1988b; [b]Mudge 1984; [c]Peterson and Meltofte 1979; [d]Smit et al. 1988a; [e]Imre 1994; [f]Pain 1991a; [g]Mauvais and Pinault 1993; [h]Guitart et al. 1994a; [i]Mateo et al. 1997b; [j]Mateo 1998; [k]Mateo et al. 1998; [l]Bonet et al. 1995; [m]Mateo et al. 2007[a]; [n]2000a; [o]Mateo and Taggart 2007.

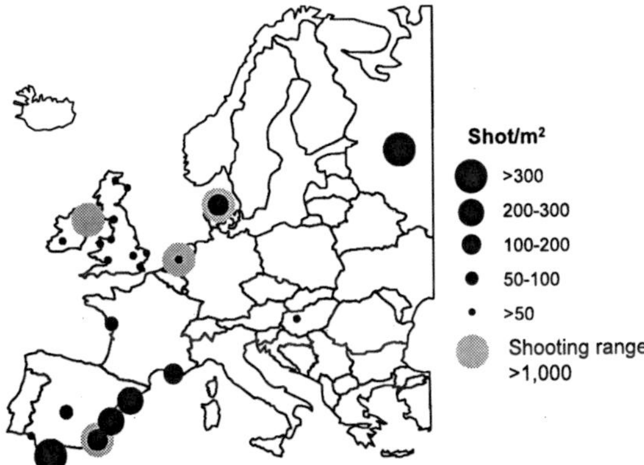

Figure 1. Densities of lead shot in wetlands due to waterfowl hunting and sport shooting.

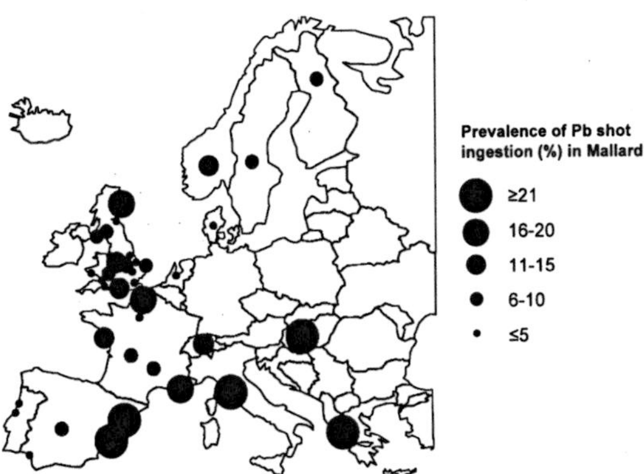

Figure 2. Prevalence of lead shot ingestion in Mallard (*Anas platyrhynchos*) in different European wetlands.

Prevalence of Lead Shot Ingestion and Mortality.— The prevalence of lead shot ingestion in waterfowl has been studied in different countries since 1957 (Tables 2a, 2b). Like lead shot densities in sediments, the highest levels have been found in the waterfowl wintering areas in the dry Mediterranean region, where birds concentrate in a limited number of wetlands intensively hunted for decades (Figure 2). A similar latitudinal trend may be observed in North America (Sanderson and Bellrose 1986).

The Mallard (*Anas platyrhynchos*) can be a good bioindicator species for lead poisoning because it is widely distributed in the world and its prevalence of lead shot ingestion is moderate to high among waterfowl species (Figure 2). The mean prevalence of lead shot ingestion in Mallards from northern Europe varies from 2.2% in Holland to 10.9% in Norway, with an overall value of 3.6% for a sample size of 8,683 shot or trapped individuals (Table 2a, Figure 2). In central and southern Europe, the prevalence of lead shot ingestion in Mallards ranges from 3.2% in Portugal to 36.4% in Greece, with an overall value for 11,239 individuals sampled of 17.3% (Table 2b; Figure 2).

In northern Europe the highest prevalence was observed in Common Goldeneyes (*Bucephala clangula*) with 13.8% of 152 sampled birds, followed by Tufted Ducks (*Aythya fuligula*) with 11.7% of 290 birds (Table 2a). The highest prevalence in these two species was found in Finland, with 32.1% for Common Goldeneye and 58.3% in Tufted Duck (reviewed in Pain 1990b). The species with the highest prevalence of lead shot ingestion in south-central Europe are the Northern Pintail (*Anas acuta*) with 45% of 598 birds, followed by the Common Pochard (*Aythya ferina*) with 24% of 507 birds (Table 2b). In the case of Mediterranean wetlands like the deltas of rivers Ebro, Rhône and Evros, the prevalence in the Northern Pintail and the Common Pochard ranges from 50 to 70% (Pain 1990a; Pain and Handrinos 1990; Mateo et al.

Table 2a. Prevalence of lead shot ingestion in waterfowl from northern Europe.

Species	United Kingdom 1957-81[a]			Holland 1977-87[b]			Denmark 1974-77[c]			Norway 1978-87[d]			Sweden 1972-94[e]			Finland 1973-1975[f]			Northern Europe		
	n	+	%	n	+	%	n	+	%	n	+	%	n	+	%	n	+	%	n	+	%
Tundra Swan *Cygnus columbianus*	516	1	0.2	-	-														516	1	0.2
Pink-footed Goose *Anser brachyrhynchus*	73	2	2.7	-	-	-	-	-	-	-	-	-	-	-	-	-	-	-	73	2	2.7
G. White-fronted Goose *Anser albifrons*	30	0	0.0	-	-	-	-	-	-	-	-	-	-	-	-	-	-	-	30	0	0.0
Greylag Goose *Anser anser*	42	3	7.1	-	-	-	-	-	-	-	-	-	-	-	-	-	-	-	42	3	7.1
Barnacle Goose *Branta leucopsis*	61	0	0.0	-	-	-	-	-	-	-	-	-	-	-	-	-	-	-	61	0	0.0
Eurasian Wigeon *Anas penelope*	574	0	0.0	-	-	-	251	4	1.6				142	4	2.8	15	0	0.0	982	8	0.8
Gadwall *Anas strepera*	61	2	3.3	5	1	20.0	-	-	-	-	-	-	-	-	-	-	-	-	66	3	4.5
Common Teal *Anas crecca*	1,188	12	1.0	-	-	-	167	0	0.0				180	0	0.0	88	0	0.0	1,623	12	0.7
Mallard *Anas platyrhynchos*	1,495	86	5.8	3,260	72	2.2	3,251	94	2.9	128	14	10,9	464	40	8.6	85	5	6.0	8,683	311	3.6
Northern Pintail *Anas acuta*	265	11	4.2	-	-	-	44	2	4.5	-	-	-	40	4	10.0	5	2	40.0	354	19	5.4
Northern Shoveler *Anas clypeata*	161	3	1.9	-	-	-	9	0	0.0	-	-	-	15	0	0.0	1	0	0.0	186	3	1.6
Common Pochard *Aythya ferina*	246	15	6.1	-	-	-	1	0	0.0	-	-	-	16	7	43.5	6	3	50.0	269	25	9.3
Tufted Duck *Aythya fuligula*	210	9	4.3	-	-	-	28	1	3.6	-	-	-	28	10	35.7	24	14	58.3	290	34	11.7
Greater Scaup *Aythya marila*	11	0	0.0	-	-	-	-	-	-	-	-	-	-	-	-	-	-	-	11	0	0.0
Common Goldeneye *Bucephala clangula*	15	1	6.7	20	1	5.0	-	-	-	-	-	-	89	10	11.2	28	9	32.1	152	21	13.8

[a]Olney 1960, 1968; Thomas 1975; Mudge 1983; Street 1983; [b]Smit et al. 1988b; Lumeij et al. 1989; Lumeij and Scholten 1989; [c]Wium-Andersen and Fransmann 1974; Petersen and Meltofte 1979; Clausen and Wolstrup 1979; [d]Pain 1990b; [e]Danell and Anderson 1975; Danell et al. 1977; Jågas 1996; [f]Danell 1980. Some of these data reviewed before in Pain 1990b.

Table 2b. Prevalence of lead shot ingestion in waterfowl from southern and central Europe.

Species	Portugal 1993-99[a]			Spain 1977-2004[b]			France 1960-2001[c]			Italy 1990[d]			Switzerland[e] 1979-81			Hungary[f]			Greece[g] 1989-90			Southern-Central Europe		
	n	+	%	n	+	%	n	+	%	n	+	%	n	+	%	n	+	%	n	+	%	n	+	%
Greylag Goose *Anser anser*	-	-	-	161	6	3.7	-	-	-	-	-	-	-	-	-	-	-	-	-	-	-	161	6	3.7
Eurasian Wigeon *Anas penelope*	-	-	-	28	1	3.6	472	20	4.2	13	1	7.7	-	-	-	-	-	-	7	0	0.0	520	22	3.7
Gadwall *Anas strepera*	-	-	-	31	3	9.7	653	25	3.8	-	-	-	11	0	0.0	-	-	-	15	0	0.0	710	28	3.2
Common Teal *Anas crecca*	-	-	-	59	10	16.9	41,131	1,958	4.8	-	-	-	52	3	5.8	-	-	-	34	5	14.7	41,276	1,976	4.8
Mallard *Anas platyrhynchos*	186	6	3.2	245	50	20.4	10,738	1,874	17.5	-	-	-	59	8	13.6	625	186	29.8	11	4	36.4	11,864	2,128	17.9
Northern Pintail *Anas acuta*	-	-	-	108	73	67.6	469	190	40.5	-	-	-	13	3	23.1	-	-	-	8	4	50.0	598	270	54.5
Northern Shoveler *Anas clypeata*	-	-	-	103	29	28.2	1,113	117	10.5	-	-	-	2	0	0.0	-	-	-	9	0	0.0	1,227	146	13.4
Garganey *Anas querquedula*	-	-	-	2	0	0.0	2,001	199	9.9	-	-	-	-	-	-	-	-	-	-	-	-	2,003	199	9.9
Red-crested Pochard *Netta rufina*	-	-	-	78	10	12.8	2	0	0.0	-	-	-	-	-	-	-	-	-	1	0	0.0	81	10	12.3
Common Pochard *Aythya ferina*	-	-	-	44	32	72.7	1,917	456	23.8	4	3	75.0	65	9	13.9	-	-	-	14	7	50.0	2,044	507	25.6
Tufted Duck *Aythya fuligula*	-	-	-	5	4	80.0	3,867	396	10.2	-	-	-	40	4	10.0	-	-	-	1	1	100	3,913	405	10.4
Common Goldeneye *Bucephala clangula*	-	-	-	-	-	-	2	2	100	-	-	-	-	-	-	-	-	-	-	-	-	4	4	100
White-headed Duck *Oxyura leucocephala*				25	8	32.0	-	-	-	-	-	-	-	-	-	-	-	-	-	-	-	25	8	32.0

[a]Rodrigues et al. 2001; [b]Llorente 1984; Guitart et al. 1994a; Mateo et al. 1997b, 1998, 2000b, 2006, 2007a; Martínez-Haro et al. 2005; [c]Hoffman 1960; Tamisier 1971; Hovette 1974; Pirot 1978; Alouche 1983; Cordel-Boudard 1983; Campredon 1984; Pirot and Taris 1987; Pain 1990a, 1991b; Mauvais and Pinault 1993; Lefranc 1993; Schricke and Lefranc 1994; Lamberet 1995; Pinault 1996; Mondain-Monval et al. 2002; [d]Tirelli et al. 1996; [e]Zuur 1982; [f]Imre 1994; [g]Pain and Handrinos 1990. Some of these data reviewed before in Pain 1990b, Duranel 1999 and Mezieres 1999.

1997b, 2000b). In a recent study, Figuerola et al. (2005) found with a meta-analysis of lead shot ingestion in 51 locations and 27 waterfowl species from North America and Europe that the prevalence in a given species was highly variable between localities, and was not consistently different between dabbling, grazing, and diving species.

Some estimations of mortality by lead poisoning in waterfowl have been done in Europe from the number of ingested shot, as described by Bellrose (1959). Mudge (1983) estimated that about 2.3% of the British population of wild Mallards might die each winter as a direct result of lead pellet ingestion, which represented at a conservative estimate about 8,000 birds. At a more local scale, Thomas (1975) estimated mortality rates in different species from the Ouse Washes (UK) ranging from 1-2.8% in Northern Shovelers (*Anas clypeta*) to 5.1-8.3% in Northern Pintails. In the Ebro Delta (Spain), mortality from lead poisoning within a mainly sedentary population of Mallards would be about 37% (9,600 birds) during the wintering season (Mateo et al. 1997b). From the data compiled in the present review (Tables 2a, 2b) and following the same calculations as the cited studies, about 975,115 waterfowl might die by lead poisoning in Europe during the wintering season from November to February from a total population of 11,228,700 of 17 species (Table 3). This percentage of waterfowl mortality by lead poisoning in Europe of 8.7% is higher than the value estimated in the USA of 2-3% (Bellrose 1959). In fact, several waterfowl species have a higher prevalence of lead shot ingestion in Europe (Table 3) than in North America (Sanderson and Bellrose 1986; Figure 3). Based on recaptures of banded ducks, Tavechia et al. (2001) found that relative monthly survival of lead-affected Mallards (with >1 ingested lead shot in the gizzard) was 19% lower than in unaffected birds over the period 1960–71 in Camargue (France). In a similar study, Guillemain et al. (2007) observed a lower survival in Common Teals (*Anas crecca*) carrying at least one lead shot in the gizzard than in Common Teals with no lead. Tavecchia et al. (2001) did not observe a negative effect on survival in Mallards with only one ingested shot, suggesting that Common Teal were more sensitive to lead poisoning.

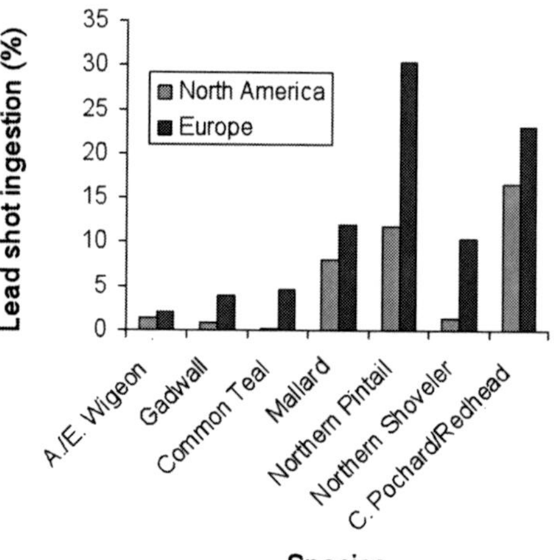

Figure 3. Comparison of the prevalence of lead shot ingestion in waterfowl species from North America (n = 171,697) and Europe (n = 75,761).

Lead poisoning in Mute Swans (*Cygnus olor*) has been observed in the UK (Simpson et al. 1979; Birkhead 1982, 1983; Poole 1986; Sears 1988; Pennycott 1998; Perrins et al. 2003; Kelly and Kelly 2004). These studies reported fatalities due to the ingestion of lead fishing weights, although shotgun pellets have been frequently found in Mute Swans (21.9%, n = 32), Bewick's Swans (*Cygnus columbianus*) (10%, n=20) and Whooper Swans (*Cygnus cygnus*) (40%, n =5) found dead in the UK by Mudge (1983). In Ireland, O'Halloran et al. (1988a, 1988b) studied 101 dead or moribund Mute Swans and found ingested lead fishing weights in 10 birds and shotgun pellets in 49 birds, in all cases with elevated lead levels in tissues; one Whooper Swan also died by lead shot ingestion. Lead poisoning by shotgun pellet ingestion was recorded in two Mute Swans from Marano Lagoon (Perco et al. 1983) and in other Italian wetlands (Di Modugno et al. 1994). Lead poisoning in Mute Swans has been detected in Sweden (7-17% of examined birds; Frank and Borg 1979, Mathiasson 1986). After the ban on the use of lead angling weights in England and Wales in 1987, the incidence of lead poisoning cases in Mute Swans started to fall and the population started to increase, although the problem was not eradicated (Sears and Hunt 1991, Perrins et al. 2003).

Lead shot ingestion has been studied in *Rallidae* species. The prevalence in Eurasian Coot has been

Table 3. Prevalence of lead shot ingestion and estimation of mortality in waterfowl wintering in Europe.

Species	Wintering Population (n)	Trend[a]	Prevalence 1957-2004		Estimated mortality[b]	
			n	(%)	n	%
Tundra Swan *Cygnus columbianus*	23,000	-3	516	0.2	45	0.2
Pink-footed Goose *Anser brachyrhynchus*	290,000	3	73	2.7	8,049	2.8
G. White-fronted Goose *Anser albifrons*	1,100,000	0	30	0.0	0	0.0
Greylag Goose *Anser anser*	390,000	3	203	4.4	17,517	4.5
Barnacle Goose *Branta leucopsis*	370,000	3	61	0.0	0	0.0
Eurasian Wigeon *Anas penelope*	1,700,000	0	1,502	2.0	34,398	2.0
Gadwall *Anas strepera*	96,000	3	776	4.0	3,885	4.0
Common Teal *Anas crecca*	730,000	-1	42,899	4.6	34,271	4.7
Mallard *Anas platyrhynchos*	3,700,000	-1	20,547	11.9	444,942	12.0
Northern Pintail *Anas acuta*	120,000	-2	952	30.4	36,905	30.8
Northern Shoveler *Anas clypeata*	200,000	-2	1,413	10.5	21,365	10.7
Red-crested Pochard *Netta rufina*	84,000	3	81	12.3	10,506	12.5
Common Pochard *Aythya ferina*	790,000	-2	2,313	23.0	184,078	23.3
Tufted Duck *Aythya fuligula*	1,200,000	-2	4,203	10.4	126,977	10.6
Greater Scaup *Aythya marila*	120,000	-3	11	0.0	0	0.0
Common Goldeneye *Bucephala clangula*	310,000	1	156	16.0	50,329	16.2
White-headed Duck *Oxyura leucocephala*	5,700	(-3)[c]	25	32.0	1,848	32.4
All species	11,228,700		75,761	-	975,115	8.7

[a] The trend value was scored from BirdLife International (2004) by considering large increase=3, moderate increase=2, slight increase=1, stable=0, slight decline=-1, moderate decline=-1, large decline=-3. [b] The estimation of mortality has been done following the method described by Bellrose (1959). The calculations were done with the mean distribution of the number of ingested pellets found in European waterfowl (Mudge 1983, Pain 1990a, Mateo et al., 1997b). The distribution was 1 shot: 47.1%, 2: 15.7%, 3: 5.4%, 4: 6.3%, 5: 3.5%, 6: 2%, >6: 19.9%. Birds with these numbers of ingested shot were assumed to have rates of mortality of 9, 23, 30, 36, 43, 50 and 75, respectively. Previously, the prevalences were corrected for the hunting bias (1.5, 1.9, 2.0, 2.1, 2.2, 2.3 and 2.4, respectively). A turn-over factor of 6 was considered. [c]White-headed Duck population fluctuates, but it is “Endangered” and the lowest value of trend has been assigned.

highly variable, with values of 0% in the Ouse Washes (UK) (Thomas 1975, Mudge, 1983), 3.6% in the Ebro Delta (Spain) (Mateo et al. 2000b), 5.1% in Switzerland (reviewed in Thomas 1982) and 14-19% in Camargue (France) (Pain 1990a, Mondain-Monval et al. 2002). The prevalence in the Moorhen (*Gallinula chloropus*) was 0-6.3% in the Ouse Washes (UK) (Thomas 1975, Mudge 1983) and in one of six from Camargue (France) (Pain 1990a). In Camargue, one of four Water Rail (*Rallus aquaticus*) had ingested lead shot (Pain 1990a). The Purple Gallinule (*Porphyrio porphyrio*) in Doñana had a rate of ingestion of 7.4% (Rodríguez and Hiraldo 1975), although a more recent study found no ingested shot in this species (Mateo et al. 2007a). In the shorebirds of the families *Scolopacidae* and *Charadridae*, lead shot ingestion has been observed in 18% of Black-tailed Godwits (*Limosa limosa*), 11% of Ruffs (*Philomachus pugnax*), 8% of Common Snipes (*Gallinago gallinago*) and 0% of Wood Sandpipers (*Tringa glareola*) from Camargue (France) (Pain 1990a, Pain et al. 1992). In the north-west of France the prevalence of lead shot ingestion in the Common Snipe was 1.8-15.6%, and in the Jack Snipe (*Lymnocryptes minimus*) it was 8.5% (Beck and Granval 1997). The Jack Snipe in Gironde (France) had a rate of shot ingestion of 21.6% (Veiga 1985). In the Ouse Washes (UK), 1.5% of Common Snipes had ingested lead shot in the gizzard (Thomas 1975). In the Albufera de València (Spain), lead shot ingestion was not detected in a sample of 30 Common Snipes (Mateo et al. 1998). Lead poisoning has been diagnosed in Black-tailed Godwits in Italy (Galasso 1976) and Spain (Mateo 1998), and Lapwings (*Vanellus vanellus*), and Avocets (*Recurvirostra avosetta*) in Spain (Guitart et al. 1994a, b).

Several fatalities by lead poisoning of Greater Flamingos (*Phoenicopterus ruber*) have been detected in southern Europe. In Spain, 22 Greater Flamingos were found dead in 1991 in Doñana (Ramos et al. 1992), 106 Greater Flamingos died between 1992 and 1994 in El Hondo and Salinas de Santa Pola (Mateo et al. 1997a), and 24% (n=41) of illegally shot Greater Flamingos from the Ebro Delta contained ingested lead shot in the gizzard (Mateo 1998). In France, lead poisoning was observed in three Greater Flamingos from the region of Marseille (Bayle et al. 1986), but low blood lead levels were found in nestlings from Camargue (Amiard-Triquet et al. 1991). More recently, lead poisoning has been diagnosed in 16 Greater Flamingos in the Po Delta (Arcangeli et al. 2007) and two Greater Flamingos from Tuscany (Ancora et al. 2008).

In some species or in some countries there is no data about lead shot ingestion, but there are studies on the accumulation of lead in waterbird livers. From these studies one can expect low exposures in Mallards and Eurasian Coots from the Gösku Delta in Turkey (Ayas and Kolankaya 1996), and a low exposure in Common Eiders (*Somateria mollissima*) in the Netherlands (Hontelez et al. 1992) because all the studied birds had <5 µg/g of lead in dry weight of liver. On the other hand, 14% of Common Eiders trapped in Finland had ≥0.2 µg/g of lead in wet weight of blood, the threshold level of subclinical poisoning in waterfowl (Franson et al. 2000), and four adult birds found dead had a combination of lesions (acid-fast intranuclear inclusion bodies in renal epithelial cells) and tissue lead residues (47.9-81.7 µg/g of lead in dry weight of liver) characteristic of lead poisoning (Hollmén et al. 1998). Similarly, lead shot ingestion may occur in the Greater Scaup (*Aythya marila*) and Common Pochard from the Szczecin Lagoon in Poland because 25% and 46%, respectively, had >10 µg/g of lead in bones (Kalisińska et al. 2007). Lead shot ingestion may also occur in Mallards from Szczecin Lagoon and Słońsk Reserve because 5.7% and 13.5% of shot birds, respectively, had >1.5 µg/g of lead in wet weight of liver (Kalisińska et al. 2004). In Italy, Tirelli et al. (1996) found >40 µg/dL of lead in blood of 54.8% of Mallards from Orbetello lagoon. Rodrigues et al. (2005) found >20 µg/dL of blood lead in 38.6% of 427 Mallards and 20.2% of 92 Common Teals from Vouga Lowlands (N Portugal) and 38.1% of 21 Mallards from Lagoa dos Patos (S Portugal).

Impact on Threatened Species and Population Trends.—Lead poisoning has been identified as an important cause of death for the White-headed Duck (*Oxyura leucocephala*), a globally endangered (EN) species according to IUCN. This species has a global population of 5,700 individuals and 75-94% of them are in Europe. The prevalence of lead shot ingestion is difficult to determine because it is not a hunted species, but the Spanish program to eradicate the Ruddy Duck (*Oxyura jamaicensis*) and their hybrids with the White-headed Ducks led to the observation of a 32% of prevalence of lead shot ingestion in the *Oxyura* genus. In addition, 73.3% of the *Oxyura* found dead in the Spanish wetlands contained lead shot in the gizzard, and 80% had >20 μg/g dry weight of lead in the liver (Mateo et al. 2001b). The continued monitoring of the White-headed Ducks accompanied by the analysis of lead isotopes confirms the importance of lead poisoning by shot ingestion as a cause of mortality (Svanberg et al. 2006).

The Marbled Teal (*Marmaronetta angustirostris*), a waterfowl species under the Global IUCN category of vulnerable (VU), is also affected by lead poisoning. As with the White-headed Duck, this is not a game species, but some birds found dead due to different causes in southern Spain between 1996 and 2001 have provided data on lead shot exposure in this species. The rate of lead shot ingestion in these fatalities was 32.9%, and 20% had >20 μg/g dry weight of lead in liver (Mateo et al. 2001b, Svanberg et al. 2006).

The Ferruginous Duck (*Aythya nyroca*) is considered near-threatened (NT) and should be monitored for the effect of lead poisoning because other species of the genus *Aythya*, like the Common Pochard and the Tufted Duck, show high prevalence in some European wetlands (Tables 2a, 2b). No information exists about lead poisoning in the Red-breasted Goose (*Branta ruficollis*) (EN) and the Lesser White-fronted Goose (*Anser erythropus*) (VU). However, cases of lead poisoning have been described in other goose species like the Greater White-fronted Goose (*Anser albifrons*), which is closely related to the Lesser White-fronted Goose (Mudge 1983, Ochiai et al. 1993).

The trends of the populations of waterfowl wintering in Europe are of marked decline for Northern Pintail and Common Pochard (BirdLife 2004, Table 3), both with the highest prevalence of lead shot ingestion in southern Europe in the 1990s (Mateo et al. 1997b). Wintering Common Pochard in northwest Europe and west Mediterranean had declined 30% and 70% during the previous two decades. In 1993, the Northern Pintail had the lowest population index in the west Mediterranean since 1969,

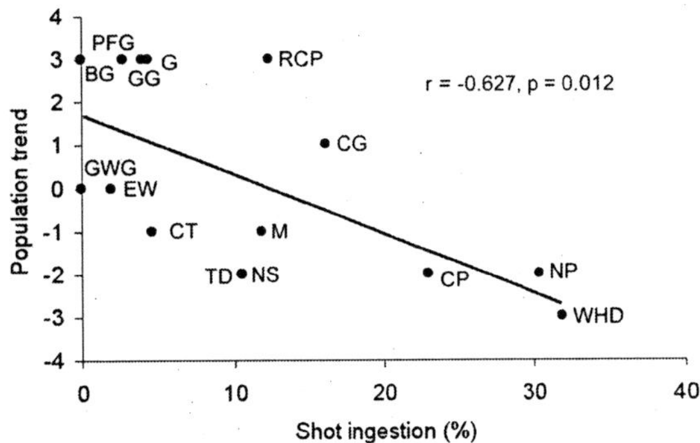

Figure 4. Correlation between the prevalence of lead shot ingestion and the trend of the wintering population in Europe of 15 species of waterfowl. Pink-footed Goose: PFG, Greater White-fronted Goose: GWG, Greylag Goose: GG, Barnacle Goose: BG, Eurasian Wigeon: EW, Gadwall: G, Common Teal: CT, Mallard: M, Northern Pintail: NP, Northern Shoveler: NS, Red-crested Pochard: RCP, Common Pochard: CP, Tufted Duck: TD, Common Goldeneye: CG, White-headed Duck: WHD.

and the east Mediterranean/Black Sea population decreased between 1967 and 1993 at a rate of -6.37% per year (Rose 1995). If we consider all the commonly hunted duck species from Europe we can observe a significant relationship between their wintering population trend and the prevalence of lead shot ingestion in them (Figure 4). Owen (1996) considers lead poisoning as a cause of the decline of the Common Pochard because females winter predominantly in far southern Europe with higher hunting pressure and lead poisoning prevalence; female survival there is only 67% of that of males.

LEAD SHOT INGESTION IN UPLAND GAMEBIRDS AND OTHERS SPECIES

Lead Shot Densities in Hunting Estates.—Lead shot densities in upland habitats has been scarcely studied. Imre (1997) found 0-1.09 shot/m^2 in four Common Pheasant hunting estates in Hungary (mean 0.46 shot/m^2). Ferrandis et al. (2008) studied shot densities on a driven shooting estate in Central Spain used for hunting of Red-legged Partridge (*Alectoris rufa*). The reported density was 7.4 shot/m^2 in 1 cm of soil depth and this is possibly low compared with other estates because the frequency of driven shootings per season in the study area only ranged from zero to two and the number of hunters spaced around 40 m apart was from six to 16 per shooting line. The densities might be higher in other more intensively driven shooting estates, where farm-reared partridges are released in large numbers just before the hunting day and driven shootings are conducted during all of the hunting season. More information is needed about lead shot densities in upland habitats in Europe, especially in intensive hunting areas.

Prevalence of Lead Shot Ingestion and Mortality.— Cases of lead poisoning in upland species have been described in Common Pheasants from the UK (Calvert 1876, Holland 1882), Denmark (Clausen and Wolstrup 1979) and Hungary (Imre 1997), Grey Partridge (*Perdix perdix*) from the UK (Keymer 1958, Anger 1971, Keymer and Stebbings 1987) and Denmark (Clausen and Wolstrup 1979), and Wood Pigeons (*Columba palumbus*) from Denmark (Clausen and Wolstrup 1979). The Game Conservancy Trust (Potts 2005) studied 1,318 Grey Partridges from 1947 to 1992 in the UK and 1.4% of birds found dead had ingested lead shot in the gizzard, with an increase in the incidence of lead poisoning from 1947-58 to 1963-92. During the second period the incidence was 4.5% in adults and 6.9% in chicks; 16 (76%) of the 21 birds that contained ingested lead shot had died as a result of lead poisoning. Clausen and Wolstrup (1979) found ingested lead shot in 1.6% of Grey Partridges from Denmark between 1971 and 1977. In the Red-legged Partridge from the UK, Butler (2005) found lead shot in the gizzard of one bird (0.16%) among 637 collected between 1955 and 1992, and in two birds (1.4%) of 144 shot in the 2001/02 season. Another study of lead shot ingestion in 437 Common Pheasants shot in 32 estates in the UK from 1996 to 2002 found an overall prevalence of 3%; the number of ingested shot were 1 (77%), 2 (15%), or 3 (8%) (Butler et al. 2005). The prevalence of lead shot ingestion in Common Pheasants (n=947) from 14 hunting estates in Hungary ranged from 0 to 23.1% (all areas=4.75%), and the number of ingested shot ranged from one to eight (Imre 1997). In Spain, Soler-Rodríguez et al. (2004) examined seven shot Red-legged Partridges and found one bird with 14 ingested lead shotgun pellets in the gizzard and 35.6 µg/g of lead in wet weight of liver. Despite the high level detected, no signs of lead poisoning were observed in this Red-legged Partridge. More recently, Ferrandis et al. (2008) examined 76 Red-legged Partridges shot in the same estate where lead shot density was studied (7.4 shot/m^2), and the prevalence of lead shot ingestion was 3.9%, with important variations between years (20% in 2004, 1.5% in 2006), with corresponding lead values in liver and bone tissues. This difference may be explained by variations in diet and grit selection between years.

A particular case of lead poisoning in two Grey-headed Woodpeckers (*Picus canus*) and one White-backed Woodpecker (*Dendrocopos leucotos*) from Sweden was reported by Mörner and Petersson (1999). The source of lead was thought to be lead pellets shot into trees and picked out by the woodpeckers during food search because holes made by shot resemble holes made by insects.

LEAD SHOT INGESTION IN BIRDS OF PREY

Cases of lead poisoning by shot or bullet ingestion have been described in 14 species of diurnal birds of prey and three species of nocturnal raptors in Europe, some of which have been near-threatened such as the White-tailed Eagle (*Haliaeetus albicilla*), or endangered, i.e., the Spanish Imperial Eagle (*Aquila adalberti*) (Figure 5). In several of these cases, lead shot ingestion can be associated with the elevated presence of embedded lead shot in the flesh of prey subjected to intensive hunting, like waterfowl (Mateo et al. 1999, 2001a) or deer (Krone et al. 2009).

Presence of Embedded Lead Shot in Prey.—The presence of embedded shot in the potential prey of raptorial birds has been mainly studied in waterfowl species for the purpose of measuring hunting pressure on a certain species and to estimate the effect on survival of embedded shot in the body. The percentage of waterfowl with embedded shot differs between species, areas with different hunting pressures, and the age of birds. These values in Europe ranged in goose species between 9.2 and 65.3%, in freshwater ducks up to 68%, and in sea ducks between 11.3% and 29.0% (Table 4). In several of these studies, waterfowl were trapped, shot with different ammunition than normally used in the area (bullets or steel), or they died by causes other than shot trauma; thus the values may represent the

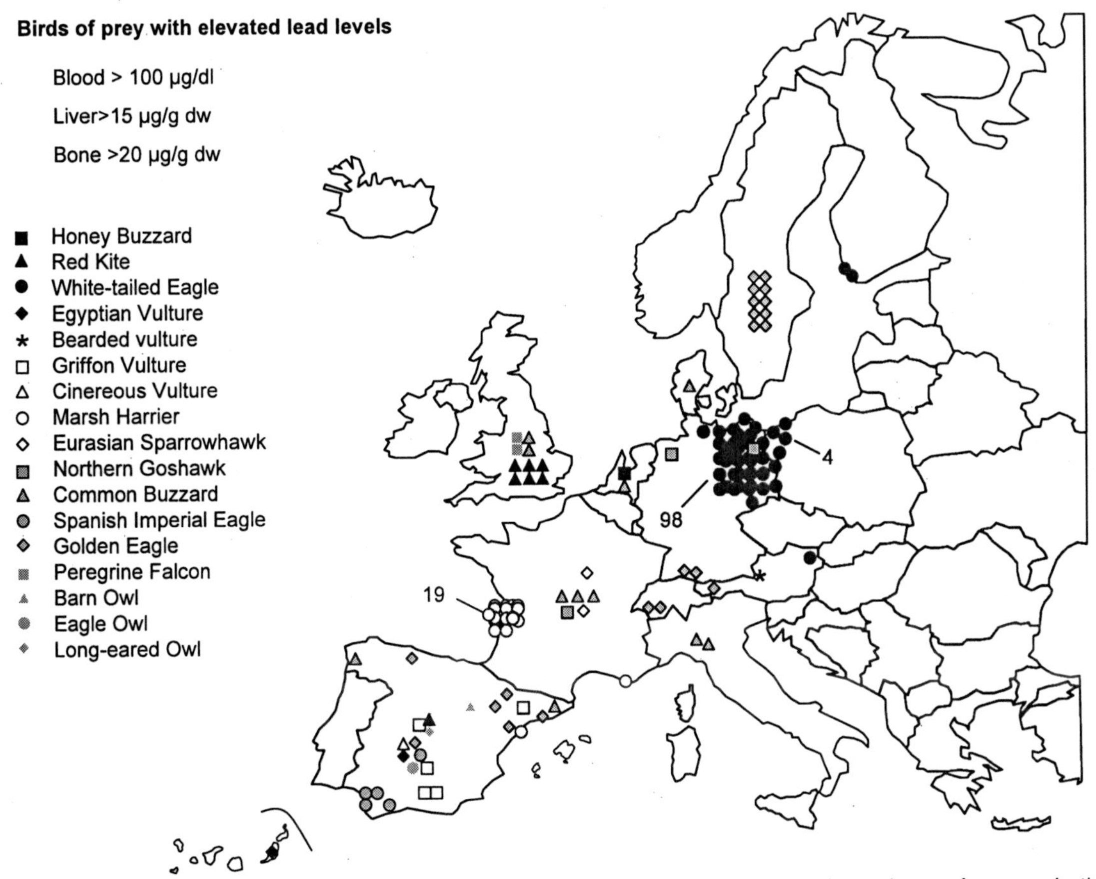

Figure 5. Birds of prey from Europe with elevated lead levels in blood, liver or bone (see references in the text). The map shows the distribution of reports by countries, and in some cases the approximate location.

Table 4. Occurrence of embedded lead shot in potential prey for European raptors.

Species	Zone/Country	M[a]	n	% with embedded shot
Pink-footed Goose	Denmark 1990-92[b]	T	413	24.6-36.0
	Denmark 1998-05[b]	T	1,904	9.2-22.2
Greylag Goose	Doñana, Spain[c]	FD	49	65.3
		T	45	44.4
Common Teal	Camargue, France[d]	T	38,909	4.4-9.6
Mallard	Camargue, France[e]	T	2740	23.4
	The Netherlands[f]	T	865	1.4-3.4 in gizzard wall (22-68 in whole body)
		S	2,859	4.9-6.4 in gizzard wall
	Doñana, Spain[c]	FD	35	14.2
Northern Shoveler	Doñana, Spain[c]	FD	17	0
Common Eider	SW Greenland[g]	FD/T	729	15.1-29.0
King Eider	SW Greenland[h]	FD	114	11.3-20.0
Eurasian Coot	Doñana, Spain[c]	FD	20	5.0
Red-legged Partridge	C Spain[i]	S	64	87.5

[a] T = trapped, FD = found dead, S = shot.
[b]Noer et al. 2007, [c]Mateo et al. 2007a, [d]Guillemain et al. 2007, [e]Tavecchia et al. 2001, [f]Lumeij and Scholten 1989, [g]Merkel et al. 2006, [h]Falk et al. 2006, [i]R. Mateo unpubl. data.

frequency of potential exposure to lead shot in birds of prey by the consumption of birds captured alive. In addition, opportunistic birds of prey would also be exposed to prey shot by hunters, dead or crippled, and not harvested by them. About 15% of geese and 19% of ducks killed by hunters are not retrieved and therefore can be consumed by scavengers with the consequent risk of lead shot ingestion (USFWS 1975). An early description of the opportunistic behaviour of birds of prey was done by Chapman and Buck (1893), when Marsh Harriers (*Circus aeruginosus*) in Doñana (Guadalquivir Marshes, Spain) were faster than were these British naturalists to recover ducks they shot.

Upland gamebirds are intensively hunted in some estates by means of driven shooting techniques, especially in places where farm-reared partridges and pheasants are released in large numbers. An undetermined number of birds, depending on the habitat, may be unretrieved by hunters, as occurs in waterfowl. The presence of lead shot embedded in the bodies of shot Red-legged Partridges in a driven shooting estate in Spain was 87.5%, with the number of embedded shot ranging from one to 22 and a mean ± SD of 4.19 ± 3.48 shot/partridge (R. Mateo unpubl. data).

Exposure to Lead and Cases of Lead Poisoning in Birds of Prey.—Probably the species most affected by lead poisoning in Eurasia has been the White-tailed Eagle, as occurred in North America with the Bald Eagle (*Haliaeetus leucocephalus*) (Pattee and Hennes 1983) or in Asia with the Steller's Sea Eagle (*Haliaeetus pelagicus*) (Kim et al. 1999). The White-tailed Eagle has been considered until recently as near-threatened (NT) in Europe, but its population has increased substantially between 1970-2000. Kenntner et al. (2001) studied 66 White-tailed Eagles found dead, sick or injured between 1993 and 2000 in Germany and Austria; 16 birds (28%) had liver lead values in the range from 5 to 62 µg/g in wet weight of tissue. In a retrospective study, Kenntner et al. (2004) diagnosed nine White-tailed Eagles as lead poisoned from 46 birds found dead in Eastern Germany from 1978 to 1998. In a more extensive compilation of cases, Kenntner et al. (2005) investigated liver and kidney tissues of 277 immature and adult White-tailed Eagles found dead or moribund in Germany between 1979 and 2005 and found levels of lead indicating lead poisoning (>5 µg/g in wet weight) in 66 (24%) of the cases originated by the ingestion of lead shot and bullet fragments. The last update for this species in Germany reports a mortality by lead poisoning of 25% (n=391) (Krone et al. 2009). Muller et al. (2001) describes two clinical cases of lead poisoning in two White-tailed Eagles from Germany. Lead poisoning has also been detected in four White-tailed Eagles from northwestern Poland (n=

Table 5. Occurrence of lead shot in pellets of raptors and carrion eaters from Europe.

Species	Zone/Country	n	%
Red Kite *Milvus milvus*	England, UK[a]	264	1.5-2.3
	Central Spain[b]	1,233	1.05
	Doñana, Spain[c,d]	962	2.2
Egyptian Vulture *Neophron percnopterus*	Ebro V., Spain[e]	327	0.0
	Canary I., Spain[f]	424	3.1
Marsh Harrier *Circus aeruginosus*	Charente-M., France[g]	459	14.8
	Ebro D., Spain[h]	521	10.7
	Doñana, Spain[i]	513	2.1
Spanish Imperial Eagle *Aquila adalberti*	Spain[j]	2,400	0.54
	Doñana, Spain[c,d]	615	4.2
Golden Eagle *Aquila chrysaetos*	Norway[k]	-	70.00
Booted Eagle *Hieraetus pennatus*	Doñana, Spain[d]	76	0.0
Peregrine Falcon *Falco peregrinus*	Doñana, Spain[d]	117	0.9
Barn Owl *Tyto alba*	Doñana, Spain[d]	50	0.0
Raven *Corvus corax*	Doñana, Spain[d]	321	0.0

[a]Pain et al. 2007; [b]García and Viñuela 1999; [c]Mateo et al. 2001a, [d]2007a; [e]Gangoso et al. 2008; [f]Donázar et al. 2002; [g]Pain et al. 1993, 1997; [h]Mateo et al. 1999; [i]González 1991; [j]González and Hiraldo 1988; [k]Pain and Amiard-Triquet 1993.

25, 16%; Falandysz et al. 1988, 2001, Kalisińska et al. 2006), two from Greenland (n=12, 16.7%; Krone et al. 2004) and two from Finland (n=11, 18.2%; Krone et al. 2006).

The second affected species, in number of cases, is the Golden Eagle (*Aquila chrysaetos*). These cases have been widely distributed across Europe, with 10 cases in Sweden (Borg 1975), six in Spain (Cerradelo et al. 1992; R. Mateo unpubl. data), two in Switzerland (Kenntner et al. 2007), two in Germany (Bezzel and Fünfstück 1995) and one in Austria (Zechner et al. 2005). The rate of lead shot ingestion was extremely high in Norway, where 70% of the regurgitated pellets contained lead shot (Table 5).

The Spanish Imperial Eagle is now a vulnerable (VU) species with a global population of 175-180 breeding pairs, and it was considered until recently as endangered (EN) (BirdLife 2004). A significant component of the population breeds in Doñana, where they feed on crippled waterfowl, resulting in the presence of lead shot in 14.7% of the regurgitated pellets in the fall-winter seasons between 1991-96 (Mateo et al. 2001a). A lower occurrence (2.8%) was observed in pellets collected in all years during 1997-2002 (Mateo et al. 2007a). Although a lower rate of lead shot ingestion may be present in upland areas (Table 5), one case of lead poisoning has been described in central Spain (Hernández 1995). Four birds, two of them from Doñana, had bone lead levels >50 µg/g of dry weight (Pain et al. 2005). The ingestion of lead shot in the Spanish Imperial Eagle in Doñana may vary among years depending on the hunting pressure on Greylag Geese that varies according to the water level in the protected areas (Mateo et al. 2007a).

Marsh Harriers in the Mediterranean wetlands frequently ingest lead shot, being opportunistic species often seen in the search of crippled waterfowl at the end of shooting days. In Charente-Maritime (France), Pain et al. (1993, 1997) found lead shot in 11.5-25% of regurgitated pellets in winter, but only in 1.4% of pellets during May and June. In Spain, the occurrence of lead shot in pellets was 10.7% (Mateo et al. 1999) in the Ebro Delta and 1.8-4.3% in Doñana (González 1991, Mateo et al. 2007a). These rates of shot ingestion were reflected in high blood lead levels (>30 µg/dL) in Charente-Maritime (33%; Pain et al. 1993, 1997), in Camargue (26%; Pain et al. 1993) and the Ebro Delta (40%; Mateo et al. 1999). One bird from Charente-Maritime was found dead with 54.9 µg/g of lead in dry weight of liver (Pain et al. 1993).

The Red Kite (*Milvus milvus*) has suffered an important decline in Europe (NT), especially due to the use of poison to kill predators; lead poisoning has also been considered as a significant cause of death in some areas. Pain et al. (2007) studied Red Kites found dead or moribund in England during the reintroduction program initiated in 1989; six of 44 birds had >15 µg/g of lead in dry weight of liver, and four of these (9%) probably died by lead poisoning. The rate of lead shot ingestion in the reintroduced Red Kites was reflected by the presence of lead shot in 1.5-2.3% of the regurgitated pellets. The presence of lead shot has been detected in 5.5% of the regurgitated pellets in Doñana at the end of the hunting season (Mateo et al. 2001a). A lower percentage was observed in pellets collected during the whole year (1.8%; Mateo et al. 2007a) in these marshes and also in other upland areas in Spain (1.05%; García and Viñuela, 1999). One Red Kite found dead in Central Spain had >20 µg/g of lead in dry weight of bone, and another three had values between 10 and 20 µg/g dry weight (Mateo et al. 2003).

Lead poisoning has been described in the four species of vultures living in Europe: Griffon Vulture (*Gyps fulvus*) (Mateo et al. 1997c), Cinereous Vulture (*Aegypius monachus*) (Hernández and Margalida 2008), Bearded Vulture (*Gypaetus barbatus*) (Hans Frey, Richard Faust Bearded Vulture Breeding Centre, Austria, pers. comm.), and Egyptian Vulture (*Neophron percnopterus*) (Rodríguez-Ramos et al. 2009). In southeastern Spain, 21 (91%) of 23 Griffon Vultures trapped alive had blood lead >20 µg/dL and two of them had >150 µg/dL (García-Fernandez et al. 2005). High blood lead levels (>20 µg/dL) had been previously observed in six other Griffon Vultures (García-Fernández et al. 1995). Egyptian Vultures from the Canary Islands (Spain) are sedentary, unlike the continental birds which migrate to Africa in winter. This difference makes the Canary Islands population more vulnerable because they feed on rabbit carcasses shot in winter. The presence of lead shot was detected in 5.3% of pellets collected in January and 1.3% of those collected in November. Moreover, five (21.7%) of 23 birds had >20 µg/dL of lead in blood, and one had >50 µg/dL (Donázar et al. 2002). Gangoso et al. (2009) found elevated blood lead levels (>20 µg/dL) in 10 (7.3%) Egyptian Vultures from the Canary Islands (n=137 nestlings and adults), and one of these birds showed a concentration of 178 µg/dL. Moreover, they analyzed bones of 28 Egyptian Vultures found dead in the Canary Islands; one had >20 µg/g of lead in dry weight.

Several studies have been conducted to evaluate lead exposure in European raptors by tissue analysis of birds admitted to wildlife rehabilitation centers. Hontelez et al. (1992) found one Common Buzzard (*Buteo buteo*) (n=28) in the Netherlands with 22.5 µg/g in dry weight of bone. Jager et al. (1996) found 11 Common Buzzards (n=80) with outlying total burden of lead in liver of >20 µg and kidney (>6 µg) in the Netherlands, and elevated liver lead levels were also found in Common Buzzards from Denmark (Clausen and Wolstrup 1979). One case of lead poisoning in a Honey Buzzard (*Pernis apivorus*) was also described in the Netherlands (Lumeij et al. 1985). Pain and Amiard-Triquet (1993) found >15 µg/g of lead in dry weight of liver in three Common Buzzards (n=85), one Northern Goshawk (*Accipiter gentilis*) (n=1), and two Eurasian Sparrowhawks (*Accipiter nisus*) (n=30) from France. In the UK, Pain et al. (1995) found two Common Buzzards (n=56) and two Peregrine Falcons (*Falco peregrinus*) (n=26) with >15 µg/g of lead in dry weight of liver, and levels >6 µg/g in six other species, including Merlin (*Falco columbarius*) (6 of 63) and Golden Eagle (2 of 7). In Germany, two Northern Goshawks among 62 birds found dead or moribund had liver lead levels of 51 and 6.5 µg/g wet weight, respectively (Kenntner et al. 2003). In Italy, Battaglia et al. (2005) found two Common Buzzards (n=18) with >20 µg/g of lead in dry weight of liver. In Spain, Pérez-López et al. (2008) found one Common Buzzard (n=44) with a liver lead level of 18.1 µg/g dry weight. Lead shot ingestion has been studied by analysis of regurgitated pellets in several additional species of birds of prey from Doñana (Spain) (Mateo et al. 2007a; Table 5).

Nocturnal raptors are also exposed to lead shot, but few studies have been done. One Eurasian Eagle Owl (*Bubo bubo*) (n=42) from Central Spain showed 185 µg/g of lead in dry weight of bone (Mateo et al. 2003) and another from southeastern Spain (n=9) had 43 µg/g of lead in dry weight of bone (García-Fernández et al. 1997). One Long-eared Owl (*Asio otus*) was found dead of lead poisoning in Madrid (Spain) close to a pigeon shooting range (Brinzal 1996), and one Barn Owl (*Tyto alba*) was found in northern Spain with 73 µg/g of lead in wet weight of liver (González et al. 1983).

REGULATORY ACTIONS ON LEAD SHOT

The use of lead shot for hunting has been regulated in 14 countries of Europe (Table 6). All except one of these countries are within the European Union (EU), but no common EU regulation exists on lead ammunition. Despite the role of the EU in the regulation of chemical substances, including lead (Thomas and Guitart 2005), the European Commission (EC), responsible for designing, implementing and managing policy and legislation within the EU, has not taken any action on lead ammunition. The Ornis Committee, that makes decisions to implement the Birds Directive (79/409/EEC) initiated by the EC, has only recommended to member states to take their own measures (Beintema 2004).

The international conventions promoting the regulation of lead shot use for hunting have been reviewed by Thomas and Owen (1996), Beintema (2004) and Thomas and Guitart (2005). The Convention on the Conservation of European Wildlife and Natural Habitats (the Bern Convention) and the Agreement on the Conservation of African-Eurasian Migratory Waterbirds (AEWA) of the Convention on the Conservation of the Migratory Species of Wild Animals (CMS, the Bonn Convention) have been the most explicit about enforcing the ban of lead shot. Paragraph 4.1.4 of the Action Plan of AEWA stated that "Parties shall endeavour to phase out the use of lead shot for hunting in wetlands by the year 2000." All the countries with some regulation about lead ammunition are parties of the Bern Convention and all but one are parties of the AEWA. Moreover, the Declaration on Risk Reduction for Lead was adopted in 1996 by the Environment Ministers from the Organisation for

Table 6. Summary of international agreements and regulations on lead ammunition in Europe by 2007. See text for explanation of agreements.

Country	International Agreements							Shot pellets		
	EU[a]	OECD[b]	CMS[c]	AEWA[d]	Bern[e]	CBD[f]	Ramsar sites[g] Surface (ha)	Waterfowl & wetlands	Upland game	Bullets
Albania	-	-	+	+	+	+	83,062	-	-	-
Andorra	-	-	-	-	+	+	-	-	-	-
Armenia	-	-	-	-	+	+	492,239	-	-	-
Austria	+	+	-	-	+	+	122,372	-	-	-
Azerbaijan	-	-	-	-	+	+	99,560	-	-	-
Belarus	-	-	+	-	-	+	283,107	-	-	-
Belgium	+	+	+	+	+	+	42,938	(+)[h]	(+)	-
Bosnia & Herzegovina	-	-	-	-	-	+	10,911	-	-	-
Bulgaria	+	-	+	+	+	+	20,306	-	-	-
Croatia	-	-	+	+	+	+	86,579	-	-	-
Cyprus	+	-	+	-	+	+	1,585	(+)	-	-
Czech Republic	+	+	+	+	+	+	54,656	-	-	-
Denmark	+	+	+	+	+	+	2,078,823	+	+	-
Estonia	+	-	-	-	+	+	225,960	-	-	-
Finland	+	+	+	+	+	+	799,518	+	-	-
France	+	+	+	+	+	+	828,803	+	-	-
Georgia	-	-	+	+	-	+	34,480	-	-	-
Germany	+	+	+	+	+	+	843,109	(+)	(+)	(+)
Greece	+	+	+	+	+	+	163,501	-	-	-
Hungary	+	+	+	+	+	+	235,430	(+)	-	-
Iceland	-	+	-	-	+	+	58,970	-	-	-
Ireland	+	+	+	+	+	+	66,994	-	-	-
Italy	+	+	+	+	+	+	59,796	-	-	-
Latvia	+	-	+	+	+	+	148,363	(+)	-	-
Liechtenstein	-	-	+	-	+	+	101	-	-	-
Lithuania	+	-	+	+	+	+	50,451	-	-	-
Luxembourg	+	+	+	+	+	+	17,213	-	-	-
Macedonia	-	-	+	+	+	+	21,616	-	-	-
Malta	+	-	-	-	+	+	16	-	-	-
Moldova	-	-	+	+	+	+	94,705	-	-	-
Monaco	-	-	+	+	+	+	10	-	-	-
Montenegro	-	-	-	-	-	+	20,000	-	-	-
The Netherlands	+	+	+	+	+	+	818,908	+	+	-
Norway	-	+	+	-	+	+	116,369	+	+	-
Poland	+	+	+	-	+	+	145,075	-	-	-
Portugal	+	+	+	+	+	+	73,784	-	-	-
Romania	+	-	+	+	+	+	682,166	-	-	-
Russia	-	-	-	-	-	+	10,323,767	-	-	-
San Marino	-	-	-	-	-	+	-	-	-	-
Serbia	-	-	+	-	+	+	53,714	-	-	-
Slovakia	+	+	+	+	+	+	40,697	-	-	-
Slovenia	+	-	+	+	+	+	8,205	-	-	-
Spain	+	+	+	+	+	+	281,768	(+)	-	-

Country	International Agreements							Shot pellets		
	EU[a]	OECD[b]	CMS[c]	AEWA[d]	Bern[e]	CBD[f]	Ramsar sites[g] Surface (ha)	Waterfowl & wetlands	Upland game	Bullets
Sweden	+	+	+	+	+	+	514,506	+	+	+
Switzerland	-	+	+	+	+	+	8,676	+	-	-
Turkey	-	+	-	-	+	+	179,482	-	-	-
Ukraine	-	-	+	+	+	+	744,651	-	-	-
United Kingdom	+	+	+	+	+	+	917,988	(+)	-	-

[a] http://europa.eu/, [b] http://www.oecd.org/, [c] http://www.cms.int/, [d] http://www.unep-aewa.org/, [e] http://europa.eu/scadplus/leg/en/lvb/l28050.htm, [f] http://www.cbd.int/, [g] http://www.ramsar.org/, [h] (+) Only implemented in some regions or in some protected areas.

Economic Co-operation and Development (OECD). The purpose of this Declaration was to advance national and co-operative efforts to reduce the risk from exposure to lead, and one of the priority actions was to restrict the use of lead shot in wetlands (Beintema 2004). Two other international agreements, the Convention on Biological Diversity (CBD) and the Convention on Wetlands (the Ramsar Convention), identify the conservation of critical habitats for migratory birds, but do not make specific recommendations to ban lead shot (Thomas and Guitart 2005).

The regulatory actions adopted by European countries have been diverse in the extent and form of application (AEWA 2008). Denmark started with the ban on the use of lead shot for target shooting over wetlands in 1981, for hunting in Ramsar areas in 1985, and for shooting over small lakes and for target shooting over agricultural land in 1986. Moreover, from 1986, a maximum of 28 g of lead in each cartridge was allowed (Clausen 1992). Denmark enforced a total ban on the use and possession of lead shot in 1996 (Kuivenhoven et al. 1998, Kanstrup 2006). In Greenland, lead shot is being used, but the ban will be implemented in the next few years (AEWA 2008). The Netherlands totally banned the use of lead shot in all its territory in 1993, and possession has been illegal since 1998 (Kuivenhoven et al. 1998, Beintema 2000). Norway banned lead shot for waterfowl hunting in 1991 and has extended this to all types of hunting (Beintema 2001, Kanstrup and Potts 2008). Finland banned lead shot for waterfowl hunting in 1996 (Scheuhammer and Norris 1996). Lead shot was banned first in Sweden for waterfowl hunting within Ramsar areas (Scheuhammer and Norris 1996), for goose and duck hunting in 1998; in 2002 the Swedish government introduced a ban on lead ammunition to be fully implemented in 2008 (2002 for wetlands, 2006 for lead shot everywhere, and 2008 for bullets) (AEWA 2005). In the Flemish region of Belgium, waterfowl hunting with lead shot has been forbidden in Ramsar areas since 1993, and in 1998 this ban was extended to all EU Bird Directive areas (Beintema 2001). A total ban on the use of lead shot was adopted in 2003 in protected areas, and this will be extended to the remainder by 2008 (AEWA 2005). In the Walloon region, the restriction applies to hunting in wetlands, although coated lead pellets ('cartouches à plomb nickelés') are still allowed (AEWA 2008). In Switzerland, a ban on the use of lead shot in shallow water areas and wetlands was introduced in 1998 (Beintema 2001). Latvia banned lead shot for waterfowl hunting at the Natural Park Lake Engure in 1998 and this was later extended to other nature reserves (Beintema 2001). In the UK, different regulations have been adopted by the different countries. In 1999, England banned shooting with lead shot on or over any area below high-water mark of ordinary spring tides, specific sites of special scientific interest, or certain waterbird species. The same regulation was adopted by Wales in 2002. Scotland banned lead shot for shooting on or over wetland areas in 2005. In Northern Ireland, there is a voluntary ban, but a statutory ban is proposed for 2008/2009 (AEWA 2008). Spain banned the use and possession of lead shot in Ramsar areas and other protected wetlands in 2001, and this was extended in 2007 to all the Natura 2000 wetlands. Before that, the use of lead shot for hunting in wetlands was initially banned in the regions of the Balearic Islands in 1995 and Castilla-La Mancha in 1999. France banned lead shot for hunting in wetlands in 2006 (J.-Y. Mondain-Monval pers. comm.). That implementation was stimulated by the experiences of North American hunters on the use of non-toxic shot (Mondain-Monval 1999). Several regions of Germany have banned lead shot for hunting waterfowl near waterbodies: Baden-Württemberg, Bavaria, Berlin, Brandenburg, Mecklenburg-West Pomerarnia, Lower Saxony, North Rhine-Westfalia and Thuringia (AEWA 2005); while Schleswig-Holstein extended the ban to waterfowl shot in terrestrial habitats (N. Kenntner pers. comm.). Moreover, the federal state Government of Brandenburg prohibited the use of any lead ammunition, including lead bullets for game hunting in the federal forests in the year 2005 (AEWA 2005, Kenntner et al. 2007). At a national scale, the German Federal Government and hunter's associations made in 1993 a recommendation to use non-toxic shot for waterfowl hunting in wetlands (Beintema 2001). Hungary has banned lead shot for hunting in Ramsar areas and other wetlands since 2005 (AEWA 2005, Kanstrup and Potts 2008). Lead shot is banned in wetlands in Cyprus (AEWA 2008). In the Russian Federation, there are some restrictions on waterfowl hunting in some areas that will reduce lead poisoning in waterbirds (Beintema 2001). Portugal prepared a ban of lead shot for all hunting in Ramsar areas for 2007/08 and plans to extend this to all waterbird hunting in 2008/09 (D. Rodrigues pers. comm.). The phasing out of lead shot use is in progress in the Czech Republic (by 2010), Slovakia (by 2015), and Croatia and Italy (AEWA 2005, 2006, 2008, Kanstrup and Potts 2008).

These advances in the regulation of the use of lead ammunition in Europe have been slow, and the result is a patchy distribution of countries where lead shot input into the wetlands has been stopped, other countries where lead shot can be used in some of the wetlands, and others with no regulation as yet (Table 6). The EU should take a stronger position on this issue to protect the African-Eurasian migratory flyway because this is a problem originated by the developed countries where lead hunting ammunition has been intensively used for decades. As a first step, the Federation of Associations for Hunting and Conservation of the EU (FACE) and BirdLife International (FACE-BirdLife 2004) have-signed an agreement as part of the EC's Sustainable Hunting Initiative under the auspices of the Birds Directive. Both organizations ask for the phasing out of the use of lead shot for hunting in wetlands throughout the EU as soon as possible, and no later than 2009. Similarly, the International Council for Game and Wildlife Conservation (CIC 2007) recommend to the authorities in countries where lead shot is still used for hunting in wetlands to begin a process of phasing out such use as soon as possible, and at the latest by 2010.

Little action has been taken to phase out lead shot for upland game hunting. Driven shooting of partridges and pheasants produces significant accumulation of lead shot in the soil of intensively hunted estates, indicating that the same regulations as in wetlands should be applied. Based on the Precautionary Principle of the Rio Declaration on Environment and Development, the input of lead into these upland hunting estates should cease before the damage has been produced (Thomas 1997), and this damage has already been revealed in several studies. In the adoption of such regulations, the exposure to humans through the consumption of lead-exposed birds (Guitart et al. 2005) or game meat with embedded lead pellets or bullet fragments (Johansen et al. 2001, 2004; Mateo et al. 2007b) should also be considered.

ACKNOWLEDGMENTS

I wish to thank N. Kenntner, J.-Y. Mondain-Monval and D. Rodrigues for information about the status of the regulatory actions adopted in their countries, and to the UNEP/AEWA Secretariat for the submission of the draft of the last update report on the use of non-toxic shot. I also thank U. Hofle for help with the English revision.

LITERATURE CITED

AEWA. 2005. Third Session of the Meeting of the Parties to the Agreement on the Conservation of African-Eurasian Migratory Waterbirds (AEWA), 23-27 October 2005, Dakar, Senegal. [Online.] Available at www.unep-aewa.org

AEWA. 2006. Fourth Meeting of the Standing Committee, 20-21 November 2006, Bonn, Germany. [Online.] Available at www.unep-aewa.org

AEWA. 2008. 8th Meeting of the Technical Committee, 3-5 March 2006, Bonn, Germany. [Online.] Available at www.unep-aewa.org

ALLOUCHE, L. 1983. Alimentation comparée du Canard Chipeau *Anas strepera* et de la Foulque Macroule *Fulica atra* pendant leur hivernage en Camargue. Rapport Diplôme d'Études Approfondies, Montpelier, France.

AMIARD-TRIQUET, C., D. J. PAIN, AND H. T. DELVES. 1991. Exposure to trace elements of Flamingos living in a Biosphere Reserve, the Camargue (France). Environmental Pollution 69:193-201.

ANCORA, S., N. BIANCHI, C. LEONZIO, AND A. RENZONI. 2008. Heavy metals in Flamingos (*Phoenicopterus ruber*) from Italian wetlands: The problem of ingestion of lead shot. Environmental Research 107:229-236.

ANGER, H. 1971. Gamebird diseases. The Game Conservancy Trust Annual Review 2:51-53.

ARCANGELI, G., A. MANFRIN, G. BINATO, R. DE NARDI, S. VOLPONI, M. VASCELLARI, F. MUTINELLI, AND C. TERREGINO. 2007. Avvelenamento da piombo in uccelli selvatici. Obiettivi & Documenti Veterinari 9:39-45.

AYAŞ, Z., AND D. KOLANKAYA. 1996 Accumulation of some heavy metals in various environments and organisms at Göksu Delta, Türkiye, 1991-1993. Bulletin of Environmental Contamination and Toxicology 56:65-72.

BATTAGLIA, A., S. GHIDINI, G. CAMPANINI, AND R. SPAGGIARI. 2005. Heavy metal contamination in Little Owl (*Athene noctua*) and Common Buzzard (*Buteo buteo*) from northern Italy. Ecotoxicology and Environmental Safety 60:61-66.

BAYLE, P., F. DERMAIN, AND G. KECK. 1986. Trois cas de saturnisme chez le Flamant Rose *Phoenicopterus ruber* dans la region de Marseille. Bulletin de la Societé Linneane de Provence 38:95-98.

BECK, N., AND P. GRANVAL. 1997. Ingestion de plombs de chasse par la Bécassine del Marais (*Gallinago gallinago*) et la Bécassine Sourde (*Lymnocryptes minimus*) dans le nord-ouest de la France. Gibier Faune Sauvage 14:65-70.

BEER, J. V., AND P. STANLEY. 1965. Lead poisoning in the Slimbridge wildfowl collection. Wildfowl 16:30-34.

BEINTEMA, N. 2001. Lead Poisoning in Waterbirds. International Update Report 2000. Wetlands International, Wageningen, The Netherlands.

BEINTEMA, N. 2004. Non-toxic shot: a path towards sustainable use of the waterbird resource. Technical Series No. 3. UNEP/AEWA Secretariat, Bonn, Germany.

BELLROSE, F. C. 1959. Lead poisoning as a mortality factor in waterfowl populations. Illinois Natural History Survey Bulletin 27:235-288.

BEYER, W. N., D. J. AUDET, G. H. HEINZ, D. J. HOFFMAN, AND D. DAY. 2000. Relation of waterfowl poisoning to sediment lead concentrations in the Coeur d'Alene River Basin. Ecotoxicology 9:207-218.

BEZZEL, E., AND H.-J. FÜNFSTÜCK. 1995. Alpine Steinadler *Aquila chrysaetos* durch bleivergiftung gefährdet? Journal für Ornithologie 136:294-296.

BIRDLIFE INTERNATIONAL. 2004. Birds in Europe: Population Estimates, Trends and Conservation Status. BirdLife International, Wageningen, The Netherlands.

BIRKHEAD, M. 1982. Causes of mortality in the Mute Swan *Cygnus olor* on the River Thames. Journal of Zoology 198:15-25.

BIRKHEAD, M. 1983. Lead levels in the blood of Mute Swans *Cygnus olor* on the River Thames. Journal of Zoology 199:59-73.

BONET, A., C. OLIVARES, M. L. PICÓ, AND S. SALES. 1995. L'acumulació de perdigons de plom al Parc Natural del Fondó d'Elx (Alacant): distribució espacial i propostes d'actuació. Butlletí de la Instució Catalana d'Història Natural 63:149-166.

BONET, A., B. TERRONES, AND J. PEÑA. 2004. El Hondo, a la cabeza en cantidad de perdigones de plomo depositados. Quercus 221:66-67.

BORG, K. 1975. Viltjukdomar. L.T.'s Forlag Helingoorg.

Brinzal. 1996. S.O.S. Venenos: Búho Chico. Quercus 124:45.

Butler, D. A. 2005. Incidence of lead shot ingestion in Red-legged Partridges (*Alectoris rufa*) in Great Britain. Veterinary Record 157:661-662.

Butler, D. A., R. B. Sage, R. A. H. Draycott, J. P. Carroll, and D. Potts. 2005. Lead exposure in Ring-necked Pheasants on shooting estates in Great Britain. Wildlife Society Bulletin 33:583-589.

Calvert, H. S. 1876. Pheasants poisoned by swallowing shot. The Field 47:189.

Campredon, P. 1984. Régime alimentaire du canard siffleur pendant son hivernage en Camargue. L'Oiseau et R.F.O. 54:189-200.

Cerradelo, S., E. Muñoz, J. To-Figueras, R. Mateo, and R. Guitart. 1992. Intoxicación por ingestión de perdigones de plomo en dos Águilas Reales. Doñana, Acta Vertebrata 19:122-127.

Chapman, A., and W. J. Buck. 1893. Wild Spain. Gurney and Jackson, London, UK.

CIC. 2007. Migratory Bird Commission Recommendation. Phasing out of Lead Shot for Hunting in Wetlands. 54th General assembly, 2-5 May 2007, Belgrade, Serbia. [Online.] Available at www.cic-wildlife.org

Clausen, B. 1992. Lead poisoning control measures in Denmark. Pages 68-70 *in* D. J. Pain (Ed.). Lead Poisoning in Waterfowl. Proceedings of an IWRB Workshop, Brussels, Belgium, 13–15 June 1991. International Waterfowl and Wetlands Research Bureau Special Publication 16, Slimbridge, UK.

Clausen, B., and C. Wolstrup. 1979. Lead poisoning in game from Denmark. Danish Review of Game Biology 11:1-22.

Cordel-Boudard, C. 1983. Le saturnisme chez les Anatidés en Dombes. Thèse Docteur Vétérinaire. Ecole Nationale Vétérinaire de Lyon, France.

Danell, K. 1980. The detrimental effects of using lead shot for shooting waterfowl. Rapport Institutionen för Viltekologi 4: 1-51.

Danell, K., and A. Andersson. 1975. Blyhagelforekomst i andmager. Statens Naturvarksverk 538:1-23.

Danell, K., A. Andersson, and V. Marcstrom. 1977. Lead shot dispersed by hunters – ingested by ducks. Ambio 6:235-237.

Del Bono, G. 1970. Il saturnismo degli ucelli acquatici. Annali della Facolta di Medicina Veterinaria de Pisa 23:102-151.

Di Modugno, G., A. Camarda, and N. Zizzo. 1994. Avvelenamento da piombo in Cigni Reali (*Cygnus olor*) di passo nella peninsola Salentina. Zootecnica International, 15-16: 90-94.

Donázar, J. A., C. J. Palacios, L. Gangoso, O. Ceballos, M. J. González, and F. Hiraldo. 2002. Conservation status and limiting factors in the endangered population of Egyptian Vulture (*Neophron percnopterus*) in the Canary Islands. Biological Conservation 107:89-97.

Duranel, A. 1999. Effets de l'ingestion de plombs de chasse sur le comportament alimentaire et la condition corporelle du Canard Colvert (*Anas platyrhynchos*). Thèse Docteur Vétérinaire. Ecole Nationale Vétérinaire de Nantes, France.

Erne, K., and K. Borg. 1969. Lead poisoning in Swedish wildlife. Ecological Research Committee Bulletin 5:31-33.

FACE-BirdLife (2004). Agreement between BirdLife International and the Federation of Associations for Hunting and Conservation of the EU on the Directive 79/409/EEC. [Online.] Available at www.face-europe.org

Falandysz, J., H. Ichihashi, K. Szymczyk, S. Yamasaki, and T. Mizera. 2001. Metallic elements and metal poisoning among White-tailed Sea Eagles from the Baltic south coast. Marine Pollution Bulletin 42:1190-1193.

Falandysz, J., B. Jakuczun, and T. Mizera. 1988. Metals and organochlorines in four female White-tailed Eagles. Marine Pollution Bulletin 19:521-526.

Falk, K., F. Merkel, K. Kampp, and S. E. Jamieson. 2006. Embedded lead shot and infliction rates in Common Eiders *Somateria mollissima* and King Eiders *S. spectabilis* wintering in southwest Greenland. Wildlife Biology 12:257-265.

Ferrandis, P., R. Mateo, F. R. López-Serrano, M. Martínez-Haro, and E. Martínez-Duro. 2008. Lead-shot exposure in Red-legged Partridge (*Alectoris rufa*) on a driven shooting estate. Environmental Science and Technology 42:6271-6277.

Figuerola, J., R. Mateo, A. J. Green, J.-Y. Mondain-Monval, H. Lefranc, and G. Mentaberre. 2005. Interspecific and spatial variability in the ingestion of grit and lead shot by waterfowl. Environmental Conservation 32:226-234.

FISHER, I. J., D. J. PAIN, AND V. G. THOMAS. 2006. A review of lead poisoning from ammunition sources in terrestrial birds. Biological Conservation 131:421- 432.

FRANK, A., AND K. BORG. 1979. Heavy metals in tissues of Mute Swans (*Cygnus olor*). Acta Veterinaria Scandinavica 20:447-465.

FRANSON, J. C., T. HOLLMÉN, R. H. POPPENGA, M. HARIO, M. KILPI, AND M. R. SMITH. 2000. Selected trace elements and organochlorines: Some findings in blood and eggs of nesting Common Eiders (*Somateria mollissima*) from Finland. Environmental Toxicology and Chemistry 19:1340-1347.

GALASSO, C. 1976. Caso di avvelenamento da piombo in una Pittima Reale (*Limosa limosa*). Rivista Italiana di Ornitologia 46:117.

GANGOSO, L., P. ÁLVAREZ-LLORET, A. B. RODRÍGUEZ-NAVARRO, R. MATEO, F. HIRALDO, AND J. A. DONÁZAR. 2009. Long-term effects of lead poisoning on bone mineralization in Egyptian Vultures (*Neophron percnopterus*) exposed to ammunition sources. Environmental Pollution 157:569–574.

GARCÍA-FERNÁNDEZ, A. J., E. MARTÍNEZ-LÓPEZ, D. ROMERO, P. MARÍA-MOJICA, A. GODINO, AND P. JIMÉNEZ. 2005. High levels of blood lead in Griffon Vultures (*Gyps fulvus*) from Cazorla Natural Park (southern Spain). Environmental Toxicology 20:459-463.

GARCÍA-FERNÁNDEZ, A. J., M. MOTAS-GUZMÁN, I. NAVAS, P. MARÍA-MOJICA, A. LUNA, AND J. A. SÁNCHEZ-GARCÍA. 1997. Environmental exposure and distribution of lead in four species of raptors in southeastern Spain. Archives of Environmental Contamination and Toxicology 33:76-82.

GARCÍA-FERNÁNDEZ, A. J., J. A. SÁNCHEZ-GARCÍA, P. JIMÉNEZ-MONTALBÁN, AND A. LUNAS. 1995. Lead and cadmium in wild birds in southeastern Spain. Environmental Toxicology and Chemistry 14:2049-2058.

GARCÍA, J. T., AND J. VIÑUELA. 1999. El plumbismo: una primera aproximación en el caso del Milano Real. Pages 213-220 *in* J. Viñuela, R. Martí, and A. Ruiz (Eds.). El Milano Real en España. Sociedad Española de Ornitología/ BirdLife, Madrid, Spain.

GIONFRIDO, J. P., AND L. B. BEST. 1999. Grit use by birds: a review. Current Ornithology 15:89-148.

GONZÁLEZ, J. L. 1991. El Aguilucho Lagunero (*Circus aeruginosus*) en España. ICONA-CSIC, Madrid, Spain.

GONZÁLEZ, L. M., AND F. HIRALDO. 1988. Organochlorine and heavy metals contamination in the eggs of the Spanish Imperial Eagle (*Aquila adalberti*) and accompanying changes in eggshell morphology and chemistry. Environmental Pollution 51:241-258.

GONZÁLEZ, M. J., M. C. RICO, M. C. FERNÁNDEZ-ACEYTUNO, L. M. HERNÁNDEZ, AND G. BALUJA. 1983. Contaminación xenobiótica del Parque Nacional de Doñana. II. Residuos de insecticidas organoclorados, bifenilos policlorados (PCBs) y metales pesados en Falconiformes y Strigiformes. Doñana, Acta Vertebrata 10:177-189.

GUILLEMAIN, M., O. DEVINEAU, J.-D. LEBRETON, J.-Y. MONDAIN-MONVAL, A. R. JOHNSON, AND G. SIMON. 2007. Lead shot and Teal (*Anas crecca*) in the Camargue, Southern France: Effects of embedded and ingested pellets on survival. Biological Conservation 137:567-576.

GUITART, R., S. MAÑOSA, V. G. THOMAS, AND R. MATEO. 1999. Perdigones y pesos de plomo: ecotoxicología y efectos para la fauna. Revista de Toxicología 16:3-11.

GUITART R, J., SERRATOSA, AND V. G. THOMAS. 2002. Lead-poisoned wildfowl in Spain: a significant threat for human consumers. International Journal of Environmental Health Research 12:301-309.

GUITART, R., J. TO-FIGUERAS, R. MATEO, A. BERTOLERO, S. CERRADELO, AND A. MARTÍNEZ-VILALTA. 1994a. Lead poisoning in waterfowl from the Ebro delta, Spain: Calculation of lead exposure thresholds for Mallards. Archives of Environmental Contamination and Toxicology 27:289-293.

GUITART, R., M. TORRA, S. CERRADELO, P. PUIG-CASADO, R. MATEO, AND J. TO-FIGUERAS. 1994b. Pb, Cd, As, and Se concentrations in livers of dead wild birds from the Ebro delta, Spain. Bulletin of Environmental Contamination and Toxicology 52:523-529.

HERNÁNDEZ, M. 1995. Lead poisoning in a free-ranging Imperial Eagle. Supplement to the Journal of Wildlife Diseases 31 (3), Newsletter.

HERNÁNDEZ, M., AND A. MARGALIDA. 2008. Pesticide abuse in Europe: effects on the Cinereous Vulture (*Aegypius monachus*) population in Spain. Ecotoxicology 17: 264-272.

HOFFMANN, L. 1960. Le saturnisme fleau de la sauvagine en Camargue. Terre et Vie 107: 120-131.

HOLLAND, G. 1882. Pheasant poisoned by swallowing shot. The Field 59:232.

HOLLMÉN, T., J. C. FRANSON, R. H. POPPENGA, M. HARIO, AND M. KILPI. 1998. Lead poisoning and trace elements in Common Eiders *Somateria mollissima* from Finland. Wildlife Biology 4:193-203.

HOLT, G., A. FROESLIE, AND G. NORHEIM. 1978. Lead poisoning in Norwegian waterfowl. Nord Veterinaermed 30:380-389.

HONTELEZ, L. C. M. P., H. M. VAN DEN DUNGEN, AND A. J. BAARS. 1992. Lead and cadmium in birds in The Netherlands : a preliminary survey. Archives of Environmental Contamination and Toxicology 23:453-456.

HOVETTE, C. 1971. Le saturnisme en Camargue. Union Internationale des Biologistes du Gibier. Actes du Xe Congrès, Paris, France.

HOVETTE, C. 1972. Le saturnisme des anatidés de la Camargue. Alauda 40:1-117.

HOVETTE, C. 1974. Le saturnisme des anatidés sauvages. L'Institut Techniche de l'Avicurture, 211:1-4.

IMRE, Á. 1994. Vadkacsák sörét eredetú ólommérgezése. Magyar Állatorvosok Lapja 49: 345-348.

IMRE, Á. 1997. Fácánok sörét eredetu ólommérgezése. Magyar Allatorvosok Lapja 119: 328-330.

JÅGAS, T. 1996. Lead levels and lead poisoning in Swedish *Anseriformes* birds. Ekotoxikologi N° 47. Uppsala Universitet, Uppsala, Sweden.

JAGER, L. P., F. V. J. RIJNIERSE, H. ESSELINK, AND A. J. BAARS. 1996. Biomonitoring with the Buzzard *Buteo buteo* in the Netherlands: Heavy metals and sources of variation. Journal für Ornithologie 137:295-318.

JOHANSEN, P., G. ASMUND, AND F. RIGET. 2001. Lead contamination of seabirds harvested with lead shot: Implications to human diet in Greenland. Environmental Pollution 112:501-504.

JOHANSEN, P., G. ASMUND, AND F. RIGET. 2004. High human exposure to lead through consumption of birds hunted with lead shot. Environmental Pollution 127:125-129.

KALISIŃSKA, E., W. SALICKI, AND A. JACKOWSKI. 2006. Six trace metals in White-tailed Eagle from northwestern Poland. Polish Journal of Environmental Studies 15: 727-737.

KALISIŃSKA, E., W. SALICKI, K. M. KAVETSKA, AND M. LIGOCKI. 2007. Trace metal concentrations are higher in cartilage than in bones of Scaup and Pochard wintering in Poland. Science of the Total Environment 388:90-103.

KALISIŃSKA, E., W. SALICKI, P. MYSŁEK, K. M. KAVETSKA, AND A. JACKOWSKI. 2004. Using the Mallard to biomonitor heavy metal contamination of wetlands in north-western Poland. Science of the Total Environment 320:145-161.

KANSTRUP, N. 2006. Non-toxic shot-Danish experiences. Page 861 *in* C. G. Boere, C. A. Galbraith, and D. A. Stroud (Eds.). Waterbirds around the World. The Stationery Office, Edinburgh, UK.

KANSTRUP, N., AND D. POTTS. 2008. Lead shot: significant new developments with relevance to all hunters. CIC-News, 4 February 2008. [Online.] Available at www.cic-wildlife.org

KELLY, A., AND S. KELLY. 2004. Fishing tackle injury and blood lead levels in Mute Swans. Waterbirds 27:60-68.

KENNTNER, N., Y. CRETTENAND, H.-J. FÜNFSTÜCK, M. JANOVSKY, AND F. TATARUCH. 2007. Lead poisoning and heavy metal exposure of Golden Eagles (*Aquila chrysaetos*) from the European Alps. Journal of Ornithology 148:173-177.

KENNTNER, N., O. KRONE, R. ALTENKAMP, AND F. TATARUCH. 2003. Environmental contaminants in liver and kidney of free-ranging Northern Goshawks (*Accipiter gentilis*) from three regions of Germany. Archives of Environmental Contamination and Toxicology 45: 128-135.

KENNTNER, N., G. OEHME, D. HEIDECKE, AND F. TATARUCH. 2004. Retrospektive untersuchung zur bleiintoxikation und exposition mit potenziell toxischen schwermetallen von Seeadlern *Haliaaetus albicilla* in Deutschland. Wogelwelt 125:63-74.

KENNTNER, N., F. TATARUCH, AND O. KRONE. 2001. Heavy metals in soft tissue of White-tailed Eagles found dead or moribund in Germany and Austria from 1993 to 2000. Environmental Toxicology and Chemistry 20:1831-1837.

KENNTNER, N., F. TATARUCH, AND O. KRONE. 2005. Risk assessment of environmental contaminants in White-tailed Sea-eagles (*Haliaeetus albicilla*) from Germany. Pages 125-127 *in* K. Pohlmeyer (Ed.). Extended Abstracts of the XXVIIth Congress of the International Union of Game Biologists, Hannover 2005. DSV-Verlag, Hamburg, Germany.

KEYMER, I. F. 1958. A survey and review of the causes of mortality in British birds and the significance of wild birds as disseminators of disease. Veterinary Record 70: 713-720.

KEYMER, I. F., AND R. S. J. STEBBINGS. 1987. Lead poisoning in a Partridge (*Perdix perdix*) after ingestion of gunshot. Veterinary Record 120:276-277.

KIM, E. Y., R. GOTO, H. IWATA, Y. MASUDA, S. TANABE, AND S. FUJITA. 1999. Preliminary survey of lead poisoning of Steller's Sea Eagle (*Haliaeetus pelagicus*) and White-tailed Sea Eagle (*Haliaeetus albicilla*) in Hokkaido, Japan. Environmental Toxicology and Chemistry 18: 448-451.

KRONE, O., N. KENNTNER, A. TRINOGGA, M. NADJAFZADEH, F. SCHOLZ, J. SULAWA, K. TOTSCHEK, P. SCHUCK-WERSIG, AND R. ZIESCHANK. 2009. Lead poisoning in White-tailed Sea Eagles: Causes and approaches to solutions in Germany. *In* R. T. Watson, M. Fuller, M. Pokras, and W. G. Hunt (Eds.). Ingestion of Lead from Spent Ammunition: Implications for Wildlife and Humans. The Peregrine Fund, Boise, Idaho, USA. DOI 10.4080/ilsa.2009.0207

KRONE, O., T. STJERNBERG, N. KENNTNER, F. TATARUCH, J. KOIVUSAARI, AND I. NUUJA. 2006. Mortality factors, helminth burden, and contaminant residues in White-tailed Sea Eagles (*Haliaeetus albicilla*) from Finland. Ambio 35:98-104.

KRONE, O., F. WILLE, N. KENNTNER, D. BOERTMANN, AND F. TATARUCH. 2004. Mortality factors, environmental contaminants, and parasites of White-tailed Sea Eagles from Greenland. Avian Diseases 48:417-424.

KUIVENHOVEN, P., J. VAN VESSEM, AND E. VAN MAANEN. 1998. Saturnisme des oiseaux d'eau. Rapport d'actualisation internationale 1997. Bulletin Mensuel de l'Office National de la Chasse 229 : 28-31.

LAMBERET, M. 1995. Enquête sur le saturnisme des Anatides en France métropolitaine. Bilan de la saison 1993/94. Rapport de stage BTA GFS – ONC.

LEFRANC, H. 1993. Enquête sur le saturnisme des Anatides en France métropolitaine. Bilan de la saison 1992/93. Rapport de stage BTA GFS – ONC.

LLORENTE, G. A. 1984. Contribución al conocimiento de la biología y la ecología de cuatro especies de anátidas en el delta del Ebro. Tesis Doctoral, Universidad de Barcelona, Barcelona, Spain.

LOCKE, L. N., AND M. FRIEND. 1992. Lead poisoning of avian species other than waterfowl. Pages 19–22 *in* D. J. Pain (Ed.). Lead Poisoning in Waterfowl. Proceedings of an IWRB Workshop, Brussels, Belgium, 13-15 June 1991. International Waterfowl and Wetlands Research Bureau Special Publication 16, Slimbridge, UK.

LUMEIJ, J.T., H. HENDRIKS, AND A. TIMMERS. 1989. The prevalence of lead shot ingestion in wild Mallards (*Anas platyrhynchos*) in the Netherlands. Veterinary Quarterly 11: 51-55.

LUMEIJ, J. T., AND H. SCHOLTEN. 1989. A comparision of two methods to establish the prevalence of lead shot ingestion in Mallards (*Anas platyrhynchos*) from The Netherlands. Journal of Wildlife Diseases 25:297-299.

LUMEIJ, J.T., W. T. WOLVEKAMP, G. M. BRON-DIETZ, AND A. J. SCHOTMAN. 1985. An unusual case of lead poisoning in a Honey Buzzard (*Pernis apivorus*). Veterinary Quarterly 7:165-168.

MACDONALD, J. W., C. J. RANDALL, H. M. ROSS, G. M. MOON, AND A. D. RUTHVEN. 1983. Lead poisoning in captive birds of prey. Veterinary Record 113:65-66.

MARTÍNEZ-HARO, M., R. GUITART, G. LLORENTE, AND R. MATEO. 2005. Seasonal and annual fluctuations of lead poisoning in waterfowl in the Ebro delta, Spain. Pages 146-147 *in* K. Pohlmeyer (Ed.). Extended Abstracts of the XXVIIth Congress of the International Union of Game Biologists, Hannover 2005. DSV-Verlag, Hamburg, Germany.

MATEO, R. 1998. La intoxicación por ingestión de perdigones de plomo en aves silvestres: aspectos epidemiológicos y propuestas para su prevención en España. Tesis Doctoral, Universitat Autònoma de Barcelona, Bellaterra, Spain.

MATEO, R., J. BELLIURE, J. C. DOLZ, J. M. AGUILAR-SERRANO, AND R. GUITART. 1998. High prevalences of lead poisoning in wintering waterfowl in Spain. Archives of Environmental Contamination and Toxicology 35:342-347.

MATEO, R., A. BONET, J. C. DOLZ, AND R. GUITART. 2000a. Lead shot densities in a site of grit ingestion for Greylag Geese *Anser anser* in Doñana (Spain). Ecotoxicology and Environmental Restoration 3:76-80.

MATEO, R., R. CADENAS, M. MÁÑEZ, AND R. GUITART. 2001a. Lead shot ingestion in two raptor species from Doñana, Spain. Ecotoxicology and Environmental Safety 48:6-10.

MATEO, R., J. C. DOLZ, J. M. AGUILAR-SERRANO, J. BELLIURE, AND R. GUITART. 1997a. An outbreak of lead poisoning in Greater Flamingos *Phoenicopterus ruber roseus* in Spain. Journal of Wildlife Diseases 33:131-134.

MATEO, R., J. ESTRADA, J-Y. PAQUET, X. RIERA, L. DOMÍNGUEZ, R. GUITART, AND A. MARTÍNEZ-VILALTA. 1999. Lead shot ingestion by Marsh Harriers *Circus aeruginosus* from the Ebro delta, Spain. Environmental Pollution 104:435-440.

MATEO, R., A. J. GREEN, C. W. JESKE, V. URIOS, AND C. GERIQUE. 2001b. Lead poisoning in the globally threatened Marbled Teal and White-headed Duck in Spain. Environmental Toxicology and Chemistry 20:2860-2868.

MATEO, R., A. J. GREEN, H. LEFRANC, R. BAOS, AND J. FIGUEROLA. 2007a. Lead poisoning in wild birds from southern Spain: A comparative study of wetland areas and species affected, and trends over time. Ecotoxicology and Environmental Safety 66:119-126.

MATEO, R., AND R. GUITART. 2000. The effects of grit supplementation and feed type on steel-shot ingestion in Mallards. Preventive Veterinary Medicine 44:221-229.

MATEO, R., R. GUITART, AND A. J. GREEN. 2000b. Determinants of lead shot, rice, and grit ingestion in ducks and coots. Journal of Wildlife Management 64:939-947.

MATEO, R., R. GUITART, AND A. MARTÍNEZ-VILALTA. 1997b. Lead shot pellets in the Ebro delta, Spain: Densities in sediments and prevalence of exposure in waterfowl. Environmental Pollution 96:335-341.

MATEO, R., R. MOLINA, J. GRÍFOLS, AND R. GUITART. 1997c. Lead poisoning in a free ranging Griffon Vulture (*Gyps fulvus*). Veterinary Record 140:47-48.

MATEO, R., M. RODRÍGUEZ-DE LA CRUZ, D. VIDAL, M. REGLERO, AND P. CAMARERO. 2007b. Transfer of lead from shot pellets to game meat during cooking. Science of the Total Environment 372:480-485.

MATEO, R., AND M. A. TAGGART. 2007. Toxic effects of the ingestion of lead polluted soil on waterfowl. Proceedings of the International Meeting of Soil and Wetland Ecotoxicology, Barcelona, Spain.

MATEO, R., M. A. TAGGART, A. J. GREEN, C. CRISTÒFOL, A. RAMIS, H. LEFRANC, J. FIGUEROLA, AND A. A. MEHARG. 2006. Altered porphyrin excretion and histopathology of Greylag Geese (*Anser anser*) exposed to soil contaminated with lead and arsenic in the Guadalquivir Marshes, SW Spain. Environmental Toxicology and Chemistry 25: 203-212.

MATEO, R., M. TAGGART, AND A. A. MEHARG. 2003. Lead and arsenic in bones of birds of prey from Spain. Environmental Pollution 126:107-114.

MATHIASSON, S. 1986. Lead in tissue and gizzards of Mute Swan *Cygnus olor* from the Swedish west coast, with remarks on other heavy metals and possible additive and synergetic effects. Vår Fågelvärld Supplement 11:111-126.

MAUVAIS, G., AND L. PINAULT. 1993. Le saturnisme des anatidés (Anatidae) sur le site du Lac de Grand-Lieu (Loire-Atlantique). Gibier Faune Sauvage 10:85-101.

MERKEL, F.R., K. FALK, AND S. E. JAMIESON. 2006. Effect of embedded lead shot on body condition of Common Eiders. Journal of Wildlife Management 70:1644-1649.

MEZIERES, M. 1999. Effets de l'ingestion de plombs de chasse sur la reproduction du Canard Colvert (*Anas platyrhynchos*). Thèse Docteur Vétérinaire. Ecole Nationale Véterinaire de Nantes, France.

MONDAIN-MONVAL, J.-Y. 1999. Programme d'education à la chasse à tir, l'approche nord-américaine. Bulletin Mensuel de l 'Office National de la Chasse 246:26-35.

MONDAIN-MONVAL, J.-Y., L. DESNOUHES, AND J.-P. TARIS. 2002. Lead shot ingestion in waterbirds in the Camargue (France). Gibier Faune Sauvage 19:237-246.

MOORE, J. L., W. L. HOHMAN, T. M. STARK, AND G. A. WEISBRICH. 1998. Shot prevalences and diets of diving ducks five years after the ban on use of lead shotshells at Catahoula Lake, Louisiana. Journal of Wildlife Management 62:564-569.

MÖRNER, T., AND L. PETERSSON. 1999. Lead poisoning in woodpeckers in Sweden. Journal of Wildlife Diseases 35:763-765.

MUDGE, G. P. 1983. The incidence and significance of ingested lead pellet poisoning in British wildfowl. Biological Conservation 27:333-372.

MUDGE, G. P. 1984. Densities and settlement rates of spent shotgun pellets in British wetland soils. Environmental Pollution 8: 299-318.

MULLER, K., O. KRONE, T. GOBEL, AND L. BRUNNBERG. 2001. Akute bleoontoxication bei zwei Seeadlern (*Haliaeetus albicilla*). Tierarztliche Praxis Ausgabe K: Kleintiere-Heimtiere 29:209-213.

NOER, H., J. MADSEN, AND P. HARTMANN. 2007. Reducing wounding of game by shotgun hunting: effects of a Danish action plan on Pink-footed Geese. Journal of Applied Ecology 44:653-662.

OCHIAI, K., K. JIN, M. GORYO, T. TSUZUKI, AND C. ITAKURA. 1993. Pathomorfologic findings of lead poisoning in White-fronted Geese *Anser albifrons.* Veterinary Pathology 30:522-528.

O'HALLORAN, J., P. F. DUGGAN, AND A. A. MYERS. 1988a. Biochemical and hematological values for Mute Swans *Cygnus olor*: Effects of acute lead poisoning. Avian Pathology 17:667-678.

O'HALLORAN, J., A. A. MYERS, AND P. F. DUGGAN. 1988b. Lead poisoning in swans and sources of contamination in Ireland. Journal of Zoology 216:211-223.

OLNEY, P. J. S. 1960. Lead poisoning in waterfowl. Wildfowl 11:123-134.

OLNEY, P. J. S. 1968. The food and feeding habits of Pochard. Biological Conservation 1: 71-76.

OWEN, M. 1996. A review of the migration strategies of the Anatidae: Challenges for conservation. Gibier Faune Sauvage 13:123-129.

PAIN, D. J. 1990a. Lead shot ingestion by waterbirds in the Camargue, France: An investigation of levels and interspecific differences. Environmental Pollution 66: 273-285.

PAIN, D. J. 1990b. Lead poisoning of waterfowl: a review. Pages 172-181 *in* G. V. T. Matthews (Ed.). Managing Waterfowl Populations. International Waterfowl and Wetlands Research Bureau. Slimbridge, UK.

PAIN, D. J. 1991a. Lead shot densities and settlement rates in Camargue marshes, France. Biological Conservation 57:273-286.

PAIN, D. J. 1991b. L'intoxication saturnine de l'avifaune: Une synthèse des travaux français. Gibier Faune Sauvage 8:79-92.

PAIN D. J. 1992. Lead poisoning in waterfowl: A review. Pages 7-13 *in* D. J. Pain (Ed.). Lead Poisoning in Waterfowl. Proceedings of an IWRB Workshop, Brussels, Belgium, 13-15 June 1991. International Waterfowl and Wetlands Research Bureau Special Publication 16, Slimbridge, UK.

PAIN, D. J., AND C. AMIARD-TRIQUET. 1993. Lead poisoning in raptors in France and elsewhere. Ecotoxicology and Environmental Safety 25:183-192.

PAIN, D. J., C. AMIARD-TRIQUET, C. BABOUX, G. BURNELEAU, L. EON, AND P. NICOLAU-GUILLAUMET. 1993. Lead poisoning in wild populations of Marsh Harrier *Circus aeruginosus* in the Camargue and Charente-Maritime, France. Ibis 135:379-386.

PAIN, D. J., C. AMIARD-TRIQUET, AND C. SYLVESTRE. 1992. Tissue lead concentrations and shot ingestion in nine species of waterbirds from the Camargue (France). Ecotoxicology and Environmental Safety 24:217-233.

PAIN, D. J., C. BABOUX, AND G. BURNELEAU. 1997. Seasonal blood lead concentrations in Marsh Harriers *Circus aeruginosus* from Charente-Maritime, France: Relationship with the hunting season. Biological Conservation 81:1-7.

PAIN, D. J., I. CARTER, A.W. SAINSBURY, R. F. SHORE, P. EDEN, M. A. TAGGART, S. KONSTANTINOS, L.A. LAKER, A.A. MEHARG, AND A. RAAB. 2007. Lead contamination and associated disease in captive and reintroduced Red Kites *Milvus milvus* in England. Science of the Total Environment 376:116-127.

PAIN, D. J., AND G. I. HANDRINOS. 1990. The incidence of ingested lead shot in ducks of the Evros delta, Greece. Wildfowl 41:167-170.

PAIN, D. J., A. A. MEHARG, M. FERRER, M. A. TAGGART, AND V. PENTERIANI. 2005. Lead concentrations in bones and feathers of the globally threatened Spanish Imperial Eagle. Biological Conservation 121:603-610.

PAIN, D. J., J. SEARS, AND I. NEWTON. 1995. Lead concentrations in birds of prey in Britain. Environmental Pollution 87:173-180.

PATTEE, O. H., AND S. K. HENNES. 1983. Bald Eagles and waterfowl: the lead shot connection. Pages 230–237 *in* 48th North American Wildlife Conference. The Wildlife Management Institute, Washington, DC, USA.

PENNYCOTT, T. W. 1998 Lead poisoning and parasitism in a flock of Mute Swans (*Cygnus olor*) in Scotland. Veterinary Record 142:13-17.

PERCO, F., S. FOCARDI, C. FOSSI, AND A. RENZONI. 1983. Intossiczione da piombo in due Cigni Reali della Laguna di Marano (Nord-Est Italia). Avocetta 7:105-116.

PÉREZ-LÓPEZ, M., M. HERMOSO DE MENDOZA, A. LÓPEZ BECEIRO, AND F. SOLER RODRÍGUEZ. 2008. Heavy metal (Cd, Pb, Zn) and metalloid (As) content in raptor species from Galicia (NW Spain). Ecotoxicology and Environmental Safety 70:154-162.

PERRINS, C. M., G. COUSQUER, AND J. WAINE. 2003. A survey of blood lead levels in Mute Swans *Cygnus olor*. Avian Pathology 32:205-212.

PETERSEN, B. D., AND H. MELTOFTE. 1979. Forekomst af blyhagl i vestjyske vådområder samt i kråsen hos danske ænder. Dansk Ornitologisk Forenings Tidsskrift 73:257-264.

PINAULT, L. 1996. Evaluation de l'exposition au plomb des canards en France: résultats d'une enquête conduite de 1992 à 1995 en sept sites. Rapport d'étude, Ecole National Vétérinaire de Nantes, France.

PIROT, J.-Y. 1978. Régime alimentaire de la Sarcelle d'été, *Anas querquedula* L., pendant son transient en Camargue. Rapport Diplôme d'Études Approfondies, Paris, France.

PIROT, J.-Y., AND J. P. TARIS. 1987. Le saturnisme des anatides hivernant en Camargue: réactualisation des données. Gibier Faune Sauvage 4:83-94.

POOLE, C. 1986. Surgical treatment of lead poisoning in a Mute Swan (*Cygnus olor*). Veterinary Record 119:501-502.

POTTS, G. R. 2005. Incidence of ingested lead gunshot in wild Grey Partridge (*Perdix perdix*) from the UK. European Journal of Wildlife Research 51:31-34.

RAMO, C., C. SÁNCHEZ, AND L. HERNÁNDEZ SAINT-AUBIN. 1992. Lead poisoning of Greater Flamingos *Phoenicopterus ruber*. Wildfowl 43:220-222.

RODRIGUES, D., M. FIGUEIREDO, AND A. FABIÃO. 2001. Mallard *Anas platyrhynchos* lead poisoning risk in central Portugal. Wildfowl 52:169-174.

RODRIGUES, D., M. FIGUEIREDO, A. FABIÃO., M. C. VAZ, G. SARMENTO, J. FRANÇA, AND J. BACELAR. 2005. Lead poisoning in Portuguese waterfowl. Pages 170-171 *in* K. Pohlmeyer (Ed.). Extended Abstracts of the XXVIIth Congress of the International Union of Game Biologists, Hannover 2005. DSV-Verlag, Hamburg, Germany.

RODRIGUEZ, R., AND F. HIRALDO. 1975. Régimen alimenticio del Calamón *Porphyrio porphyrio* en las Marismas del Guadalquivir. Doñana Acta Vertebrata 2:201-213.

RODRIGUEZ-RAMOS, J., V. GUTIERREZ, U. HÖFLE, R. MATEO, L. MONSALVE, E. CRESPO, J. M. BLANCO. 2009. Lead in Griffon and Cinereous Vultures in Central Spain: Correlations between clinical signs and blood lead levels. Extended Abstract *in* R. T. Watson, M. Fuller, M. Pokras, and W. G. Hunt (Eds.). Ingestion of Lead from Spent Ammunition: Implications for Wildlife and Humans. The Peregrine Fund, Boise, Idaho, USA. DOI 10.4080/ilsa.2009.0213

ROSE, P. M. 1995. Western Palearctic and South-West Asia Waterfowl Census 1994. IWRB Publication 35.

SANDERSON, G. C. 1992. Lead poisoning mortality. Pages 14-18 *in* D. J. Pain (Ed.). Lead Poisoning in Waterfowl. Proceedings of an IWRB Workshop, Brussels, Belgium, 13-15 June 1991. International Waterfowl and Wetlands Research Bureau Special Publication 16, Slimbridge, UK.

SANDERSON, G. C., AND F. C. BELLROSE. 1986. A review of the problem of lead poisoning in waterfowl. Illinois Natural History Survey Special Publication 4:1-34.

SCHEUHAMMER, A. M., AND S. L. NORRIS. 1996. The ecotoxicology of lead shot and lead fishing weights. Ecotoxicology 5:279-295.

SCHRICKE, V., AND H. LEFRANC. 1994. Enquête sur le saturnisme des anatides en France métropolitaine. Bilan de la saison 1992/93. Bulletin Mensuel de l'Office National de la Chasse 192:2-15.

SEARS, J. 1988. Regional and seasonal variations in lead poisoning in the Mute Swan *Cygnus olor* in relation to the distribution of lead and lead weights in the Thames area, England. Biological Conservation 46:115-134.

SEARS, J., AND A. HUNT. 1991. Lead poisoning in Mute Swans, *Cygnus olor*, in England. Wildfowl Supplement 1:383-388.

SILEO, L., AND S. I. FEFER. 1987. Paint chip poisoning of Laysan Albatross at Midway atoll. Journal of Wildlife Diseases 23:432-437.

SIMPSON, V. R., A. E. HUNT, AND M. C. FRENCH. 1979. Chronic lead poisoning in a herd of Mute Swans. Environmental Pollution 18:187-202.

SMIT, T., T. BAKHUIZEN, C. P. H. GAASENBEEK, AND L. G. MORAAL. 1988a. Voorkomen van loodkorrels rondjachthutten en kleiduivenbanen. Limosa 61:183-186.

SMIT, T., T. BAKHUIZEN, AND L. G. MORAAL. 1988b. Metallisch lood als bron van loodvergiftiging in Nederland. Limosa 61:175-178.

SOLER-RODRÍGUEZ, F., A. L. OROPESA-JIMÉNEZ, J. P. GARCÍA-CAMBERO, AND M. PÉREZ-LÓPEZ. 2004. Lead exposition by gunshot ingestion in Red-legged Partridge (*Alectoris rufa*). Veterinary and Human Toxicology 46:133-134.

STREET, M. 1983. The assessment of mortality resulting from the ingestion of spent lead shot by Mallard wintering in south-east England. *In* Proceedings of the XI Congress of the International Union of Game Biologists. Trujillo, Spain.

SVANBERG, F., R. MATEO, L. HILLSTRÖM, A. J. GREEN, M. A. TAGGART, A. RAAB, AND A. A. MEHARG. 2006. Lead isotopes and lead shot ingestion in the globally threatened Marbled Teal (*Marmaronetta angustirostris*) and White-headed Duck (*Oxyura leucocephala*). Science of the Total Environment 370:416-424.

TAMISIER, A. 1971. Régime alimentaire des Sarcelles d'hiver, *Anas crecca* L., en Camargue. Alauda 39:261-311.

TAVECCHIA, G., R. PRADEL, J.-D. LEBRETON, A. R. JOHNSON, AND J.-Y. MONDAIN-MONVAL. 2001. The effect of lead exposure on survival of adult Mallards in the Camargue, southern France. Journal of Applied Ecology 38:1197-1207.

THOMAS, G. J. 1975. Ingested lead pellets in waterfowl at the Ouse washes. England 1968-1973. Wildfowl 26:43-48.

THOMAS, G. J. 1982. Lead poisoning in waterfowl. Pages 260-268 *in* Managing Wetlands and their Birds: A Manual of Wetland and Waterfowl Management. International Waterfowl and Wetlands Research Bureau, Slimbridge, UK.

THOMAS, V. G. 1997. Attitudes and issues preventing bans on toxic lead shot and sinkers in North America and Europe. Environmental Values 6:185-199.

THOMAS, V. G., AND R. GUITART. 2005. Role of international conventions in promoting avian conservation through reduced lead toxicosis: Progression towards a non-toxic agenda. Bird Conservation International 15:147-160.

THOMAS, V. G., AND M. OWEN. 1996. Preventing lead toxicosis of European waterfowl by regulatory and non-regulatory means. Environmental Conservation 23:358-364.

TIRELLI, E., N. MAESTRINI, S. GOVONI, E. CATELLI, AND R. SERRA. 1996. Lead contamination in the Mallard (*Anas platyrhynchos*) in Italy. Bulletin of Environmental Contamination and Toxicology 56:729-733.

TROST, R. E. 1981. Dynamics of grit selection and retention in captive Mallards. Journal of Wildlife Management 45:64-73.

USFWS. 1975 Issuance of annual regulations permitting the sport hunting of migratory birds. Final Environmental Statement. U. S. Fish and Wildlife Service, Washington, DC, USA.

VEIGA, J. 1985. Contribution à l'etude du regimen alimentaire de la Bécassine Sourde *Lymnocriptes minimus*. Gibier Faune Sauvage 2:75-84.

WIUM-ANDERSEN, S., AND N. E. FRANSMANN. 1974. Dor andefulge af at spise blyhagl? Feltornithologen 16:14.

ZECHNER, L., T. STEINECK, AND F. TATARUCH. 2005. Bleivergiftung bei einem Steinadler (*Aquila chrysaetos*) in der Steiermark. Egretta 47:157-158.

ZUUR, B. 1982. Zum vorkommen von bleischrotkornern im magen von wasservölgen am Untersee. Ornithologische Beobachter 79:97-103.

A GLOBAL UPDATE OF LEAD POISONING IN TERRESTRIAL BIRDS FROM AMMUNITION SOURCES

DEBORAH J. PAIN[1], IAN J. FISHER[2], AND VERNON G. THOMAS[3]

[1]*Wildfowl and Wetlands Trust, Slimbridge, Gloucestershire, GL2 7BT, UK*
E-mail: debbie.pain@wwt.org.uk

[2]*Royal Society for the Protection of Birds, The Lodge, Sandy, Bedfordshire SG19 2DL, UK*

[3]*Department of Integrative Biology, College of Biological Science, University of Guelph, Guelph, ON N1G 2W1, Canada.*

ABSTRACT.—Lead poisoning mortality, through the ingestion of spent shot, is long established in waterfowl, and more recently in raptors and other avian taxa. Raptors (vultures, hawks, falcons, eagles and owls) are exposed to lead from spent ammunition (shot, bullets, or fragments from either) while feeding on game species, and other avian taxa are exposed when feeding in shot-over areas, including shooting ranges. Here we review the published literature on ingestion of and poisoning by lead from ammunition in terrestrial birds. We briefly discuss methods of evaluating exposure to and poisoning from ammunition sources of lead, and the use of lead isotopes for confirming the source of lead. Documented cases include 33 raptor species and 30 species from *Gruiformes*, *Galliformes* and various other avian taxa, including ten Globally Threatened or Near Threatened species. Lead poisoning is of particular conservation concern in long-lived slow breeding species, especially those with initially small populations such as the five Globally Threatened and one Near Threatened raptor species reported as poisoned by lead ammunition in the wild. Lead poisoning in raptors and other terrestrial species will not be eliminated until all lead gunshot and rifle bullets are replaced by non-toxic alternatives. *Received 29 May 2008, accepted 24 July 2008.*

PAIN, D. J., I. J. FISHER, AND V. G. THOMAS. 2009. A global update of lead poisoning in terrestrial birds from ammunition sources. *In* R. T. Watson, M. Fuller, M. Pokras, and W. G. Hunt (Eds.). Ingestion of Lead from Spent Ammunition: Implications for Wildlife and Humans. The Peregrine Fund, Boise, Idaho, USA. DOI 10.4080/ilsa.2009.0108

Key words: Ammunition, global, lead, poisoning, terrestrial birds.

TERRESTRIAL BIRDS AT RISK

LEAD POISONING IN WILDFOWL AND WADERS from the ingestion of spent lead gunshot has been extensively studied, documented and reviewed over the last half century (Bellrose 1959, Kaiser and Fry 1980, Hall and Fisher 1985, Veiga 1985, Pain 1990a, 1990b, 1991a, 1992, 1996, Locke and Friend 1992, Sharley et al. 1992, Scheuhammer and Norris 1995, Beck and Granval 1997). Poisoning from the ingestion of lead from both gunshot and bullets by terrestrial birds initially received less attention, but today a considerable and increasing literature exists. While ingestion of discharged lead from firearms has been recorded in a diverse range of terrestrial birds, two groups of birds are particularly susceptible.

One group includes birds that are exposed in a similar way to wildfowl and waders, while feeding in any shot over areas where spent lead is deposited. Such areas include upland and lowland game shooting areas, shooting ranges including clay-pigeon shoots, and areas where species such as squirrels and other rodents are shot as pests. Exposure is likely to be higher in species that, like wildfowl, deliberately ingest grit to help break down food and/or ingest food items similar in appearance to lead shot (such as seeds) and may therefore mistakenly ingest lead. Many species fall into this category, but species are predominantly *Gruiformes* and *Galliformes* (Table 1). The second group at risk comprise species that ingest lead from bullets or shot while preying upon or scavenging game or other species of mammals or birds (such as 'pest' species, e.g., Knopper et al. 2006) that have been shot but survived, or killed but unretrieved. Species at risk are predominantly raptors, herein defined as vultures (New and Old World), hawks, falcons, eagles and owls (Table 1). Here, risk of exposure is likely to be particularly high in species that feed more frequently on hunter-shot quarry and have a propensity to scavenge. Species are also more likely to be exposed if there is intensive hunting within their foraging range (e.g. Mateo et al. 1998a, Wayland and Bollinger 1999) and during the hunting season (Elliott et al. 1992, Pain et al. 1997).

One of the earliest reports of lead poisoning from lead shot ingestion in a non-wildfowl species involved a Ring-necked Pheasant (*Phasianus colchicus*) in the UK (Calvert 1876), and a recent study from British shooting estates found an overall shot ingestion rate of 3%, with correspondingly elevated bone lead levels (Butler et al. 2005). Evidence for lead poisoning in this species therefore spans 125 years. Species of *Galliformes* and *Gruiformes* have been reported to have ingested shot and/or suffered lead poisoning through exposure to lead shot in a wide range of countries across Europe, and North America (Table 1).

In areas where lead ammunition is used, all raptors that feed on game or other hunted species are potentially at risk from lead poisoning. Species that take live prey, such as the Goshawk (*Accipiter gentilis*), are exposed through ingesting animals that have been shot but survived carrying shot in their flesh. In some cases exposure rates can be high, e.g., 25% and 36% of first year and adult live-trapped Pink-footed Geese (*Anser brachyrhynchus*) carried shot (Noer and Madsen 1996). However, species that scavenge, or both hunt and scavenge, are likely to be at particular risk from lead ingestion through feeding on injured or hunter-killed but unretrieved prey. Season and local hunting intensity play an important part in determining the level of dietary lead exposure (Pain et al. 1993, Mateo 1998a). For example, of Bald Eagles (*Haliaeetus leucocephalus*) found sick, injured or dead with significantly elevated lead exposure in British Columbia, the greatest number were found between January and March when they were feeding heavily on wintering waterfowl (Elliott et al. 1992). The prevalence of Bald Eagles with elevated lead exposure was found to be higher in areas of high waterfowl hunting intensity than areas with low hunting intensity (Wayland and Bollinger 1999). Similarly, Pain et al. (1997) found a significantly higher prevalence of elevated blood lead concentrations in Marsh Harriers (*Circus aeruginosus*) in France during than outside the hunting season.

Raptors are at risk both from prey killed with shotguns, such as wildfowl, rabbits and rodents, and with rifles, such as deer and other large game. The prevalence of exposure to lead from bullets is likely to be higher in larger raptors, and poisoning from bullet fragments has been reported in wild Bald Eagles, Golden Eagles (*Aquila chrysaetos*), Steller's Sea Eagles (*Haliaeetus pelagicus*), White-tailed Eagles (*H. albicilla*), and California Condors (*Gymnogyps californianus*) (Table 1).

Both captive and supplementary-fed wild birds may also be at risk if food includes animals that have been killed with lead ammunition, even when shot and bullets have been removed (e.g., Pain et al. 2007). Until recently, the extent to which lead shot and bullets fragment while passing through the tissues of prey was not realised. However, radiographs have shown that both shot and bullets undergo considerable fragmentation and fragments may be both distant from the wound canal, and too small to be visually detected (e.g., Pain et al. 2007). Many authors have now shown that feeding on species killed by lead gunshot or bullets can present a significant risk to scavengers, whether or not they are exposed to whole shot or bullets (Scheuhammer et al. 1998, Sergeyev and Shulyatieva 2005, Hunt et al. 2006, Knopper et al. 2006, Pain et al. 2007). For example, Scheuhammer et al. (1998) found that 11% (92 of 827) of pectoral muscle pools from hunter-killed game (predominantly waterfowl) had elevated lead concentrations, in excess of 0.5 µg g^{-1} ww (mean of 12± 38 µg g^{-1}ww). While no lead could be detected visibly in the samples, radiography showed tiny embedded metal fragments.

EXPOSURE TO LEAD GUNSHOT AND BULLETS

Lead exposure is usually measured through the presence of lead shot or lead fragments in pellets regurgitated by raptors (e.g., Platt 1976, Pain et al. 1997, 2007), or the presence of shot, bullets or fragments in the alimentary tracts of birds post-mortem (e.g. Church et al. 2006 and numerous authors in Table 1).

The regurgitation of ingested shot in pellets illustrates that not all exposure to lead need be fatal, and regurgitation will undoubtedly reduce the amount of lead absorbed and the level of poisoning in raptors. However, repeated exposure and regurgitation of lead shot can result in poisoning (Pattee et al. 1981). In areas of high exposure, the likelihood of picking up shot (and of repeated exposure) increases. For example, as Bald Eagles and their waterfowl prey become concentrated in smaller areas, the likelihood of eagles picking up shot increases (Pattee and Hennes 1983). Consequently, exposure measured through the presence of shot in regurgitated pellets is likely to give a good indication of the likelihood of a species to become poisoned. Pain et al. (1997) found that the incidence of shot in regurgitated Marsh Harrier pellets dropped from up to 25% during the hunting season to just 1.4% after, in parallel with a significant reduction in blood lead concentrations in the birds outside the hunting season. Other studies have similarly found both the presence of lead shot in regurgitated pellets and evidence of poisoning and mortality from lead in wild birds (e.g., Red Kites [*Milvus milvus*] in the UK, Pain et al. 2007).

The presence of lead particles in the alimentary tracts of birds at post mortem is a widely used indicator of exposure, and both the number of lead shot or fragments found and the degree of abrasion or dissolution may provide supplementary evidence to help in the diagnosis of lead poisoning. Occasionally, large numbers of shot are ingested, and in Spain, where Golden Eagles have been found with lead toxicosis from shot ingestion, one bird had 40 pellets in its proventriculus (Cerradelo et al. 1992).

It is possible that birds that have not ingested lead from ammunition but carry shot-in lead in their tissues may also have elevated blood lead concentrations, as has been found in humans that have been shot with lead bullets (McQuirter et al. 2004), although these are likely to be far lower than in birds that have ingested lead fragments.

EVALUATING LEAD POISONING

Lead is not an essential element, and is a non-specific poison affecting all body systems. Lead poisoned birds often exhibit a distended proventriculus, green watery faeces, weight loss, anaemia and drooping posture, among other signs (Redig et al. 1980, Reiser and Temple 1981, Franson et al. 1983, Custer et al. 1984, Sanderson and Bellrose 1986, Friend 1987, Mateo 1998b). Lead affects the nervous system, kidneys and the circulatory system, resulting in a range of sub-lethal physiological, biochemical and behavioural changes (Scheuhammer 1987). There is no true 'no-effect' threshold level for lead; for example even at extremely low concentrations lead depresses the activity of the blood enzyme delta aminolevulinic acid, essential for haemoglobin production (Grasman and Scanlon 1995, Redig et al. 1991).

Once ingested, lead ammunition or fragments are readily dissolved in the acidic conditions in the intestine, and in birds that ingest grit to aid digestion (e.g., some *Gruiformes* and *Galliformes*) abrasion of the lead may accelerate this process. Directly following absorption the highest lead concentrations tend to be found in the blood, from which lead is transported around the body and deposited in soft tissue and bone. Blood lead generally has a half-life of around two weeks in birds (e.g., 14 days in the California Condor, Fry et al. 2009 this volume), whereas liver and kidney tissues generally retain elevated lead concentrations for weeks to several

Table 1. Lead shot ingestion and poisoning.

Species	Status	Evidence	Countries	References
Common Raven (*Corvus corax)*	LC	Poisoning	USA	Craighead and Bedrosian 2008
Chukar (*Alectoris chukar*)	LC	Ingestion	USA	Hanspeter and Kerry 2003
Grey Partridge (*Perdix perdix*)	LC	Ingestion and Poisoning	Denmark, UK	Clausen and Wolstrup 1979, Keymer and Stebbings 1987, Potts, 2005
Common Pheasant (*Phasianus colchicus)*	LC	Ingestion and Poisoning	Denmark, UK, USA	Calvert 1876, Elder 1955, Clausen and Wolstrup 1979, NWHL 1985, Dutton and Bolen 2000, Butler et al. 2005.
Wild Turkey (*Meleagris gallopavo*)	LC	Ingestion	USA	Stone and Butkas 1978
Scaled Quail (*Callipepla squamata*)	LC	Ingestion	USA	Campbell 1950
Northern Bobwhite Quail (*Colinus virginianus*)	NT	Ingestion	USA	Stoddard 1931, Keel et al. 2002
Great Horned Owl (*Bubo virginianus*)	LC	Poisoning	Canada	Clark and Scheuhammer 2003
Eurasian Eagle Owl (*B. bubo*)	LC	Poisoning	Spain	Mateo et al. 2003
Snowy Owl (*Nyctea scandiaca*)	LC	Poisoning	Captive	MacDonald et al. 1983
Long-eared Owl (*Asio otus*)	LC	Poisoning	Spain	Brinzal 1996
Rock Pigeon (*Columba livia*)	LC	Ingestion	USA , Belgium	Dement et al. 1987, Tavernier et al. 2004
Common Wood-pigeon (*C. palumbus*)	LC	Poisoning	Denmark	Clausen and Wolstrup 1979
Mourning Dove (*Zenaida macroura*)	LC	Ingestion	USA	Locke and Bagley 1967, Lewis and Legler 1968, Best et al. 1992, Schulz et al. 2002
Sandhill Crane (*Grus canadensis*)	LC	Ingestion	USA	Windingstad et al. 1984, NWHL 1985
Whooping Crane (*G. americana*)	EN	Poisoning	USA	Hall and Fisher 1985
Clapper Rail (*Rallus longirostris*)	LC	Ingestion	USA	Jones 1939
King Rail (*R. elegans*)	LC	Ingestion	USA	Jones 1939
Virginia Rail (*R. limicola*)	LC	Ingestion	USA	Jones 1939
Sora (*Porzana carolina*)	LC	Poisoning	USA	Jones 1939, Artman and Martin 1975, Stendell et al. 1980
Common Moorhen (*Gallinula chloropus*)	LC	Ingestion	Europe, USA	Jones 1939, Locke and Friend 1992
Common Coot (*Fulica atra*)	LC	Ingestion	France	Pain 1990a
American Coot (*F. americana*)	LC	Ingestion	USA	Jones 1939

Species	Status	Evidence	Countries	References
American Woodcock (*Scolopax minor*)	LC	Ingestion	Canada	Scheuhammer et al. 1999, 2003
Ruffed Grouse (*Bonasa umbellus*)		Ingestion	Canada	Rodrigue et al. 2005
California Gull (*Larus californicus*)	LC	Ingestion	USA	Quortrup and Shillinger 1941
Glaucous-winged Gull (*L. glaucescens*)	LC	Ingestion	USA	NWHL 1985
Herring Gull (*L. argentatus*)	LC	Ingestion	USA	NWHL 1985
European Honey-buzzard (*Pernis apivorus*)	LC	Unknown (ingestion or shot)	Netherlands	Lumeiji et al. 1985
Red Kite (*Milvus milvus*)	LC	Ingestion or poisoning	Germany, Spain, UK, Captive	Mateo 1998a, Mateo et al. 2001, 2003, Pain et al. 1997, Kenntner et al. 2005
Bald Eagle (*Haliaeetus leucocephalus*)	LC	Poisoning, shot and bullets	Canada, USA	Platt 1976, Jacobson et al. 1977, Kaiser et al. 1980, Pattee and Hennes 1983, Reichel et al. 1984, Frenzel and Anthony 1989, Craig et al. 1990, Langelier et al. 1991, Elliott et al. 1992, Gill and Langelier 1994, Scheuhammer and Norris 1996, Wayland and Bollinger 1999, Miller et al. 2000, 2001, Clark and Scheuhammer 2003, Wayland et al. 2003
Steller's Sea-eagle (*H. pelagicus*)	VU	Poisoning, bullets	Japan	Kim et al. 1999, Iwata et al. 2000, Kurosawa 2000
White-rumped Vulture (*Gyps bengalensis*)	CR	Poisoning (origin unknown)	Pakistan	Oaks et al. 2004
Eurasian Griffon (*G. fulvus*)	LC	Poisoning	Spain	Mateo et al. 1997, Guitart 1998, Mateo et al. 2003, Garcia-Fernandez et al. 2005
Egyptian Vulture (*Neophron percnopterus)*	EN	Poisoning	Canary Islands; Iberian peninsula	Donazar et al. 2002, Rodriguez-Ramos et al. 2009, this volume
Cinereous Vulture *(Aegypius monachus)*	NT	Poisoning	Spain	Hernandez and Margalida 2008
Eastern Marsh-harrier (*Circus spilonotus)*		Ingestion	Japan	Hirano et al. 2004
Western Marsh-harrier (*Circus aeruginosus*)	LC	Poisoning	France, Germany, Spain	Pain et al. 1993, 1997, Mateo et al., 1999, Kenntner et al. 2005
Northern Harrier (*C. cyaneus*)	LC	Ingestion	Canada, USA	Martin and Barrett 2001, Martin et al. 2003
Eurasian Sparrowhawk (*Accipiter nisus*)	LC	Ingestion	France, Captive	MacDonald et al. 1983, Pain and Amiard-Triquet 1993
Sharp-shinned Hawk (*A. striatus*)	LC	Ingestion	Canada, USA	Martin and Barrett 2001
Cooper's Hawk (*A. cooperii*)	LC	Ingestion	Canada, USA	Snyder et al. 1973, Martin and Barrett 2001
Northern Goshawk (*A. gentilis*)	LC	Poisoning	Canada, France, Germany, USA, Captive	Stehle 1980, Pain and Amiard-Triquet 1993, Martin and Barrett 2001, Kenntner et al. 2003, 2005

Species	Status	Evidence	Countries	References
Red-tailed Hawk (*Buteo jamaicensis*)	LC	Poisoning	Canada, USA	Franson et al. 1996, Martin and Barrett 2001, Clark and Scheuhammer 2003, Martina et al. 2008
Common Buzzard (*B. buteo*)	LC	Poisoning	France, Germany; UK, Italy, Captive	Stehle 1980, MacDonald et al. 1983, Pain and Amiard-Triquet 1993, Pain et al. 1995, Kenntner et al. 2005, Battaglia et al. 2005
Rough-legged Buzzard (*B. lagopus*)	LC	Poisoning	USA	Locke and Friend 1992
Spanish Imperial Eagle (*Aquila adalberti*)	VU	Poisoning	Spain	González and Hiraldo 1988, Hernández 1995, Mateo 1998a, Mateo et al. 2001, Pain et al. 2004
Golden Eagle (*A. chrysaetos*)	LC	Poisoning	Canada, Germany, Spain, Switzer-land, USA	Bloom et al. 1989, Craig et al. 1990, Pattee et al. 1990, Cerradelo et al. 1992, Bezzel and Fünfstück 1995, Scheuhammer and Norris 1996, Wayland and Bollinger 1999, Clark and Scheuhammer 2003, Wayland et al. 2003, Kenntner et al. 2007
White-tailed Eagle (*Haliaeetus albicilla*)	LC	Ingestion and Poison-ing – shot and bullets	Greenland, Po-land; Austria, Germany, Ja-pan, Finland	Falandysz et al. 1988, Kim et al. 1999, Iwata et al. 2000, Kurosawa 2000, Kenntner et al. 2001, 2004, 2005, Krone et al. 2004, 2006, Kalisinska et al. 2006
American Kestrel (*Falco sparverius*)	LC	Ingestion	Canada, USA	Martin and Barrett 2001
Laggar Falcon (*F. jugger*)	NT	Poisoning	Captive	MacDonald et al. 1983
Prairie Falcon (*F. mexicanus*)	LC	Poisoning	Captive	Benson et al. 1974, Stehle 1980
Peregrine Falcon (*F. peregrinus*)	LC	Poisoning	UK, Captive	MacDonald et al. 1983, Pain et al. 1995
Turkey Vulture (*Cathartes aura*)	LC	Ingestion	Canada, USA	Wiemeyer et al. 1986, Martin et al. 2003, Martina et al. 2008
California Condor (*Gymnogyps californianus*)	CR	Poisoning, bullets	USA	Wiemeyer et al. 1983, 1986, 1988, Janssen et al. 1986, Bloom et al. 1989, Pattee et al. 1990, Meretsky et al. 2000, Snyder and Snyder 2000
Andean Condor (*Vultur gryphus*)	NT	Poisoning	Captive	Locke et al. 1969
King Vulture (*Sarcorhampus papa*)	LC	Poisoning	Captive	Decker et al. 1979
White-throated Sparrow (*Zonotrichia albicollis*)	LC	Ingestion	USA	Vyas et al. 2000
Dark-eyed Junco (*Junco hyemalis*)	LC	Ingestion	USA	Vyas et al. 2000
Brown-headed Cowbird (*Molothrus atar*)	LC	Ingestion	USA	Vyas et al. 2000
Yellow-rumped Warbler (*Dendroica coronata*)		Poisoning	USA	Lewis et al. 2001
Brown Thrasher (*Toxostoma rufum*)		Poisoning	USA	Lewis et al. 2001
Solitary Vireo (*Vireo solitarius*)		Poisoning	USA	Lewis et al. 2001

Countries are given for wild birds; captive birds are simply listed as captive, and were poisoned through shot present in their feed. Conservation status is from BirdLife International (2004) and is coded as: CR—Critically Endangered, EN—Endangered , NT—Near Threatened, LC—Least Concern. Evidence of poisoning indicates tissue lead concentrations indicative of poisoning (e.g., Franson 1996) with the source of poisoning most likely to be lead gunshot and/or diagnosis of lead poisoning in one or more individuals; 'Ingestion' indicates evidence of ingestion of shot usually in the absence of tissue analysis. Rare cases of lead poisoning believed to be from non-ammunition sources are excluded.

Table 2. Guidelines for interpretation of tissue lead concentrations in *Falconiformes* and *Galliformes* (ppm ww).

	Falconiformes (ppm ww)			Galliformes (ppm ww)		
	Blood	Liver	Kidney	Blood	Liver	Kidney
Sub clinical	0.2–1.5	2–4	2–5	0.2–3	2–6	2–20
Toxic	>1	>3	>3	>5	>6	>15
Compatible with death	>5	>5	>5	>10	>15	>50

Adapted from Franson (1996)

months following absorption. Once deposited in bone, lead is far less mobile, and bone lead concentrations tend to remain elevated for months to years, reflecting lifetime exposure (Pain 1996). Several authors have suggested guidelines to help interpret tissue lead concentrations in different avian taxa (Franson 1996, Pain 1996, Table 2). Differences in sensitivity exist both among and within taxa (e.g., Table 2, Carpenter et al. 2003), and other factors, such as the duration and level of exposure, may influence the tissue lead concentrations at which effects are observed. There is no simple way of relating tissue lead concentrations to effect. However, the general conclusions on tissue lead levels associated with sub-lethal poisoning, toxicity, and death from lead poisoning that have been drawn from the vast field and experimental data reported in the published literature are useful guidelines (Table 2, Franson 1996).

The definition of 'background' lead concentrations is rather more complex and will differ with the context of the question being asked and the population under study. Natural levels of exposure to lead in the environment no longer exist as lead resulting from anthropogenic emissions is ubiquitous. Concentrations in the environment distant from emission sources are generally described as 'background.' In remote areas most birds would be expected to have <2 and certainly <5 μg/dL blood lead from exposure to background levels. However, populations of scavengers that feed in urban areas or at rubbish dumps may have median population blood lead concentrations far higher than this as they will be exposed to a mix of anthropogenic sources of lead. These higher concentrations could still usefully be defined as 'background for this population' when attempting to distinguish between low level chronic exposure from a mix of general environmental sources, and acute exposure, for example through the ingestion of lead from ammunition.

The diagnosis of lead poisoning is usually based upon clinical signs of poisoning in combination with blood lead concentrations in live birds. In dead birds, diagnosis is based upon tissue (liver or kidney) lead concentrations and clinical signs of poisoning, sometimes, but not necessarily, in combination with evidence of exposure to lead. Clinical signs of poisoning and/or evidence of exposure alone are insufficient to positively diagnose lead poisoning and tissue lead analysis is required. The presence of ingested lead objects in the intestine, while a good indicator, is not diagnostic of poisoning as little absorption may have occurred if objects have been ingested very recently. Similarly, the absence of lead objects in the intestine does not mean that birds have not been lead poisoned, as lead objects may be regurgitated or passed through after considerable absorption has occurred, or may be ground down and/or totally dissolved and absorbed.

Clinical signs are more frequently recorded in cases of chronic poisoning, whereas acutely poisoned birds may die rapidly in apparently good body condition (e.g., Gill and Langelier 1994). Liver and kidney lead concentrations are most useful in lead poisoning diagnosis post-mortem. Bone lead concentrations are less useful as, in cases of acute poisoning, lead poisoned birds may die rapidly with comparatively low bone lead concentrations, and conversely, high bone lead concentrations may result from chronic lifetime exposure in the absence of lead toxicity. Except in cases of acute exposure, bone lead concentrations have been shown to be positively correlated with age in waterfowl populations (Stendell et al. 1979, Clausen et al. 1982).

IDENTIFYING THE SOURCE OF LEAD

There are relatively few sources of lead to which exposure is sufficient to result in tissue concentrations indicative of toxicity (Table 2) or to result in mortality. By far the majority of cases in birds result from the ingestion of shot or bullet fragments (Table 1), or anglers' lead weights in certain water birds (Scheuhammer and Norris 1995). Rare cases result from point source exposure, e.g., to mine waste (Henny et al. 1991, 1994, Sileo et al. 2001) or industrial pollution (Bull et al. 1983), or the ingestion of lead in paint (Sileo and Fefer 1987). In cases where lead poisoning has been diagnosed and lead shot and/or bullet fragments are found in the bird's intestine, the likely source is self-evident. In most cases where no lead from ammunition is present, knowledge of feeding habits or areas of the species is often sufficient to conclude that lead from ammunition is the likely source.

One additional tool to help identify likely sources of lead contamination is the examination of the distribution of blood or soft tissue lead concentrations in a population of birds. Blood and soft tissue lead concentrations in birds that have simply been exposed to generally distributed anthropogenic lead (e.g., from atmospheric deposition) should follow a relatively normal distribution. Dramatic outliers from this, or distributions particularly skewed to higher lead concentrations, suggest elevated lead ingestion from an additional source in a proportion of birds. Distributions are likely to be normal with a few elevated outliers in birds that infrequently ingest lead ammunition, and heavily skewed to elevated concentrations, with more very elevated outliers in populations that more frequently ingest lead ammunition. Although the examination of population lead distributions is a useful indicator and can provide supporting evidence, and while lead from ammunition is by far the most frequently reported source of exposure, most distributions could potentially result from other forms of exposure, and the examination of tissue lead distributions should simply be used as a tool.

Recently, lead isotopes have been used to help source lead exposure (e.g., Meharg et al. 2002). Ratios of lead isotopes, $Pb^{206, 207, 208}$, can vary with the origin of the lead. For example, lead in background sources such as soil and air may have markedly different isotopic ratios from certain industrial sources, or from lead mined in different geographical regions. The utility of lead isotopes for distinguishing sources of lead contamination in wild birds was illustrated by Scheuhammer and Templeton (1998). They compared isotopic ratios in lead-exposed waterfowl and eagles and found them to be similar to those from lead shot pellets, whereas juvenile Herring Gulls (*Larus argentatus*) had isotopic ratios within the range characterising lead from gasoline combustion. Stable lead isotope ratios were used by Martin and Barrett (2001) to determine the source of lead exposure to wildlife on the north shore of Lake Erie. While none of the migrating birds sampled had lead levels indicating lead poisoning, one or more individuals of the following species were found to have levels indicative of sub-lethal lead exposure, with the lead isotope ratios for most of the samples falling within the range of gunshot pellets: American Kestrel (*Falco sparverius*), Sharp-shinned Hawk (*Accipiter striatus*), Cooper's Hawk, Northern Goshawk, Northern Harrier (*Circus cyaneus*), and Red-tailed Hawk (*Buteo jamaicensis*). More recently, Pain et al. (2007) found lead isotope ratios in the livers of Red Kites in the UK to be consistent with ratios in lead pellets, and Church et al. (2006) found that free-flying California Condors with relatively high blood lead concentrations had isotope values consistent with lead from ammunition and distinct from those of pre-release condors with low blood lead concentrations.

In most cases of poisoning from lead ammunition sources, there is sufficient knowledge of the feeding habits of birds, and their proclivity to ingest lead shot and bullet fragments to negate the need for isotopic studies. Isotopic studies may be particularly useful in those rare cases when the source of lead is unclear or where unusual patterns of tissue lead concentrations are found.

CASE STUDIES

Many lead poisoning studies are based upon knowledge of a species' feeding habits and ranges, the likelihood of shot and bullet fragment ingestion, and tissue lead concentrations, or upon the presence of ingested shot or bullet fragments. While these factors are generally sufficient to evaluate lead poisoning from ammunition sources, many other studies have included a much wider range of data. A few illustrative examples are outlined here. Evidence for lead poisoning of Marsh Harriers in southern Europe came from (1) knowledge of the propensity of Marsh Harriers to scavenge waterfowl, (2) temporal variation in the proportion of regurgitated pellets containing lead shot, with far higher incidence during the hunting season and, (3) temporal variation in the proportion of birds with elevated blood lead concentrations, with a far higher incidence during than outside the hunting season (Pain et al. 1993, 1997, Mateo et al. 1999). In the UK, evidence for lead shot poisoning of Red Kites included (1) knowledge of propensity to scavenge game and other shot species, (2) the presence of lead gunshot in regurgitated pellets, (3) tissue lead concentrations consistent with lead poisoning mortality at post mortem, and (4) lead isotope values in the tissues of dead birds with elevated lead concentrations consistent with those of lead shot (Pain et al. 1997). By far the most comprehensively studied species is the California Condor, for which evidence for lead poisoning from spent shot and bullets includes all of the above. It includes: post-mortem evidence and/or evidence of ammunition ingestion, moribund birds with clinical signs of lead poisoning and elevated blood lead levels, elevated blood lead levels in free-flying birds, temporal and spatial correlations between big game hunting seasons and areas and elevated lead levels in condors, and lead isotope ratios in the tissues of exposed condors consistent with those of lead shot and bullets (Church et al. 2006 and authors in Table 1). This provides only a small sample of the numerous studies that combine a wide range of evidence of lead poisoning from ammunition sources in individual species.

Species Affected by Lead Poisoning

Cases of lead shot ingestion and/or poisoning in terrestrial birds are documented in Table 1. This currently includes 33 raptor species and 30 species from *Gruiformes, Galliformes* and various other avian taxa. Twenty-eight raptor species and all species from other taxa were free-flying, and an additional ten raptor species were poisoned accidentally in captivity from lead shot in food carcasses. This was usually hunter-killed rabbits, squirrels and other small mammals, including an example where shot were manually removed from prey, but where small lead fragments likely remained (Locke et al. 1969, Benson et al. 1974, Decker et al. 1979, Jacobsen et al. 1977, Stehle 1980, MacDonald et al. 1983, Pain et al. 1997). While most cases of poisoning are in raptors, *Gruiformes* or *Galliformes*, several species from other taxa have been shown to have ingested shot, including pigeons, doves, gulls, and a range of passerines.

In addition to published cases reviewed for this study, records of many other species poisoned by lead ammunition exist at veterinary schools and similar institutions (M. Pokras pers. comm.).

Physiological and Population Impacts of Lead Poisoning

Lead poisoning through the ingestion of shot and bullet fragments by terrestrial birds causes a wide range of sub-lethal impacts affecting physiology and behaviour, and when exposure is sufficiently high or of sufficient duration results in mortality. Experimental studies in birds have shown that, among other effects, lead can impair blood synthesis, immune function and reproduction (Grandjean 1976, Kendall et al. 1981, Kendall and Scanlon 1981, Veit et al. 1982, Edens and Garlich 1983, Scheuhammer 1987, Grasman and Scanlon 1995, Redig et al. 1991). Consequently, free-living birds exhibiting sub-lethal poisoning will likely be more susceptible to disease, starvation and predation, and an increased probability of death from other causes (Scheuhammer and Norris 1996). Although the extent to which survival is reduced by sub-lethal lead poisoning is difficult to quantify in field studies, it is clearly undesirable in any species.

Mortality from lead poisoning is of greatest concern in long-lived, slow-breeding species, characteristics typical of many raptors, and particularly species with populations that are either naturally small or have been reduced by other factors. Such species are particularly vulnerable to increases in adult mortality, and all age classes can be affected by lead poisoning.

Lead poisoning in the Bald Eagle, a species whose population was depressed through losses due to pesticide abuses in the 1950s and 1960s, is well documented (Elliott et al. 1992, Pattee and Hennes 1983, Wayland and Bollinger 1999, Table 1). In Canada and the USA, approximately 10–15% of recorded post-fledging mortality in Bald and Golden Eagles was attributed to the ingestion of lead shot from prey animals (Scheuhammer and Norris 1996). Elliott et al. (1992) found that 14% of 294 sick, injured. or dead Bald Eagles in British Columbia (1988 to 1991) were lead-poisoned and an additional 23% sub-clinically exposed, with the majority of exposure occurring during January to March, when the birds were feeding heavily on wintering wildfowl.

Of particular concern is lead poisoning in globally threatened species. To date, eight globally threatened (Critically Endangered, Endangered or Vulnerable) or Near Threatened (BirdLife International 2008) free-living species have been reported as lead-poisoned (Table 1), of which six are raptors. While there were isolated reports in two cases, the Endangered Whooping Crane (*Grus americana*) and Critically Endangered Oriental White-backed Vulture (*Gyps bengalensis*), multiple reports exist for the other species. In addition, two captive Near Threatened species have been reported with lead poisoning from ammunition sources: Laggar Falcon (*Falco jugger*) and Andean Condor (*Vultur gryphus*). High levels of adult mortality occurred in globally Vulnerable Steller's Sea-eagles and White-tailed Eagles (Least Concern) in Hokkaido, Japan, until a ban on the use of lead bullets in this area (Kurosawa 2000). Population modelling indicated that in the absence of preventative measures Steller's Sea Eagle populations would decline as a result of lead poisoning, and that populations were most sensitive to changes in adult mortality (Ueta and Masterov 2000). The Vulnerable Spanish Imperial Eagle (*Aquila adelberti*) is an example of a species with a small population, estimated at 200 pairs (BirdLife International 2008) that cannot sustain high, especially adult, mortality (Ferrer 2001, Ferrer et al. 2003). The most significant impacts of lead poisoning on any terrestrial species have been on the Critically Endangered California Condor. Lead poisoning from lead shot and bullet fragments was a major factor in the decline of this species, which became extinct in the wild, and has seriously hampered reintroduction efforts (Meretsky et al. 2000, Cade 2007).

In addition to lead poisoning in globally threatened species, many of the species in Table 1 are of national or regional conservation concern (e.g., Gregory et al. 2002 for the UK).

Not only does lead poisoning cause unnecessary mortality in many terrestrial avian species, and present a significant problem at a population level to a few, it also severely compromises the welfare of large numbers of wild birds, both terrestrial and aquatic. Sainsbury et al. (1995) investigated the scale and severity of welfare issues in wild European birds and mammals. They described the nature and level of harm caused (pain, stress and fear), the duration of effects and numbers of individuals affected, and found lead poisoning through shot ingestion to be among the most significant human activities that severely compromise the welfare of large numbers of animals.

REMOVING THE THREAT OF LEAD POISONING IN TERRESTRIAL BIRDS

Lead poisoning from ammunition affects the populations of several globally threatened species, and hampers the reintroduction of the critically endangered California Condor. In many species it causes mortality or sub-lethal effects which may influence survival. Limited legislative steps have been taken to reduce the threat of lead poisoning to terrestrial birds, including nationwide legislation banning the use of lead shot for all waterfowl hunting in the USA from 1991, aimed at preventing secondary lead toxicosis of the Bald Eagle (Anderson 1992), and a ban on the use of lead bullets for hunting on Hokkaido, Japan, where secondary poisoning affected many Steller's Sea Eagles (Matsuda 2003). More recently, and in the face of considerable opposition from the gun lobby, the State of California passed Assembly Bill 821 in 2007 to restrict the use of lead shot and bullets in the range of the California Condor. Such legislation is an important step towards reducing the risk of lead poisoning in terrestrial birds; policing and compliance are also essential, and much remains to be done in these areas.

The risk to terrestrial and water birds, of lead poisoning from ammunition is just one of many reasons for replacing lead in all ammunition with non-toxic alternatives. Other reasons have been covered elsewhere in this volume, and include the following:

- Lead is a highly toxic heavy metal with no known biological function that affects humans and other animals at the lowest measurable concentrations. Lead has already been banned from most uses that could result in human and wildlife exposure (e.g., from paint, and as an anti-knocking agent and octane booster in gasoline/petrol).
- The use of lead ammunition results in huge quantities of lead deposited in the environment annually (Thomas 1997), where it is extremely persistent in contaminating soils and waterways (e.g., Sorvari et al. 2006) and augmenting potential future risks to wildlife inhabiting shot-over areas.
- Environmental clean-up following the deposition of lead ammunition is both difficult and costly.
- Some lead from ammunition sources is or becomes bioavailable and may be bioaccumulated by edaphic organisms (Migliorini et al. 2004), and puts at risk vegetation, invertebrates and other organisms (Bennett et al. 2007).
- Lead poisoning may severely compromise the welfare of large numbers of wild animals.
- Lead from ammunition can present a human health risk, e.g., at firing ranges and to people who consume game shot with lead (Scheuhammer et al. 1998, Tsuji et al. 1999, Guitart et al. 2002, Johansen et al. 2004, Gustavsson and Gerhardsson 2005). While not all people that consume game will necessarily develop elevated blood lead concentrations (e.g., Haldimann et al. 2002), those that frequently consume game such as subsistence hunters appear particularly at risk (e.g., Bjerregaard et al. 2004) as may be people who cook game using recipes of low pH that facilitate lead dissolution (e.g., Mateo et al. 2007).

There are few reasons to continue with the use of lead ammunition. Non-toxic shot regulations have been shown to be successful at reducing the incidence of lead poisoning (e.g., Anderson et al. 2000, Stevenson et al. 2005) and are unlikely to result in increased crippling rates (Schultz et al. 2006). Cade (2007) reviewed the compelling evidence for the impacts of lead poisoning from ammunition on the California Condor population, and concluded that the use of non-toxic shot and bullets would be highly efficacious for hunting, economically feasible, and ethically the right thing to do.

While previous regulatory emphasis has been placed on wetland habitats or waterfowl hunting, there is now an urgent and compelling need to replace all lead gunshot and rifle bullets for shooting game, pest species, and targets with non-toxic alternatives. The wide extent of primary and secondary lead poisoning has been demonstrated across a large range of bird species and continents wherein hunting has been practiced. The burden of proof that lead from spent gunshot and bullets causes lead toxicosis has already been achieved for wetland species, upland species, and avian predators and scavengers. More recently, the risks to humans have been documented, especially those who regularly eat hunted game meat (see the collection of papers in this volume). Thomas and Guitart (2005) indicated that several global treaties and agreements exist that could authorize the transition to non-toxic shot and bullets in many countries, and presage the passing of binding legislation to support bans on the use of lead products. The removal of lead shot, bullets and fishing weights from sporting use is also completely consistent with the Precautionary Principle (Matsuda 2003), and does not appear compromised by considerations of economic costs, commercial availability, or ballistic effectiveness. The transition to use of non-toxic lead substitutes is vital to ensure the sustainable use of wildlife, and to reduce the risks of lead poisoning from lead bullets and shot to human health.

LITERATURE CITED

ANDERSON, W. L. 1992. Legislation and lawsuits in the United States and their effects on non-toxic shot regulations. Pages 56–60 *in* D.J. Pain (Ed.). Lead Poisoning in Waterfowl. Proceedings of an IWRB Workshop, Brussels, Belgium, 13–15 June 1991. International Waterfowl and Wetlands Research Bureau Special Publication 16, Slimbridge, UK.

ANDERSON, W. L., S. P. HAVERA, AND B. W. ZERCHER. 2000. Ingestion of lead and non-toxic shotgun pellets by ducks in the Mississippi flyway. Journal of Wildlife Management 64:848–857.

ARTMANN, J. W., AND E. M. MARTIN. 1975. Incidence of ingested lead shot in Sora Rails. Journal of Wildlife Management 39:514–519.

BATTAGLIA, A., S. GHIDINI, G. CAMPANINI, AND R. SPAGGIARI. 2005. Heavy metal contamination in Little Owl (*Athene noctua*) and Common Buzzard (*Buteo buteo*) from northern Italy. Ecotoxicology and Environmental Safety 60:61–66.

BECK, N., AND P. GRANVAL. 1997. Ingestion de plombs de chasse par la Bécassine del Marais (*Gallinago gallinago*) et la Bécassine Sourde (*Lymnocryptes minimus*) dans le nord-ouest de la France. Gibier Faune Sauvage 14:65–70.

BELLROSE, F. C. 1959. Lead poisoning as a mortality factor in waterfowl populations. Illinois Natural History Survey Bulletin 27:235–288.

BENSON, W. W., B. PHARAON, AND P. MILLER. 1974. Lead poisoning in a bird of prey. Bulletin of Environmental Contamination Toxicology 11:105–108.

BENNETT, J. R., C. A. KAUFMAN, I. KOCH, J. SOVA, AND K. J. REIMER. 2007. Ecological risk assessment of lead contamination at rifle and pistol ranges using techniques to account for site characteristics. Science of the Total Environment 374:91–101.

BEST, T. L., T. E. GARRISON, AND C. G. SCHMITT. 1992. Availability and ingestion of lead shot by Mourning Doves (*Zenaida macroura*) in Southeastern New Mexico. The South-eastern Naturalist 37:287–292.

BEZZEL, E., AND H. J. FÜNFSTÜCK. 1995. Alpine Steinadler (*Aquila chrysaetos*) druch Bleivergiftung gefährdet? Journal für Ornithologie 136: 294–296.

BIRDLIFE INTERNATIONAL. 2008. Species factsheets downloaded from http://www.birdlife.org on 29 May 2008.

BJERREGAARD, P., P. JOHANSEN, G. MULVAD, H. S. PEDERSEN, AND J. C. HANSEN. 2004. Lead sources in human diet in Greenland. Environmental Health Perspectives 112:1496–1498.

BLOOM, P. H., J. M. SCOTT, O. H. PATTEE, AND M. R. SMITH. 1989. Lead contamination of Golden Eagles (*Aquila chrysaetos*) within the range of the Californian Condor (*Gymnogyps californianus*). Pages 481–482 *in* B-U. Meyburgh and R.D. Chancellor (Eds.). Raptors in the Modern World. Proceedings of the 3rd World Conference on Birds of Prey and Owls, Eliat, Israel, 22–27 March 1987. World Working Group on Birds of Prey, Berlin, London, and Paris.

BRINZAL. 1996. S.O.S. venenos: Búho Chico. Quercus 124:45.

BULL, K. R., W. J. EVERY, P. FREESTONE, J. R. HALL, AND D. OSBORN. 1983. Alkyl lead pollution and bird mortalities on the Mersey estuary, UK 1979–1981. Environmental Pollution 31: 239–259.

BUTLER, D. A., R. B. SAGE, R. A. H. DRAYCOTT, J. P. CARROLL, AND D. POTTS. 2005. Lead exposure in Ring-Necked Pheasants on shooting estates in Great Britain. Wildlife Society Bulletin 33: 583–589.

CADE, T. J. 2007. Exposure of California Condors to lead from spent ammunition. Journal of Wildlife Management 71:2125–2133.

CALVERT, J. H. 1876. Pheasants poisoned by swallowing shots. The Field 47:189.

CAMPBELL, H. 1950. Quail picking up lead shot. Journal of Wildlife Management 14:243–244.

CARPENTER, J. W., O. H. PATTEE, S. H. FRITTS, B. A. RATTNER, S. N. WIEMEYER, J. A. ROYLE, AND M. R. SMITH. 2003. Experimental lead poisoning in Turkey Vultures (*Cathartes aura*). Journal of Wildlife Diseases 39:96–104.

CERRADELO, S., E. MUÑOZ, J. TO-FIGUERAS, R. MATEO, AND R. GUITART. 1992. Intoxicación por ingestión de perdigones de plomo en dos Águilas Reales. Doñana, Acta Vertebrata 19: 122–127.

CHURCH, M. E., R. GWIAZDA, R. W. RISEBOROUGH, K. SORENSON, C. P. CHAMBERLAIN, S. FARRY, W. HEINRICH, B. A. RIDEOUT, AND D. R. SMITH. 2006. Ammunition is the principal source of lead accumulated by California Condors reintroduced to the wild. Environmental Science and Technology 40: 6143–6150.

CLARK, A. J., AND A. M. SCHEUHAMMER. 2003. Lead poisoning in upland-foraging birds of prey in Canada. Ecotoxicology 12:23–30.

CLAUSEN, B., AND C. WOLSTRUP. 1979. Lead poisoning in game from Denmark. Danish Review of Game Biology 11:22.

CLAUSEN, B., K. ELVESTAD, AND O. KARLOG. 1982. Lead burden in Mute Swans from Denmark. Nordisk Veterinaer Medicine 34:83–91.

CRAIG, T. H., J. W. CONNELLY, E. H. CRAIG, AND T. L. PARKER. 1990. Lead concentrations in Golden and Bald Eagles. Wilson Bulletin 102: 130–133.

CRAIGHEAD, D., AND B. BEDROSIAN. 2008. Blood lead levels of Common Ravens with access to big-game offal. Journal of Wildlife Management 72:240–245.

CUSTER, T. W., J. C. FRANSON, AND O. H. PATTEE. 1984. Tissue lead distribution and hematologic effects in American Kestrels (*Falco sparverius*) fed biologically incorporated lead. Journal of Wildlife Diseases 20: 39–43.

DECKER, R. A., A. M. MCDERMID, AND J. W. PRIDEAUX. 1979. Lead poisoning in two captive King Vultures. Journal of the American Medical Association 175:1009.

DEMENT, S. H., J. J. CHISOLM, JR., M. A. ECKHAUS, AND J. D. STRANDBERG. 1987. Toxic lead exposure in the urban Rock Dove. Journal of Wildlife Diseases 23:273–278.

DONAZAR, J. A., C. J. PALACIOS, L. GANGOSO, O. CEBALLOS, M. J. GONZALEZ, AND F. HIRALDO. 2002. Conservation status and limiting factors in the endangered population of Egyptian Vulture (*Neophron percnopterus*) in the Canary Islands. Biological Conservation 107:89–97.

DUTTON, C. S., AND E. G. BOLEN. 2000. Fall diet of a relict pheasant population in North Carolina. Journal of the Elisha Mitchell Scientific Society 116:41–48.

EDENS, F. W., AND J. D. GARLICH. 1983. Lead-induced egg production decrease in leghorn and Japanese quail hens. Poultry Science 64:1757–1763.

EISLER, R. 1988. Lead hazards to fish, wildlife, and invertebrates: a synoptic review. United States Fish and Wildlife Service Biological Report 85: 1–4.

ELDER, W. H. 1955. Fluoroscope measures of hunting pressure in Europe and North America. Transactions of the North American Wildlife Conference 20: 298–322.

ELLIOTT, J. E., K. M. LANGELIER, A. M. SCHEUHAMMER, P. H. SINCLAIR, AND P. E. WHITEHEAD. 1992. Incidence of lead poisoning in Bald Eagles and lead shot in waterfowl gizzards from British Columbia 1988–91. Canadian Wildlife Service Program Note No. 200. Canadian Wildlife Service, Quebec, Canada.

FALANDYSZ, J., B. JAKUCZUN, AND T. MIZERA. 1988. Metal and organochlorines in four female White-tailed Eagles. Marine Pollution 19:521–526.

FERRER, M. 2001. The Spanish Imperial Eagle. Lynx Editions, Barcelona, Spain.

FERRER, M., V. PENTERIANI, J. BALBONTÍN, AND M. PANDOLFI. 2003. The proportion of immature breeders as a reliable early warning signal of population decline: evidence from the Spanish Imperial Eagle in Doñana. Biological Conservation 114:463–466.

FRANSON, J. C. 1996. Interpretation of tissue lead residues in birds other than waterfowl. Pages 265–279 *in* W. N. Beyer, G. H. Heinz, and A. W. Redmon-Norwood (Eds.). Environmental Contaminants in Wildlife: Interpreting Tissue Concentrations. SETAC, CRC/Lewis Publishers, Boca Raton, Florida, USA.

FRANSON, J. C., L. SILEO, O. H. PATTEE, AND J. F. MOORE. 1983. Effects of chronic dietary lead in American Kestrels (*Falco spaverius*). Journal of Wildlife Diseases 19:110–113.

FRANSON, J. C., N. J. THOMAS, M. R. SMITH, A. H. ROBBINS, S. NEWMAN, AND P. C. MCCARTIN. 1996. A retrospective study of post-mortem findings in Red-tailed Hawks. Journal of Raptor Research 30: 7–14.

FRENZEL, R. W., AND R. G. ANTHONY. 1989. Relationship of diets and environmental contaminants in wintering Bald Eagles. Journal of Wildlife Management 53:792–802.

FRIEND, M. 1987. Field guide to wildlife diseases. United States Fish and Wildlife Service, Washington, DC, USA.

FRY, M., K. SORENSON, J. GRANTHAM, J. BURNETT, J. BRANDT, AND M. KOENIG. 2009. Lead Intoxication Kinetics in Condors from California. Abstract *in* R. T. Watson, M. Fuller, M. Pokras, and W. G. Hunt (Eds.). Ingestion of Lead from Spent Ammunition: Implications for Wildlife and Humans. The Peregrine Fund, Boise, Idaho, USA. DOI 10.4080/ilsa.2009.0301

GARCIA-FERNANDEZ, A. J., E. MARTINEZ-LOPEZ, D. ROMERO, P. MARIA-MOJICA, A. GODINO, AND P. JIMENEZ. 2005. High levels of blood lead in Griffon Vultures (*Gyps fulvus*) from Cazorla Natural Park (southern Spain). Environmental Toxicology 20: 459–463.

GILL, E., AND K. M. LANGELIER. 1994. Acute lead poisoning in a Bald Eagle secondary to bullet ingestion. Canadian Veterinary Journal 35:303–304.

GONZÁLEZ, L. M., AND F. HIRALDO. 1988. Organochlorine and heavy metals contamination in the eggs of the Spanish Imperial Eagle (*Aquila adalberti*) and accompanying changes in eggshell morphology and chemistry. Environmental Pollution 51:241–258.

GRANDJEAN, P. 1976. Possible effect of lead on eggshell thickness in kestrels 1874–1974. Bulletin of Environmental Contamination and Toxicology 16:101–106.

GRASMAN, K. A. AND P. F. SCANLON. 1995. Effects of acute lead ingestion and diet on antibody and T-cell-mediated immunity in Japanese quail. Archives of Environmental Contamination and Toxicology 28:161–167.

GREGORY, R. D., N. I. WILKINSON, D. G. NOBLE, J. A. ROBINSON, A. F. BROWN, J. HUGHES, D. PROCTER, D. W. GIBBONS, AND C. GALBRAITH. 2002. The population status of birds in the United Kingdom, Channel Islands and Isle of Man: an analysis of conservation concern 2002–2007. British Birds 95:410–48.

GUITART, R. 1998. Plumbismo en buitres. Quercus 146:47.

GUITART, R., J. SERRATOSA, AND V. G. THOMAS. 2002. Lead-poisoned wildfowl in Spain: a significant threat for human consumers. International Journal of Environmental Health Research 12:301–309.

GUSTAVSSON, P., AND L. GERHARDSSON. 2005. Intoxication from an accidentally ingested lead shot retained in the gastrointestinal tract. Environmental Health Perspectives 113:491–493.

HALDIMANN, M., A. BAUMGARTNER, AND B. ZIMMERLI. 2002. Intake of lead from game meat—a risk to consumers' health? European Food Research and Technology 215:375–379.

HALL, S. L., AND F. M. FISHER. 1985. Lead concentrations in tissues of marsh birds and relationships of feeding habits and grit preference to spent shot ingestion. Bulletin of Environmental Contamination and Toxicology 35:1–8.

HANSPETER, W., AND R. P. KERRY. 2003. Fall diet of Chukars (*Alectoris chukar*) in Eastern Oregon and discovery of ingested lead pellets. Western North American Naturalist 63:402–405.

HENNY, C. J., L. J. BLUS, D. J. HOFFMAN, AND R. A. GROVE. 1994. Lead in hawks, falcons and owls downstream from a mining site on the Coeur d'Alene river, Idaho. Environmental Monitoring and Assessment 29:267–288.

HENNY, C. J., L. J. BLUS, D. J. HOFFMAN, R. A. GROVE, AND J. S. HATFIELD. 1991. Lead accumulation and Osprey production near a mining site on the Coeur d'Alene river, Idaho. Archives of Environmental Contamination and Toxicology 21:415–424.

HERNÁNDEZ, M. 1995. Lead poisoning in a free-ranging Imperial Eagle. Supplement to the Journal of Wildlife Diseases 31, Newsletter.

HERNÁNDEZ, M., AND A. MARGALIDA. 2008 Pesticide abuse in Europe: Effects on the Cinereous Vulture (*Aegypius monachus*) population in Spain. Ecotoxicology 17:264–272.

HIRANO, T., I. KOIKE, AND C. TSUKAHARA. 2004. Lead shots retrieved from the pellets of Eastern Marsh Harriers wintering in Watarase Marsh, Tochigi Prefecture, Japan. Japanese Journal of Ornithology 53:98–100.

HUNT, W. G., W. BURNHAM, C. N. PARISH, K. K. BURNHAM, B. MUTCH, AND J. L. OAKS. 2006. Bullet fragments in deer remains: Implications for lead exposure in avian scavengers. Wildlife Society Bulletin 34:167–170.

IWATA, H., M. WATANABE, E.-Y. KIM, R. GOTOH, G. YASUNAGA, S. TANABE, Y. MASUDA, AND S. FUJITA. 2000. Contamination by chlorinated hydrocarbons and lead in Steller's Sea Eagle and White-tailed Sea Eagle from Hokkaido, Japan. Pages 91–106 *in* M. Ueta and M. J. McGrady (Eds.). First Symposium on Steller's and White-tailed Sea Eagles in East Asia. Wild Bird Society of Japan, Tokyo, Japan.

JACOBSON, E., J. W. CARPENTER, AND M. NOVILLA. 1977. Suspected lead toxicosis in a Bald Eagle. Journal of the American Veterinary Medical Association 171:952–954.

JANSSEN, D. L., J. E. OOSTERHUIS, J. L. ALLEN, M. P. ANDERSON, D. G. KELTS, AND S. N. STANLEY. 1986. Lead poisoning in free-ranging California Condors. Journal of the American

Veterinary Medical Association 189:1115–1117.

JOHANSEN, P., G. ASMUND, AND F. RIGET. 2004. High human exposure to lead through consumption of birds hunted with lead shot. Environmental Pollution 127:125–129.

JOHNSGARD, P. A. 1983. Sandhill Crane. Pages 171–184 *in* Cranes of the World. Croom Helm, London, Canberra.

JONES, J. C. 1939. On the occurrence of lead shot in stomachs of North American gruiformes. Journal of Wildlife Management 3:353–357.

KAISER, G. W., AND K. FRY. 1980. Ingestion of lead shot by Dunlin. The Murrelet 61:37.

KAISER, T. E., W. L. REICHEL, L. N. LOCKE, E. CROMARTIE, A. J. KRYNITSKY, T. G. LAMONT, B. M. MULHERN, R. M. PROUTY, C. J. STAFFORD, AND D. M. SWINEFORD. 1980. Organochlorine pesticide, PCB, and PBB residues and necropsy data for Bald Eagles from 29 states, 1975–77. Pesticide Monitoring Journal 14:145–149.

KALISINSKA, E., W. SALICKI, AND A. JACKOWSKI. 2006. Six trace metals in White-tailed Eagle from northwestern Poland. Polish Journal of Environmental Studies 15:727–737.

KEEL, M. K., W. R. DAVIDSON, G. L. DOSTER, AND L. A. LEWIS. 2002. Northern Bobwhite and lead shot deposition in an upland habitat. Archives of Environmental Contamination and Toxicology 43:318–322.

KENDALL, R. J., AND P. F. SCANLON. 1981. Effects of chronic lead ingestion on reproductive characteristics of Ringed Turtle Doves (*Streptopelia risoria*) and on tissue lead concentrations of adults and their progeny. Environmental Pollution A26:203–213.

KENDALL, R. J., H. P. VEIT, AND P. F. SCANLON. 1981. Histological effects and lead concentrations in tissues of adult male Ringed Turtle Doves (*Streptopelia risoria*) that ingested lead shot. Journal of Toxicology and Environmental Health 8:649–658.

KENNEDY, S., J. P. CRISLER, E. SMITH, AND M. BUSH. 1977. Lead poisoning in Sandhill Cranes. Journal of the American Veterinary Medical Association 171:955–958.

KENNTNER, N., O. KRONE, R. ALTENKAMP, AND F. TATARUCH. 2003. Environmental contaminants in liver and kidney of free-ranging Northern Goshawks (*Accipiter gentilis*) from three regions of Germany. Archives of Environmental Contamination and Toxicology 45:128–135.

KENNTNER, N., G. OEHME, D. HEIDECKE, AND F. TATARUCH. 2004. Retrospektive Untersuchung zur Bleiintoxikation und Exposition mit potenziell toxischen Schwermetallen von Seeadlern (*Haliaeetus albicilla*) in Deutschland. Vogelwelt 125:63–75.

KENNTNER, N., F. TATARUCH, AND O. KRONE. 2001. Heavy metals in soft tissue of White-tailed Eagles found dead or moribund in Germany and Austria from 1993 to 2000. Environmental Toxicology and Chemistry 20: 1831–1837.

KENNTNER, N., F. TATARUCH, AND O. KRONE. 2005. Risk assessment of environmental contaminants in White–tailed Sea Eagles (*Haliaeetus albicilla*) from Germany. Pages 125–127 *in* K. Pohlmeyer (Ed.). Extended Abstracts of the XXVIIth Congress of the International Union of Game Biologists, Hannover 2005. DSV Verlag, Hamburg, Germany.

KENNTNER, N., Y. CRETTENAND, H. J. FÜNFSTÜCK, M. JANOVSKY, AND F. TATARUCH. 2007. Lead poisoning and heavy metal exposure of Golden Eagles (*Aquila chrysaetos*) from the European Alps. Journal of Ornithology 148:173–177.

KEYMER, I. F., AND R. S. STEBBINGS. 1987. Lead poisoning in a Partridge (*Perdix perdix*) after ingestion of gunshot. Veterinary Records 21: 276–277.

KIM, E-Y., R. GOTO, H. IWATA, S. TANABE, Y. MASUDA, AND S. FUJITA. 1999. Preliminary survey of lead poisoning of Steller's Sea Eagle (*Haliaeetus pelagicus*) and White-tailed Eagle (*Haliaeetus albicilla*) in Hokkaido, Japan. Environmental Toxicology and Chemistry 18:448–451.

KNOPPER, L. D., P. MINEAU, A. M. SCHEUHAMMER, D. E. BOND, AND D. T. MCKINNON. 2006. Carcasses of shot Richardson's Ground Squirrels may post lead hazards to scavenging hawks. Journal of Wildlife Management 70:295–299.

KRONE, O., F. WILLE, N. KENNTNER, D. BOERTMANN, AND F. TATARUCH. 2004. Mortality factors, environmental contaminants, and parasites of White-tailed Sea Eagles from Greenland. Avian Diseases 48:417–424.

KRONE, O., T. STIERNBERG, N. KENNTNER, F. TATARUCH, J. KOIVUSAARI, AND I. NUUJA. 2006.

Mortality factors, helminth burden, and contaminant residues in White-tailed Sea Eagles (*Haliaeetus albicilla*) from Finland. Ambio 35:98–104.

KUROSAWA, N. 2000. Lead poisoning in Steller's Sea Eagles and White tailed Sea Eagles. Pages 107–109 *in* M. Ueta and M. J. McGrady (Eds.). First Symposium on Steller's and White-tailed Sea Eagles in East Asia. Wild Bird Society of Japan, Tokyo, Japan.

LANGELIER, K. M., C. E. ANDRESS, T. K. GREY, C. WOOLDRIDGE, R. J. LEWIS, AND R. MARCHETTI. 1991. Lead poisoning in Bald Eagles in British Columbia. Canadian Veterinary Journal 32: 108–109.

LEWIS, J. C., AND E. LEGLER. 1968. Lead shot ingestion by Mourning Doves and incidence in soil. Journal of Wildlife Management 32:476–482.

LEWIS, L. A., R. J. POPPENGA, W. R. DAVIDSON, J. R. FISCHER, AND K. A. MORGAN. 2001. Lead toxicosis and trace element levels in wild birds and mammals at a firearms training facility. Archives of Environmental Contamination and Toxicology 41:208–214.

LOCKE, L. N., AND G. E. BAGLEY. 1967. Lead poisoning in a sample of Maryland Mourning Doves. Journal of Wildlife Management 31: 515–518.

LOCKE, L. N., G. E. BAGLEY, D. N. FRICKIE, AND L. T. YOUNG. 1969. Lead poisoning and Aspergillosis in an Andean Condor. Journal of the American Veterinary Medical Association 155:1052–1056.

LOCKE, L. N., AND M. FRIEND. 1992. Lead poisoning of avian species other than waterfowl. Pages 19–22 *in* D.J. Pain (Ed.). Lead Poisoning in Waterfowl. Proceedings of an IWRB Workshop, Brussels, Belgium, 13–15 June 1991. International Waterfowl and Wetlands Research Bureau Special Publication 16, Slimbridge, UK.

LUMEIJI, J. T., W. T. C. WOLVEKAMP, G. M. BRON-DIETZ, AND A. J. H. SCHOTMAN. 1985. An unusual case of lead poisoning in a Honey Buzzard (*Pernis apivorus*). The Veterinary Quarterly 7:165–168.

MACDONALD, J. W., C. J. RANDALL, H. M. ROSS, G. M. MOON, AND A. D. RUTHVEN. 1983. Lead poisoning in captive birds of prey. Veterinary Record 113:65–66.

MARTIN, P. A., AND G. C. BARRETT. 2001. Exposure of terrestrial raptors to environmental lead: determining sources using stable isotope ratios. Page 84 *in* International Association for Great Lakes Research Conference Program and Abstracts 44, University of Wisconsin—Green Bay, Green Bay, 10–14 June 2001. IAGLR, Ann Arbor, Michigan, USA.

MARTIN, P. A., D. CAMPBELL, AND A. SCHEUHAMMER. 2003. Lead exposure in terrestrial foraging raptors in southern Ontario, 1999–2001. Page 269 *in* International Association for Great Lakes Research Conference Program and Abstracts 46, DePaul University, Chicago, 22–26 June 2003. IAGLR, Ann Arbor, Michigan, USA.

MARTINA, P. A., D. CAMPBELL, K. HUGHES, AND T. MCDANIEL. 2008. Lead in the tissues of terrestrial raptors in southern Ontario, Canada 1951–2001. Science of the Total Environment 391:96–103. Art No. 10.1016/j.scitotenv.2007.11.012\ISSN 0048–9697 2008.

MATEO, R. 1998a. Ingestión de perdigones de plomo en rapaces de las marismas del Guadalquivir. Pages 159–170 *in* La Intoxicación por Ingestión de Perdigones de Plomo en Aves Silvestres: Aspectos Epidemiológicos y Propuestas para su Prevención en España. Doctoral Thesis, Universitat Autònoma de Barcelona, Spain.

MATEO, R. 1998b. La intoxicación por ingestión de objetos de plomo en aves: una revisión de los aspectos epidemiológicos y clínicos. Pages 5–44 *in* La Intoxicación por Ingestión de Perdigones de Plomo en Aves Silvestres: Aspectos Epidemiológicos y Propuestas para su Prevención en España. Doctoral Thesis, Universitat Autònoma de Barcelona, Spain.

MATEO, R., R. CADENAS, M. MANEZ, AND R. GUITART. 2001. Lead shot ingestion in two raptor species from Doñana, Spain. Ecotoxicology and Environmental Safety 48:6–10.

MATEO, R., J. ESTRADA, J-Y. PAQUET, X. RIERA, L. DOMÍNGUEZ, R. GUITART, AND A. MARTÍNEZ-VILALTA. 1999. Lead shot ingestion by Marsh Harriers (*Circus aeruginosus*) from the Ebro Delta, Spain. Environmental Pollution 104:435–440.

MATEO, R., R. MOLINA, J. GRÍFOLS, AND R. GUITART. 1997. Lead poisoning in a free-ranging Griffon Vulture (*Gyps fulvus*). The Veterinary Record 140: 47–48.

MATEO, R., M. TAGGART, AND A. A. MEHARG. 2003. Lead and arsenic in bones of birds of prey from Spain. Environmental Pollution 126:107–114.

MATEO, R., M. RODRIGUEZ-DE LA CRUZ, D. VIDAL, M. REGLERO, AND P. CAMARERO. 2007. Transfer of lead from shot pellets to game meat during cooking. Science of the Total Environment 372:480–485.

MATSUDA, H. 2003. Challenges posed by the precautionary principle and accountability in ecological risk assessment. Environmetrics 14: 245–254.

MCQUIRTER, J. L., S. J. ROTHENBERG, G. A. DINKINS, V. KONDRASHOV, M. MANOLO, AND A. C. TODD. 2004. Change in blood lead concentration up to 1 year after a gunshot wound with a retained bullet. American Journal of Epidemiology 159:683–692.

MEHARG, A. A., D. J. PAIN, R. M. ELLAM, R. BAOS, V. OLIVE, A. JOYSON, N. POWELL, A. J. GREEN, AND F. HIRALDO. 2002. Isotopic identification of the sources of lead contamination for White Storks (*Ciconia ciconia*) in a marshland ecosystem (Doñana, S.W. Spain). Science of the Total Environment 300: 81–86.

MERETSKY, V. J., N. F. R. SNYDER, S. R. BEISSINGER, D. A. CLENDENEN, AND J. W. WILEY. 2000. Demography of the California Condor: implications for reestablishment. Conservation Biology 14:957–967.

MIGLIORINI, M., G. PIGINO, N. BIANCHI, F. BERNINI, AND C. LEONZIO. 2004. The effects of heavy metal contamination on the soil arthropod community of a shooting range. Environmental Pollution 129:331–340.

MILLER, M. J. R., M. E. WAYLAND, E. H. DZUS, AND G. BORTOLOTTI. 2000. Availability and ingestion of lead shotshell pellets by migrant Bald Eagles in Saskatchewan. Journal of Raptor Research 34:167–174.

MILLER, M. J. R., M. E. WAYLAND, E. H. DZUS, AND G. BORTOLOTTI. 2001. Exposure of migrant Bald Eagles to lead in prairie Canada. Environmental Pollution 112:153–162.

MUDGE, G. P. 1992. Options for alleviating lead poisoning: a review and assessment of alternatives to the use of non-toxic shot. Pages 23–25 *in* D.J. Pain (Ed.). Lead Poisoning in Waterfowl. Proceedings of an IWRB Workshop, Brussels, Belgium, 13–15 June 1991. International Waterfowl and Wetlands Research Bureau Special Publication 16, Slimbridge, UK.

NATIONAL WILDLIFE HEALTH LABORATORY (NWHL). 1985. Lead poisoning in non-waterfowl avian species. United States Fish and Wildlife Service Unpublished Report.

NOER, H., AND J. MADSEN. 1996. Shotgun pellet loads and infliction rates in Pink-footed Geese *Anser brachyrhynchus*. Wildlife Biology 2:65–82.

OAKS, J. L., M. GILBERT, M. Z. VIRANI, R. T. WATSON, C. U. METEYER, B. A. RIDEOUT, H. L. SHIVAPRASAD, S. AHMED, M. J. I. CHAUDHRY, M. ARSHAD, S. MAHMOOD, A. ALI, AND A. A. KHAN. 2004. Diclofenac residues as the cause of vulture decline in Pakistan. Nature 2317 Letters: 1–4.

PAIN, D. J. 1990a. Lead shot ingestion by waterbirds in the Camargue, France: an investigation of levels and interspecific differences. Environmental Pollution 66:273–285.

PAIN, D. J. 1990b. Lead poisoning of waterfowl: a review. Pages 172–181 *in* G. Matthews (Ed.). Managing Waterfowl Populations. The International Waterfowl and Wetlands Research Bureau, Slimbridge, UK.

PAIN, D. J. 1991a. Lead poisoning in birds: an international perspective. Acta XX Congressus Internationalis Ornithologici, 2343–2352.

PAIN, D. J. 1991b. Why are lead-poisoned waterfowl rarely seen? The disappearance of waterfowl carcasses in the Camargue, France. Wildfowl 42:118–122.

PAIN, D. J. 1992. Lead poisoning in waterfowl: a review. Pages 7–13 *in* D.J. Pain (Ed.). Lead Poisoning in Waterfowl. Proceedings of an IWRB Workshop, Brussels, Belgium, 13–15 June 1991. International Waterfowl and Wetlands Research Bureau Special Publication 16, Slimbridge, UK.

PAIN, D. J. 1996. Lead in waterfowl. Pages 251–262 *in* W. M. Beyer, G. H. Heinz, and A. W. Redman-Norwood (Eds.). Environmental Contaminants in Wildlife: Interpreting Tissue Concentrations. Lewis Publishers, Boca Raton, Florida, USA.

PAIN, D. J., AND C. AMIARD-TRIQUET. 1993. Lead poisoning in raptors in France and elsewhere. Ecotoxicology and Environmental Safety 25: 183–192.

PAIN, D. J., C. AMIARD-TRIQUET, C. BAVOUX, G. BURNELEAU, L. EON, AND P. NICOLAU-GUILLAUMET. 1993. Lead poisoning in wild populations of Marsh Harrier (*Circus aeruginosus*) in the Camargue and Charente-Maritime, France. Ibis 135:379–386.

PAIN, D. J., C. BAVOUX, AND G. BURNELEAU. 1997. Seasonal blood lead concentrations in Marsh Harriers (*Circus aeruginosus*) from Charente-Maritime, France: relationship with hunting season. Biological Conservation 81:1–7.

PAIN, D. J., A. A. MEHARG, M. FERRER, M. TAGGART, AND V. PENTARIANI. 2004. Lead concentrations in bones and feathers of the globally threatened Spanish Imperial Eagle. Biological Conservation 121:603–610.

PAIN, D. J., J. SEARS, AND I. NEWTON. 1995. Lead concentrations in birds of prey in Britain. Environmental Pollution 87:173–180.

PAIN, D. J., I. CARTER, A. W. SAINSBURY, R. F. SHORE, P. EDEN, M. A. TAGGART, S. KONSTANTINOS, L. A. WALKER, A. A. MEHARG, AND A. RAAB. 2007. Lead contamination and associated disease in captive and reintroduced Red Kites *Milvus milvus* in England. Science of the Total Environment 376:116–127.

PATTEE, O. H., P. H. BLOOM, J. M. SCOTT, AND M. R. SMITH. 1990. Lead hazards within the range of the California Condor. The Condor 92: 931–937.

PATTEE, O. H., AND S. K. HENNES. 1983. Bald Eagles and waterfowl: the lead shot connection. Pages 230–237 *in* 48th North American Wildlife Conference 1983. The Wildlife Management Institute, Washington, DC, USA.

PATTEE, O. H., S. N. WIEMEYER, B. M. MULHERN, L. SILEO, AND J. W. CARPENTER. 1981. Experimental lead-shot poisoning in Bald Eagles. Journal of Wildlife Management 45:806–810.

PLATT, J. B. 1976. Bald Eagles wintering in the Utah desert. American Birds 30: 783–788.

POTTS, G. R. 2005. Incidence of ingested lead gunshot in wild Grey Partridges. European Journal of Wildlife Research 51:31–34.

QUORTRUP, E. R., AND J. E. SHILLINGER. 1941. 3,000 wild bird autopsies on western lake areas. Journal of the American Veterinary Medical Association 99:382–397.

REDIG, P. T., E. M. LAWLER, S. SCHWARTZ, J. L. DUNNETTE, B. STEPHENSON, AND G. E. DUKE. 1991. Effects of chronic exposure to sub-lethal concentrations of lead acetate on heme synthesis and immune function in Red-tailed Hawks. Archives of Environmental Contamination and Toxicology 21:72–77.

REDIG, P. T., C. M. STOWE, D. M. BARNES, AND T. D. ARENT. 1980. Lead toxicosis in raptors. Journal of the American Veterinary Association 177:941–943.

REICHEL, W. L., S. K. SCHMELING, E. CROMARTIE, T. E. KAISER, A. J. KRYNITSSKY, T. G. LAMONT, B. M. MULHERN, R. M. PROUTY, D. J. STAFFORD, AND D. M. SWINEFORD. 1984. Pesticide, PCB and lead residues and necropsy data from Bald Eagles for 32 states—1978–81. Environmental Monitoring and Assessment 4:395–403.

REISER, M. H., AND S. A. TEMPLE. 1981. Effect of chronic lead ingestion on birds of prey. Pages 21–25 *in* J. E. Cooper and A. G. Greenwood (Eds.). Recent Advances in the Study of Raptor Diseases. Chiron Publications, Keighley, UK.

RODRIGUE, J., R. MCNICOLL, D. LECLAIR, AND J. F. DUCHESNE. 2005. Lead concentrations in Ruffed Grouse, Rock Ptarmigan, and Willow Ptarmigan in Quebec. Archives of Environmental Contamination and Toxicology 49:97–104.

RODRIGUEZ-RAMOS, J., V. GUTIERREZ, U. HÖFLE, R. MATEO, L. MONSALVE, E. CRESPO, J. M. BLANCO. 2009. Lead in Griffon and Cinereous Vultures in Central Spain: Correlations between clinical signs and blood lead levels. Extended abstract *in* R. T. Watson, M. Fuller, M. Pokras, and W. G. Hunt (Eds.). Ingestion of Lead from Spent Ammunition: Implications for Wildlife and Humans. The Peregrine Fund, Boise, Idaho, USA. DOI 10.4080/ilsa.2009.0213

SAINSBURY, A. W., P. M. BENNETT, AND J. K. KIRKWOOD. 1995. The welfare of free-living wild animals in Europe: harm caused by human activities. Animal Welfare 4:183–206.

SANDERSON, G. C., AND F. C. BELLROSE. 1986. A review of the problem of lead poisoning in waterfowl. Illinois Natural History Survey, Special Publication 4:1–34.

SCHEUHAMMER, A. M. 1987. The chronic toxicity of aluminium, cadmium, mercury, and lead in birds: a review. Environmental Pollution 46: 263–295.

SCHEUHAMMER, A. M., D. E. BOND, N. M. BURGESS, AND J. RODRIGUE. 2003. Lead and stable isotope ratios in soil, earthworms and bones of American Woodcock (*Scolopax minor*) from eastern Canada. Environmental Toxicology and Chemistry 22:2585–2591.

SCHEUHAMMER, A. M., AND S. L. NORRIS. 1995. A review of the environmental impacts of lead shotshell ammunition and lead fishing weights in Canada. Canadian Wildlife Service, Occasional Paper 88.

SCHEUHAMMER, A. M., AND S. L. NORRIS. 1996. The ecotoxicology of lead shot and lead fishing weights. Ecotoxicology 5:279–295.

SCHEUHAMMER, A. M., J. A. PERRAULT, E. ROUTHIER, B. M. BRAUNE, AND G. D. CAMPBELL. 1998. Elevated lead concentrations in edible portions of game birds harvested with lead shot. Environmental Pollution 102:251–257.

SCHEUHAMMER, A. M., C. A. ROGERS, AND D. BOND. 1999. Elevated lead exposure in American Woodcock (*Scolopax minor*) in eastern Canada. Archives of Environmental Contamination and Toxicology 36:334–340.

SCHEUHAMMER, A. M., AND D. M. TEMPLETON. 1998. Use of stable isotope ratios to distinguish sources of lead exposure in birds. Ecotoxicology 7:37–42.

SCHULZ, J. H., J. J. MILLSPAUGH, B. E. WASHBURN, G. R. WESTER, J. T. LANIGAN, AND J. C. FRANSON. 2002. Spent-shot availability and ingestion on areas managed for Mourning Doves. Wildlife Society Bulletin 30:112–120.

SCHULZ, J. H., P. I. PADDING, AND J. J. MILLSPAUGH. 2006. Will Mourning Dove crippling rates increase with nontoxic-shot regulations? Wildlife Society Bulletin 34:861–865.

SERGEYEV, A. A., AND N. A. SHULYATIEVA. 2005. Meat quality of birds after using lead shot. Pages 472–473 *in* K. Pohlmeyer (Ed.). Extended Abstracts of the XXVIIth Congress of the International Union of Game Biologists, Hannover 2005. DSV Verlag, Hamburg, Germany.

SHARLEY, A. J., L. W. BEST, J. LANE, AND P. WHITEHEAD. 1992. An overview of lead poisoning in Australian waterfowl and implications for management. Pages 73–77 *in* D.J. Pain (Ed.). Lead Poisoning in Waterfowl. Proceedings of an IWRB Workshop, Brussels, Belgium, 13–15 June 1991. International Waterfowl and Wetlands Research Bureau Special Publication 16, Slimbridge, UK.

SILEO, L., L. H. CREEKMORE, D. J. AUDET, M. R. SNYDER, C. U. METEYER, J. C. FRANSON, L. N. LOCKE, M. R. SMITH, AND D. L. FINLEY. 2001. Lead poisoning of waterfowl by contaminated sediment in the Coeur d'Alene River. Archives of Environmental Contamination and Toxicology, 41:364–368.

SILEO, L., AND S. I. FEFER. 1987. Paint chip poisoning of Laysan Albatross at Midway atoll. Journal of Wildlife Diseases 23:432–437.

SNYDER, N. F., AND H. A. SNYDER. 2000. The California Condor. Academic Press, New York, USA.

SNYDER, N. F., H. A. SNYDER, J. L. LINCER, AND R. T. REYNOLDS. 1973. Organochlorines, heavy metals and the biology of North American accipiters. Biosciences 23:300–305.

SORVARI, J., R. ANTIKAINEN, AND O. PYY. 2006. Environmental contamination at Finnish shooting ranges—the scope of the problem and management options. Science of the Total Environment 366:21–31.

STEHLE, V. S. 1980. Orale Bleivergiftung bei Greifvögeln (Falconiformes)—Vorläufige Mitteilung. Kleintier Prakis 25:309–310.

STENDELL, R. C., R. I. SMITH, K. P. BURNHAM, AND R. E. CHRISTENSEN. 1979. Exposure of waterfowl to lead: a nationwide survey of residues in wing bones of seven species, 1972–1973. U.S. Fish and Wildlife Service Special Scientific Report Wildlife 223:1–12.

STENDELL, R. C., J. W. ARTMANN, AND E. MARTIN. 1980. Lead residues in Sora Rails from Maryland. Journal of Wildlife Management 44:525–527.

STEVENSON, A. L., A. M. SCHEUHAMMER, AND H. M. CHAN. 2005. Effects of nontoxic shot regulations on lead accumulation in ducks and American Woodcock in Canada. Archives of Environmental Contamination and Toxicology 48:405–413.

STODDARD, H. L. 1931. The Bobwhite Quail, its habits, preservation and increase. Charles Scribner's Sons, New York, USA.

STONE, W. B., AND S. A. BUTKAS. 1978. Lead poisoning in a wild turkey. New York Fish and Game Journal 25:169.

SZYMCZAK, M. R. 1978. Steel shot use on a goose hunting area in Canada. Wildlife Society Bulletin 6:217–225.

TAVERNIER, P., S. ROELS, K. BAERT, K. HERMANS, F. PASMANS, AND K. CHIERS. 2004. Lead intoxication by ingestion of lead shot in racing pigeons (*Columba livia*). Vlaams Diergeneeskundig Tijdschrift 73:307–309.

THOMAS, V. G. 1997. The environmental and ethical implications of lead shot contamination of rural lands in north America. Journal of Agricultural and Environmental Ethics 10:41–54.

THOMAS, V. G., AND R. GUITART. 2005. Role of international conventions in promoting avian conservation through reduced lead toxicosis: progression towards a non-toxic agenda. Bird Conservation International 15:147–160.

TSUJI, L. J. S., E. NIEBOER, J. D. KARAGATZIDES, R. M. HANNING, AND B. KATAPATUK. 1999. Lead shot contamination in edible portions of game birds and its dietary implications. Ecosystem Health 5:183–192. Art No. ISSN 1076-2825.

UETA, M., AND V. MASTEROV. 2000. Estimation by a computer simulation of population trend of Steller's Sea Eagles. Pages 111–116 *in* M. Ueta and M. J. McGrady (Eds.). First Symposium on Steller's and White-tailed Sea Eagles in East Asia. Wild Bird Society of Japan, Tokyo, Japan.

VEIGA, J. 1985. Contribution à l'étude du regime alimentaire de la Bécassine Sourde (*Lymnocryptes minimus*). Gibier Faune Sauvage 1:75–84.

VEIT, H. P., R. J. KENDALL, AND P. F. SCANLON. 1982. The effect of lead shot ingestion on the testes of adult Ringed Turtle Doves (*Streptophelia risoria*). Avian Diseases 27:442–452.

VON KUSTER, D. D., AND D. SCHNEBERGER. 1992. Diet of the Great Horned Owl in central Saskatchewan. Blue Jay 50: 195–200.

VYAS, N. B., J. W. SPANN, G. H. HEINZ, W. N. BEYER, J. A. JAQUETTE, AND J. M. MENGELKOCH. 2000. Lead poisoning of passerines at a trap and skeet range. Environmental Pollution 107:159–166.

WAYLAND, M., AND T. BOLLINGER. 1999. Lead exposure and poisoning in Bald Eagles and Golden Eagles in the Canadian prairie provinces. Environmental Pollution 104:341–350.

WAYLAND, M., L. K. WILSON, J. E. ELLIOTT, J. E. MILLER, T. BOLLINGER, M. MCADIE, K. LANGELIER, J. KEATING, AND J. M. W. FROESE. 2003. Mortality, morbidity and lead poisoning of eagles in western Canada 1986–1998. Journal of Raptor Research 37:8–18.

WIEMEYER, S. N., R. M. JUREK, AND J. F. MOORE. 1986. Environmental contaminants in surrogates, foods, and feathers of California Condors (*Gymnogyps californianus*). Environmental Monitoring and Assessment 6:91–111.

WIEMEYER, S. N., A. J. KRYNITSKY, AND S. R. WILBUR. 1983. Environmental contaminants in tissues, foods, and feces of California Condors. Pages 4727–4739 *in* S. R. Wilbur and J. A. Jackson, (Eds.). Vulture Biology and Management. University of California Press, Los Angeles, California, USA.

WIEMEYER, S. N., J. M. SCOTT, M. P. ANDERSON, P. H. BLOOM, AND C. J. STAFFORD. 1988. Environmental contaminants in California Condors. Journal of Wildlife Management 52:238–247.

WINDINGSTAD, R. M., S. M. KERR, L. N. LOCKE, AND J. J. HUNT. 1984. Lead poisoning of Sandhill Cranes (*Grus canadensis*). Prairie Naturalist 16:21–24.

WOBESTER, G., AND A. G. WOBESTER. 1992. Carcass disappearance and estimation of mortality in a simulated die-off of small birds. Journal of Wildlife Diseases 28:548–554.

GUNSHOT WOUNDS: A SOURCE OF LEAD IN THE ENVIRONMENT

RICHARD K. STROUD[1] AND W. GRAINGER HUNT[2]

[1]*US Fish & Wildlife Service, National Forensics Laboratory*
1490 East Main Street, Ashland, OR 97520-1310 USA

[2]*The Peregrine Fund, 5668 West Flying Hawk Lane, Boise ID 96056 USA*
E-mail: grainger@peregrinefund.org

ABSTRACT.—Ingested lead shotgun pellets and rifle bullet fragments have been shown to be an important source of lead poisoning in water birds, raptors, avian scavengers, and even seed-eating birds. Ingestion of spent lead shotgun pellets by waterfowl and secondary ingestion by Bald Eagles (*Haliaeetus leucocephalus*) scavenging on waterfowl led to the change of hunting regulations that prohibit the use of toxic lead pellets for waterfowl hunting in the United States. However, bullets containing toxic lead are still widely used to hunt large game animals and "varmints" and are a source of lead in the environment available to wildlife.

Basic bullet materials available to the bullet manufacturer include lead alloys, lead with external copper wash, lead core with copper jacket, pure copper, and bismuth. Lead and bismuth are highly frangible, whereas pure copper bullets tend to remain intact after impact. Bullet fragmentation increases the degree of lead contamination in tissue ingested by scavengers feeding on hunter-killed animal remains. Modern bullet design, velocity, composition, and bone impact are significant factors in the character and distribution of lead particles in carcasses, gut piles, and wound tissue left in the field by hunters. Prior to the 1900s, bullets were made entirely of lead. Their velocities were relatively slow (<2000 feet per second), and their tendency to fragment was accordingly lower than that of modern ammunition. Development of smokeless powder in the 1890s increased bullet speeds above 2000 feet (610 m) per second, causing lead bullets to melt in the barrels and produce fouling which reduced accuracy. Copper jacketed lead-core bullets were therefore developed, which permitted velocities that may exceed 3000 or even 4000 ft/sec in modern firearms. Standard hunting bullets now typically travel at 2600 to 3100 ft/sec, speeds highly conducive to fragmentation. Plastic-tipped "hollow-point bullets" used for varmint hunting are actually designed to completely fragment, leaving the entire mass of the bullet to contaminate the carcass.

Hunters value bullets that are accurate and have adequate impact and destructive power to humanely kill the target animal. To decrease the incidence of lead exposure in wildlife and in humans that consume game animals, alternatives to traditional lead-based bullets have been and are being developed. These bullets must be proven to be both as accurate and as lethal as traditional lead bullets in producing humane kills before they are accepted for general use by the hunting community. Additionally, they must be reasonably priced. *Received 8 September 2008, accepted 19 November 2008.*

STROUD, R. K., AND W. G. HUNT. 2009. Gunshot wounds: A source of lead in the environment. *In* R. T. Watson, M. Fuller, M. Pokras, and W. G. Hunt (Eds.). Ingestion of Lead from Spent Ammunition: Implications for Wildlife and Humans. The Peregrine Fund, Boise, Idaho, USA. DOI 10.4080/ilsa.2009.0109

Key words: Ammunition, ballistics, bullet fragmentation, lead poisoning, hunting, wildlife.

THE TOXICOLOGY AND THE PATHOLOGICAL EFFECTS of ingested lead projectiles are well documented in waterfowl, birds of prey (Locke and Thomas 1996), and avian scavengers such as the endangered California Condor (*Gymnogyps californianus*) (Parish et al. 2007). The institution of non-toxic shot regulations for waterfowl hunting throughout the United States and Canada has made much progress in reducing the occurrence of lead poisoned waterfowl and scavengers that feed on waterfowl carrying lead shot. In recent years, another form of ammunition lead has been recognized as a source of exposure in scavenging birds and even potentially in humans, namely, lead fragments emanating from lead-core rifle bullets used in big game hunting (Hunt et al. 2006) and "varmint hunting" (Knopper el al. 2006, Pauli and Buskirk 2007). The leading cause of death in the California Condor reintroduction effort in Arizona is lead poisoning from scavenged carcasses of big game animals killed with lead core bullets (Cade 2007).

Currently, most high velocity bullets manufactured for centerfire rifles have a copper jacket overlaying a lead core. They are engineered to mushroom in the wound track of the target animal, and in doing so, they tend to disperse lead particles into the wound tissue (Hunt et al. 2006). Such fragmentation is generally thought undesirable for big game hunting because of reduced penetration and meat wastage, but varmint shooters prefer bullets that fragment. In either case, scavenging birds feed on the tissues of bullet-killed animals and ingest the lead fragments dispersing from the bullet path. Acute lead poisoning may result and it is often lethal; moreover, there are questions about possible long-lasting sublethal manifestations, especially from repeated exposure (see Douglas-Stroebel 1992, Gangoso et al. in press).

Some of the lead available to scavenging birds such as eagles and condors may come from the use of lead shot for upland game and other types of shotgun hunting not currently covered in the nontoxic shot waterfowl hunting regulations. However, by far the greater potential sources are rifle-killed mammal carcasses, including "varmints" killed and left in the field (Knopper et al. 2006) and the gut piles and unrecovered carcasses of big game animals killed during the fall hunting seasons. Hunt et al. (2007) and Green et al. (2009, this volume), show the relationship of condor blood lead levels with deer hunting seasons and condor movements into deer hunting areas, demonstrating a direct effect of the use of standard centerfire bullets containing lead cores (Parish et al. 2007).

As in the substitution of nontoxic metals for lead in shotgun shells used for waterfowl hunting, alternative technologies are being developed for environmentally safer rifle bullets for use in other types of hunting. Acceptance of these alternatives by the hunting public is desirable, but may require substantial research and educational outreach to validate and explain the impact of this source of lead in the environment, the functionality and the efficiency of the new forms of bullets, and the potential health implications for humans who eat game animals contaminated with lead fragments.

Our objective is to familiarize the reader with the mechanisms that contribute to the dispersion of lead and copper bullet fragments in a carcass of a big game animal. Understanding the factors that promote bullet fragmentation is basic to the further development of alternative bullets with decreased potential for poisoning wildlife and humans who eat hunter harvested game animals.

DEVELOPMENT OF MODERN HUNTING BULLETS

The history of the development of lead core bullets and their increasing tendency to fragment into smaller particles with a greater potential for ingestion began near the close of the 19th Century with the development of modern high velocity centerfire rifles that employed smokeless gunpowder. Prior to that, black powder rifles used solid lead bullets that traveled at less than 1600 ft/sec (488 m/sec). These relatively slow-moving bullets simply deformed or sometimes broke apart into relatively large pieces when they struck the target animal. Today, shotgun slugs, travelling at these slower velocities, are used in many near-urban areas for deer hunting, and black powder rifles have resurged in popularity for big game hunting throughout the country. Because of the lower velocity and construction of bullets used, bullets of neither weapon fragment to the degree exhibited by higher velocity centerfire rifle bullets.

The development of bullets with a copper jacket over a lead core used in modern high velocity hunting rifles followed the development of smokeless powders that produced greater volumes of gases and whose burning rates could be adjusted to specific needs by the manufacturer. The original intent of the copper jacket over the lead projectile was to prevent the lead fouling of the rifling of the gun barrel, which decreased the accuracy of the rifle. This occurred after a few shots due to the "melting" of the lead from the bullet by the higher pressures and barrel friction produced by the larger volumes of gas. The combination of smokeless powder and copper jacketing produced bullets that travelled at much higher velocities and with flatter trajectories than those of earlier rifles. The original bullets for use with smokeless powder were developed for the military and were fully-jacketed, meaning that the lead core was completely encased in copper.

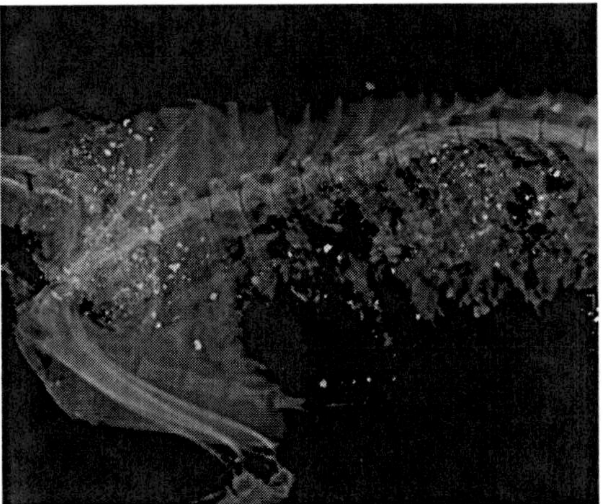

Figure 1. Radiograph of "lead snowstorm" in rifle-killed coyote.

Soft-nosed or semijacketed bullets with lead cores were soon developed for hunting large game mammals such as deer and elk. These bullets were designed to mushroom so as to maximize the wounding potential and provide for quicker, more humane kills. The incomplete jacket exposed a portion of the lead core on the nose of the bullet to the forces of tissue impact. Fragmentation was an emergent property of greater velocities and mushrooming designs. Later, however, hollow point bullets were designed specifically for the "varmint shooter" to fragment completely when striking small thin-skinned animals such as rodents, rabbits and coyotes. Plastic or metallic inserts in the nose of the bullet are often incorporated into the design to streamline the bullet and to act as a wedge to facilitate expansion or fragmentation on impact. The striking of bone early in the wound path may cause additional malformation and fragmentation of the bullet. The result from soft points and hollow points is a "snowstorm" of small to minute, irregular metal fragments throughout the wound channel (Fig. 1). The remaining, partially fragmented, deformed base of the soft-point bullet may or may not exit into the environment.

Hunting cartridges are now available in a wide variety of sizes and calibers and contain bullets of various weights and designs. Each was developed for a different perceived need of the hunting community. For example, a 55-grain (3.6 g) bullet for a .22-caliber rifle barrel is available in the ubiquitous .22-caliber "long rifle" rimfire cartridge and with a normal velocity of <1300 ft/sec, but also for the very high velocity .223 centerfire magnum cartridge whose bullet travels up to 4000 ft/sec. The lower velocity bullet leaves relatively few fragments in the wound compared to the numerous very small fragments of the high velocity bullet that literally "explodes" when the target animal is hit. These high velocity small caliber bullets are used primarily for shooting "varmints," the carcasses of which are usually left in the field and available to scavengers (Knopper et al. 2006, Pauli and Buskirk 2007).

Cartridges used for big game hunting are generally loaded with larger bullets weighing 100–225 grains and with muzzle velocities of 2500–3100 ft/sec, depending on the intended use. Most are designed to mushroom yet retain most of their original weight as they penetrate the tissues of the game animal. Virtually all modern lead-based bullets fragment to some extent, most of them shedding 30% or more of their original mass. As a consequence, the abandoned gut piles of rifle-killed big game animals, as well as carcasses lost to wounding, typically contain numerous bullet fragments accessible to scavengers (Hunt et al. 2006).

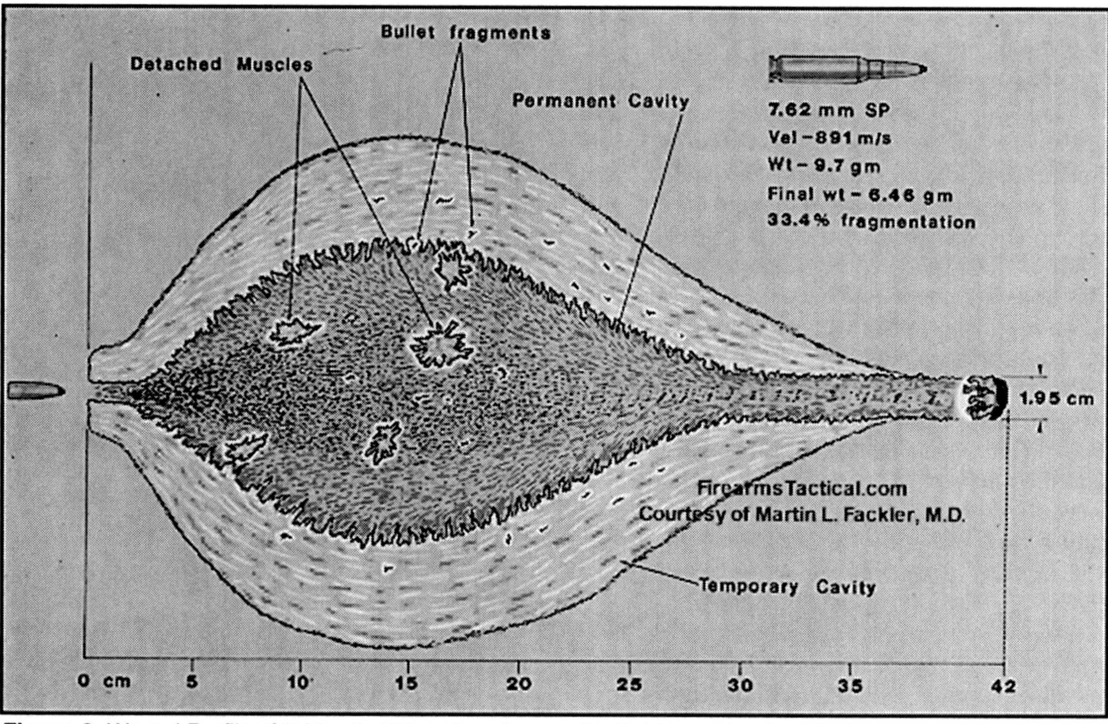

Figure 2. Wound Profile of 7.62 cal, 150 grain soft-nosed bullet (from Fackler 1986).

WOUND BALLISTICS: THE MECHANISM OF BULLET WOUND PRODUCTION

The study of ballistics includes the factors that influence the flight of the bullet from the end of the gun barrel to the target. Hunters strive to kill humanely. To this end, the two essential aspects of bullet performance are accuracy and lethality. The former proceeds from the ignition of gunpowder in the cartridge case to the pressurization of gases propelling the bullet through the rifling of the barrel. These events send the bullet spinning through the air as unerringly as possible to the precise point of aim, namely a center of vital processes, the destruction of which will produce a quick and painless death. Wound ballistics, in contrast, is the study of the impact of the bullet and its penetration of tissue. The velocity and mass of the bullet determine the amount of kinetic energy potentially imparted to the target animal (Fackler 1986). The relevant velocity therefore is that of the bullet as it strikes the target, not the velocity at which the bullet leaves the muzzle of the gun. Distance to the target decreases the velocity to some extent, and this decrease may be proportionately greater for larger projectiles or those not aerodynamically designed.

$$\text{Kinetic Energy} = \text{Mass} \times \text{Velocity}^2$$

Velocity is exponential, whereas the mass of the bullet is linear in the equation that describes the amount of kinetic energy available within the bullet to transfer to the target animal. Therefore, the velocity of the bullet has proportionally much more influence than the weight of the bullet in the wounding process. The mass retained by the bullet after initial penetration of the skin and partial fragmentation is an important factor in the depth of penetration of the bullet along the wound track; a bullet that loses mass through fragmentation shows less penetration, whereas the tendency of the bullet to mushroom, tumble, and fragment while traveling through the tissue are factors influencing the amount of tissue destruction. All these factors are influenced by bullet construction, namely, the internal components and the thickness and extent of coverage of the copper jacket.

The biological significance of the smaller fragments from higher velocity bullets versus larger bullet fragments in wound tissue or animal carcasses is that smaller irregular lead particles have a higher surface area relative to the mass of the fragment; they are therefore more likely to be quickly eroded by the gastric acids of the animal ingesting them and thus more efficiently absorbed within the intestines. Additionally, large fragments may have a greater likelihood than smaller ones of being regurgitated or avoided altogether by raptors.

Wounds in the carcass of an animal hit by a high velocity bullet are produced by three separate mechanisms. As the semijacketed or soft-nosed bullet passes through tissue, it disperses its energy, a product of the mass and velocity, into the surrounding tissue. This mechanism produces a radiating tissue expansion perpendicular to the path of the bullet called the "temporary wound cavity" (Figure 2). Tissue is torn by this wave of energy that is often referred to as hydrostatic pressure. Tissue such as liver, which has poor elasticity, tears and fragments whereas lung tissue is highly elastic and may expand and contract without tearing. This may be visualized in a slow-motion film of a bullet entering a tissue-simulating gel block and showing characteristics like ripples in a pool around a pebble striking the water. As the bullet passes through tissue and starts to mushroom and break apart, secondary projectiles consisting of bullet fragments and disrupted bone fragments cause additional damage alongside the bullet path by lacerating tissue. The third mechanism is the crushing of tissue along the wound path by the mushroomed and tumbling bullet itself. This is referred to as the "permanent wound cavity."

Translating this into a wound path through a carcass is more complicated. Differences in elasticity in various tissues account for variable wound path damage profiles. A bullet that fragments (soft-nosed) compared to a non-fragmenting bullet (solid or fully jacketed) of equal mass and velocity has a larger permanent wound cavity that, in theory, would cause a more extensive wound if it penetrated to the same degree (Fackler 1986). Factors making the study of animal wounds additionally complex are the sizes of target animals and the specific tissues encountered along the wound path. Impacted bone, for example, may increase the degree of fragmentation and produce secondary projectiles (bone fragments). In summary, evaluation of any given bullet's effectiveness in killing the targeted animal is often a very subjective process.

Generally speaking, the slower the projectile, the narrower the wound path. A slower-moving shotgun slug will not cause the same large diameter temporary wound cavity and exit-wound as a high velocity centerfire bullet as it passes through a carcass of similar size, even though the shotgun slug may be significantly larger in mass and could cause a larger permanent wound channel than a smaller high velocity bullet. For this and for reasons of accuracy and trajectory, hunters have increasingly preferred higher velocity bullets. Some hunters argue that the tendency of such bullets to fragment makes them kill more humanely than those that only mushroom. However, in addition to their greatly reduced penetration, the reduced bullet core may divert into nonvital areas. Finally, there is a greater potential for meat wastage though the production of bloodshot, lead-infused tissue. Accordingly, there have been numerous design innovations to prevent fragmentation and maintain bullet mass. Among these, are (1) "bonding" techniques whereby the copper jacket is soldered or otherwise made to adhere to the lead core, (2) copper partitions placed between the rear of the bullet and its mushrooming nose, (3) reductions of the lead core and its encapsulation in a thin steel sheath, and (4) construction of bullets made entirely of copper.

FUTURE DIRECTIONS

Copper bullets are now available that do not tend to fragment on impact and do not contain the lead core common to most hunting bullets. These bullets are designed to mushroom and expend most of their kinetic energy within the carcass so as to humanely and effectively kill large game animals. Interestingly, the development of copper expanding bullets had little to do with environmental or health considerations, but rather with the desire among big game hunters for deep-penetrating non-fragmenting bullets. Reports on the accuracy and killing performance of copper expanding bullets in hunting have been highly favorable (McMurchy 2003, Towsley 2005, Sullivan et al. 2007). The Barnes Bullet Company has been the pioneer of this technology, but other companies are following suit. The use of these bullets in areas of high use by endangered California Condors with seasonal big game hunting and varmint shooting has promise in reducing the exposure to lead fragments in their diet and those of other scavengers (Sieg et al. 2009, this volume). Copper bullets also show promise in reducing the contamination of game meat for human consumption (Cornatzer et al. 2009, this volume). Other technologies may eventually have application for hunting, including variations of the so-called "green bullet" developed by the military. This bullet substitutes a compressed metallic (nonlead) matrix for the lead bullet core. These bullets have been extensively used at practice shooting ranges and for some urban military combat situations.

Education of the hunting public concerning the impacts on wildlife of lead in the environment is an essential component of wildlife conservation (Friend 2009, this volume). The research showing that steel shot was a reasonable substitute for lead shot for waterfowl was extensive, and once proven scientifically, there followed an educational program to inform the hunting public on how to use steel shot effectively for waterfowl hunting. Voluntary use of copper bullets in behalf of condors in Arizona is promising, and follow-up questionnaires are showing that hunters are finding the new copper expanding bullets comparable in lethality and accuracy for Mule Deer and elk (Sullivan et al. 2007). Any significant changes in hunting regulations designed to reduce toxic lead availability to wildlife should be preceded by research into wound ballistics to validate that non-lead bullets are effective and humane in killing the targeted animal and accurate in existing rifles. Educational programs should also explain the negative impacts of toxic lead in the environment as well as the potential for human health impacts, especially to fetuses and children.

LITERATURE CITED

CADE, T. 2007. Exposure of California Condors to lead from spent ammunition. Journal of Wildlife Management 71:2125–2133.

CORNATZER, W. E., E. F. FOGARTY, AND E. W. CORNATZER. 2009. Qualitative and quantitative detection of lead bullet fragments in random venison packages donated to the Community Action Food Centers of North Dakota, 2007. *In* R. T. Watson, M. Fuller, M. Pokras, and W. G. Hunt (Eds.). Ingestion of Lead from Spent Ammunition: Implications for Wildlife and Humans. The Peregrine Fund, Boise, Idaho, USA. DOI 10.4080/ilsa.2009.0111

DOUGLAS-STROEBEL, E. K., G. L. BREWER, AND D. J. HOFFMAN. 2005. Effects of lead-contaminated sediment and nutrition on Mallard duckling behavior and growth. Journal of Toxicology and Environmental Health 68:113–128.

FACKLER, M. L. 1986. Ballistic Injury. Annals of Emergency Medicine 15:1451–1455.

FRIEND, M., J. C. FRANSON, AND W. L. ANDERSON. 2009. Biological and societal dimensions of lead poisoning in birds in the USA. *In* R. T. Watson, M. Fuller, M. Pokras, and W. G. Hunt (Eds.). Ingestion of Lead from Spent Ammunition: Implications for Wildlife and Humans. The Peregrine Fund, Boise, Idaho, USA. DOI 10.4080/ilsa.2009.0104

GANGOSO, L., P. ALVAREZ-LLORET, A. A. B. RODRIGUEZ-NAVARRO, R. MATEO, F. HIRALDO, AND J. A. DONAZAR. 2008. Long-term effects of lead poisoning on bone mineralization in vultures exposed to ammunition sources. Environmental Pollution, DOI: 10.1016/j.envpol.2008.09.015.

GREEN, R. E., W. G. HUNT, C. N. PARISH, AND I. NEWTON. 2009. Effectiveness of action to reduce exposure of free-ranging California Condors in Arizona and Utah to lead from spent ammunition. *In* R. T. Watson, M. Fuller, M. Pokras, and W. G. Hunt (Eds.). Ingestion of Lead from Spent Ammunition: Implications for Wildlife and Humans. The Peregrine Fund, Boise, Idaho, USA. DOI 10.4080/ilsa.2009.0218

HUNT, W. G., W. BURNHAM, C. N. PARISH, K. BURNHAM, B. MUTCH, AND J. L. OAKS. 2006. Bullet fragments in deer remains: Implications for lead exposure in scavengers. Wildlife Society Bulletin 34:168–171.

HUNT, W. G., C. N. PARISH, S. C. FARRY, T. G. LORD, R. SIEG. 2007. Movements of introduced California Condors in Arizona in relation to lead exposure. Pages 79–96 *in* A. Mee, L. S. Hall and J. Grantham (Eds.). California Condors in the 21st Century. Series in Ornithology, no. 2. American Ornithologists Union, Washington, DC, and Nuttall Ornithological Club, Cambridge, Massachusetts, USA.

LOCKE, L. N., AND N. J . THOMAS. 1996. Lead poisoning in raptors and waterfowl. Pages 108–117 *in* A. Fairbrother, L. N. Locke, and G. L. Hoff (Eds.). Noninfectious Diseases of Wildlife, second edition. Iowa State University Press, Ames, Iowa, USA.

KNOPPER, L. D., P. MINEAU, A. M. SCHEUHAMMER, D. E. BOND, AND D. T. MCKINNON. 2006. Carcasses of shot Richardson's ground squirrels may pose lead hazards to scavenging hawks. Journal of Wildlife Management 70 (1):295–299.

MCMURCHY, I. 2003. Barnes XLC bullets. American Hunter 31 (January):70–71.

PARISH, C. N., W. R. HEINRICH, AND W. G. HUNT. 2007. Lead exposure, diagnosis, and treatment in California Condors released in Arizona. Pages 97–108 *in* A. Mee, L. S. Hall and J. Grantham (Eds.). California Condors in the 21st Century. Series in Ornithology, no. 2. American Ornithologists Union, Washington, DC, and Nuttall Ornithological Club, Cambridge, Massachusetts, USA.

PAULI, J. N. AND S. W. BUSKIRK. 2007. Recreational shooting of prairie dogs: A portal for lead entering wildlife food chains. Journal of Wildlife Management 71(1)103–108.

SIEG, R., K. SULLIVAN, AND C. N. PARISH. 2009. Voluntary lead reduction efforts within the northern Arizona range of the California Condor. *In* R. T. Watson, M. Fuller, M. Pokras, and W. G. Hunt (Eds.). Ingestion of Lead from Spent Ammunition: Implications for Wildlife and Humans. The Peregrine Fund, Boise, Idaho, USA. DOI 10.4080/ilsa.2009.0309

SULLIVAN, K., R. SIEG, AND C. PARISH. 2007. Arizona's efforts to reduce lead exposure in California Condors. Pages 100–122 *in* A. Mee, L. S. Hall and J. Grantham (Eds.). California Condors in the 21st Century. Series in Ornithology, no. 2. American Ornithologists Union, Washington, DC, and Nuttall Ornithological Club, Cambridge, Massachusetts, USA.

TOWSLEY, B. M. 2005. The hunting bullet redefined. American Rifleman 153 (December):34.

HUMAN EXPOSURE TO LEAD FROM AMMUNITION IN THE CIRCUMPOLAR NORTH

LORI A. VERBRUGGE[1], SOPHIE G. WENZEL[1], JAMES E. BERNER[2], AND ANGELA C. MATZ[3]

[1]*Alaska Department of Health and Social Services, Division of Public Health, Section of Epidemiology, 3601 C Street Suite 540, Anchorage, AK 99503, USA.*
E-mail: lori.verbrugge@alaska.gov

[2]*Alaska Native Tribal Health Consortium, Division of Community Health Services, 4000 Ambassador Drive, Anchorage, AK 99508, USA.*

[3]*US Fish and Wildlife Service, 101 – 12th Avenue, Box 19, Room 110, Fairbanks, AK 99701, USA.*

ABSTRACT.—Circumpolar subsistence cultures use firearms, including shotguns and rifles, for hunting game for consumption. Lead shot is still used for waterfowl and seabird hunting in many subsistence areas (despite lead shot bans) because it is inexpensive, readily available, and more familiar than non-toxic or steel shot, which shoot differently. Here we review published literature on lead concentrations and lead isotope patterns from subsistence users in the circumpolar North, indicating that elevated lead exposure is associated with use of lead ammunition. Mechanisms of exposure include ingestion of lead dust, ammunition fragments, and shot pellets in harvested meat, and inhalation of lead dust during ammunition reloading. In Alaska, ammunition-related lead exposures have also been attributed to the use of certain indoor firing ranges, and the melting and casting of lead to make bullets. Since there is no safe lead exposure limit, especially for children, use of lead shot and bullets in subsistence cultures results in unnecessary and potentially harmful lead exposure. In order for lead ammunition to be feasibly phased out, alternatives must be affordable and readily available to subsistence hunters. Community outreach, including describing the harmful effects of even small amounts of lead, especially in children and women of child-bearing age, and training on the different shot patterns, velocities, and distances inherent in using shot and bullet materials other than lead, will also be necessary to promote acceptance of alternatives to lead ammunition. *Received 15 September 2008, accepted 3 October 2008.*

VERBRUGGE, L. A., S. G. WENZEL, J. E. BERNER, AND A. C. MATZ. 2009. Human exposure to lead from ammunition in the circumpolar north. *In* R.T. Watson, M. Fuller, M. Pokras, and W.G. Hunt (Eds.). Ingestion of Lead from Spent Ammunition: Implications for Wildlife and Humans. The Peregrine Fund, Boise, Idaho, USA. DOI 10 .4080/ilsa.2009.0110

Key words: Alaska, ammunition, arctic, game, human, hunters, lead, subsistence, waterfowl.

HUMANS IN THE NORTH have been exposed to lead from many of the same sources as in temperate regions. In the 20th century, the greatest exposure was inhalation of atmospherically transported lead produced from leaded gasoline. Other atmospheric sources included combustion of other fossil fuels, particularly coal, non-ferrous metal production (mining, smelting), and waste incineration (AMAP

[3] The findings and conclusions in this document are those of the author(s) and do not necessarily represent the views of the US Fish and Wildlife Service.

2004, AMAP 2002). Lead leachate from lead solder used in food cans may have poisoned the crews of *Erebus* and *Terror*, the ships of the 1850s Franklin expedition to the North Pole (Bayliss 2002). Interestingly, lead solder for canning wasn't banned in the United States until 1995 (Federal Register 60(123): 33106-9), and may still be used elsewhere. Ingestion of lead-based paint chips by children remains an issue worldwide, although in abatement with regulation of leaded paint.

With control of these lead sources, however, blood lead levels in humans have dropped over the past few decades. A phase-out of leaded gas beginning in the 1980s, for example, resulted in a substantial decline in lead levels in humans in North America (Pirkle et al. 1994) and Greenland (Hansen et al. 1991), as well as in snow from Greenland (Robinson 1981) and in the Arctic ice pack. The prevalence of blood lead levels ≥10 μg/dL dropped from over 80% before 1980 to less than 10% in the 1990s (Pirkle et al.1998).

Still, some northern populations, especially indigenous peoples dependent upon subsistence foods, continue to have elevated blood lead levels. A primary source is thought to be lead from ammunition, by ingestion of lead fragments in game shot with lead, inhalation of fumes from home production of shot or sinkers (as in rural areas in Russia; AMAP 2004), and inhalation of dust or particles during prolonged shooting. In fact, the Arctic Monitoring and Assessment Programme stated:

Lead levels in Arctic indigenous peoples have declined since the implementation of controls on lead emissions. Concentrations of lead in blood currently reported are below a level of concern, however, continued monitoring is warranted because of the potent effects of lead on neurological development in the fetus and children (AMAP 1998).

This is still valid. In addition, recent data have shown that lead shot can be a significant source of human exposure (AMAP 2003).

Lead is exceptionally dense, making it ideal for projectiles. It is also relatively soft, which allows it to be formed, even in home environments, into a variety of bullet and shot gauges. This malleability also results in fracturing of the shot and bullets. The latter can leave macro- and microscopic traces of lead on average 15 cm from bullet pathways in meat (Hunt et al. 2006) and spread over an average of 24 cm and up to 45 cm apart (Hunt et al. 2009). Therefore, even if game is carefully cleaned and damaged meat discarded, embedded and invisible fragments of lead may still contaminate the meat (Stroud and Hunt 2009, Hunt et al. 2009).

In this paper we review data on lead concentrations in people living in the circumpolar north and evaluate lead from ammunition as an important source for current lead exposure. We conclude that exposure to lead from ammunition is unnecessary and potentially harmful to Arctic indigenous populations.

Review of Lead Toxicology

Absorption.—Lead can enter the human body through three main routes of exposure: eating, breathing, or being shot. The third route has obvious health consequences and will not be discussed further.

People can ingest lead that is present in their immediate environment, such as dust, or that is in food or water. Leachate from lead solder use in canned foods has already been discussed. Wild game that has been shot with lead ammunition can contain lead fragments, particles or dust that is consumed along with the meat. Lead can also be ingested if people handle lead products such as fishing sinkers, and then fail to wash their hands before eating food. Children often ingest lead when they mouth lead-containing toys or objects, or suck their fingers after touching lead objects or lead-containing dust or soil.

In humans, the percentage of lead that is absorbed into the bloodstream after oral ingestion is influenced by several factors, including age. Gastrointestinal absorption of water-soluble lead appears to be higher in children than in adults (ATSDR 2007). Estimates derived from dietary balance studies indicate that children (ages two weeks to eight years) absorb approximately 40–50% of ingested water-soluble lead, while non-fasting adults absorb only 3–10% of ingested water-soluble lead (ATSDR

2007). Nutritional status also affects gastrointestinal absorption of lead; fasting status increases lead absorption. The presence of food in the gastrointestinal tract lowers lead absorption, especially if calcium or phosphate is present in the meal. Children who have calcium or iron deficiencies have a higher absorption of lead from the gastrointestinal tract (ATSDR 2007).

Exposure to lead through inhalation can occur in a variety of ways. When lead is melted to make fishing sinkers, ammunition or other products, especially in a home environment, dangerous levels of lead fumes can be produced and inhaled. Lead can also be inhaled on dust particles, contaminated soils, or via occupational exposure in manufacturing and mining. When leaded gasoline is combusted, tetraalkyl lead is an inhalable byproduct.

Amounts and patterns of deposition of particulate aerosols in the respiratory tract are affected by the size of the inhaled particles, age-related factors that determine breathing patterns (e.g., nose vs. mouth breathing), airway geometry, and airstream velocity within the respiratory tract (ATSDR 2007). Absorption of deposited lead is influenced by particle size and solubility. Larger particles (>2.5 microns) that are deposited in the upper airways can be transferred by mucociliary transport into the esophagus and swallowed. Smaller particles (<1 micron) can be deposited deeper into the lungs including the alveolar region, where intimate contact with the bloodstream enhances absorption (ATSDR 2007).

Distribution and Excretion.—The excretory half-life of lead in blood is approximately 30 days for adult humans (ATSDR 2007). Lead that is retained by the body is mostly stored in bone, where it is assimilated due to its chemical similarity to calcium (AMAP 2002). Lead can be mobilized from bone and released into the bloodstream during the process of bone resorption. Mobilization of bone lead can occur during pregnancy and lactation, and after menopause due to osteoporosis (ATSDR 2007). Lead in a pregnant mother's blood is effectively transferred to the fetus, and maternal lead can also be transferred to infants during breastfeeding (ATSDR 2007).

Toxicity.—Lead poses a greater risk to children than to adults for several reasons. Lead is more toxic to children than to adults because the nervous system of children is still developing. Also, children absorb a greater percentage of the lead they are exposed to (ATSDR 2007), and children are often exposed to more lead than adults. Children play outdoors and sometimes engage in hand-to-mouth behaviors that increase their exposure potential. The crawling and mouthing behaviors of older infants and young toddlers place them at particular risk for exposure; blood lead levels (BLLs) in children typically peak at the age of two years for this reason (American Academy of Pediatrics 2005). Children are shorter than are adults; this means they breathe dust, soil, and vapors close to the ground. A child's lower body weight and higher intake rate results in a greater dose of hazardous substance per unit of body weight. If toxic exposure levels are high enough during critical growth stages, the developing body systems of children can sustain permanent damage. Children's brains are developing rapidly during the first six years of life, which is why exposure to a chemical like lead that targets the brain is most devastating at that critical time.

Lead can delay or impair brain development in children and adversely affect IQ, and impair a child's ability to learn. Lead can also cause anemia and impaired metabolism of vitamin D. The Centers for Disease Control and Prevention (1997) recognized BLLs of ≥10 µg/dL in children aged ≤6 years as levels of concern, and based on studies since then, the CDC now recognizes that 10 µg/dL does not define a lower threshold for the harmful effects of lead (Brown 2007). Multiple studies have shown that as blood lead concentrations increase, IQ decreases, for example, by 7.4 points as blood lead increased from 1 to 10 µg/dL in children up to five years old (Canfield et al. 2003), and with significantly higher rates of intellectual decrement in children with maximal BLL <7.5 µg/dL than ≥7.5 µg/dL (Lanphear et al. 2005). Thus, BLLs less than 10 µg/dL are clearly harmful, and there is growing consensus that there is no "safe" level of lead exposure. Other adverse health effects associated with relatively low BLLs in children include delayed sexual maturation, increased blood pressure, depressed renal glomerular filtration rate, and inhibition of pathways in heme synthesis (ATSDR 2007).

As BLLs rise in children, the harmful health effects of lead become more severe. A child exposed to a large amount of lead may develop anemia, kidney damage, colic, muscle weakness, and brain damage, which can ultimately kill the child (ATSDR 2007). Such symptoms of clinical lead poisoning are commonly observed in children with BLLs of 45 μg/dL or higher; children with BLLs of 70 μg/dL or higher should be hospitalized immediately for treatment (Centers for Disease Control and Prevention 2002).

Studies have reported adverse health effects in adults with blood lead levels between 25–40 μg/dL, including hypertension, subtle or sub-clinical central nervous system deficits, and adverse reproductive outcomes (Centers for Disease Control and Prevention 2002). Lead exposure is clearly related to elevated blood pressure, and may also cause negative clinical cardiovascular outcomes and impaired performance on cardiovascular function tests (Navas-Acien et al. 2007). Cardiovascular and renal effects have been seen in adults chronically exposed to lead at levels <5 μg/dL in blood, and no lower threshold has been established for any lead-cardiovascular association (Navas-Acien et al. 2007).

At high levels of lead exposure, the brain and kidney in adults or children can be severely damaged, and death can result. High levels of lead exposure may also cause miscarriage in pregnant women, and affect testicular hormones in men. Other symptoms of lead poisoning in adults include colic, anemia, and muscle weakness. Clinical symptoms of lead poisoning can occur in adults with BLLs above 40 μg/dL (ATSDR 2007).

Human Exposure to Lead in the Arctic

Research on human lead exposure in the Arctic in the last decade has linked elevated lead exposure to use of lead shot or bullets for hunting. Other lead exposures of prior importance have largely been controlled, such as lead-based paints, lead in drinking water, and lead from gasoline. Leaded gasoline was phased out from North American use in the 1980s, with subsequent declines in environmental levels, including blood lead in humans (AMAP 1998, AMAP 2003, Van Oostdam et al. 2003). The exception may be in northern Russia, where industrial contamination from mining and smelting of lead ores, and use of lead-containing gasoline, continues (AMAP 2003). However, populations in Russia who practice subsistence hunting, such as people on the Kola Peninsula, are probably also exposed to lead from ammunition (AMAP 2003, Odland et al. 1999).

Specific studies of lead exposure from lead shot began decades ago with documentation of residual (embedded or ingested) lead in waterfowl. Embedded lead shot were found in 18–45% of waterfowl, depending upon the species, tested in the USA, Canada, and Western Europe in the 1950s (Elder 1955). In Canada in the 1980s, 15% of 227 pooled breast muscle samples from waterfowl harvested with lead shot had lead concentrations >0.5 mg/kg (Canadian Wildlife Service unpublished data, cited in Scheuhammer and Norris 1995), and Frank (1986) found lead concentrations, some >100 μg/kg, in tissues of waterfowl harvested with lead shot. These fragments, confirmed by radiographs and ranging in size from dust to 1–2 mm, resulted from collision of shot with bone. In the mid-1990s, Hicklin and Barrow (2004) used fluoroscopy on live Canada Geese (*Branta canadensis*), American Black Ducks (*Anas rubripes*), Mallards (*A. platyrhynchos*) and Common Eiders (*Somateria mollissima*) from eastern Canada. Twenty-five percent of 1,624 birds had embedded shot, most of which was assumed to be lead. From 15–29%, depending upon age, of over 700 Common Eiders collected in western Greenland after colliding with boats or drowning in fishing nets had embedded lead shot in them (Merkel et al. 2006). It is clear that both micro- and macroscopic lead particles remain in avian meat that has been shot with lead pellets (Scheuhammer et al. 1998) and in large mammals shot with lead-based rifle bullets (Hunt et al. 2006). Therefore, lead from ammunition is a potential public health concern for indigenous peoples (Tsuji et al. 1999) and others who depend on wild game for food.

In a study specifically designed to examine the link between lead shot use for subsistence hunting of birds and potential human exposure, Johansen et al. (2001) x-rayed 50 Thick-billed Murre (*Uria lomvia*) carcasses bought from hunters in Greenland. The birds had been harvested with lead shot, and had an average of 3.7 lead pellets per carcass (range 0–12). There was no correlation between the number of pellets and the lead concentration in meat, which ranged from 0.0074–1.63 ppm wet weight, although most lead found in the breast meat was from pellets that had gone through the meat and left fragments. The authors concluded that even after pellets were removed, lead shot fragmented to fine dust upon collision with bone, resulting in substantially greater (although variable) lead concentrations in murres shot with lead compared to those shot with steel. They estimated a potential dose of 50 μg of lead from eating one bird. An estimated 200,000 murres are harvested annually in Greenland, in addition to other seabirds and waterfowl. The authors concluded that using lead shot to hunt birds could be a significant public health concern (Johansen et al. 2001).

A variety of raptor species have been exposed to or poisoned by lead from predating or scavenging lead-shot game (Hunt et al. 2006) and waterfowl (Pattee and Hennes 1983, Elliott et al. 1992, Pain et al. 1993, Kendall et al. 1996, Miller et al. 1998, Mateo et al. 1999, Samour and Naldo 2002, Pain et al. 2009). Therefore, it is not surprising that people who consume game shot with lead can also have elevated blood lead levels. Numerous studies at both the population and individual levels have implicated and linked lead ammunition to elevated blood lead levels and clinical symptoms in northern peoples.

For example, blood lead levels were monitored in 50 male hunters in Greenland before, during, and after the bird-hunting season in order to establish the association between bird consumption and blood lead concentrations (Johansen et al. 2006). Frequency of reported bird consumption was strongly associated with measured BLLs in the hunters, and eider meals were more important than murre meals as a lead source in the blood. Mean BLLs (12.8 μg/dL) were more than eight times higher in the group reporting more than 30 bird meals per month than in the group reporting no bird consumption (1.5 μg/dL).

At the population level, the Dene/Métis and bird-hunting Inuit in Canada averaged from 3.1–5.0 μg/dL of lead in maternal blood, compared to 1.9–2.2 μg/dL among Caucasians and other Inuit (Van Oostdam et al. 2003). However, 3.4% and 2.2% of the blood samples from the Inuit and Dene/Métis women, respectively, exceeded the 10.0 μg/dL Canadian Action Level (Walker et al. 2001). In Greenland, blood lead levels in Inuit mothers averaged 3.1–5.0 μg/dL, similar to the Canadian Inuit and Dene/Métis (AMAP 2003). In Siberia, indigenous women had average blood lead levels of 2.1–3.2 μg/dL, while non-indigenous women, who presumably obtained a smaller proportion, if any, of their food from hunting, averaged 0.02–0.04 μg/dL (AMAP 2003). In Nunavik (Arctic Quebec), adult Inuit blood lead levels were elevated and were related to age, smoking and, in particular, daily consumption of waterfowl (Dewailly et al. 2001). Blood lead, adjusted for age and sex, was associated with seabird consumption in Greenland (Bjerregaard et al. 2004). In that study, Greenlanders who reported consuming sea birds several times a week had a blood lead level >50% higher than those who reported eating sea birds only a few times a month or less.

Lead shot exposure and effects have also been documented at the individual level in northern humans. For example, Madsen et al. (1988) noted that lead shot in the appendix were often seen in lower abdominal x-rays in Denmark, and those with lead in the appendix had greater blood lead concentrations. Of 132 randomly selected radiographic charts from a hospital serving six native Cree communities in Northern Ontario (1990–1995), 15% showed lead shot in the gastrointestinal system (Tsuji and Nieboer 1997). Sixty-two patients in one Newfoundland hospital had from 1–200 lead shot in their appendices (Reddy 1985), and Hillman (1967), Greensher et al. (1974), Durlach et al. (1986), and Gustavsson and Gerhardsson (2005) all documented clinical symptoms resulting from lead shot in human appendices. In the USA in 2005, Cox and Pesola (2005) published a radiograph from an Alaska Native elder with an appendix full of shot, and stated "buckshot is commonly seen in Alaskan natives."

Using lead isotopes to identify the source of lead when blood lead is elevated combines population and individual assessments. This method was used by Tsuji et al. (2008) to definitively document lead from ammunition—both shot and bullets—as a source of lead in First Nations Cree in northern Ontario. Lead isotope signatures of southern Ontario urban dwellers were different from those of northern First Nations people, who depended upon subsistence foods. Lead from ammunition had a separate signature from that found on lichens and, significantly, isotope signatures of First Nations people overlapped with that of lead from ammunition. Levesque et al. (2003) used a similar approach to identify the source of lead in cord blood of Nunavik Inuit infants born from 1993–96. Although mobilization of maternal bone lead resulted in less definite signatures than those documented by Tsuji et al. (2008), there was still a strong suggestion that the source of elevated cord blood lead, found in approximately 7% of Inuit newborns, was lead from ammunition. There were also signature differences between Inuit infants from Nunavik in northern Quebec, and Caucasian infants from southern Quebec. In Alaska, recent lead isotope data from blood of Alaska Natives from Bethel on the Yukon-Kuskokwim Delta and Barrow on the North Slope, regions where subsistence waterfowl hunts occur, showed signatures that overlapped with those of shot (Alaska Native Tribal Health Consortium, unpubl. data).

Blood Lead Surveillance in Alaska.—Alaska regulations require laboratories and health care providers to report all blood lead test results ≥10 µg/dL to the Alaska Division of Public Health, Section of Epidemiology; however, most laboratories report all BLL results (Section of Epidemiology 2008b). The Section of Epidemiology maintains a blood lead surveillance database of all reported blood lead levels from Alaskans (>26,000 records as of August 2008), and conducts individual case follow-up activities for all elevated BLLs.

In Alaska, the majority of adults with BLLs ≥25 µg/dL were males who worked in the metal ore mining industry (State of Alaska 2008a). Across all age groups, the majority (81%) of known non-occupational elevated lead exposures involved people exposed on indoor firing ranges, followed by children who were born or adopted from abroad (10%), and people casting lead as a hobby (3.4%) (State of Alaska 2008b).

Major lead sources for children aged <6 years in the contiguous United States are lead-contaminated dust and soil and deteriorated lead-based paint (Brown 2007), but these exposure sources are not frequently encountered in Alaska. The majority of Alaska children aged <6 years with elevated BLLs obtained their lead exposures abroad (State of Alaska 2008b). Many of the other sources of non-occupational lead exposure in Alaskans reflect the hunting and fishing, outdoor lifestyle of Alaska. Lead ammunition or lead fishing sinkers are commonly implicated as the primary exposure source of elevated BLLs in Alaska.

Elevated BLLs have been attributed to use of indoor firing ranges in Alaska (Lynn et al. 2005, Verbrugge 2007). Students shooting on high school rifle teams that used the problematic indoor shooting ranges were among the persons with elevated BLLs. Inadequate ventilation systems and improper maintenance practices at indoor firing ranges were documented at several ranges with lead exposure problems. The cleaning practice of dry sweeping is particularly hazardous, and should never be performed in indoor ranges. Elevated lead exposures have also occurred among Alaskans who hand reload ammunition, and among sportsmen who melt lead to cast their own bullets (State of Alaska 2008b). In June 2001 an adult Alaskan male suffered acute lead poisoning as a result of inhaling lead dust and fumes while melting and casting lead to make fishing sinkers (State of Alaska 2001). The patient had a BLL of 133 µg/dL and exhibited symptoms of fatigue, stomach pain with gastric upset for several months, and a fever of 102°F for 10 days. The patient was hospitalized and received chelation therapy, and his BLL subsequently declined. The State of Alaska has not yet investigated whether consumption of game shot with lead may also be causing elevated lead exposures in Alaska, although this has recently been added to the list of potential risk factors under consideration during follow-ups for elevated BLLs.

REDUCING LEAD EXPOSURE IN CIRCUMPOLAR PEOPLE

In the circumpolar north, many indigenous peoples and other rural inhabitants depend on wild game for subsistence. In Alaska and elsewhere, scientists have documented the nutritional value of traditional foods such as fish, marine and terrestrial mammals, wild birds, and plants (Egeland et al. 1998, Nobmann et al. 1992). In many rural northern communities, wage-paying jobs are limited and market food is not available or is expensive. Further, wild foods are often nutritionally superior to market foods, which have high levels of processed sugars and fats. Subsistence food gathering is essential if people are to have enough healthy food. Traditional foods represent not just a critical food source, but also an integral part of Native culture and a way of life that has existed for many generations. Risk reduction strategies for lead exposure from ammunition must account for the need for inexpensive shot that is easy to use for subsistence hunting—a niche that is still being filled by purchased and reloaded lead shot in much of the North.

Risk reduction strategies that have been suggested for reducing lead exposure from use of lead shot include culture-specific outreach (see Tsuji 1998) to lead shot users and sellers, with the goal of voluntary behavior changes; capacity-building, which trains community members in outreach regarding lead shot risks and non-lead shot shooting techniques; and regulation, both from within and outside of subsistence communities (Tsuji 1999, AMAP 2003). Some are more successful than others; for example, regulation is often most effective if it is community-generated. Enforcement from outside the community, especially with the large distances and relatively low human population densities in Arctic regions, can be inefficient on broad scales.

After Inuit from Nunavik were found to have high cord blood lead levels, lead shot bans (Dallaire et al. 2003) and public health intervention (Levesque et al. 2003) resulted in "marked" and "significant" decreases in cord blood lead concentrations, from an average of 0.20 µmol/L before the ban in 1999 to 0.12 µmol/L after the ban (Dallaire et al. 2003). In the Mushkegowuk Territory of northern Ontario, collaborative health education outreach with direct community involvement was essential to changing attitudes about the safety of lead shot and inspiring behavioral change (Tsuji et al. 1999). In Alaska, outreach to food preparers, school-age children, and hunters about the risk of lead exposure from lead shot to human and bird health, resulted in two community-generated injunctions on the use of lead shot in areas covering 83 million acres (2.4 million ha) and numerous subsistence communities on the North Slope and Yukon-Kuskokwim Delta.

Reducing lead exposure from other sources, which may not be as widespread as the use of lead ammunition, could respond well to targeted outreach and regulation. For example, as the Alaskan examples illustrate, lead should not be melted and formed into shot or sinkers in home environments. In indoor shooting ranges, ventilation systems must be built correctly and correctly maintained, dry sweeping should be prohibited, and blood lead testing for regular users such as rifle teams should be performed at the beginning and end of each shooting season.

CONCLUSION

Since bans on lead in gasoline, instituted primarily in the 1980s and 1990s, lead levels in northern hemisphere humans have generally declined. A notable exception is the blood lead levels of Arctic indigenous peoples who rely on subsistence foods. In many cases, elevated blood lead levels in the Arctic have been associated with ingestion of lead from spent ammunition, primarily shot, although lead from fragmented bullets in big game may have been overlooked as a source until recently (Hunt et al. 2006, Tsuji et al. 2008, Hunt et al. 2009, Titus et al. 2009). Other cases of harmful lead exposure have resulted indirectly from use of lead in ammunition or for fishing (indoor firing ranges, home melting and manufacture of lead sinkers, shot, or bullets, and home reloading). Because subsistence populations by definition hunt much of their food, and because this food is important economically, nutritionally, and socially (Titus et al. 2009), an inexpensive source of ammunition is required. Lead is relatively inexpensive, but use of lead in ammunition comes with risks to humans, especially children, which do not occur with non-lead substitutes. Many approaches to reducing lead exposure have been proposed or implemented. For example, human health agencies can work with ammunition manufacturers and sellers to reduce the availability of lead ammunition, facilitate the availability of inexpensive non-toxic alternatives, and offer training on the different shot patterns, velocities, and distances inherent in using materials other than lead. The most effective means of reducing lead exposure have included community-based outreach and education on the dangers of lead from ammunition to both humans and the environment. These approaches have achieved positive behavioral changes, and may result in subsistence hunters and their families choosing to use non-toxic shot and bullets for their subsistence needs.

LITERATURE CITED

AMAP. 1998. AMAP Assessment Report: Arctic Pollution Issues. Arctic Monitoring and Assessment Programme (AMAP), Oslo, Norway.

AMAP. 2002. AMAP Assessment 2002: Arctic Pollution. Arctic Monitoring and Assessment Programme (AMAP), Oslo, Norway.

AMAP. 2003. AMAP Assessment 2002: Human Health in the Arctic. Arctic Monitoring and Assessment Programme (AMAP), Oslo, Norway.

AMAP. 2004. Persistent Toxic Substances, Food Security and Indigenous Peoples of the Russian North. Arctic Monitoring and Assessment Programme (AMAP), Oslo, Norway.

AMERICAN ACADEMY OF PEDIATRICS, COMMITTEE ON ENVIRONMENTAL HEALTH. 2005. Lead exposure in children: prevention, detection, and management. Pediatrics 116:1036–1046.

ATSDR. 2007. Toxicological profile for lead. Agency for Toxic Substances and Disease Registry, Atlanta, Georgia, USA.

BAYLISS, R. 2002. Sir John Franklins's last arctic expedition: a medical disaster. Journal of the Royal Society of Medicine 95:151–153.

BJERREGAARD, P., P. JOHANSEN, G. MULVAD, H. S. PEDERSEN, AND J. C. HANSEN. 2004. Lead sources in human diet in Greenland. Environmental Health Perspectives 112:1496–1498.

BROWN, M. J. 2007. Interpreting and managing blood lead levels <10 μg/dL in children and reducing childhood exposures to lead: recommendations of CDC's Advisory Committee on Childhood Lead Poisoning Prevention. MMWR 56 (RR08):1–14,16.

CANFIELD, R. L., C. R. HENDERSON, JR., D. A. CORY-SLECHTA, C. COX, T. A. JUSKO, AND B. P. LANPHEAR. 2003. Intellectual impairment in children with blood lead concentrations below 10 μg per deciliter. New England Journal of Medicine 348:1517–1526.

CENTERS FOR DISEASE CONTROL AND PREVENTION. 1997. Screening young children for lead poisoning: guidance for state and local public health officials. US Department of Health and Human Services, Atlanta, Georgia, USA.

CENTERS FOR DISEASE CONTROL AND PREVENTION. March 2002. Managing elevated blood lead levels among young children: recommendations from the Advisory Committee on Childhood Lead Poisoning Prevention. US Department of Health and Human Services, Atlanta, Georgia, USA.

CENTERS FOR DISEASE CONTROL AND PREVENTION. Adult blood lead epidemiology and surveillance—United States, 1998–2001. *In:* Surveillance Summaries, December 13, 2002. MMWR 51 (No. SS-11):1–10.

COX, W. M., AND G. R. PESOLA. 2005. Buckshot ingestion. New England Journal of Medicine 353:26.

DALLAIRE, F., E. DEWAILLY, G. MUCKLE, AND P. AYOTTE. 2003. Time trends of persistent organic pollutants and heavy metals in umbilical cord blood of Inuit infants born in Nunavik (Quebec, Canada) between 1994 and 2001. Environmental Health Perspectives 111:1660–1664.

DEWAILLY, E., P. AYOTTE, S. BRUNEAU, G. LEBEL, P. LEVELLOIS, AND J.P. WEBER. 2001. Exposure of the Inuit population of Nunavik (Arctic Québec) to lead and mercury. Archives of Environmental Health 56:350–357.

DURLACH, V., F. LISOVOSKI, A. GROSS, G. OSTERMANN, AND M. LEUTENEGGER. 1986. Appendicectomy in an unusual case of lead poisoning. Lancet i(8482):687–688.

EGELAND, G. M., L. A. FEYK, AND J. P. MIDDAUGH. January 15, 1998. The use of traditional foods in a healthy diet in Alaska: Risks in perspective. State of Alaska Epidemiology Bulletin.[Online.] Available at http://www.epi.hss.state.ak.us/bulletins/docs/rr1998_01.pdf. Accessed August 21, 2008.

ELDER, W. H. 1955. Fluoroscope measures of hunting pressure in Europe and North America. Transactions of the North American Wildlife Conference 20: 298–322.

ELLIOTT, J. E., K. M. LANGELIER, A. M. SCHEUHAMMER, P. H. SINCLAIR, AND P. E. WHITEHEAD. 1992. Incidence of lead poisoning in Bald Eagles and lead shot in waterfowl gizzards from British Columbia, 1988–91. Canadian Wildlife Service Progress Notes No. 200, June 1992. Canadian Wildlife Service, Ottawa, Canada.

FRANK, A. 1986. Lead fragments in tissues from wild birds: A cause of misleading results. Science of the Total Environment 54:275–281.

GREENSHER, J., H. C. MOFENSON, C. BALAKRISHNAN, AND A. ALEEM. 1974. Lead poisoning from ingestion of lead shot. Pediatrics 54:641.

GUSTAVSSON, P., AND L. GERHARDSSON. 2005. Intoxication from an accidentally ingested lead shot retained in the gastrointestinal tract. Environmental Health Perspectives 113:491–493.

HANSEN, J. C., T. G. JENSEN, AND U. TARP. 1991. Changes in blood mercury and lead levels in pregnant women in Greenland 1983–1988. Pages 605–607 *in* B. Postl, P. Gilbert, J. Goodwill, M. E. K. Moffatt, J. D. O'Neil, P. A. Sarsfield, and T. K. Young (Eds.). Proceedings of the 8th International Congress on Circumpolar Health, 20–25 May 1990, White Horse, Yukon, Canada. University of Manitoba Press, Winnipeg, Canada.

HICKLIN, P. W., AND W. R. BARROW. 2004. The incidence of embedded shot in waterfowl in Atlantic Canada and Hudson Strait. Waterbirds 27:41–45.

HILLMAN, F. E. 1967. A rare case of chronic lead poisoning: Polyneuropathy traced to lead shot in the appendix. Industrial Medicine and Surgery 36:388–398.

HUNT, W. G., W. BURNHAM, C. N. PARISH, K. K. BURNHAM, B. MUTCH, AND J. L. OAKS. 2006. Bullet fragments in deer remains: Implications for lead exposure in avian scavengers. Wildlife Society Bulletin 34:167–170.

HUNT, W. G., R. T. WATSON, J. L. OAKS, C. N. PARISH, K. K. BURNHAM, R. L. TUCKER, J. R. BELTHOFF, AND G. HART. 2009. Lead bullet fragments in venison from rifle-killed deer: Potential for human dietary exposure. *In* R. T. Watson, M. Fuller, M. Pokras, and W. G. Hunt (Eds.). Ingestion of Lead from Spent Ammunition: Implications for Wildlife and Humans. The Peregrine Fund, Boise, Idaho, USA. DOI 10.4080/ilsa.2009.0112

JOHANSEN, P., G. ASMUND, AND F. RIGET. 2001. Lead contamination of seabirds harvested with lead shot—implications to human diet in Greenland. Environmental Pollution 112:501–504.

JOHANSEN, P., H. S. PEDERSEN, G. ASMUND, AND F. RIGET. 2006. Lead shot from hunting as a source of lead in human blood. Environmental Pollution 142:93–97.

KENDALL, R. J., T. E. LACHER, JR., C. BUNCK, B. DANIEL, C. DRIVER, C. F. GRUE, F. LEIGHTON, W. STANSLEY, P. G. WATANABE, AND M. WHITWORTH. 1996. An ecological risk assessment of lead shot exposure in non-waterfowl avian species: Upland game birds and raptors. Environmental Toxicology and Chemistry 15:4–20.

LANPHEAR, B. P., R. HORNUNG, J. KHOURY, K. YOLTON, P. BAGHURST, D. C. BELLINGER, R. L. CANFIELD, K. N. DIETRICH, R. BORNSCHEIN, T. GREENE, S. J. ROTHENBERG, H. L. NEEDLEMAN, L. SCHNAAS, G. WASSERMAN, J. GRAZIANO, AND R. ROBERTS. 2005. Low-level environmental lead exposure and children's intellectual function: an international pooled analysis. Environmental Health Perspectives 113:894–899.

LEVESQUE, B., J-F. DUCHESNES, C. GARIEPY, M. RHAINDS, P. DUMAS, A. M. SCHEUHAMMER, J-F. PROUL, S. DERY, G. MUCKLE, F. DALLAIRE, AND E. DEWAILLY. 2003. Monitoring of umbilical cord blood lead levels and sources assessment among the Inuit. Occupational and Environmental Medicine 60:693–695.

LYNN, T., S. ARNOLD, C. WOOD, L. CASTRODALE, J. MIDDAUGH, AND M. CHIMONAS. 2005. Lead exposure from indoor firing ranges among students on shooting teams—Alaska, 2002–2004. MMWR 54(23):577–579.

MADSEN, H. H., T. SKJØDT, P. J. JORGENSEN, AND P. GRANDJEAN. 1988. Blood lead levels in patients with lead shot retained in the appendix. Acta Radiology 29:745–746.

MATEO, R., J. ESTRADA, J.-Y. PAQUET, X. RIERA, L. DOMINGUEZ, R. GUITART, AND A. MARTINEZ-VILALTA. 1999. Lead shot ingestion by Marsh Harriers *Circus aeruginosus* from the Ebro delta, Spain. Environmental Pollution 104:435–440.

MERKEL, F. R., K. FALK, AND S. E. JAMIESON. 2006. Effect of embedded lead shot on body condition of Common Eiders. Journal of Wildlife Management 70:1644–1649.

MILLER, M. J., M. RESTANI, A. R. HARMATA, G. R. BORTOLOTTI, AND M. E. WAYLAND. 1998. A comparison of blood lead levels in Bald Eagles from two regions on the Great Plains of North America. Journal of Wildlife Diseases 34:704–714.

NAVAS-ACIEN, A., E. GUALLAR, E. K. SILBERGELD, AND S. J. ROTHENBERG. 2007. Lead exposure and cardiovascular disease—a systematic review. Environmental Health Perspectives 115:472–482.

NOBMANN, E. D., T. BYERS, A. P. LANIER, J. H. HANKIN, AND M. Y. JACKSON. 1992. The diet of Alaska Native adults: 1987–1988. American Journal of Clinical Nutrition 55:1024–1032.

ODLAND, J. O., I. PERMINOVA, N. ROMANOVA, Y. THOMASSEN, L. J. S. TSUJI, J. BROX, AND E. NIEBOER. 1999. Elevated blood lead concentrations in children living in isolated communities of the Kola Peninsula, Russia. Ecosystem Health 5:75–81.

PAIN, D. J., AND C. AMIARD-TRIQUET. 1993. Lead poisoning of raptors in France and elsewhere. Ecotoxicology and Environmental Safety 25:183–192.

PAIN, D. J., I. J. FISHER, AND V. G. THOMAS. 2009. A global update of lead poisoning in terrestrial birds from ammunition sources. *In* R. T. Watson, M. Fuller, M. Pokras, and W. G. Hunt (Eds.). Ingestion of Lead from Spent Ammunition: Implications for Wildlife and Humans. The Peregrine Fund, Boise, Idaho, USA. DOI 10.4080/ilsa.2009.0108

PATTEE, O. H., AND S. K. HENNES. 1983. Transactions of the 48th North American Wildlife and Natural Resources Conference. 1983:230–237.

PIRKLE, J.L., D.J. BRODY, AND E.W. GUNTER. 1994. The declines in blood lead levels in the United States: the National Health and Nutrition Examination Surveys. Journal of the American Medical Association 272:284–291.

PIRKLE, J. L., R. B. KAUFMANN, D. J. BRODY, T. HICKMAN, E. W. GUNTER, AND D. C. PASCHAL. 1998. Exposure of the USA population to lead, 1991–1994. Environmental Health Perspectives 106:745–750.

REDDY, E. R. 1985. Retained lead shot in the appendix. Canadian Journal of the Association of Radiologists 36:47–48.

ROBINSON, J. 1981. Lead in Greenland snow. Ecotoxicology and Environmental Safety 5:24–37.

SAMOUR, J. H. AND J. NALDO. 2002. Diagnosis and therapeutic management of lead toxicosis in falcons in Saudi Arabia. Journal of Avian Medicine and Surgery 16:16–20.

SCHEUHAMMER, A. M., AND S. L. NORRIS. 1995. A review of the environmental impacts of lead shotshell ammunition and lead fishing weights in Canada. Occasional Paper No. 88, Canadian Wildlife Service, Ottawa, Canada.

SCHEUHAMMER, A. M., J. A. PERRAULT, E. ROUTHIER, B. M. BRAUNE, AND G. D. CAMPBELL. 1998. Elevated lead concentrations in edible portions of game birds harvested with lead shot. Environmental Pollution 102:251–257.

SECTION OF EPIDEMIOLOGY, Division of Public Health, Department of Health and Social Services, State of Alaska. January 2008. Conditions reportable to public health. Anchorage, Alaska, USA. [Online.] Available at http://www.epi.hss.state.ak.us/pubs/conditions/ConditionsReportable.pdf. Accessed August 21, 2008.

STATE OF ALASKA EPIDEMIOLOGY BULLETIN, November 19, 2001. Cottage industry causes acute lead poisoning. [Online.] Available at http://www.epi.hss.state.ak.us/bulletins/docs/b2001_17.htm. Accessed August 21, 2008.

STATE OF ALASKA EPIDEMIOLOGY BULLETIN. January 23, 2008a. Adult blood lead epidemiology and surveillance: Occupational exposures – Alaska, 1995–2006. [Online.] Available at http://www.epi.hss.state.ak.us/bulletins/docs/b2008_02.pdf. Accessed August 21, 2008.

STATE OF ALASKA EPIDEMIOLOGY BULLETIN. March 7, 2008b. Blood lead epidemiology and surveillance: non-occupational exposures in adults and children—Alaska, 1995–2006. [Online.] Available at http://www.epi.hss.state.ak.us/bulletins/docs/b2008_07.pdf. Accessed August 21, 2008.

STROUD, R. K., AND W. G. HUNT. 2009. Gunshot wounds: A source of lead in the environment. *In* R. T. Watson, M. Fuller, M. Pokras, and W. G. Hunt (Eds.). Ingestion of Lead from Spent Ammunition: Implications for Wildlife and

Humans. The Peregrine Fund, Boise, Idaho, USA. DOI 10.4080/ilsa.2009.0109

TITUS, K., T. L. HAYNES, AND T. F. PARAGI. 2009. The importance of Moose, Caribou, deer and small game in the diet of Alaskans. *In* R. T. Watson, M. Fuller, M. Pokras, and W. G. Hunt (Eds.). Ingestion of Lead from Spent Ammunition: Implications for Wildlife and Humans. The Peregrine Fund, Boise, Idaho, USA. DOI 10.4080/ilsa.2009.0312

TSUJI, L. J. S. 1998. Mandatory use of non-toxic shotshell for harvesting of migratory game birds in Canada: cultural and economic concerns. Canadian Journal of Native Studies 18:19–36.

TSUJI, L. J. S., AND E. NIEBOER. 1997. Lead pellet ingestion in First Nation Cree of the Western James Bay region of northern Ontario, Canada: Implications for a nontoxic shot alternative. Ecosystem Health 3:54–61.

TSUJI, L. J. S., E. NIEBOER, AND J. D. KARAGATZIDES. 1999. Lead and the environment: An approach to educating adults. Journal of American Indian Education 38:25–38.

TSUJI, L. J. S., B. C. WAINMAN, I. D. MARTIN, C. SUTHERLAND, J.-P. WEBER, P. DUMAS, AND E. NIEBOER. 2008. The identification of lead ammunition as a source of lead exposure in First Nations: The use of lead isotope ratios. Science of the Total Environment 393:291–298.

VAN OOSTDAM, J., S. DONALDSON, M. FEELEY, AND N. TREMBLAY. 2003. Canadian Arctic Contaminants Assessment Report II: Human Health. Ottawa, Canada.

VERBRUGGE, L. A. 2007. Health consultation – Interior Alaska indoor shooting range. ATSDR, Public Health Service, US Department of Health and Human Services, Atlanta, Georgia, USA. [Online.] Available at http://www.atsdr.cdc.gov/HAC/pha/InteriorAlaskaIndoorShootingRange/InteriorAlaskaShootingRange061807.pdf. Accessed August 21, 2008.

WALKER, J., J. VAN OOSTDAM, AND E. MCMULLEN. 2001. Human contaminant trends in Arctic Canada: Northwest Territories and Nunavut environmental contaminants exposure baseline. Final Technical Report. Department of Health and Social Services, Government of the Northwest Territories. Yellowknife, Northwest Territories, Canada.

THE IMPORTANCE OF MOOSE, CARIBOU, DEER, AND SMALL GAME IN THE DIETS OF ALASKANS

KIMBERLY TITUS[1], TERRY L. HAYNES[2] AND THOMAS F. PARAGI[2]

[1]*Alaska Department of Fish and Game, Division of Wildlife Conservation, P.O. Box 115526, Juneau, AK 99811, USA.* E-mail: kim.titus@alaska.gov

[2]*Alaska Department of Fish and Game, Division of Wildlife Conservation, 1300 College Road, Fairbanks, AK 99701, USA.*

ABSTRACT.—With a statewide human population of about 677,000 (2006–2007 estimate) and at least 84,000 licensed resident hunters, many Alaskans rely on wild game for a significant part of their total diet. Even within Anchorage, the largest city with 283,000 residents, many families consume wild-taken fish (primarily salmon and halibut) and game (Moose, Caribou, deer) even if they did not harvest these resources themselves. We demonstrate through information from hunter harvest reports and subsistence sharing patterns that thousands of Alaskans depend on wild game. For example, some 29,000 hunters kill about 7,300 Moose annually in Alaska. Each harvested Moose and Caribou yields about 256 and 78 kg of edible meat, respectively. This meat is shared across households. In rural Alaska, reliance on ungulate meat is illustrated by communities such as Nikolai and Akiachak, where about 100 kg of Moose and Caribou meat are consumed per person annually. Small game, marine mammals, and waterfowl harvested with firearms also contribute to the local diet. The high levels of terrestrial wildlife harvest are allowed under both state and federal subsistence laws that provide a preference for Alaskan residents (under state law) and rural residents (under federal law). Specific regulations authorize long seasons and liberal bag limits for ungulates such as deer (up to six per person per season) and Caribou (five per day) in some areas. Sixty percent of the households in rural Alaska harvest game animals and 86% of these rural households consume wild game, attesting to the importance of wild foods. Alaskans consider the harvest of wild game as a healthy and cost-effective way to obtain protein as a food source. The extent to which Alaskans who harvest and consume high levels of game meat are at risk to lead exposure from spent ammunition has not been determined. *Received 15 September 2008, accepted 17 October 2008.*

TITUS, K., T. L. HAYNES, AND T. F. PARAGI. 2009. The importance of Moose, Caribou, deer and small game in the diet of Alaskans. *In* R. T. Watson, M. Fuller, M. Pokras, and W. G. Hunt (Eds.). Ingestion of Lead from Spent Ammunition: Implications for Wildlife and Humans. The Peregrine Fund, Boise, Idaho, USA. DOI 10.4080/ilsa.2009.0312

Key words: Alaska, *Alces alces*, Caribou, diet, food, hunting, *Odocoileus hemionus sitkensis*, *Rangifer tarandus*, Moose, Sitka Black-tailed Deer, wild game.

HARVEST OF WILD FISH AND GAME for human consumption is a widely practiced activity in Alaska. These activities are particularly important in rural areas of Alaska, although even in urban areas there are many families that routinely eat fish and game throughout the year. Depending on the area and time of year in Alaska, harvest of ungulates can be very important to some families and communities. Moose (*Alces alces*), Caribou (*Rangifer tarandus*) and Sitka Black-tailed Deer (*Odocoileus hemionus sitkensis*) are the most important big game species harvested. Additionally, in some rural villages, harvest of small game, waterfowl and other birds is locally important, both in the late summer/fall and also during the spring. During the fall and winter months, ungulate harvest can represent all of the red meat protein consumed by many Alaskans.

In addition to harvest for human consumption, harvest of wild game is culturally important to Alaska Natives, including the sharing of Moose, Caribou and deer that is essential for potlatches and other ceremonial events. These harvests are culturally different from those practiced by many non-Alaska Native hunters, although sharing of fish and game resources is common among all Alaskans. Lead ingestion from spent shot in game birds and bullet fragments in big game has been linked to elevated blood lead levels in subsistence hunters in the Canadian Arctic (Tsuji et al. 2008) and may, therefore, be relevant in Alaska also.

Here we review the magnitude of wild game in the diets of Alaskans. We pay particular attention to the laws, regulations and practices that protect this lifestyle. Although the amount of lead from spent ammunition that is ingested by humans while consuming wild game is unknown, we document substantial intake in the amount of game meat consumed annually by Alaskans.

METHODS

Human Population Trends.—We reviewed human population data available online through the State of Alaska—Department of Labor and Workforce Development (http://www.labor.state.ak.us/home.htm) and the Department of Commerce, Community and Economic Development (http://www.commerce.state.ak.us/dca/commdb/CF_COMDB.htm) where a variety of databases examine the demographic patterns of Alaska's human population.

Hunters and Hunter Harvest Patterns.—We evaluated the location and types of hunters using two primary information sources. First, for rural Alaskans, we used subsistence survey information based on household surveys. These surveys are conducted by subsistence resource specialists from the Alaska Department of Fish and Game (ADFG) who have extensive knowledge of rural Alaska, and have gained the trust of locals to report their subsistence harvests accurately. Second, we evaluated total harvest of Moose, Caribou and deer from mandatory hunter harvest reports and non-mandatory surveys that are conducted annually by ADFG. We chose selected examples of the harvest of Moose, Caribou and deer by community, region, and statewide, where relevant. Both general hunting and subsistence survey results are inexact in Alaska. General hunting surveys are based on Moose and Caribou harvest ticket reports that are mailed in by hunters. Deer harvest is estimated by extrapolation from random hunter surveys based on the harvest tickets they obtain before hunting deer. Subsistence surveys estimate community harvest of fish, game and other wild resources through confidential interviews of households using methods that have been established over the past few decades.

Relevant Laws and Regulations.—State general and subsistence hunting regulations can be found at www.wildlife.alaska.gov and federal subsistence hunting regulations can be found at http://alaska.fws.gov/asm/index.cfml. Alaska has a few key laws and regulations that are designed to protect and maintain the subsistence way of life. These legal protections can be found in both federal and state laws and regulations. Relevant federal regulations include the Marine Mammal Protection Act (MMPA) and the Alaska National Interests Land Conservation Act (ANILCA). The MMPA provides an exclusive preference for Alaska Natives to harvest marine mammals for food and cultural purposes. For harvest of terrestrial fish and wildlife, ANILCA provides a preference for rural residents to harvest fish and wildlife on federal public lands in Alaska. Urban residents of Anchorage, the Matanuska Valley, Kenai Peninsula, Fairbanks, Ketchikan and Juneau and variously defined surrounding areas, especially along the road system, may be restricted from hunting certain species and are ineligible to federal lands in times of game shortage relative to defined rural needs.

The State of Alaska has strong laws and regulations protecting the harvest and use of fish and wildlife for subsistence purposes. First, the common use clause in the Alaska Constitution (Article 8, Section 3) recognizes that all Alaskans shall have equal access to subsistence resources. The state subsistence law provides a preference for all state residents, and elaborates on methods for choosing among subsistence users when there is not enough fish or game to provide for all uses. The net effect of these federal and state laws is a system whereby Alaska residents have the opportunity to harvest wildlife in numbers unlike most of the USA. In times when big game populations are insufficient to provide reasonable opportunity for harvest by Alaskans, nonresident hunting is closed. Large areas of interior Alaska have had chronically low Moose populations and closed nonresident hunting seasons for several years.

Practical aspects of harvesting wild game in Alaska are also governed by strict regulations associated with the retrieval of game. Wanton waste of game meat is a serious offense, and hunters must bring all of the meat from the field before antlers/horns or other trophies can be retrieved. In addition, in many areas of the state, all of the four quarters and rib cage of harvested Moose and/or Caribou must be salvaged from the field with the "bone-in." These "bone-in" regulations were enacted largely through the insistence of local hunters and are designed to minimize waste of game meat by reducing the surface area of meat exposed to bacteria in the field.

Finally, the State of Alaska has an Intensive Management (IM) law requiring the Alaska Board of Game (BOG; appointed seven person Board that sets wildlife regulations) to identify Moose, Caribou and deer populations that need to be managed to provide sufficient game for high levels of human harvest and consumption. Most deer herds, 11 of 32 Caribou herds, and Moose range covering 40% of the state have been identified by the BOG as requiring management under the IM law. This law still requires following the sustained yield principle which is set forth in the Alaska Constitution (Article 8, Section 4). In practice this law has resulted in subsequent regulations that impose management regimes to satisfy the dependence of Alaskans on wild game. The IM law has resulted in management of predators and culling of wolves and/or bears in some instances, with the objective of increasing ungulates for harvest, especially in rural areas. The predator management aspects of the IM law have remained controversial, particularly the culling of wolves (*Canis lupus*) with the use of aircraft. A full review of this topic, including the science of predator management and human/social issues can be found in the National Research Council (1997) and other reviews such as Decker et al. (2006) and Titus (in press).

Alaska also has some other regulations that protect the importance of the harvest of wild game. For example, there is a proxy hunting regulation allowing a younger hunter to harvest game for an elderly or disabled hunter. There are also federal regulations under ANILCA for designated hunters and a provision in state regulations for community harvest of game. Both are designed to recognize and legalize the fact that a single hunter may harvest multiple big game animals for use by others in her or his community.

RESULTS

Human Population Patterns.—About 80% of Alaskans live in urban areas, including 260,000 in the Anchorage area (Division of Subsistence 2000). Generally, these areas are on the main road system (Figure 1). About 20% of Alaskans live in rural areas, which generally means off the road system (Division of Subsistence 2000). This rural population is split nearly equally between Alaska Natives (51%) and those who are not Alaska Native (49%).

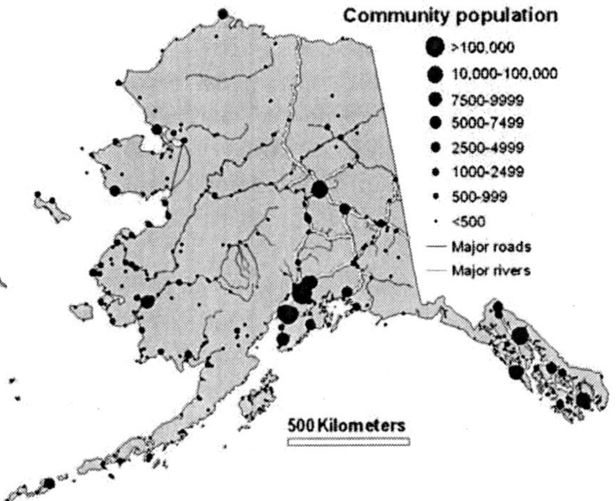

Figure 1. Location of Alaska's cities, towns and villages relative to the road system, major rivers and coastline.

Many rural communities in Alaska not connected to the road system were established along rivers or in coastal areas because of an historic dependence on marine, aquatic or terrestrial subsistence resources. Some communities have been in existence for thousands of years for this reason, especially in coastal areas where marine resources (fish and marine mammals) were harvested seasonally. Harvest of terrestrial resources, especially Caribou, often resulted in the establishment of communities where migratory Caribou could be depended on seasonally. This is the case for a community like Anaktuvuk Pass in the Brooks Range. Residents of this community seasonally depend on the migration and subsequent harvest of the Western Arctic Caribou Herd as a food resource.

Hunters and Hunter Harvest Patterns.—Alaska averaged about 84,000 resident hunters annually during 1998–2007. This accounts for about 12% of the residents. These numbers do not reflect the fact that in some rural areas of Alaska, hunting licenses are difficult to obtain and some hunters don't purchase licenses.

Moose are among the most widely distributed species in the state, with a population estimate of ~160,000–200,000 animals. From 1987–2007, an annual mean of 29,000 Moose hunters harvested 7,260 animals annually. Although they are highly coveted as a trophy species by some, they are far more important as a food resource for thousands of Alaskans (e.g., Boertje et al. 2009). There are a number of Moose herds across the state and densities vary. They are intensively managed for human harvest in some areas, including the harvest of cow Moose (Boertje et al. 2007). Moose hunting regulations are complex and are designed to provide a high sustained yield and the best possible opportunity for local residents. Some areas have winter hunting seasons that are designed to provide food during the winter for local subsistence users. Because of the large size of a Moose (Table 1), many Moose hunters participate in a hunt as a team and the meat is shared among those in the hunting party.

In addition to the general meat salvage requirements for big game, there are very strict salvage requirements in some Game Management Units (GMUs). For example, in GMU 13, the Nelchina Basin, salvage requirements include the heart, liver and all edible meat. In seven GMUs, mostly in interior Alaska, all meat of the front quarters, hindquarters, and ribs must remain naturally attached to the bone until transported from the field or processed for human consumption. For Tier II (State subsistence hunts that are only available for Alaska residents 10 years of age or older) Moose taken in some parts of the state, antler destruction is required to remove any trophy value from the antlers (e.g., typically by cutting across mid-palm on one antler).

Table 1. Approximate weight (kg) of Moose, Caribou and deer harvested in Alaska and the respective amounts of boned-out meat.

Species	Live	Carcass	Boned-out meat
Moose	750	450	255
Caribou	225	136	77
Sitka Black-tailed Deer	90	55	32

Caribou occur in 32 herds in Alaska, and their abundance can vary widely over time. Presently, the Western Arctic Caribou Herd is the largest in the state, in recent years estimated as large as 450,000 animals. When Caribou numbers are high in a herd, hunting has minimal impact on the population, and hunting regulations are often liberal. Annual statewide Caribou harvest is ~27,000, with nearly 75% of statewide harvest from the Western Arctic and Teshekpuk herds. Harvest reporting is not required for the large northern herds, so the biologists estimate harvest through hunter contacts and their own experience. Caribou are harvested year-round in some parts of rural, northern Alaska. In GMU 26A (northwest Alaska north of the Brooks Range) there is a resident 365 day bull Caribou season (and cows from 1 July to 15 May) with a bag limit of five per day. Snowmachines and boats are often used for hunting. Like for Moose, in many areas the front quarters, hindquarters and sometimes the ribs of Caribou must be salvaged intact with the meat attached. One State subsistence Caribou hunt requires salvage of the head, hide, heart, liver, kidneys and all edible meat.

Sitka Black-tailed Deer occur naturally in southeast Alaska, and transplanted populations are widely hunted in Prince William Sound and on Kodiak and Afognak islands. In Southeast Alaska, over 11,000 deer are harvested annually, nearly all by local residents. There is a long season and liberal bag limits in northern Southeast Alaska (GMU 4—Admiralty, Baranof, and Chichagof islands), an area without wolves that serves to allow for higher deer numbers in the absence of a key predator. The state season for all hunters is from 1 August–31 December with a four deer bag limit, and the season for federally qualified subsistence hunters is from 1 August –31 January with a six deer bag limit. Locals annually harvest about 2,400 deer in the Kodiak area and about 2,500 from Prince William Sound. These areas also have long seasons and liberal bag limits. State proxy hunting regulations and federal designated hunter regulations allow for higher harvests in many circumstances.

Small Game Harvest.—Statewide, the numbers of small game (Arctic Hares *Lepus arcticus,* Snowshoe Hares *L. americanus*, migratory birds, and resident gallinaceous birds) harvested are unknown. In certain communities and by some individuals, the amount of harvest can be quite high. Across much of Alaska, hare hunting has no closed season and no bag limits. Grouse and ptarmigan hunting regulations vary, but bag limits are often 10, 15 or 20 per day with twice that typically as a possession limit. Seasons are typically long (e.g., 10 August–31 March).

Community Harvest Patterns.—Surveys of community harvest patterns reveal that many of the 270 rural communities in Alaska harvest significant amounts of wild game. For example, in Nuiqsut on the Arctic coastal plain, about 125 kg of wild game are harvested annually per capita, and over 80% of this is represented by Caribou (Table 2). Survey results from 2002 in interior Alaska in Nikolai, showed that 114 kg of wild game are harvested annually per capita, and about 80% of this is represented in the harvest of about 38 Moose in that community. In southeast Alaska, where deer are the primary big game species harvested by residents, the Admiralty Island community of Angoon harvests 370 deer annually, representing over 80% of the annual harvest of 28 kg of edible meat per capita obtained by hunting. In contrast to communities like Nuiqsut and Nikolai, fish (salmon and halibut) make up a significant amount of the annual per capita portion of the protein diet for many coastal Alaska community residents.

Overall, during the 1990s wild food harvests in Alaska represented 170 kg per person per year in rural Alaska and 10 kg per person in urban Alaska (Division of Subsistence 2000). Depending on the region of the state, from 79–92% of the households in rural Alaska participate in and use game harvested with a firearm as part of their subsistence activities (Division of Subsistence 2000). Since subsistence foods are widely shared, most residents of rural communities make use of wild game, where present. Although these statistics are for rural Alaskans where the consumption is widespread, harvest and use of wild game in the diet of urban Alaskans should not be underestimated in some instances. Even among urban households, Moose, Caribou and deer are widely shared among family and friends. A large bull Moose provides hundreds of kilograms of meat and sharing is likely the norm rather than the exception.

DISCUSSION

Alaskans harvest significant amounts of wild game annually and in some areas of the state this wild game represents nearly all of the non-fish protein consumed annually in many households. Even in some urban households, Moose, Caribou and/or deer are the primary red meat consumed. A full understanding of the vitamin, contaminant, and potentially the lead levels in the wild game that is harvested is lacking (e.g., Ballew et al. 2006). In some areas of Alaska, the consumption of ungulates is associated with culture, custom and in a few cases, meat salvage requirements. In addition to the frontquarters, hindquarters, ribs, tenderloin, neck and miscellaneous meat, in many cases, the liver, heart, and other animal parts may be consumed. The effect of this consumption on ingestion of lead from spent ammunition is unknown and may be negligible in many cases. In some instances the harvest method may reduce the potential for ingesting spent lead from ammunition. For example, many Caribou are harvested in the autumn along rivers, such as the Kobuk, by shooting the animal in the head using a .22 caliber rifle at close range from a boat pulled alongside the swimming animal, which is a

Table 2. Selected wild food resources harvested with firearms in rural Alaska communities.[a]

Community/Year	Moose	Caribou	Deer	Black Bear	Migratory Birds	Ptarmigan	Grouse	Hare	Beaver[b]	Seals	Walrus	Weight (kg) of edible meat per capita obtained by hunting[c]
Akiachak (1998)												
No. harvested	106	374		36	16,103	5,450	146	2,338	433	70		
kg per capita	66	39		5	25	5	<1	5	6	9		159
Angoon (1996)												
No. harvested			370		144					63		
kg per capita			23		<1					4		28
Barrow (1989)												
No. harvested	40	1,656			12,539	329				440	101	
kg per capita	3	29			4	<1				5	11	52
Cordova (1997)												
No. harvested	98		1,441	43	4,056	813	1,017	2,443	145	212		
kg per capita	10		11	<1	<1	<1	<1	<1	<1	1		24
Kodiak (1993)												
No. harvested	57		2,165		4,140			1,785				
kg per capita	2		7		<1			<1				10
Nikolai (2002)												
No. harvested	38	17		18	395	51	363		81			
kg per capita	90	11		5	3	<1	1		3			114
Nuiqsut (1993)												
No. harvested	9	672			2,238	973				109		
kg per capita	6	103			5	1				11		125
Shungnak (2002)												
No. harvested	11	403		2	1,001	264		32	52			
kg per capita	11	100		<1	5	<1		<1	2			119

[a]Source: Alaska Department of Fish and Game, Division of Subsistence, Community Subsistence Information System [http://www.subsistence.adfg.state.ak.us/]
[b]Beaver can be harvested legally with steel traps, snares, and firearms. Legal harvest methods vary by season and from region to region.
[c]Actual amounts may be higher, as not all wildlife species hunted with firearms are included in this table.

legal method in parts of Alaska. Marine mammals, especially seals, are harvested by shooting them in the head, and this part of the animal is not consumed. In other instances, the harvest of Moose results in hundreds of kg of meat, and large portions of the animal are far away from the location of the bullet wound so that only a small proportion of the meat could be contaminated with lead residue.

Virtually all of the big game harvested in Alaska is with the use of lead ammunition. Copper bullet ammunition is seldom used. Popular rifle calibers for Moose, Caribou and deer include .30-06, .338, .375, and .300 magnum. In some cases, such as among many deer hunters, rifle and bullet size are larger than generally advised for the big game species. This is because many deer hunters are hunting in the presence of Brown Bears (*Ursus arctos*) in coastal areas where bear densities are high. As a result, hunters often shoot deer with large loads, and poor shot placement may result in highly damaged meat and many bone fragments. Whether this results in greater dispersal of lead fragments and higher lead ingestion would depend on the location of the shot (e.g., neck, shoulder, ribs), how much damaged meat is discarded, and perhaps the type of "bone-in" and salvage regulation in the hunt area (for Moose and Caribou).

Aside from cultural reasons, the costs of food, fuel (gasoline), and transportation are significant impediments to many rural Alaskans consuming any meat other than wild game. Food is expensive in the 270 bush communities across the state (Division of Subsistence 2000). As of the fall of 2008, gasoline may exceed $10.00 per gallon in some bush communities. Many Alaskans dislike beef, chicken, and other processed foods.

Aside from big game, in many parts of rural Alaska, the harvest of migratory birds can be a significant portion of the diet seasonally (Table 2). Migratory birds are harvested both in the spring and fall. In addition, resident game birds (i.e., grouse and ptarmigan) are harvested in high numbers by some households. Human lead levels associated with the harvest of large numbers of birds is largely unknown, but one might presume that the non-toxic shot now required for waterfowl hunting would reduce lead exposure from this pathway (Johansen et al. 2006, Tsuji et al. 2008). The number of lead shotgun shells still being used in remote villages is unknown, as is the prevalence of lead ammunition to harvest grouse and ptarmigan.

The Alaska Board of Game recently passed regulations restricting the use of lead shot for all bird hunting in GMU 26 (North Slope). In GMU 18 (Yukon-Kuskokwim Delta centered on Bethel), the taking of game under either a hunting or trapping license with lead shot size T or 0.20 or smaller while in the possession of lead shot, is prohibited. In part, these regulations are designed to reduce the amount of lead shotgun shells in rural villages and force suppliers to provide non-toxic alternatives. These regulations may have the benefit of reducing human exposure to lead via consumption of birds (e.g., Johansen et al. 2001) and other small game, like hares. These regulatory actions, designed initially for bird conservation, may have the broader benefit of helping reduce lead exposure in humans (Rattner et al. 2008).

ACKNOWLEDGMENTS

We thank the many dedicated staff from the Division of Subsistence, Alaska Department of Fish and Game, who conducted the numerous community harvest surveys over the past few decades. Combined with hunter harvest surveys, these provide the basis for understanding the importance and consumption of wild game to Alaskans.

LITERATURE CITED

BALLEW, C., A. R. TZILKOWSKI, K. HAMRICK, AND E. D. NOBMANN. 2006. The contribution of subsistence foods to the total diet of Alaska Natives in 13 rural communities. Ecology of Food and Nutrition 45:1–26.

BOERTJE, R. D., K. A. KELLIE, C. T. SEATON, M. A. KEECH, D. D. YOUNG, B. W. DALE, L. G. ADAMS, AND A. R. ALDERMAN. 2007. Ranking Alaska Moose nutrition: Signals to begin liberal antlerless harvests. Journal of Wildlife Management 71:1494–1506.

BOERTJE, R. D., M. A. KEECH, D. D. YOUNG, K. A. KELLIE, AND C. T. SEATON. 2009. Managing for elevated yield of Alaska Moose. Journal of Wildlife Management 73. In press.

DECKER, D. J., C. A. JACOBSON, AND T. L. BROWN. 2006. Situation-specific "impact dependency" as a determinant of management acceptability: insights from wolf and Grizzly Bear management in Alaska. Wildlife Society Bulletin 34:426–432.

DIVISION OF SUBSISTENCE. 2000. Subsistence in Alaska: A year 2000 update. [Online.] Available at www.subsistence. adfg.state.ak.us/download/ subupd00.pdf. Division of Subsistence, Alaska Department of Fish and Game. Juneau, Alaska, USA.

JOHANSEN, P., G. ASMUND, AND F. RIGET. 2001. Lead contamination of seabirds harvested with lead shot—implications to human diet in Greenland. Environmental Pollution 112:501–504.

JOHANSEN, P., H. S. PEDERSEN, G. ASMUND, AND F. RIGET. 2006. Lead shot from hunting as a source of lead in human blood. Environmental Pollution 142:93–97.

NATIONAL RESEARCH COUNCIL. 1997. Wolves, bears and their prey in Alaska. National Academy Press. Washington, DC, USA.

RATTNER, B. A., J. C. FRANSON, S. R. SHEFFIELD, C. I. GODDARD, N. J. LEONARD, D. STAND, AND P. J. WINGATE. 2008 Sources and implications of lead ammunition and fishing tackle on natural resources. Wildlife Society Technical Review. The Wildlife Society, Bethesda, Maryland, USA.

TITUS, K. In press. Intensive management of wolves and ungulates in Alaska. Transactions of the North American Wildlife and Natural Resources Conference. Portland, Oregon, USA.

TSUJI, L. J. S., B. C. WAINMAN, I. D. MARTIN, J. P. WEBER, C. SUTHERLAND, E. N. LIBERDA, AND E. NIEBOER. 2008. Elevated blood-lead levels in First Nation people of northern Ontario, Canada: policy implications. Bulletin of Environmental Contamination and Toxicology 80:14–18.

LEAD BULLET FRAGMENTS IN VENISON FROM RIFLE-KILLED DEER: POTENTIAL FOR HUMAN DIETARY EXPOSURE

W. GRAINGER HUNT[1], RICHARD T. WATSON[1], J. LINDSAY OAKS[2], CHRIS N. PARISH[1], KURT K. BURNHAM[1], RUSSELL L. TUCKER[3], JAMES R. BELTHOFF[4], AND GARRET HART[5]

[1]*The Peregrine Fund, 5668 W. Flying Hawk Lane, Boise, ID 83709, USA.*
E-mail: grainger@peregrinefund.org

[2]*Washington Animal Disease Diagnostic Laboratory, Pullman, WA 99164-7034, USA.*

[3]*Department of Veterinary Clinical Sciences, Washington State University, Pullman, WA 99164, USA.*

[4]*Department of Biology, Boise State University, 1910 University Drive, Boise, ID 83725, USA.*

[5]*School of Earth & Environmental Sciences, Washington State University, Pullman, WA 99164, USA.*

ABSTRACT.—Human consumers of wildlife killed with lead ammunition may be exposed to health risks associated with lead ingestion. This hypothesis is based on published studies showing elevated blood lead concentrations in subsistence hunter populations, retention of ammunition residues in the tissues of hunter-killed animals, and systemic, cognitive, and behavioral disorders associated with human lead body burdens once considered safe. Our objective was to determine the incidence and bioavailability of lead bullet fragments in hunter-killed venison, a widely-eaten food among hunters and their families. We radiographed 30 eviscerated carcasses of White-tailed Deer (*Odocoileus virginianus*) shot by hunters with standard lead-core, copper-jacketed bullets under normal hunting conditions. All carcasses showed metal fragments (geometric mean = 136 fragments, range = 15–409) and widespread fragment dispersion. We took each carcass to a separate meat processor and fluoroscopically scanned the resulting meat packages; fluoroscopy revealed metal fragments in the ground meat packages of 24 (80%) of the 30 deer; 32% of 234 ground meat packages contained at least one fragment. Fragments were identified as lead by ICP in 93% of 27 samples. Isotope ratios of lead in meat matched the ratios of bullets, and differed from background lead in bone. We fed fragment-containing venison to four pigs to test bioavailability; four controls received venison without fragments from the same deer. Mean blood lead concentrations in pigs peaked at 2.29 µg/dL (maximum 3.8 µg/dL) 2 days following ingestion of fragment-containing venison, significantly higher than the 0.63 µg/dL averaged by controls. We conclude that people risk exposure to lead from bullet fragments when they eat venison from deer killed with standard lead-based rifle bullets and processed under normal procedures. At risk in the U.S. are some ten million hunters, their families, and low-income beneficiaries of venison donations. *Reproduced with permission from PLoS ONE 4(4): e5330.*[6]

[6] Reproduced in accordance with the Creative Commons Attribution License with permission of the authors from: Hunt, W. G., R. T. Watson, J. L. Oaks, C. N. Parish, K. K. Burnham, R. L. Tucker, J. R. Belthoff, and G. Hart. 2009. Lead bullet fragments in venison from rifle-killed deer: potential for human dietary exposure. PLoS ONE 4(4): e5330. doi: 10.1371/journal.pone.0005330.

HUNT, W. G., R. T. WATSON, J. L. OAKS, C. N. PARISH, K. K. BURNHAM, R. L. TUCKER, J. R. BELTHOFF, AND G. HART. 2009. Lead bullet fragments in venison from rifle-killed deer: potential for human dietary exposure. Reproduced *in* R. T. Watson, M. Fuller, M. Pokras, and W. G. Hunt (Eds.). Ingestion of Lead from Spent Ammunition: Implications for Wildlife and Humans. The Peregrine Fund, Boise, Idaho, USA. DOI 10.4080\ilsa.2009.0112

Key words: Bullet fragmentation, bush meat, game meat, lead, lead exposure, venison.

LEAD HAS BEEN IMPACTING the health of humankind since the Romans began mining it 2500 years ago, and despite early knowledge of its harmful effects, exposure to lead from a wide variety of sources persists to this day (Warren 2000). Government-based guidelines for acceptable degrees of exposure prior to the 1970s were based upon thresholds of overt toxicity and on apparent acceptance that norms in lead concentrations in a society enveloped in lead-permeated exhaust fumes and lead paint must somehow reflect organic tolerance. Medical science has since concluded that virtually no level of lead exposure can be considered harmless in consideration of its many sublethal, debilitating, and often irreversible effects (Needleman 2004). Lead quantities formerly regarded as trivial are associated with permanent cognitive damage in children (Lanphear et al. 2005), including those prenatally exposed (Schnaas et al. 2006). Lead is associated with impaired motor function (Cecil et al. 2008), attentional dysfunction (Braun et al. 2006), and even criminal behavior (Needleman et al. 2002, Wright et al. 2008). Release of lead stores from bone exposes fetuses during pregnancy (Tellez-Rojo et al. 2004), and adults late in life (Schwartz and Stewart 2007, Shih et al. 2007). Lead is implicated in reduced somatic growth (Hauser et al. 2008), decreased brain volume (Cecil et al. 2008), spontaneous abortion (Borja-Aburto et al. 1999), nephropathy (Ekong et al. 2006), cancer, and cardiovascular disease (Menke et al. 2006, Lustberg and Silbergeld 2002).

Ingested residues of lead ammunition are a recently identified pathway of lead exposure to human consumers of gun-killed game animals. An analysis of North Dakota residents showed that recent (≤ 1 mo) consumers of game meat had higher covariate-adjusted blood lead concentrations than those with a longer interval (> 6 mo) since last consumption (Iqbal 2008). Studies have linked elevated blood lead concentrations of subsistence hunters in northern Canada, Alaska, Greenland, and elsewhere to consumption of shotgun-killed birds (Hanning et al. 2003, Levesque et al. 2003, Johansen et al. 2004, 2006, Bjerregaard et al. 2004, Tsuji et al. 2008a, 2008b, 2008c; see Burger et al.1998, Mateo et al. 2007). The hypothesis that rifle bullet fragments are an additional source of human lead exposure is suggested by radiographic studies of deer killed with standard lead-based bullets, which show hundreds of small metal fragments widely dispersed around wound channels (Hunt et al. 2006, Dobrowolska and Melosic 2008, Krone et al. 2009). The possibility of inadvertent lead contamination in prepared meat consumed by hunters and their families is noteworthy, considering the millions of people who hunt big game in the USA (USFWS and USCB 2006) and the thousands of deer annually donated to food pantries for the poor (Cornatzer et al. 2009, Avery and Watson 2009). In this report, we test two hypotheses: (1) that fragments of lead from rifle-bullets remain in commercially processed venison obtained under normal hunting conditions in the USA, and (2) humans absorb lead when they eat venison containing bullet fragments.

MATERIALS AND METHODS

Ethics Statement.—Nine licensed hunters provided the deer carcasses analyzed in this study, and obtained them during the established hunting season and in accordance with normal practices as permitted under the authority of the Wyoming Game and Fish Commission, Cheyenne, Wyoming. The latter institution also granted permission to the authors to convey the processed meat from each carcass to the Washington Animal Disease Diagnostic Laboratory at Washington State University, Pullman, for analysis. The Washington State University Institutional Animal Care and Use Committee approved the lead bioavailability experiment involving eight swine.

Deer Collection.—Hunters used conventional center-fire hunting rifles to kill 30 White-tailed Deer (*Odocoileus virginianus*) under normal hunting conditions in Sheridan County, Wyoming in November 2007. All bullets were of 7-mm Remington Magnum caliber and of identical mass (150 grains, 9720 mg); cartridges were of a single brand reported in local mass-market vendor interviews as the most widely sold to deer hunters. Bullets consisted of a lead core (68% of mass) and a copper jacket (32%); lead was exposed only at the 1.7-mm-diameter tip of the bullet. Reported shot distances averaged 116 m (range = 25–172 m). All deer were eviscerated according to the hunters' normal practice. Weights of 29 eviscerated deer averaged 33.8 kg (SD = 7.1). We recorded the positions of bullet entry and exit wounds; 26 deer (87%) were shot in the thorax, and some portion of the projectile exited the animal in 92% of shots. We removed the skin and head, and we excised from each animal a ≥4 cm section of tibia for isotope analyses and a ≥30 g sample of muscle (shank) along the tibia to determine background lead levels in each deer.

Carcass Radiography.—We radiographed with conventional veterinary equipment the area of the wound channel (lateral view) of eviscerated deer and adjusted exposures to maximize contrast. We included along the margin of each radiograph a strip of clear plastic tape containing arrayed samples of lead bullet fragments (obtained by shooting through light plastic jugs filled with water), comparably-sized samples of bone fragments, and locally-obtained sand and gravel; only the lead fragments were clearly visible in the radiographs at the applied settings. We scanned radiographs into digital format and counted unambiguous metal fragments under 400% magnification. We did not attempt to distinguish between copper and lead in fragment counts.

Commercial Processing.—We transported each deer carcass to a different commercial meat processing plant in 22 towns throughout Wyoming and requested normal processing into boneless steaks and ground meat in 2-pound (0.91 kg) packages; we retrieved the processed, frozen, and packaged meat usually within 4 days.

Radiography of Processed Meat.—We used digital radiography (EDR6 Digital Radiography, Eklin Medical Systems, Santa Clara, California) and fluoroscopy (MD3 Digital Fluoroscopy, Philips Medical Systems, Best, Netherlands) to scan all the thawed ground meat packages (N = 234); we scanned an additional 49 loin steak packages from 16 carcasses in which radiography had revealed fragments near the spine. We unwrapped every package showing visible radiodense fragments in a subsample of 13 deer, flattened the meat to c. 1-cm thickness on a light plastic plate, and rescanned. We marked the vicinity of each visible fragment with a stainless steel needle and then used a 2.8-cm diameter plastic tube as a "cookie-cutter" to obtain samples of meat with radiodense fragments.

Analysis of Metal Samples.—Each of the fragment-containing meat samples was weighed and then divided into approximately 5-g subsamples, each of which was completely digested in a known volume of concentrated nitric acid. Inductively coupled plasma (ICP) analysis was then used to measure the concentrations of lead and copper in each subsample. The lower detection limit for both metals was 2 μg/g. The analysis was performed commercially by the Analytical Sciences Laboratory, University of Idaho, Moscow, where quality management conforms with applicable Federal Good Laboratory Practices (40 CFR Part 160); the Laboratory is accredited through the American Association of Veterinary Laboratory Diagnosticians, which stipulates ISO 17025 quality assurance measures.

Lead Isotope Analysis.—We analyzed bullet, bone, and meat samples for lead isotope compositions. Bullet fragments were cleaned in dilute (1M) HCl, leached with 2 ml of 7M HNO_3, and then removed from the acid leachate. The leachate was then dried and treated with 2 drops of 14M HNO_3. Bone and meat samples were digested in 14M HNO_3, dried and treated with 2 drops of 14M HNO_3. Lead was separated using standard HBr and HCl on an anion-exchange column (Bio Rad, AG 1X8). Isotope compositions were determined with a ThermoFinnigan Neptune MC-ICPMS at the Washington State University GeoAnalytical Laboratory. Reproducibility of the lead standard (NBS-981), run before, during, and after the samples, was <0.012% (2 SE, n = 4) for Pb^{206}/Pb^{204}, and <0.018% for Pb^{208}/Pb^{204}.

Lead concentrations in the procedural blanks were negligibly small.

Bioavailability Experiment.—We tested the bioavailability of ingested bullet fragments by feeding processed venison known by radiography to contain radiodense fragments to pigs. The latter were considered a good model for the absorption of lead from the human gastrointestinal tract (USEPA 2007). We used eight female Yorkshire/Landrace and Berkshire/Duroc cross-bred pigs, 70–82 days of age and weighing 28.2–32.7 kg (mean 30.3 kg) at the termination of the experiment. All were initially fed 1.36 kg of standard pelleted pig grower ration divided into two meals per day, then acclimated for 7 days to consuming cooked ground commercial beef patties mixed with the pellet ration. We gradually increased the amount of ground meat from 113 g per meal to 500 g, as pellet amounts were correspondingly decreased. We withheld all food for 24 hours prior to the venison feeding trial.

Ground venison and venison steaks from four deer were used in the feeding trial. Each of the eight pigs consumed 1.26–1.54 kg of meat over two feedings 24 hours apart on days 0 and 1 of the experiment; no pig consumed meat from more than one deer. Four pigs received venison containing fluoroscopically visible metal fragments. The total amount of lead fed to each pig was unknown, but quantitative analysis of similar packages from other deer in the study showed 0.2–168 mg (median 4.2 mg) of lead. The four control pigs were simultaneously fed equivalent amounts of venison with no fluoroscopically visible fragments from the same four deer. We assessed background levels of lead in each deer from shank meat, collected well away from any potential bullet contamination. All venison for the test and control pigs was either already ground, or finely chopped if steaks, and cooked in a microwave oven until brown. For feeding, we mixed the cooked venison in a bowl with small amounts of pig ration to improve palatability. We verified that all meat was eaten, and we monitored the pigs for signs of illness.

We collected anticoagulated blood samples (2 ml whole blood in EDTA) from each pig at 1 hour prior to feeding venison on day 0, and on days 1, 2, 3, 4, 7 and 9 after feeding venison, and stored the samples at 4°C until testing. Lead levels were determined by inductively coupled plasma mass spectrometry (ICP-MS) with a lower detection limit of 0.5 µg/dL; we assigned all values below the detection limits as 0.5 µg/dL. We compared mean blood lead concentrations between control pigs and test pigs on days 0 through 9 using 2-way ANOVA with repeated measures and restricted maximum likelihood (REML) estimation; we performed linear group contrasts for each day. A single outlier datum among control pigs on day 4 (6.8 µg/dL) was an order of magnitude higher than a retest of the same sample (0.54 µg/dL); the latter was consistent with all other control samples. We omitted both results from statistical analysis, resulting in a sample of three rather than four control pigs on day 4. We used JMP (SAS Institute, Cary, NC, USA, Vers. 7.0.1) for all statistical analyses.

RESULTS

Bullet Fragments in Venison.—Wound radiographs of all 30 eviscerated deer showed metal fragments (median = 136 fragments, range = 15–409) and offered a measure of fragment dispersion, albeit two-dimensional. Extreme distance between fragment clusters in standard radiographs averaged 24 cm (range ± SD = 5–43 ± 9 cm), and maximum single fragment separation was 45 cm. Radiography revealed visible metal fragments in the ground meat of 24 (80%) of the 30 deer. At least one fragment was visible in radiographs of 74 (32%) of 234 packages of ground meat; 160 (68%) revealed no fragments, 46 (20%) had one, 16 (7%) had two, and 12 (5%) showed 3-8 fragments. An average of 32% of ground meat packages (N = 3–15 packages, mean 7.8) per deer showed metal fragments (range = 0–100% of packages). The ground meat derived from one deer showed more fragments (N = 42) than counted in the radiograph of the carcass (N = 31), and two ground meat packages (2 deer) each contained a single shotgun pellet which had not been detected on the carcass radiographs. No relationship was apparent between the number of metal fragments counted in carcasses and those subsequently counted in ground meat from the same individual (correlation coefficient 0.06). In the aggregate, we observed 155 metal particles in the ground meat packages, 3.1% of the 5074 we counted in the carcasses. Of 16 deer carcasses with metal frag-

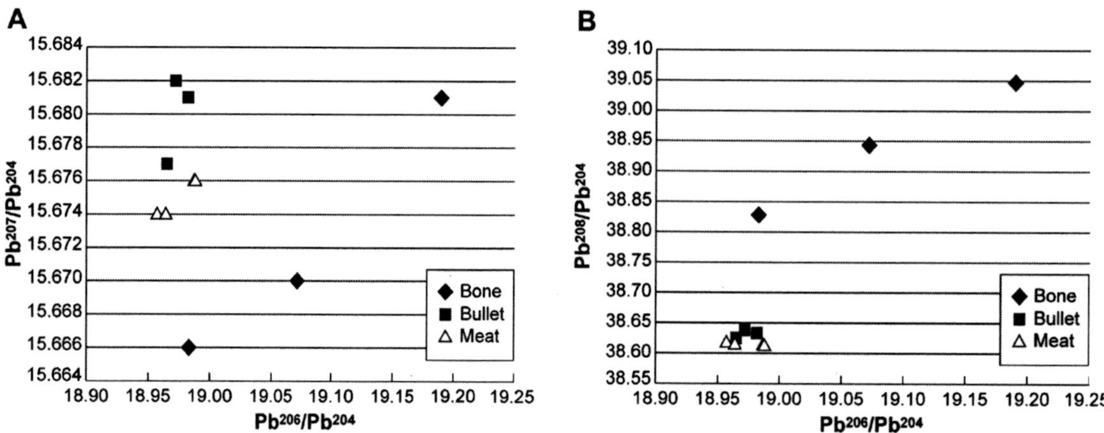

Figure 1. Plots of lead isotope ratios in ground meat samples containing radiodense fragments from four deer. Ratios from lead-in-meat samples clustered with those of unfired bullets but were distinct from bone lead ratios. Note that there are four meat data points (open triangles) in each graph, but two have almost identical positions and are superimposed.

ments near the spine, four (25% of selected deer, 8% of 49 packages) showed fragments in processed loin steaks (1–9 fragments). Additional fragments may have occurred in 220 unscanned packages of steaks derived from all animals.

ICP analysis of radiodense fragments excised from ground meat packages from 13 deer identified lead in 25 (93%) of 27 samples; aggregate lead fragment mass per package averaged 17.2 mg (range ± SD = 0.2–168 ± 39.8 mg) or 0.03% of the lead component of bullet mass. Nine samples contained copper at levels above background values, including the two samples with no detectable lead. Lead concentrations in unprocessed muscle tissue collected from the shank and well away from the bullet path of the same 13 deer were all below the detection limit of 2.0 µg/g and served as internal controls for measures of lead in ground meat.

The ratio of lead isotopes 206/204 plotted against 207/204 ratios (Figure 1a) and 208/204 ratios (Figure 1b) showed that meat samples with elevated lead levels from four deer, and lead from bullets from the same boxes (N = 3) supplying the bullets used to kill those deer, formed tight clusters distinct from ratios of background lead in tibial bone. Variation in the bone ratios apparent in Figure 1 likely represent long term, cumulative lead exposure encompassing varied sources of natural and anthropogenic lead.

Bioavailability Experiment.—All the pigs consumed all the venison provided to them within 2 hours. None of the experimental animals showed any signs of lead toxicosis or other illness for the duration of the experiment; none exhibited vomiting or diarrhea which might have affected gastrointestinal physiology or retention times in the stomach or intestines.

Blood lead concentrations in the four control pigs ranged from below the level of ICP-MS detection (0.5 µg/dL) to 1.2 µg/dL throughout the experiment (mean ± SD = 0.63 ± 0.19 µg/dL; Figure 2). Blood lead concentrations in pigs fed metal fragment-containing venison ranged from below the level of detection to 1.4 µg/dL on day 0, immediately prior to feeding venison. The 2-way ANOVA revealed a significant interaction between treatment (feeding venison either with fragments or no fragments) and day ($F_{6,35.32}$ = 3.413, P = 0.009; Figure 2). Mean blood lead concentrations in the pigs fed fragment-containing venison were significantly elevated above those of control pigs on days 1, 2 and 3 post-exposure (linear contrast: $F_{1,39.79}$ = 10.39, P = 0.003, $F_{1,39.79}$ = 17.76, P = 0.0001, and $F_{1,39.79}$ = 14.71, P = 0.0004, respectively; Figure 2); the maximum observed value was 3.8 µg/dL. Blood lead concentrations did not differ (P >0.05) between the control pigs and exposed pigs on days 0, 4, 7 and 9 (Figure 2).

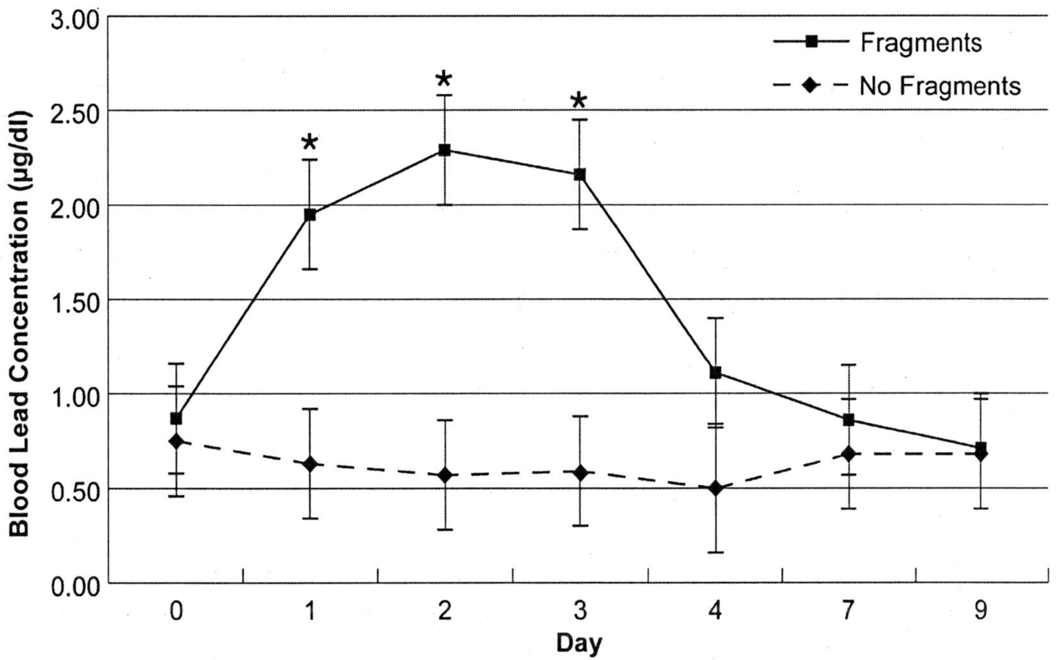

Figure 2. Mean blood lead concentrations observed during swine feeding experiment. Mean (± SE) blood lead concentrations (µg/dL) in four pigs fed venison containing radiographically dense fragments (Fragments) compared with four control pigs fed venison without visible fragments (No Fragments) on days 0 and 1. Asterisks indicate days when means differed significantly between test and control groups.

DISCUSSION

Our findings show that people risk exposure to lead when they eat venison from deer killed with standard lead-based rifle bullets and processed under normal commercial procedures. Evidence includes a high proportion (80%) of deer showing at least one bullet fragment in one or more ground meat packages, a substantial frequency of contamination (32% of all ground meat packages), a majority (93%) of assayed fragments identified as lead, isotopic homogeneity of bullet lead with that found in the meat, and increased blood lead concentrations in swine fed fragment-containing venison. Considering that all the carcasses we brought to the processors contained fragments (15-409 fragments counted in radiographs), the high rate of removal evident in the ground meat implies meticulous care on the part of the processors to avoid contamination, but an apparent inability of 80% of them to do so entirely. We conclude that, in a majority of cases, one or more consumers of a hunter-killed, commercially-processed deer will consume bullet lead.

We interpret the absorption of lead into the bloodstream of all four test pigs as clear evidence of the bioavailability of lead from ingested bullet fragments (Figure 2), and we infer that human consumption of venison processed under prevailing standards of commerce results in increased blood lead concentrations. The rate of bioavailability cannot be calculated from our experiment because the exact amounts of lead in the meat packages were unknown. Rather, we directed our test at the condition experienced by human consumers of venison from rifle-killed deer of variable amounts of lead patchily distributed as fragments in ground meat or steak.

Depuration of lead in blood does not imply its excretion, but rather the sequestration of a substantial proportion in soft tissues and ultimately in bone from which it may eventually be mobilized, as during pregnancy (Tellez-Rojo et al. 2004) or in old age (Schwartz and Stewart 2007). The observed elevations in blood lead concentrations, while not considered overtly toxic, would nevertheless contribute to cumulative lead burdens, and would be additive with further meals of contaminated venison. Observed blood lead concentrations of up to 3.8 µg/dL, and daily means of 2.3 and 2.2 µg/dL in the experimental animals, do approach what is considered significant with respect to adverse effects in humans by contemporary assessments (Gilbert and Weiss 2006, Levin et al. 2008). Whereas the CDC advisory level for intervention in individual children is 10 µg/dL in blood (CDC 1991), studies now associate as little as 2 µg/dL with increased risk of cardiovascular mortality in adults (Menke et al. 2006) and impaired cognitive function in children (Jusko et al. 2008). Hauser et al. (2008) detected an impact threshold of 5 µg/dL on male maturation rates, and Lanphear et al. (2005) concluded that "...lead exposure in children who have maximal blood lead concentrations <7.5 µg/dL is associated with intellectual deficits." These latter values would appear attainable with the repeated consumption of venison possible among deer hunting families, especially those incurring additional exposure from other sources.

Factors that may influence dietary lead exposure from spent lead bullets include the frequency and amount of venison consumption, degree of bullet fragmentation, anatomical path of the bullet, the care with which meat surrounding the bullet wound is removed, and any acidic treatments of the meat that would dissolve lead, i.e., coating the hanging carcass with vinegar or the use of acidic marinades in cooking. Exposure to lead from spent bullets is easily preventable if health-minded hunters use lead-free copper bullets now widely available and generally regarded as fully comparable to lead-based bullets for use in hunting (Carter 2007). The potential for toxic exposure to copper from these bullets is presumably insignificant because little or no fragmentation occurs (Hunt et al. 2006), and there is no meat wastage from having to discard tissue suspected of contamination.

Fragmenting lead bullets have been in use for hunting since the early 1900s (Stroud and Hunt 2009). Although hunter numbers have diminished slightly in recent years, there were 10.7 million big game hunters in the United States in 2006, the majority of whom still use lead-based bullets (USFWS 2006, Watson and Avery 2009). Many state wildlife agencies annually issue multiple deer harvest permits to individuals, effectively offering venison as a year-round protein staple for some families; game meat is the principal source of protein for a considerable proportion of Alaska's population (Titus et al. 2009). Hunter-donated venison to food pantries and shelters for low income families in most states produced an estimated minimum of 9 million venison meals associated with the 2007/08 hunting season (Avery and Watson 2009). With these concerns, we anticipate that health sciences will further examine the bioavailability of lead from bullets and shot, the epidemiology of exposure, and the possible consequences among hunters, their families, and others who consume venison.

ACKNOWLEDGEMENTS

The data were collected as part of The Peregrine Fund's California Condor Restoration Project, which is supported by the U.S. Fish and Wildlife Service, Arizona Game and Fish Department, Bureau of Land Management, The Charles Engelhard Foundation, Liz Claiborne and Art Ortenberg Foundation, Nina Mason Pulliam Charitable Trust, Grand Canyon Conservation Fund, National Fish and Wildlife Foundation, Jane Smith Turner Foundation, and other important donors. We thank P. and L. Widener, R. Berry, P. Jenny, B. Mutch, A. Montoya, P. Juergens, B. Oakleaf, R. Green, T. Hunt, A. Siedenstrang, the Wyoming Game and Fish Department, the University of Idaho Analytical Sciences Laboratory-Holm Research Center, and The Peregrine Fund Research Library for help with this project. S. McGeehan and T. Case performed laboratory analyses for heavy metals. D. Lewis, G. Turner, G. Van Orden, and J. Luft provided care for the pigs and performed blood collection.

LITERATURE CITED

AVERY, D., AND R. T. WATSON. 2009. Distribution of venison to humanitarian organizations in the U.S. and Canada. *In* R. T. Watson, M. Fuller, M. Pokras, and W. G. Hunt (Eds.). Ingestion of

Lead from Spent Ammunition: Implications for Wildlife and Humans. The Peregrine Fund, Boise, Idaho, USA. DOI:10.4080/ilsa.2009.0114.

BJERREGAARD, P., P. JOHANSEN, G. MULVAD, H. S. PEDERSEN, AND J. C. HANSEN. 2004. Lead sources in human diet in Greenland. Environmental Health Perspectives 112:1496-1498. DOI: 10.1289/ehp.7083.

BORJA-ABURTO, V. H., I. HERTZ-PICCIOTTO, M. R. LOPEZ, P. FARIAS, C. RIOS, AND J. BLANCO. 1999. Blood lead levels measured prospectively and risk of spontaneous abortion. American Journal of Epidemiology 150:590-597.

BRAUN, J. M., R. S. KAHN, T. FROEHLICH, P. AUINGER, AND B. P. LANPHEAR. 2006. Exposure of environmental toxicants and attention deficit hyperactivity disorder in U.S. children. Environmental Health Perspectives 114:1904-1909. DOI: 10.1289/ehp.9478.

BURGER, J., R. A. KENNAMER, I. L. BRISBIN, JR., AND M. GOCHFELD. 1998. A risk assessment for consumers of Mourning Doves. Risk Analysis 18:563-573.

CARTER, A. 2007. Xtraordinary. Shooting Illustrated Magazine, July 2007. National Rifle Association Publication. Fairfax, Virginia, USA.

CDC (Centers for Disease Control and Prevention USA). 1991. Preventing lead poisoning in young children. [Online.] Available at http://www.cdc.gov/nceh/lead/publications/prevleadpoisoning.pdf. Accessed 2008 November 3.

CECIL, K. M., C. J. BRUBAKER, C. M. ADLER, K. N. DIETRICH, M. ALTAYE, J. C. EGELHOFF, S. WESSEL, I. ELANGOVAN, R. HORNUNG, K. JARVIS, AND B. P. LANPHEAR. 2008. Decreased brain volume in adults with childhood lead exposure. PLoS Med 5:741-750.
DOI: 10.1371/journal.pmed.0050112.

CORNATZER, W. E., E. F. FOGARTY, AND E. W. CORNATZER. 2009. Qualitative and quantitative detection of lead bullet fragments in random venison packages donated to the Community Action Food Centers of North Dakota, 2007. *In* R. T. Watson, M. Fuller, M. Pokras, and W. G. Hunt (Eds.). Ingestion of Lead from Spent Ammunition: Implications for Wildlife and Humans. The Peregrine Fund, Boise, Idaho, USA. DOI 10.4080/ilsa.2009.0111.

DOBROWOLSKA, A., AND M. MELOSIK. 2008. Bullet-derived lead in tissues of the Wild Boar (*Sus scrofa*) and Red Deer (*Cervus elaphus*). European Journal of Wildlife Research 54:231–235. DOI: 10.1007/s10344-007-0134-y.

EKONG, E. B., B. G. JAAR, AND V. M. WEAVER. 2006. Lead-related nephrotoxicity: A review of the epidemiologic evidence. Kidney International 70:2074-2084.
DOI: 10.1038/sj.ki.5001809.

GILBERT, S. G. AND B. WEISS. 2006. A rationale for lowering the blood lead action level from 10 to 2 μg/dL. NeuroToxicology 27:693–701. DOI: 10.1016/j.neuro.2006.06.008.

HANNING, R. M., R. SANDHU, A. MACMILLAN, L. MOSS, L. J. S. TSUJI, AND E. NIEBOER. 2003. Impact on blood Pb levels of maternal and early infant feeding practices of First Nation Cree in the Mushkegowuk Territory of northern Ontario, Canada. Journal of Environmental Monitoring 5:241-245. DOI: 10.1039/b208220a.

HAUSER, R., O. SERGEYEV, S. KORRICK, M. M. LEEM, B. REVICH, E. GITIN, J. S. BURNS, AND P. L. WILLIAMS. 2008. Association of blood lead levels with onset of puberty in Russian boys. Environmental Health Perspectives 116:976-980. DOI: 10.1289/ehp.10516.

HUNT, W. G., W. BURNHAM, C. N. PARISH, K. K. BURNHAM, B. MUTCH, AND J. L. OAKS. 2006. Bullet fragments in deer remains: Implications for lead exposure in avian scavengers. Wildlife Society Bulletin 34:167-170. Also DOI: 10.4080/ilsa.2009.0123.

IQBAL, S. 2008. Epi-Aid Trip Report: Assessment of human health risk from consumption of wild game meat with possible lead contamination among the residents of the State of North Dakota. National Center for Environmental Health, Centers for Disease Control and Prevention: Atlanta, Georgia, USA. [Online.] Available at http://www.rmef.org/NR/rdonlyres/F07627AA-4D94-4CBC-B8FD-4F4F18401303/0/ND_report.pdf.
Accessed 2009 March 23.

JOHANSEN, P., G. ASMUND, AND F. RIGET. 2004. High human exposure to lead through consumption of birds hunted with lead shot. Environmental Pollution 127:125-129.
DOI: 10.1016/S0269-7491(03)00255-0.

JOHANSEN, P., H. S. PEDERSEN, G. ASMUND, AND F. RIGET. 2006. Lead shot from hunting as a source of lead in human blood. Environmental Pollution 142:93-97. DOI: 10.1016/j.envpol.2005.09.015.

JUSKO, T. A., C. R. HENDERSON, B. P. LANPHEAR, D. A. COREY-SLECHTA, AND P. J. PARSONS. 2008. Blood lead concentrations <10 μg/dL and child intelligence at 6 years of age. Environmental Health Perspectives 116:243-248. DOI: 10.1289/ehp.10424.

KRONE, O., N. KENNTNER, A. TRINOGGA, M. NADJAFZADEH, F. SCHOLZ, J. SULAWA, K. TOTSCHEK, P. SCHUCK-WERSIG, AND R. ZIESCHANK. 2009. Lead poisoning in White-tailed Sea Eagles: Causes and approaches to solutions in Germany. *In* R. T. Watson, M. Fuller, M. Pokras, and W. G. Hunt (Eds.). Ingestion of Lead from Spent Ammunition: Implications for Wildlife and Humans. The Peregrine Fund, Boise, Idaho, USA. DOI: 10.4080/ilsa.2009.0207.

LANPHEAR, B. P., R. HORNUNG, J. KHOURY, K. YOLTON, P. BAGHURST, D. C. BELLINGER, R. L. CANFIELD, K. N. DIETRICH, R. BORNSCHEIN, T. GREENE, S. J. ROTHENBERG, H. L. NEEDLEMAN, L. SCHNAAS, G. WASSERMAN, J. GRAZIANO, AND R. ROBERTS. 2005. Low-level environmental lead exposure and children's intellectual function: An international pooled analysis. Environmental Health Perspectives 113:894-899. DOI: 10.1289/ehp.7688.

LEVESQUE, B., J. F. DUCHESNE, C. GARIEPY, M. RHAINDS, P. DUMAS, A. M. SCHEUHAMMER, J. F. PROULX, S. DERY, G. MUCKLE, F. DALLAAIRE, AND E. DEWAILLY. 2003. Monitoring of umbilical cord blood lead levels and sources assessment among the Inuit. Occupational and Environmental Medicine 60:693-695.

LEVIN, R., M. J. BROWN, M. E. KASHTOCK, D. E. JACOBS, E. A. WHELAN, J. RODMAN, M. R. SCHOCK, A. PADILLA, AND T. SINKS. 2008. U.S. lead exposures in U.S. children, 2008: Implications for prevention. Environmental Health Perspectives 116:1285-1293. DOI: 10.1289/ehp.11241.

LUSTBERG, M., AND E. SILBERGELD. 2002. Blood lead levels and mortality. Archives of Internal Medicine 162:2443-2449.

MATEO, R., M. RODRÍGUEZ-DE LA CRUZ, D. VIDAL, M. REGLERO, AND P. CAMARERO. 2007. Transfer of lead from shot pellets to game meat during cooking. Science of the Total Environment 372:480-485. DOI: 10.1016/j.scitotenv.2006.10.022.

MENKE, A., P. MUNTNER, V. BATUMANN, E. SILBERGELD, AND E. GUALLAR. 2006. Blood lead below 0.48 μmol/L (10μg/dL) and mortality among US adults. Circulation 114:1388. DOI: 10.1161/circulationaha.106.628321.

NEEDLEMAN, H. L. 2004. Lead poisoning. Annual Review of Medicine 55:209–222.

NEEDLEMAN, H. L., C. MCFARLAND, R. B. NESS, S. E. FIENBERG, AND M. J. TOBIN. 2002. Bone lead levels in adjudicated delinquents: A case control study. Neurotoxicology and Teratology 24:711-717.

SCHNAAS, L., S. J. ROTHENBERG, M. F. FLORES, S. MARTINEZ, C. HERNANDEZ, E. OSORIO, S. R. VELASCO, AND E. PERRONI. 2006. Reduced intellectual development in children with prenatal lead exposure. Environmental Health Perspectives 114:791-797. DOI: 10.1289/ehp.8552.

SCHWARTZ, B. S., AND W. F. STEWART. 2007. Lead and cognitive function in adults: A questions and answers approach to a review of the evidence for cause, treatment, and prevention. International Review of Psychiatry 19: 671–692. DOI: 10.1080/09540260701797936.

SHIH, R. A., H. HU, M. G. WEISSKOPH, AND B. S. SCHWARTZ. 2007. Cumulative lead dose and cognitive function in adults: a review of studies that measured both blood lead and bone lead. Environmental Health Perspectives 115:483-492. DOI: 10.1289/ehp.9786.

STROUD, R. K., AND W. G. HUNT. 2009. Gunshot wounds: A source of lead in the environment. *In* R. T. Watson, M. Fuller, M. Pokras, and W. G. Hunt (Eds.). Ingestion of Lead from Spent Ammunition: Implications for Wildlife and Humans. The Peregrine Fund, Boise, Idaho, USA. DOI 10.4080/ilsa.2009.0109.

TELLEZ-ROJO, M. M., M. HERNANDEZ-AVILA, H. LAMADRID-FIGUERUA, D. SMITH, L. HERNANDEZ-CARDENA, A. MERCADO, A. ARO, J. SCHWARTZ, AND H. HU. 2004. Impact of bone lead and bone resorption on plasma and whole blood lead levels during pregnancy. American Journal of Epidemiology 160:668-678. DOI: 10.1093/aje/kwh271.

TITUS, K., T. L. HAYNES, AND T. F. PARAGI. 2009. The importance of moose, caribou, deer and small game in the diet of Alaskans. *In* R. T. Watson, M. Fuller, M. Pokras, and W. G. Hunt (Eds.). Ingestion of Lead from Spent Ammunition: Implications for Wildlife and Humans. The Peregrine Fund, Boise, Idaho, USA. DOI: 10.4080/ilsa.2009.0312.

TSUJI, L. J. S., B. C. WAINMAN, I. D. MARTIN, C. SUTHERLAND, J-P. WEBER, P. DUMAS, AND E. NIEBOER. 2008a. The identification of lead ammunition as a source of lead exposure in First Nations: The use of lead isotope ratios. Science of the Total Environment 393:291-298. DOI: 10.1016/j.scitotenv.2008.01.022.

TSUJI, L. J. S., B. C., WAINMAN, I. D. MARTIN, C. SUTHERLAND, J-P. WEBER, P. DUMAS, AND E. NIEBOER. 2008b. Lead shot contribution to blood lead of First Nations people: The use of lead isotopes to identify the source of exposure. Science of the Total Environment 405:180-185. DOI: 10.1016/j.scitotenv.2008.06.048.

TSUJI, L. J. S., B. C. WAINMAN, I. D. MARTIN, J-P. WEBER, C. SUTHERLAND, E. N. LIBERDA, AND E. NIEBOER. 2008c. Elevated blood-lead levels in First Nation People of northern Ontario Canada: Policy implications. Bulletin of Environmental Contamination and Toxicology 80:14-18. DOI: 10.1007/s00128-007-9281-9.

USEPA (United States Environmental Protection Agency). 2007. Estimation of relative bioavailability of lead in soil and soil-like materials using in vivo and in vitro methods. Office of Solid Waste Emergency Response 9285.7-77. [Online.] Available at http://www.epa.gov/superfund/health/contaminants/bioavailability/lead_tsd_main.pdf. Accessed 2008 November 3.

USFWS AND USCB [United States Fish and Wildlife Service and United States Census Bureau]. 2006. 2006 National survey of fishing, hunting, and wildlife-associated recreation. [Online.] Available at http://library.fws.gov/nat_survey2006_final.pdf. Accessed 2008 Nov 3.

WARREN, C. 2000. Brush with Death: A Social History of Lead Poisoning. Johns Hopkins University Press, Baltimore, Maryland, USA.

WATSON, R. T., AND D. AVERY. 2009. Hunters and anglers at risk of lead exposure in the United States. *In* R. T. Watson, M. Fuller, M. Pokras, and W. G. Hunt (Eds.). Ingestion of Lead from Spent Ammunition: Implications for Wildlife and Humans. The Peregrine Fund, Boise, Idaho, USA. DOI: 10.4080/ilsa.2009.0117.

WRIGHT, J. P., K. N. DIETRICH, M. D. RIS, R. W. HORNUNG, S. D. WESSEL, B. P. LANPHEAR, M. HO, AND M. N. RAE. 2008. Association of prenatal and childhood blood lead concentrations with criminal arrests in early adulthood. PLoS Medicine 5:732-740. DOI: 10.1371/journal.pmed.0050101.

QUALITATIVE AND QUANTITATIVE DETECTION OF LEAD BULLET FRAGMENTS IN RANDOM VENISON PACKAGES DONATED TO THE COMMUNITY ACTION FOOD CENTERS OF NORTH DAKOTA, 2007

WILLIAM E. CORNATZER, EDWARD F. FOGARTY, AND ERIC W. CORNATZER

University of North Dakota School of Medicine, Southwest Campus, Bismarck, ND 58501, USA. E-mail: doccornatzer@qwestoffice.net

ABSTRACT.—We studied randomly selected ground venison packages donated to the Community Action Food Centers of North Dakota by the Hunters For The Hungry Association. These packages were studied by high resolution computerized tomography imaging and x-ray fluoroscopy for qualitative detection of metal fragments. Quantitative measurements of lead levels in both randomly selected and fluoroscopic image guided site-specific subsamples from packages were performed. This study documented a health risk from lead exposure to humans consuming venison. *Received 30 July 2008, accepted 30 October 2008.*

CORNATZER, W. E., E. F. FOGARTY, AND E. W. CORNATZER. 2009. Qualitative and quantitative detection of lead bullet fragments in random venison packages donated to the Community Action Food Centers of North Dakota, 2007. *In* R. T. Watson, M. Fuller, M. Pokras, and W. G. Hunt (Eds.). Ingestion of Lead from Spent Ammunition: Implications for Wildlife and Humans. The Peregrine Fund, Boise, Idaho, USA. DOI 10.4080/ilsa.2009.0111

Key words: Computed tomography imaging, health risk, humans, lead, venison.

STUDIES OF LEAD TOXICITY in the diet of California Condors (*Gymnogyps californianus*) have shown there are small particles of lead in the tissues of deer shot with high velocity rifle bullets (Hunt et al. 2006). Reports from Canada and Greenland have shown a statistically significant correlation between elevated serum lead levels in people and the consumption of wild game killed with lead bullets (Bjerregaard et al. 2004, Dewailly et al. 2001, Tsuji et al. 2008). Preliminary research presented at The Peregrine Fund's Board of Directors meeting in May of 2007 (Parish pers. comm.) showed small metal fragments in processed venison. Based on these data, we conjecture that there might be lead fragments from rifle bullets in venison consumed by the general population.

METHODS AND MATERIALS

One hundred, one-pound ground venison packages were randomly selected from the Community Action Food Pantry program in North Dakota. The venison had been donated by the Hunters for the Hungry Program in the fall of 2007. The sample of 100 was selected from a total of 15,250 donated one-pound packages. High definition CT scan and fluoroscopy were performed on the sample for qualitative detection of metal fragments. In conjunction with the North Dakota Health Department, fifteen of the 100 randomly selected packages were punch biopsied in a blind fashion yielding 4-g tissue biopsies; this gave 15 random sub-samples from within the randomly selected packages. These specimens were sent to the University of Iowa Hygienic Laboratory for flame absorption atomic spectrometry to detect and quantify the mass of lead in sub-samples. An additional five samples obtained from among the 100 CT screened packages using fluoroscopic image-guided retrieval of metal-containing venison were also submitted for analysis. These image-guided biopsies yielded a maximum of four grams of combined ground venison and metal (Figure 1).

Osirix® DICOM® workstation software running on Mac OS X® was used for visual analysis of CT image data for Hounsfield unit assessments of suspected lead fragments. Objects having Hounsfield unit measurements over 1500 were considered suspicious for metal fragments. Color look-up tables from Osirix® were used for color encoding of CT data.

RESULTS

Qualitative analysis of the randomly selected ground venison samples showed 59 packages out of the 100 had one or more visible metal fragments on high definition computed tomography (Figure 2). Quantitative analysis with flame absorption atomic spectrometry of the fifteen random blind biopsies showed one sample with 120 ppm lead (1 ppm = 1 milligram/kilogram). All five fluoroscopic image-guided biopsies showed elevated lead concentrations varying from 4,200 to 55,000 ppm lead dry weight (Table 1).

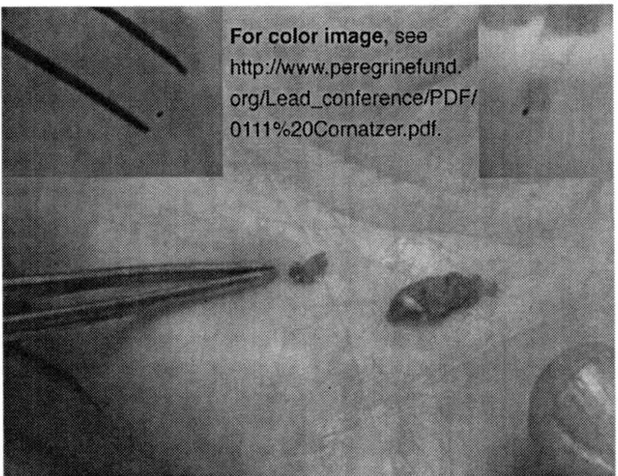

Figure 1. Fluoroscopic image-guided fragment sampling. The fluoroscopic image in upper left shows forceps approaching a metal fragment within a package of ground venison. The upper right image shows a retrieved metal fragment embedded within a small volume of ground venison contained in a glass test tube. Photograph shows forceps pointing to a metal fragment embedded in ground venison.

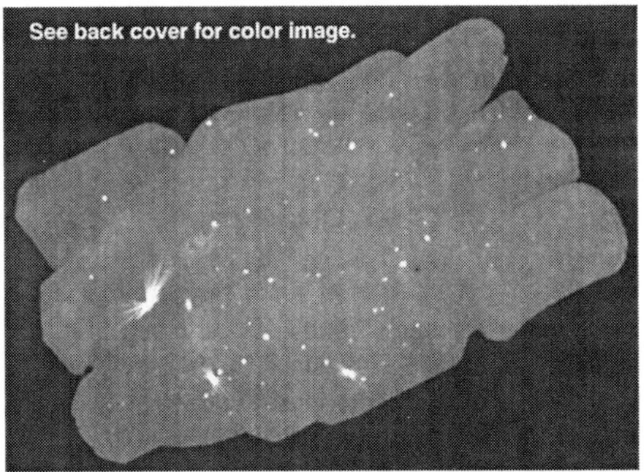

Figure 2. High definition computed tomography (CT) image of *ca.* 20 one-pound venison packages. Bright spots are metal fragments embedded in the tissue.

Table 1. Lead concentrations of five venison samples retrieved by fluoroscopically guided biopsy.

Sample	Lead Concentration (ppm or mg/kg dry weight)
1	52000
2	34000
3	4200
4	55000
5	9700

DISCUSSION

Our study has shown that 59% of 100 randomly selected packages of ground venison donated to the Community Action Food Pantry in North Dakota in the fall of 2007 were contaminated with lead fragments. Venison is a common dietary staple for many families throughout the United States. Lead has been shown to be a major health threat and in children there is no safe minimum threshold of lead exposure. Sources of dietary lead vary from country to country. In the United States, paint chips, dust, jewelry, toys, lead-based gasoline, and lead plumbing (Markowitz 2007) have been identified as sources of lead exposure in the past. Our study reveals lead-based ammunition residues in venison as a source of lead exposure among the USA population that is largely unrecognized as a threat to human health, other than among subsistence hunters of the circumpolar north including Alaska and Canada (Tsuji et al. 2008, Verbrugge et al. 2009, this volume).

ACKNOWLEDGMENTS

We thank the assistance of Terry Dwelle, M.D. and Sandy Washik of the North Dakota State Health Department, Steven Pickard, M.D., Field Officer of the CDC in the State of North Dakota, and Craig Lambrecht, M.D., Bismarck, North Dakota.

LITERATURE CITED

BJERREGAARD, P., P. JOHANSEN, G. MULVAD, H. PEDERSEN, AND J. C. HANSEN. 2004. Environment lead sources in human diet in Greenland. Environmental Health Perspective 112 (15):1496-1498.

DEWAILLY, E. P., S. AYOTT, S. BRUNEAU, G. LEBEL, P. LEVALLOS, AND J. P. WEBER. 2001. Exposure of the Inuit population of Nunivik (Arctic Quebec) to lead and mercury. Archives of Environmental Health 56:350-357.

HUNT, W. G., W. BURNHAM, C. N. PARISH, K. BURNHAM, B. MUTCH, AND J. L. OAKS. 2006. Bullet fragments in deer remains: implications for lead exposure in scavengers. Wildlife Society Bulletin 34:168-171.

MARKOWITZ, M. 2007. Lead Poisoning. Pages 2913–2918 *in* R. M. Kliegman, E. Behrman, H. B. Jenson, and B. F. Stanton (Eds.). Nelson Textbook of Pediatrics, 18th ed. W. B. Saunders Co., Philadelphia, Pennsylvania, USA.

TSUJI, L. J. S., B. C. WAINMAN, I. D. MARTIN, C. SUTHERLAND, J.-P. WEBER, P. DUMAS, AND E. NIEBOER. 2008. The identification of lead ammunition as a source of lead exposure in First Nations: The use of lead isotope ratios. Science of the Total Environment 393:291-298.

VERBRUGGE, L. A., S. G. WENZEL, J. E. BERNER, AND A. C. MATZ. 2009. Human exposure to lead from ammunition in the circumpolar north. *In* R. T. Watson, M. Fuller, M. Pokras, and W. G. Hunt, (Eds.). Ingestion of lead from spent ammunition: Implications for wildlife and humans. The Peregrine Fund, Boise, Idaho, USA. DOI 10.4080/ilsa.2009.0110

DISTRIBUTION OF VENISON TO HUMANITARIAN ORGANIZATIONS IN THE USA AND CANADA

DOMINIQUE AVERY AND RICHARD T. WATSON

The Peregrine Fund, 5668 West Flying Hawk Lane, Boise, ID 83709, USA.
E-mail: rwatson@peregrinefund.org

ABSTRACT.—Hunter organizations throughout the United States provide donations of wild game meat to food banks, shelters, and other humanitarian organizations. These programs provide the needy with a source of low cost, high quality protein. To understand the scale of the programs operating around the country, we conducted a survey of each of the hunter donation programs to determine the amount of venison and other game donated annually. We attempted to identify and question every venison donation program in the country. Survey participants (n = 103) were identified from program websites and by other program participants if no website existed. Surveys were sent to participants via e-mail, and non-responses were contacted by phone to increase response rates. Venison donation programs operate in all 50 states and in at least four Canadian provinces. Figures for the total pounds donated in the 2007/2008 hunting season were received for 74% (n=75) of programs. They reported providing an average of 34,943 lb (SD=96,028) of hunted game meat annually. The meat donated to all responding programs totaled 2,655,730 lb, which provided approximately 10,623,000 meals annually. Many of the programs reported starting only within the last two years and expressed their hope to increase their donations of game meat in the future. Other studies in this conference show that consumers of game meat hunted with lead bullets or shot, which would include beneficiaries of game meat donations, may be at risk of lead exposure. *Received 16 May 2008, accepted 29 July 2008.*

AVERY, D., AND R. T. WATSON. 2009. Distribution of venison to humanitarian organizations in the USA and Canada. *In* R. T. Watson, M. Fuller, M. Pokras, and W. G. Hunt (Eds.). Ingestion of Lead from Spent Ammunition: Implications for Wildlife and Humans. The Peregrine Fund, Boise, Idaho, USA. DOI 10.4080/ilsa.2009.0114

Key words: Donation, humanitarian organization, hunter, lead ingestion, venison.

THE US DEPARTMENT OF AGRICULTURE estimated that 11% of all USA households are food insecure at some time during the year, and 4% experience very low food security (Nord et al. 2007). Between 3.5 million and 6.1 million adults, and 211,000 to 467,000 children, are hungry at some time during the year. Emergency food assistance from food pantries, churches, or food banks is sought by 2.8% of all USA households annually. Part of this need for food is met by hunter organizations which provide donations of wild game meat to food banks, shelters, and other humanitarian organizations. Game meat donation programs provide the needy with a source of low cost, high quality protein which they suggest would not otherwise be received due to the high cost of beef and other meat.

METHODS

We conducted a survey of hunter donation programs in North America to determine the amount of venison and other game donated annually. We attempted to identify and question every program. Programs were identified from websites and by other program participants if no website existed. Surveys were sent to all programs via e-mail, and non-responses were contacted by phone to increase response rates. Information for many organizations was obtained from a national coordinator rather than the specific chapter.

Our original survey consisted of 10 questions, including two about where donations are distributed and the number of women and children receiving game meat. The survey was subsequently revised to exclude this information due to the lack of response. Most food banks, shelters, or other organizations receive the game meat as a general food donation. The hunter donation programs are typically not affiliated with the food banks, so they are not able to provide details of the recipients. Many donation programs supply multiple sources which would make the collection of this information difficult. Additionally, many food banks or shelters do not keep statistics on recipient age or gender since they serve anyone in need.

RESULTS

Summary statistics describing the scope of game meat donation programs in North America are provided in Table 1. Of the states that responded to our survey (n=46) 34 had more than one game meat donation program run by different organizations. Most programs served a few cities or counties (n=57), but some provided a statewide service (n=15). The earliest program began in 1990 and the number of programs has increased steadily since then. Responding programs (35% of the total contacted) reported donating 2.24 million lb (1.02 million kg) of meat, estimated to provide almost 9 million meals in the 2007/08 hunting season. Extrapolation to all 212 known programs suggests that about 25 million meals were provided in the 2007/08 season from donated game meat. The high variance in donation amounts between responding programs produced a 95% confidence interval of this estimate from 6.6 to 44 million meals, the lower limit of which is known to be false because just 35% of programs reported providing at least 9 million meals. Most programs (72%) donated less than 500 lb of meat, while a few (11%) donated over 50,000 lb and one, the Hunters Helping the Hungry sponsored by West Virginia Division of Natural Resources Wildlife Resources Section, donated 588,937 lb of meat in the 2007/08 hunting season, its sixteenth.

DISCUSSION

Hunter donation programs make a substantial contribution to providing food to the needy through humanitarian organizations. The large variance in the amount of meat donated by game meat donation programs makes extrapolation from the responding sample to all known programs imprecise but estimates ranged from a minimum known of 9 million meals to a maximum 95% confidence limit of 44 million meals, with a mean of 25 million meals. Other studies show that consumers of game meat hunted with lead shot or with lead bullets fired from high-powered rifles are at some risk of lead exposure (Johansen et al. 2004, 2006, Tsuji et al. 2008, Verbrugge et al. 2009, Cornatzer et al. 2009, Hunt et al. 2009). Beneficiaries of donated game meat share this risk, and may be at higher risk if they subsist on this source of protein. Hunting method (e.g., rifle, shotgun, bow), ammunition preferences (e.g., lead, non-lead), and variables associated with butchering (e.g., care, skill, cross contamination with other hunters' deer) are additional factors likely to affect the level of risk to lead exposure from hunted game meat. Some of these factors vary somewhat predictably by region (e.g., prevalence of high-powered rifles for deer hunting in western states), while others vary unpredictably between individual hunters and butchers. Without a detailed understanding of these variables, it is impossible to reliably predict the specific risks of lead exposure in donated game meat to all beneficiaries. Women and children are at a higher risk of health effects from lead exposure (Canfield 2003, Kosnett 2009) and the mean annual estimate of 25 million meals of donated meat implies a substantial national rate of lead exposure among children, the group at greatest risk. Game meat donation programs can avoid the risk of lead exposure in the beneficiaries of these worthy humanitarian programs by accepting meat only from hunters who use non-lead ammunition and bow-hunters.

Table 1. Descriptive statistics of game meat donation programs surveyed in the 2007/08 hunting season.

Meat donation programs		212 programs
USA states and Canadian provinces with donation programs		46 USA states 4 Canadian provinces
Number and proportion of programs that responded to our survey		75 (35%)
Amount of game meat donated by responding programs:	Imperial units (as reported)	Metric units
Range =	40 to 588,937 lb	18 to 267,138 kg
Mean =	29,890 lb	13,558 kg
Standard Deviation =	96,028 lb	43,558 kg
Median =	2,200 lb	998 kg
Sum of game meat donated by responding programs =	2,241,730 lb	1,016,830 kg
Estimated number of meals provided by responding programs, assuming one pound of meat makes four meals		8,967,000 meals
Extrapolated number of meals provided by all 212 programs in North America in the 2007/08 hunting season (±95% CI)		25,347,000 (±18,735,700) meals (= ca. 6.6 to 44 million meals)

LITERATURE CITED

CANFIELD, R. L., J. HENDERSON, D. A. CORY-SLECHTA, C. COX, T. A. JUSKO, AND B. P. LANPHEAR. 2003. Intellectual impairment in children with blood lead concentrations below 10 µg per deciliter. The New England Journal of Medicine 348:1517–1526.

CORNATZER, W. E., E. F. FOGARTY, AND E. W. CORNATZER. 2009. Qualitative and quantitative detection of lead bullet fragments in random venison packages donated to the Community Action Food Centers of North Dakota, 2007. *In* R. T. Watson, M. Fuller, M. Pokras, and W. G. Hunt (Eds.). Ingestion of Lead from Spent Ammunition: Implications for Wildlife and Humans. The Peregrine Fund, Boise, Idaho, USA. DOI 10.4080/ilsa.2009.0111

HUNT, W. G., R. T. WATSON, J. L. OAKS, C. N. PARISH, K. K. BURNHAM, R. L. TUCKER, J. R. BELTHOFF, AND G. HART. 2009. Lead bullet fragments in venison from rifle-killed deer: Potential for human dietary exposure. *In* R. T. Watson, M. Fuller, M. Pokras, and W. G. Hunt (Eds.). Ingestion of Lead from Spent Ammunition: Implications for Wildlife and Humans. The Peregrine Fund, Boise, Idaho, USA. DOI 10.4080/ilsa.2009.0112

JOHANSEN, P., G. ASMUND, AND F. RIGET. 2004. High human exposure to lead through consumption of birds hunted with lead shot. Environmental Pollution 127:125–129.

JOHANSEN, P., H. S. PEDERSON, G. ASMUND, AND F. RIGET. 2006. Lead shot from hunting as a source of lead in human blood. Environmental Pollution 142:93–97.

KOSNETT, M. J. 2009. Health effects of low dose lead exposure in adults and children, and preventable risk posed by the consumption of game meat harvested with lead ammunition. *In* R. T. Watson, M. Fuller, M. Pokras, and W. G. Hunt (Eds.). Ingestion of Lead from Spent Ammunition: Implications for Wildlife and Humans. The Peregrine Fund, Boise, Idaho, USA. DOI 10.4080/ilsa.2009.0103

NORD, M., M. ANDREWS, AND S. CARLSON. 2007. Household Food Security in the United States, 2006. Economic Research Report No. (ERR-49) 66 pp. United States Department of Agriculture.

TSUJI, L. J. S., B. C. WAINMAN, I. D. MARTIN, C. SUTHERLAND, J-P WEBER, P. DUMAS, AND E. NIEBOER. 2008. The identification of lead ammunition as a source of lead exposure in First Nations: The use of lead isotope ratios. Science of the Total Environment 393:291–298.

VERBRUGGE, L. 2009. Commentary. *In* R. T. Watson, M. Fuller, M. Pokras, and W. G. Hunt, (Eds.). Ingestion of Lead from Spent Ammunition: Implications for Wildlife and Humans. The Peregrine Fund, Boise, Idaho, USA. DOI 10.4080/ilsa.2009.0320

REGULATION OF LEAD-BASED AMMUNITION AROUND THE WORLD

DOMINIQUE AVERY AND RICHARD T. WATSON

The Peregrine Fund, 5668 West Flying Hawk Lane, Boise, ID 83709, USA.
E-mail: rwatson@peregrinefund.org

ABSTRACT.—The use of lead shot and bullets has been regulated in many countries around the world. Using published literature, we compiled a summary of the extent, type, reason, and date for establishing ammunition legislation in each country where it exists. We documented 29 countries with regulations on lead ammunition. The types of bans varied widely and ranged from partial, voluntary restrictions of the use of lead shot to a total ban on the use and import of lead ammunition. The most common restriction (n=14) was the ban of lead shot for hunting of waterfowl over wetlands. The reason for the ban of lead ammunition was most often due to concerns over populations of waterfowl or avian scavengers. Many countries created legislation in response to the African-Eurasian Waterfowl Agreement's (AEWA) recommendation for the use of nontoxic shot over wetlands. Other countries, such as Liberia, banned the use of lead ammunition after a military coup. A timeline demonstrates the momentum with which this issue is gaining ground with most of the regulations taking place in the past 15 years and further regulations under discussion in many areas. An accumulating body of evidence shows that a reduction in the use of lead for hunting also benefits wildlife and humans who consume wild game. *Received 16 May 2008, accepted 25 July 2008.*

AVERY, D., AND R. T. WATSON. 2009. Regulation of lead-based ammunition around the world. *In* R. T. Watson, M. Fuller, M. Pokras, and W. G. Hunt (Eds.). Ingestion of Lead from Spent Ammunition: Implications for Wildlife and Humans. The Peregrine Fund, Boise, Idaho, USA. DOI 10.4080/ilsa.2009.0115

Key words: Ammunition, country, lead, regulation, state, world.

THE USE OF LEAD SHOT AND BULLETS is regulated in many countries around the world. The types of regulation vary widely and range from partial, voluntary restrictions of the use of lead shot to a total ban on the use and import of lead ammunition. Most countries have implemented regulations due to concerns about the health of migratory waterfowl and avian scavengers. Existing regulations are being strengthened and new ones implemented due to accumulating evidence of the adverse health effects in wildlife and humans of lead from spent ammunition.

METHODS

We used the internet to search for reports and peer-reviewed articles on lead ammunition regulation. We compiled a summary of the extent, type, reason, and date for establishing lead-based ammunition legislation in each country where it exists. Data that could not be verified for accuracy were excluded.

RESULTS

Our search yielded 29 countries that have implemented voluntary or legislative restrictions on the use of lead ammunition (Table 1). Two counties have banned all forms of lead ammunition. Six countries have a partial ban on the use of lead bullets in addition to full bans on lead shot. Four countries have banned the use of lead shot for all hunting. Fourteen countries and Australian territories have banned the use of lead shot in wetlands or for waterfowl hunting. Two countries have voluntary or recommended restrictions in place. Eleven countries and Australian territories have a partial ban on lead shot. Twenty-five states of the United States have implemented regulations on the use of lead shot in addition to the Federal guidelines. Seven countries have implemented increasingly strict regulations on lead ammunition over time.

DISCUSSION

Many countries created legislation in response to the African-Eurasian Waterfowl Agreement's (AEWA) recommendation for the use of nontoxic shot over wetlands (Beintema 2001). Concern about populations of avian scavengers have prompted bans in several countries, such as Japan, and in the United States lead ammunition has been banned in portions of California used by condors. Liberia, where lead shot was banned due to a military coup, was the only country to ban lead ammunition for reasons other than health of wildlife or humans.

Increasingly strict regulation imposed on the use of lead ammunition is a growing trend internationally. A timeline (Table 2) demonstrates the momentum with which this issue is gaining ground with most of the regulations taking place in the past 15 years and further regulations under discussion in many areas. Scandinavian countries have led the way in a full ban on lead ammunition, with Denmark banning lead in 2000 and Sweden scheduled to implement a full ban in 2008.

Evidence of lead exposure in Arctic subsistence hunters who continue to use lead shot (Dewailly et al. 2001, Johansen et al. 2003) suggests that the ban on behalf of eagles has benefited humans as well. Countries worldwide are responding to an accumulating body of evidence that shows that the reduction in the use of lead-based ammunition for hunting benefits wildlife and humans who consume wild game.

LITERATURE CITED

BEINTEMA, N. 2001. Lead poisoning in waterbirds. International Update Report 2000. Wetlands International and UNEP/African-Eurasian Waterbird Agreement Secretariat, Bonn, Germany.

DEWAILLY, E., P. AYOTT, S. BRUNEAU, G. LEBEL, P. LEVALLOIS, AND J. P. WEBER. 2001. Exposure of the Inuit population of Nunavik (Arctic Quebec) to lead and mercury. Archives Environmental Health 56:350–357.

JOHANSEN, P., G. ASMUND, AND F. RIGET. 2004. High human exposure to lead through consumption of birds hunted with lead shot. Environmental Pollution 127:125–129.

Table 1. Comparison of types of lead-based ammunition regulation worldwide in 2008. Asterisk indicates states or other sub-regions of countries.

Country or State	Recommended use of nontoxic shot	Partial ban on lead shot	Ban on lead shot in wetlands or for waterfowl	Ban on lead shot for all hunting	Partial ban on lead ammunition	Ban on all forms of lead ammunition	Ban on hunting	Nontoxic shot regulations in addition to Federal
Austria	Banned prior to 2002							
Australia								
*Capital Territory, AU							Hunting ban on native wildlife	
*Western Australia, AU							Hunting ban on duck and quail	
*South Australia, AU		Banned during duck season, 1998		1993				
*Northern Territory, AU		Banned in hunting reserves, 1998						
*Queensland, AU	2001	Banned at three sites					Hunting ban on duck and quail, 2005	
*Tasmania, AU			2004					
*New South Wales, AU		Ban for duck hunting					Hunting ban on duck	
*Victoria, AU		Banned for duck hunting, 1993	1995					
Denmark	1985		1993	1996		Ban on the import of lead ammunition, 2000		
Belgium		Banned in Ramsar sites, 1993	1998	Ban considered for 2008				
Canada			1997	1999 lead shot banned for hunting game birds				

Country or State	Recommended use of nontoxic shot	Partial ban on lead shot	Ban on lead shot in wetlands or for waterfowl	Ban on lead shot for all hunting	Partial ban on lead ammunition	Ban on all forms of lead ammunition	Ban on hunting	Nontoxic shot regulations in addition to Federal
Cyprus			1993					
Finland			1996					
France			2006					
Germany	1993	Ban in 10 states						
Ghana							Hunting ban in wetlands and irrigation sites	
Hungary			2005					
India							All hunting banned	
Israel							Most wetlands closed to hunting–must use lead shot	
Italy			Proposed date unknown					
Japan					Partial ban on lead ammunition for deer, 2000			
Kenya								
Latvia		Banned in wetland SPA's, 2000						
Liberia			Military coup banned lead shot, 1980					
Malaysia		Date unknown						
Malta		Banned in two wetlands						

Country or State	Recommended use of nontoxic shot	Partial ban on lead shot	Ban on lead shot in wetlands or for waterfowl	Ban on lead shot for all hunting	Partial ban on lead ammunition	Ban on all forms of lead ammunition	Ban on hunting	Nontoxic shot regulations in addition to Federal
Mauritania						Ban on all lead for large game and sport hunting 1975		
Netherlands				1993	Banned for clay pigeon shooting, 2004			
Norway			1991	2005				
Poland	Recommended							
Portugal			Proposed for 2008					
Russia		Some restrictions for wetlands						
South Africa		Partial ban on lead shot for waterfowl						
Spain		Banned in Ramsar sites in 1994	2001					
Sweden			2002		Banned for clay pigeon shooting, 2002	2008		
Switzerland			1998					
Great Britain								
*England		Voluntary ban in 1995	1999					
*Scotland			2005					
*Wales		Banned in SSSI wetlands 2002						
New Zealand		10 or 12 gauge shot banned, 2006						
United States			1991					
*Tejon Ranch, CA						2008		
*Camp Roberts, CA						2007		
*Alabama								

Country or State	Recommended use of nontoxic shot	Partial ban on lead shot	Ban on lead shot in wet-lands or for waterfowl	Ban on lead shot for all hunting	Partial ban on lead ammunition	Ban on all forms of lead ammunition	Ban on hunting	Nontoxic shot regu-lations in addition to Federal
*Alaska								Yes
*Arizona								
*Arkansas								
*California					Banned in Condor range 2008			Yes
*Colorado								
*Connecticut								
*Delaware								
*Florida								
*Georgia								
*Hawaii								
*Idaho								
*Illinois								Yes
*Indiana								
*Iowa								Yes
*Kansas								Yes
*Kentucky								Yes
*Louisiana								Yes
*Maine								Yes
*Maryland								Yes
*Massachusetts								Yes
*Michigan								Yes
*Minnesota								Yes
*Mississippi								
*Missouri								Yes
*Montana								
*Nebraska								Yes
*Nevada								
*New Hampshire								

Country or State	Recommended use of nontoxic shot	Partial ban on lead shot	Ban on lead shot in wet-lands or for waterfowl	Ban on lead shot for all hunting	Partial ban on lead ammunition	Ban on all forms of lead ammunition	Ban on hunting	Nontoxic shot regu-lations in addition to Federal
*New Jersey								Yes
*New Mexico								Yes
*New York								Yes
*North Carolina								Yes
*North Dakota								Yes
*Ohio								Yes
*Oklahoma								
*Oregon								Yes
*Pennsylvania								
*Rhode Island								
*South Carolina								
*South Dakota								Yes
*Tennessee								
*Texas								
*Utah								Yes
*Vermont								
*Virginia								
*Washington								Yes
*Wyoming								Yes

Table 2. Regulation of lead ammunition over time.

Date	Country and type of regulation
1975	Mauritania hunting laws prohibit use of toxic ammunition for large game and sport hunting.
1980	Liberia bans lead shot due to military coup.
1985	Denmark hunters initiate use of nontoxic shot.
1989	
1990	
1991	USA bans the use of lead shot over wetlands.
	Norway bans lead shot in wetlands for hunting of all ducks, geese, and waders.
1992	
1993	South Australia, Australia bans the use of lead shot.
	Victoria, Australia bans the use of lead shot during duck season.
	Denmark bans the use of lead shot over wetlands.
	Cyprus bans the use of lead shot over wetlands.
	Germany bans the use of lead shot over wetlands in 8 Lander and recommends voluntary use of nontoxic shot over all wetlands.
	Belgium bans the use of lead shot over Ramsar wetlands.
	February-Netherlands bans the use of lead shot for hunting over wetlands.
1994	
1995	Victoria, Australia bans the use of lead shot for duck hunting.
	Netherlands bans the use of lead shot in all hunting.
	UK instills voluntary use of nontoxic shot over wetlands.
1996	Denmark bans the use of lead shot in all hunting.
	Finland bans the use of lead shot over wetlands.
1997	Canada bans the use of lead shot for hunting migratory game birds near water.
1998	Switzerland bans the use of lead shot for hunting over wetlands and shallow water areas.
	Belgium bans the use of lead shot over all wetlands.
	Northern Territory, Australia bans the use of lead shot during duck season.
1999	England prohibits use of lead shot over wetlands and for all waterfowl.
	Canada bans the use of lead shot for hunting all migratory game birds (with a few exceptions).
2000	Japan bans the use of lead bullets for deer hunting in Hokkaido.
	Latvia bans the use of lead shot over wetland special protected areas.
	Spain bans the use of lead shot at Ramsar sites.
	Denmark bans the import of all lead products including ammunition.
2001	Queensland, Australia instills voluntary ban on the use of lead shot over wetlands.
	1 June-Spain bans the use of lead shot over all wetlands.
2002	Sweden bans the use of lead shot over wetlands.
	Wales bans the use of lead shot over wetland sites of special scientific interest.
	Sweden bans the use of lead shot for clay pigeons.
2003	
2004	Netherlands bans the use of lead shot for clay pigeons.
	Tasmania, Australia bans the use of lead shot over public wetlands and Crown Land.
2005	Hungary bans the use of lead shot over wetlands.
	31 March-Scotland bans the use of lead shot over wetlands.
	Norway bans the use of lead shot for all hunting.
2006	New Zealand bans 10 and 12 gauge shot for waterfowl near water.
	France bans the use of lead shot over wetlands.
2007	Camp Roberts, California, USA bans all lead ammunition for hunting.
	Fort Hunter Liggett, California, USA bans lead ammunition for hunting.
2008	Tejon Ranch, California, USA bans all lead ammunition for hunting.
	Camp Roberts, California, USA bans use of all lead shot and ammunition for hunting.
	California, USA bans the use of lead ammunition when taking big game and coyotes in the California Condor range in California.
	Sweden enacts a total ban on lead shot and ammunition.
	Belgium considers a total ban on the use of lead shot.
	Portugal proposes a ban on the use of lead shot in wetlands.

HUNTERS AND ANGLERS AT RISK OF LEAD EXPOSURE IN THE UNITED STATES

RICHARD T. WATSON AND DOMINIQUE AVERY

The Peregrine Fund, 5668 West Flying Hawk Lane, Boise, ID 83709, USA
E-mail: rwatson@peregrinefund.org

ABSTRACT.—Lead (Pb) is toxic and known to have neurological and cognitive development effects at low levels in children, is associated with increased mortality from heart attack and stroke at levels >2 µg/dL blood lead level in adults, and has other well established and serious health effects in people and wildlife. We used the *US Fish and Wildlife Service 2006 National Survey of Fishing, Hunting, and Wildlife-Associated Recreation* to assess the numbers and proportions of state populations that may be at risk of lead exposure from lead-based ammunition and fishing gear. In 2006, 12.5 million people (6% of the population) aged 16 years and older in the United States hunted on 220 million days, of which 45% were urban residents and 55% were rural. An estimated 1.6 million children aged 6 to 15 years hunted. Although 28% of hunters reported using bow and arrows, 93% used a rifle or shotgun, and 20% used muzzle-loaders, indicating some hunters used more than one method. Big game hunting, such as deer and elk, was most popular (10.7 million hunters on 164 million days), followed by small game hunting, such as squirrels and rabbits (4.8 million hunters on 52 million days), migratory bird hunting, such as waterfowl and doves (2.3 million hunters on 20 million days), and other animals, such as raccoons and groundhogs (1.1 million hunters on 15 million days). Texas had the largest number of hunters (979,000), followed by Pennsylvania (933,000) and Michigan (721,000). Montana (19%) had the largest proportion of hunters among the state's population, followed by North Dakota (17%), and Wisconsin and South Dakota (each 15%). Lead was made illegal for hunting waterfowl in 1992, but it continues to be the favored metal for bullets and shot used for big game hunting, small game hunting, upland bird hunting (e.g., doves), and varmint (other animals) hunting. In addition to lead exposure from handling ammunition (e.g., hunters who load their own ammunition), lead exposure can occur through inhalation of vapor upon firing, and from ingestion of game meat contaminated with bullet fragments and shot. In 2006, 30 million people (13% of the population) aged 16 years and older in the United States fished on 517 million days. Freshwater anglers numbered 25.4 million and they fished on 433 million days. Saltwater anglers numbered 7.7 million and they fished on 86 million days. Texas had the largest number of anglers (2.3 million), followed by Florida (1.9 million) and California (1.6 million). Minnesota (28%) had the largest proportion of anglers among the state's population, followed by Alaska (27%), and Wyoming (24%). Lead exposure in anglers can occur in handling and making lead sinkers, and through accidental ingestion. Frequent loss of fishing gear annually contaminates aquatic ecosystems with lead, and causes wildlife mortality and potential sub-lethal effects. Non-lead substitutes for lead ammunition and fishing gear are available and recommended to benefit human and wildlife health. *Received 21 May 2008, accepted 28 July 2008.*

WATSON, R.T. AND D. AVERY. 2009. Hunters and anglers at risk of lead exposure in the United States. *In* R. T. Watson, M. Fuller, M. Pokras, and W. G. Hunt (Eds.). Ingestion of Lead from Spent Ammunition: Implications for Wildlife and Humans. The Peregrine Fund, Boise, Idaho, USA. DOI 10.4080/ilsa.2009.0117

Key words: Angler, exposure, hunter, lead, risk.

LEAD (PB) IS TOXIC and known to have neurological and cognitive development effects at low levels in children (Canfield et al. 2003), is associated with increased mortality from heart attack and stroke at levels ≥2 μg/dL blood lead level in adults (Menke et al. 2006), and has other well established and serious health effects in people and wildlife (Kosnett 2009, Pokras 2009, this volume). Lead shot was banned for hunting waterfowl in the USA in 1991 due to its effects on waterfowl populations and secondary poisoning of avian scavengers, including Bald Eagles (*Haliaeetus leucocephalus*), but it continues to be the favored metal for bullets and shot used for big game hunting (e.g., deer, elk, moose and caribou), small game hunting (e.g., rabbits), upland bird hunting (e.g., doves), and varmint hunting (e.g., coyote, ground squirrel, and prairie dogs). In addition to lead exposure from handling ammunition (e.g., hunters who load their own ammunition), lead exposure can occur through inhalation of vapor upon firing, and from ingestion of game meat contaminated with bullet fragments and shot (Johansen et al. 2004, 2006, Tsuji et al. 2008). Lead exposure in anglers can occur in handling and making lead sinkers, and through accidental ingestion (St. Clair and Benjamin 2008).

METHODS

We used the *US Fish and Wildlife Service 2006 National Survey of Fishing, Hunting, and Wildlife-Associated Recreation* to assess the numbers and proportions of state populations that may be at risk of lead exposure from lead-based ammunition and fishing gear.

RESULTS

In 2006, 12.5 million people (6% of the population) aged 16 years and older in the United States hunted on 220 million days, of which 45% were urban residents and 55% were rural. An estimated 1.6 million children aged 6 to 15 years hunted. Although 28% of hunters reported using a bow and arrows, 93% used a rifle or shotgun, and 20% used muzzle-loaders, indicating some hunters used more than one method. Big game hunting, such as deer and elk, was most popular (10.7 million hunters on 164 million days), followed by small game hunting, such as squirrels and rabbits (4.8 million hunters on 52 million days), migratory bird hunting, such as waterfowl and doves (2.3 million hunters on 20 million days), and other animals, such as raccoons and groundhogs (1.1 million hunters on 15 million days).

Texas had the largest number of hunters (979,000), followed by Pennsylvania (933,000) and Michigan (721,000, Figure 1). Montana (19%) had the largest proportion of hunters among the state's population, followed by North Dakota (17%), and Wisconsin and South Dakota (each 15%, Figure 2).

In 2006, 30 million people (13% of the population) aged 16 years and older in the United States fished on 517 million days. Freshwater anglers numbered 25.4 million and they fished on 433 million days. Saltwater anglers numbered 7.7 million and they fished on 86 million days. Texas had the largest number of anglers (2.3 million), followed by Florida (1.9 million) and California (1.6 million, Figure 3). Minnesota (28%) had the largest proportion of anglers among the state's population, followed by Alaska (27%), and Wyoming (24%, Figure 4).

DISCUSSION

Hunting is popular throughout the USA, and in some states, such as Alaska where 11% of the population hunts, much of the state's rural poor subsist on hunting (Verbrugge et al. 2009, Titus et al. 2009, this volume). In at least 14 other states ≥11% of the population hunt, and although annual hunting limits vary by state, those states where it is most popular could reasonably expect significant dietary intake of game meat among some sectors of the population and therefore potentially significant lead exposure from lead-based ammunition.

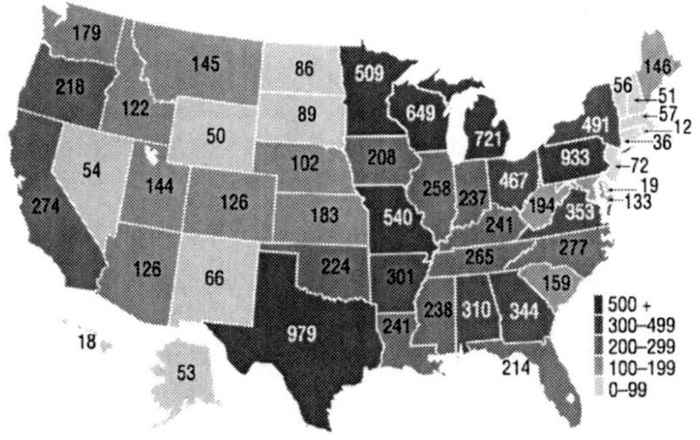

Figure 1. The number (in thousands) of hunters in each state in 2006.

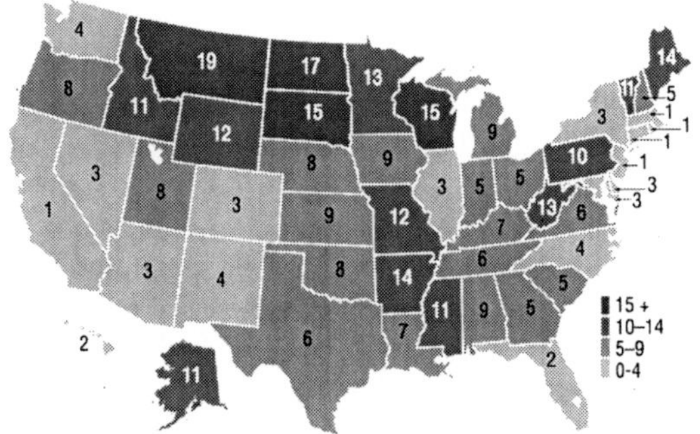

Figure 2. The proportion (%) of the population in each state that hunted in 2006.

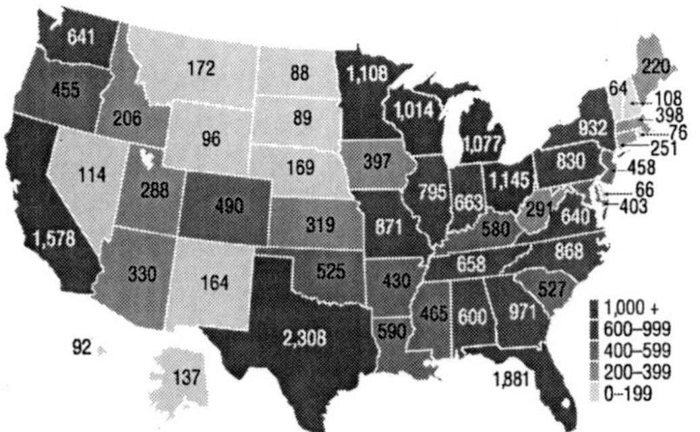

Figure 3. The number (in thousands) of anglers in each state in 2006.

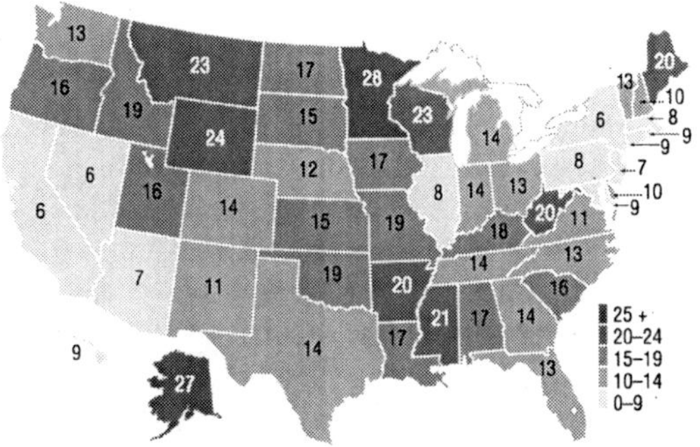

Figure 4. The proportion (%) of the population in each state that fished in 2006.

The number of people exposed to lead from ingestion of contaminated game meat is unknown, but likely includes hunters, their families and friends, and the beneficiaries of donated game meat to humanitarian organizations (Avery and Watson 2009, Cornatzer et al. 2009, Hunt et al. 2009, Verbrugge et al. 2009, this volume) and therefore could exceed the number of hunters by several orders of magnitude. The frequency and amount of lead exposure from ingestion of contaminated game meat is unknown, but likely to be greater in people who consume game meat frequently, such as beneficiaries of donated game meat (Avery and Watson 2009, this volume) and others who subsist on this source of protein (Verbrugge et al. 2009, Titus et al. 2009, this volume). An assessment of the health effects of lead exposure from game meat shot with lead-based ammunition should target frequent or long-term consumers using both blood lead concentration to measure short-term exposure (the half-life for lead in blood is about 35 days, Gordon et al. 2002) and K-shell X-ray fluorescence of bone to measure long-term exposure (Hu et al. 1995). States shown in Figure 2 with a high proportion of the population engaged in hunting, such as Montana, North and South Dakota, Wisconsin, Arkansas, and Maine would be good candidates for further study, as would states with large absolute numbers of hunters, such as Texas, Pennsylvania, Michigan, and Wisconsin (Figure 1).

The large number of hunters (about 12.5 million), who collectively hunt annually on 220 million days, causes concern about the annual mass of lead from unrecovered bullets and shot accumulating in the environment. Unrecovered game and offal is a significant source of lead exposure in scavenging and predatory wildlife resulting in mortality and potential sub-lethal effects (Hunt et al. 2006, Knopper et al. 2006, Pauli and Buskirk 2007, Pokras et al. 2009, Rattner et al. 2009, this volume). Consumption of unrecovered shot is a source of lead exposure to game birds that can lead to secondary exposure in predators and humans (Franson et al. 2009, Schulz et al. 2009, Bingham et al. 2009, this volume).

The number of people exposed to lead from lead-based fishing gear, and the frequency and amount of lead exposure is unknown. Frequent loss of fishing gear annually contaminates aquatic ecosystems with lead, and causes wildlife mortality and potential sub-lethal effects (Friend et al. 2009, Pokras et al. 2009, Rattner et al. 2009, this volume). States in which these effects may be greatest are illustrated in Figure 4 as those with the highest proportion of the population engaged in fishing, such as Alaska, Minnesota, Wisconsin, Wyoming, and Montana.

Large numbers of people are at risk of lead exposure from recreational and subsistence hunting. Although there is special concern for children and women in their child-bearing years, hunters, their families and friends, and beneficiaries of donated game meat may also be at risk. Large amounts of lead are annually deposited in aquatic and terrestrial ecosystems from unrecovered lead bullets, shot and fishing gear, where it can accumulate to appreciable mass over time. Hunters and anglers can avoid the risk of lead exposure by using non-lead ammunition and fishing gear to the benefit of both human and wildlife health.

LITERATURE CITED

AVERY, D., AND R. T. WATSON. 2009. Distribution of venison to humanitarian organizations in the USA and Canada. *In* R. T. Watson, M. Fuller, M. Pokras, and W. G. Hunt (Eds.). Ingestion of Lead from Spent Ammunition: Implications for Wildlife and Humans. The Peregrine Fund, Boise, Idaho, USA. DOI 10.4080/ilsa.2009.0114

BINGHAM, R. J., R. T. LARSEN, J. A. BISSONETTE, AND J. T. FLINDERS. 2009. Causes and consequences of ingested lead pellets in Chukars. Extended abstract *in* R. T. Watson, M. Fuller, M. Pokras, and W. G. Hunt (Eds.). Ingestion of Lead from Spent Ammunition: Implications for Wildlife and Humans. The Peregrine Fund, Boise, Idaho, USA. DOI 10.4080/ilsa.2009.0204

CANFIELD, R. L., J. HENDERSON, D. A. CORY-SLECHTA, C. COX, T. A. JUSKO, AND B. P. LANPHEAR. 2003. Intellectual impairment in children with blood lead concentrations below 10 μg per deciliter. The New England Journal of Medicine 348:1517–1526.

CORNATZER, W. E., E. F. FOGARTY, AND E. W. CORNATZER. 2009. Qualitative and quantitative detection of lead bullet fragments in random venison packages donated to the community action food centers of North Dakota, 2007. Extended abstract *in* R. T. Watson, M. Fuller, M. Pokras, and W. G. Hunt (Eds.). Ingestion of Lead from Spent Ammunition: Implications for Wildlife and Humans. The Peregrine Fund, Boise, Idaho, USA. DOI 10.4080/ilsa.2009.0111

FRANSON, J. C., S. P. HANSEN, AND J. H. SCHULZ. 2009. Ingested shot and tissue lead concentrations in Mourning Doves. *In* R. T. Watson, M. Fuller, M. Pokras, and W. G. Hunt (Eds.). Ingestion of Lead from Spent Ammunition: Implications for Wildlife and Humans. The Peregrine Fund, Boise, Idaho, USA. DOI 10.4080/ilsa.2009.0202

FRIEND, M., J. C. FRANSON, AND W. L. ANDERSON. 2009. Biological and societal dimensions of lead poisoning in birds in the USA. *In* R. T. Watson, M. Fuller, M. Pokras, and W. G. Hunt (Eds.). Ingestion of Lead from Spent Ammunition: Implications for Wildlife and Humans. The Peregrine Fund, Boise, Idaho, USA. DOI 10.4080/ilsa.2009.0104

GORDON, J. N., A. TAYLOR, AND P. N. BENNETT. 2002. Lead poisoning: case studies. British Journal of Clinical Pharmacology 53:451–458.

HU, H., A. ARO, AND A. ROTNITZKY. 1995. Bone lead measured by x-ray fluorescence: epidemiologic methods. Environmental Health Perspectives 103, Supp. 1:105–110.

HUNT, W. G., W. BURNHAM, C. N. PARISH, K. BURNHAM, B. MUTCH, AND J. L. OAKS. 2006. Bullet fragments in deer remains: implications for lead exposure in scavengers. Wildlife Society Bulletin 34:168–171.

HUNT, W. G., R. T. WATSON, J. L. OAKS, C. N. PARISH, K. K. BURNHAM, R. L. TUCKER, J. R. BELTHOFF, AND G. HART. 2009. Lead bullet fragments in venison from rifle-killed deer: Potential for human dietary exposure. *In* R. T.

Watson, M. Fuller, M. Pokras, and W. G. Hunt (Eds.). Ingestion of Lead from Spent Ammunition: Implications for Wildlife and Humans. The Peregrine Fund, Boise, Idaho, USA. DOI 10.4080/ilsa.2009.0112

JOHANSEN, P., G. ASMUND, AND F. RIGET. 2004. Environmental Pollution 127:125–129.

JOHANSEN, P., H. S. PEDERSON, G. ASMUND, AND F. RIGET. 2006. Lead shot from hunting as a source of lead in human blood. Environmental Pollution 142:93–97.

KNOPPER, L. D., P. MINEAU, A. M. SCHEUHAMMER, D. E. BOND, AND D. T. MCKINNON. 2006. Carcasses of shot Richardson's Ground Squirrels may pose lead hazards to scavenging hawks. Journal of Wildlife Management 70(1): 295–299.

KOSNETT, M. J. 2009. Health effects of low dose lead exposure in adults and children, and preventable risk posed by the consumption of game meat harvested with lead ammunition. *In* R. T. Watson, M. Fuller, M. Pokras, and W. G. Hunt (Eds.). Ingestion of Lead from Spent Ammunition: Implications for Wildlife and Humans. The Peregrine Fund, Boise, Idaho, USA. DOI 10.4080/ilsa.2009.0103

MENKE, A., P. MUNTNER, V. BATUMAN, E. K. SILBERGELD, AND E. GUALLAR. 2006 Blood lead below 0.48 μmol/L (μg/dL) and mortality among US adults. Circulation: 114:1388–1394.

PAULI, J. N., AND S. W. BUSKIRK. 2007. Recreational shooting of Prairie Dogs: A portal for lead entering wildlife food chains. Journal of Wildlife Management 71(1):103–108.

POKRAS, M. A., AND M. R. KNEELAND. Understanding lead uptake and effects across species lines: A conservation medicine based approach. *In* R. T. Watson, M. Fuller, M. Pokras, and W. G. Hunt (Eds.). Ingestion of Lead from Spent Ammunition: Implications for Wildlife and Humans. The Peregrine Fund, Boise, Idaho, USA. DOI 10.4080/ilsa.2009.0101

RATTNER, B. A., J. C. FRANSON, S. R. SHEFFIELD, C. I. GODDARD, N. J. LEONARD, D. STANG, AND P. J. WINGATE. 2009. Technical review of the sources and implications of lead ammunition and fishing tackle on natural resources. Extended abstract *in* R. T. Watson, M. Fuller, M. Pokras, and W. G. Hunt (Eds.). Ingestion of Lead from Spent Ammunition: Implications for Wildlife and Humans. The Peregrine Fund, Boise, Idaho, USA. DOI 10.4080/ilsa.2009.0106

SCHULZ, J. H., X. GAO, J. J. MILLSPAUGH, AND A. J. BERMUDEZ. 2009. Acute lead toxicosis and experimental lead pellet ingestion in Mourning Doves. Extended abstract *in* R. T. Watson, M. Fuller, M. Pokras, and W. G. Hunt (Eds.). Ingestion of Lead from Spent Ammunition: Implications for Wildlife and Humans. The Peregrine Fund, Boise, Idaho, USA. DOI 10.4080/ilsa.2009.0203

ST. CLAIR, W. S., AND J. BENJAMIN. 2008. Lead intoxication from ingestion of fishing sinkers: a case study and review of the literature. Clinical Pediatrics 47(1):66–70.

TITUS, K., T. L. HAYNES, AND T. F. PARAGI. 2009. The importance of Moose, Caribou, deer and small game in the diets of Alaskans. *In* R. T. Watson, M. Fuller, M. Pokras, and W. G. Hunt (Eds.). Ingestion of Lead from Spent Ammunition: Implications for Wildlife and Humans. The Peregrine Fund, Boise, Idaho, USA. DOI 10.4080/ilsa.2009.0312

TSUJI, L. J. S., B. C. WAINMAN, I. D. MARTIN, C. SUTHERLAND, J-P WEBER, P. DUMAS, AND E. NIEBOER. 2008. The identification of lead ammunition as a source of lead exposure in First Nations: the use of lead isotope ratios. Science of the Total Environment 393:291–298.

US DEPARTMENT OF THE INTERIOR, FISH AND WILDLIFE SERVICE, AND US DEPARTMENT OF COMMERCE, US CENSUS BUREAU. 2006 National Survey of Fishing, Hunting, and Wildlife-Associated Recreation. [Online.] Available at: http://wsfrprograms.fws.gov/Subpages/NationalSurvey/nat_survey2006_final.pdf. Accessed 1 May 2008.

VERBRUGGE, L. A., S. G. WENZEL, J. E. BERNER, AND A. C. MATZ. 2009. Human exposure to lead from ammunition in the circumpolar north. *In* R. T. Watson, M. Fuller, M. Pokras, and W. G. Hunt (Eds.). Ingestion of Lead from Spent Ammunition: Implications for Wildlife and Humans. The Peregrine Fund, Boise, Idaho, USA. DOI 10.4080/ilsa.2009.01

LEAD ISOTOPES INDICATE LEAD SHOT EXPOSURE IN ALASKA-BREEDING WATERFOWL

ANGELA MATZ[1] AND PAUL FLINT[2]

[1]*US Fish and Wildlife Service, Environmental Contaminants Program, Fairbanks, AK 99701, USA.*

[2]*US Geological Survey - Alaska Science Center, 4210 University Dr., Anchorage, AK 99508, USA.*

ABSTRACT.—Although use of lead shot in waterfowl hunting has been banned in the United States since 1991, lead concentrations with possible population-level effects continue to be documented in waterfowl breeding in Alaska, including threatened Spectacled Eiders (*Somateria fischeri*). The presumed source is ingested lead shot, which waterfowl consume intentionally or incidentally while feeding in wetlands containing spent shot. Lead shot is still used in many parts of rural Alaska for subsistence waterfowl hunting. Further, legal use of lead shot for upland game hunting may occur in waterfowl breeding habitats. Availability of spent shot may be prolonged by permafrost, which frequently underlies wetlands used for breeding and retards the sinking of shot beyond the reach of feeding waterfowl. Exposure to lead from shot can be documented using radiographs or dissection, but these methods are cumbersome or applicable only post mortem, respectively. Analysis of blood for total lead and lead isotope ratios (e.g., ^{206}Pb /^{207}Pb) is a simpler and more efficient technique. Lead isotope ratios vary geographically, and lead products such as shot can have distinct, ore-specific signatures. We compared lead isotope ratios from shot and breeding and wintering area sediments to those in blood from Spectacled, King (*S. spectabilis*), and Common Eiders (*S. mollissima*) and Long-tailed Ducks (*Clangula hyemalis*). Birds were sampled on the Yukon-Kuskokwim Delta and the North Slope of Alaska. We also analyzed bird blood for total lead concentrations. Isotopic signatures from birds with relatively high blood lead concentrations were most similar to the isotopic signatures of lead shot, while signatures from birds with low blood lead concentrations closely matched those of local sediments. Further, lead concentrations in sediment samples were very low making sediments an unlikely source for high blood concentrations. Therefore, spent lead shot is available and consumed by breeding waterfowl in Alaska. Although exposure may result from previously used shot, current lead shot use combined with the persistence of lead shot in Alaskan wetlands mandates that management, including outreach and law enforcement, be directed at entirely eliminating the use of lead shot for subsistence hunting.

MATZ, A., AND P. FLINT. 2009. Lead isotopes indicate lead shot exposure in Alaska-breeding waterfowl. Abstract *in* R.T. Watson, M. Fuller, M. Pokras, and W.G. Hunt (Eds.). Ingestion of Lead from Spent Ammunition: Implications for Wildlife and Humans. The Peregrine Fund, Boise, Idaho, USA. DOI 10.4080/ilsa.2009.0113

Key words: Ammunition, isotope, lead, shot, waterfowl, wildlife.

INGESTED SHOT AND TISSUE LEAD CONCENTRATIONS IN MOURNING DOVES

J. CHRISTIAN FRANSON[1], SCOTT P. HANSEN[1,3], AND JOHN H. SCHULZ[2]

[1]*US Geological Survey, National Wildlife Health Center, 6006 Schroeder Road, Madison, WI 53711, USA.*
E-mail: jfranson@usgs.gov

[2]*Missouri Department of Conservation, Resource Science Center, 1110 South College Ave., Columbia, MO 65201, USA.*

ABSTRACT.—A more complete understanding of nonhunting and harvest mortality for Mourning Doves (*Zenaida macroura*) will be critical to improving regional and national harvest management decisions. Poisoning from ingested lead shot is of particular concern in Mourning Doves, which are often hunted on managed shooting fields where lead shot densities can be high, potentially increasing the risk of lead exposure. Previous studies of lead exposure in Mourning Doves have been local in scope and sample sizes have varied widely among areas. We provide an evaluation of lead exposure in 4,884 hunter-harvested Mourning Doves from Arizona, Georgia, Missouri, Oklahoma, Pennsylvania, South Carolina, and Tennessee. Overall, the frequency of ingested lead pellets in gizzards of doves on hunting areas where the use of lead shot was permitted was 2.5%, although we found a high degree of variability among locations. On areas where nontoxic shot was required, 2.4% of Mourning Doves had ingested steel shot. Hatch year (HY) doves had a greater frequency of ingested lead and steel pellets than after hatch year (AHY) birds, suggesting that they either ingested pellets more frequently or that young birds with ingested shot were preferentially harvested over older birds with ingested pellets. In doves without ingested lead pellets, bone lead concentrations were lower on an area requiring the use of nontoxic shot than on areas allowing the use of lead shot. *Received 3 June 2008, accepted 8 August 2008.*

FRANSON, J. C., S. P. HANSEN, AND J. H. SCHULZ. 2009. Ingested shot and tissue lead concentrations in Mourning Doves. *In* R. T. Watson, M. Fuller, M. Pokras, and W. G. Hunt (Eds.). Ingestion of Lead from Spent Ammunition: Implications for Wildlife and Humans. The Peregrine Fund, Boise, Idaho, USA. DOI 10.4080/ilsa.2009.0202

Key words: Ingested lead pellets, ingested steel pellets, lead exposure, lead poisoning, Mourning Dove, *Zenaida macroura*.

THE MOURNING DOVE National Strategic Harvest Management Plan (National Plan) provides a long-range vision for Mourning Dove management by development, implementation, and continuous improvement of harvest strategies based on mechanistic population models. The National Plan was adopted in 2003 by the four Flyway Councils, in 2004 by the Association of Fish and Wildlife Agencies (AFWA), and Migratory Shore and Upland Game Bird Working Group, and subsequently published by the US Fish and Wildlife Service (US Fish and Wildlife Service 2005).

The use of quantitative population models that synthesize knowledge of life history parameters, and the effects of intrinsic and extrinsic factors, has

[1] Use of trade or product names does not imply endorsement by the US Government.
[3] Current address: Wisconsin Department of Natural Resources, 110 S. Neenah Avenue, Sturgeon Bay, Wisconsin 54235, USA.

a long history in wildlife management (Shenk and Franklin 2001). This modeling approach provides a framework for tracking population change as a function of changes in factors impacting life history. Thus, the model fulfills a dual role of providing a rigorous context for harvest management and improving our understanding of the population dynamics of the species by evaluating vital population rates as functions of extrinsic factors such as amount of available breeding habitat, climatic conditions, nonhunting mortality, or hunting pressure. Within this context, a better understanding of specific causes of nonhunting mortality and their relative importance for Mourning Doves contributes critical information to the modeling process and requires empirically based input data. Poisoning from ingested spent lead shot in Mourning Doves has been identified as a conservation and management issue, with a need for better understanding of its potential population effects (Mirarchi and Baskett 1994, Tomlinson et al. 1994). Although the magnitude of lead exposure and poisoning in Mourning Doves is unknown, a risk assessment of lead shot exposure in upland birds and raptors concluded that Mourning Doves are particularly likely to ingest spent lead shot (Kendall et al. 1996), and Schulz et al. (2006b) have suggested that the number poisoned may approach the number harvested on an annual basis.

In the United States, Mourning Doves are commonly hunted in selectively managed fields. In 2007, 5–6 shots per dove bagged were reported for managed shooting fields in Missouri (Missouri Department of Conservation, unpublished report), and an earlier study in Tennessee found as many as eight shotshells were expended per bird taken (Lewis and Legler 1968). Lead shot densities of greater than 860,000 pellets per hectare have been reported from heavily hunted fields after the end of the dove hunting season (Best et al. 1992). In several regional studies, ingested lead shot were found in 0.3% to 6.4% of Mourning Doves (Castrale 1991, Kendall et al. 1996, Schulz et al. 2002). Kendall et al. (1996) suggested that a conservative estimate of the frequency of ingested lead shot in upland game birds, based largely on data for Mourning Doves, is about 3%.

Lead poisoning in birds is typically a chronic disease which often results in the development of clinical signs that include weakness, emaciation, and anemia. Although the relationship between ingested lead shot and toxicological effects depends on many factors, including species, diet, body condition, and environmental factors, one #7.5 or #8 ingested lead pellet may cause acute mortality in Mourning Doves (Buerger et al. 1986, Schulz et al. 2006a). Even in the absence of direct mortality from lead toxicosis, one ingested lead pellet may result in sublethal effects on physiology and behavior that can lead to death from starvation, predation, or disease (Scheuhammer and Norris 1996, Schulz et al. 2006a). Liver is the tissue often used as an indicator of recent lead exposure. Liver lead concentrations of 2 parts per million (ppm) wet weight (about 6 ppm dry weight) or greater are generally considered elevated in birds, including Mourning Doves, while concentrations of 6 ppm wet weight (about 20 ppm dry weight) or greater are potentially toxic (Franson 1996, Pain 1996). In a recent experimental lead shot dosing study in Mourning Doves, surviving birds had a mean liver lead concentration of 3.4 ppm, wet weight, while the mean of those that died was 49.2 ppm (Schulz et al. 2006a). Bone lead concentrations are often used as a measure of chronic exposure, because lead is lost from bone very slowly (Sanderson and Bellrose 1986, Pain 1996). Doves experimentally dosed with lead shot may accumulate up to 400–500 ppm dry weight of lead in bone (Kendall et al. 1983), whereas bone lead concentrations in apparently normal doves from rural areas usually average less than 50 ppm dry weight (Kendall and Scanlon 1979, 1982).

To evaluate the prevalence of lead exposure in Mourning Doves, we studied the frequency of ingested lead shot and lead concentrations in liver and wing bones in a sample of hunter-harvested birds from locations in seven states where lead shot was permitted for hunting. In addition, we examined hunter-harvested doves from two areas where nontoxic shot was required for hunting Mourning Doves. We studied the frequency of ingested pellets at both of these locations and liver and bone lead concentrations at one.

METHODS

Sample Collection and Processing.—With the assistance of personnel from State natural resource agencies, we examined 4,884 dove carcasses (after the breast was removed in most cases) collected from cooperating hunters during 1998–2000 in Arizona, Georgia, Missouri, Oklahoma, Pennsylvania, South Carolina, and Tennessee. Carcasses were shipped to the USGS National Wildlife Health Center (NWHC), Madison, Wisconsin. Age (i.e., hatch year (HY) and after hatch year (AHY)) was determined for 4,758 birds by plumage characteristics (Mirarchi 1993, Schulz et al. 1995). Sex was determined for 4,535 birds by visual examination of gonads, and the gizzard, liver, and one wing were removed from each carcass. All gizzards were individually identified and radiographed in groups, and those that contained metallic densities were examined visually for the recovery of metal. Pellets in the gizzard lumen that had penetrated the gizzard muscle as a result of gunshot were differentiated from ingested pellets by confirming the presence of entry wounds in the gizzard, by finding feathers enveloping the pellets and carried into the gizzard lumen, and by the examination of pellets for deformities (in the case of lead shot) indicating that they had contacted hard tissues. Steel pellets were identified by visual examination and the use of a magnet. Nonmagnetic pellets were evaluated for characteristics of lead (such as softness; tendency to deform or be cut, rather than to flake; appearance of a sheen on a cut or scratched surface) by pressing the pellet with the tip of a stainless steel scalpel blade or scissors.

Lead Analysis.—Livers and wing bones (radius/ulna) were examined visually and those with damage indicative of penetration by pellets were trimmed or discarded. Livers were stored frozen at –20ºC until analysis. Wing bones were stored in paper envelopes until preparation for lead analysis, when skin, feathers, and flesh were removed. Lead concentrations were measured in all livers and wing bones (except four wings damaged by pellets) from doves with ingested lead shot. In addition, we analyzed livers and wing bones from 1,989 and 691, respectively, doves without ingested lead or steel pellets. Although the majority of these samples came from areas where lead shot was permitted, we also analyzed livers and wing bones from one area (Hackberry Flat in Oklahoma) that required the use of nontoxic shot. Livers and bones for lead analysis from birds without ingested pellets were selected at random from within individual shipments, such that all states were represented in the sample. Lead analysis of tissues was done by atomic absorption spectroscopy (AAS) at the Illinois Department of Agriculture Animal Disease Laboratory, Centralia, Illinois and the NWHC. Most (95%) of the livers and all wing bones were analyzed for lead by graphite furnace AAS (Varian SpectrAA 220 FS, Varian, Inc., Palo Alto, CA or Thermo Jarrell Ash Scan 1 Thermo Jarrell Ash, Franklin, MA), using methods described by Ihnat (1999). Samples were dried to constant weight, digested in 1–5 ml trace metal grade nitric acid, and diluted with reagent grade deionized water to 50 ml. For the remaining 5% of the livers, samples were prepared and analyzed for lead according to Franson and Smith (1999), using flame AAS (Thermo Jarrell Ash Scan 1, Thermo Jarrell Ash, Franklin, MA). Quality assurance and quality control procedures for all analyses included the preparation of one sample spiked with a known amount of lead per each batch of 12–24 samples. Lower limits of detection were 0.05 ppm dry weight (Illinois Department of Agriculture) and 0.15 ppm dry weight (NWHC) for graphite furnace AAS and 0.70 ppm dry weight for flame AAS (NWHC). Mean recoveries from spiked samples were 100.5% for graphite furnace AAS and 100.1% for flame AAS. Average moisture content for livers and bones were 70.3% and 7.9%, respectively. All results are expressed as ppm on dry weight basis.

Statistics.—We calculated frequencies of liver and bone concentrations of ≥6 ppm and ≥20 ppm, respectively, as indicators of lead exposure (Franson 1996, Pain 1996). All liver data (from graphite furnace and flame AAS) were used in these calculations and in data summaries from doves with ingested lead shot. However, because of the much higher limit of detection, we excluded liver flame AAS data (n = 89) from calculations of summary statistics and comparisons of liver concentrations in doves without ingested lead shot. For those analyses, we used only the graphite furnace AAS data (n = 1,900). For analysis of bone data, we excluded three samples that fell below the NWHC detection limit of 0.15 ppm dry weight, yielding n = 688. Thus, our lower limit of detection was effectively 0.05 ppm dry weight and we assigned a concentration of 0.025 ppm dry weight to those samples below the lower limit of detection (8.0 % of liver samples and 0.6 % of wing bones). We used nonparametric statistics as these methods require only the relative order of the lead concentrations and because samples below the detection limit of 0.05 ppm were assigned a value of 0.025 ppm, they were consistently ranked lowest (Helsel 2005). We used the Wilcoxon rank sums test to evaluate differences in lead concentrations by age and sex, Fisher's exact test to compare categorical data, Spearman correlation coefficient to compare lead concentrations in liver and bone, and factorial analysis of variance (ANOVA) to compare bone lead concentrations in doves without ingested shot by age and area (lead shot or nontoxic shot) (SAS Institute, Inc. 1996).

RESULTS

Samples Collected.—Most Mourning Dove carcasses (4,229) came from areas where the use of lead shot was permitted, but 655 were collected at two areas where nontoxic shot was required (Tables 1, 2), and the overall frequency of ingested pellets (lead and steel) was 2.5%. The majority of dove carcasses (2,832 from lead shot areas and 580 from nontoxic shot areas) were collected from September 1 through September 7, with the remainder (1,397 from lead shot areas and 75 from nontoxic shot areas) collected from September 8 through December 24 (Table 3). Of the dove carcasses for which age and sex were determined, 69.9% were HY, 30.1% were AHY, 55.5% were males, and 44.5% were females.

Ingested Pellets on Lead Shot Areas.—Combining results from all areas where lead shot was allowed for hunting, we found 106 (2.5%) doves with ingested lead pellets (Table 1). Frequencies of ingested lead pellets ranged from 0% at many of the areas sampled to 19.9% and 13.3% at Gila Valley and Yuma Valley, Arizona, respectively (Table 1). The number of ingested lead pellets per dove ranged from one, in 42% of birds with ingested lead pellets, to 43 (Table 3). Gizzards of two doves collected from lead shot areas (one each from Northampton County, Pennsylvania and Lake Wallace, South Carolina) contained ingested steel pellets, but no ingested lead pellets. All doves with ingested lead pellets, except one, were HY birds. The frequency of ingested lead pellets did not differ by sex or by early versus late sampling period (Table 4).

Ingested Pellets on Nontoxic Shot Areas.—At two of the areas sampled, nontoxic shot was required because of the presence of relatively newly constructed and restored wetlands. Ingested pellets were found in 16 of 655 (2.4%) Mourning Doves on these areas (Table 2). All 16 birds had ingested one or more steel pellets and gizzards of two doves contained ingested steel and lead pellets. The number of ingested steel pellets ranged from one to 23 (Table 3). Of the 12 doves having ingested shot for which age was determined, all were HY. The frequency of ingested steel pellets was similar for males and females and none of the doves collected after September 7 had ingested shot in their gizzards (Table 4).

Liver and Bone Lead Concentrations.—Liver and bone lead concentrations of doves with ingested lead pellets (Table 5) did not differ by sex ($P = 0.4524$ and 0.1055, respectively). Lead concentrations in tissues of birds with ingested lead pellets were not compared by age because all but one were HY birds. Liver and bone lead concentrations were significantly correlated in birds with ($r = 0.328$, n = 103, $P = 0.0007$) and without ($r = 0.337$, n = 598, $P < 0.0001$) ingested lead pellets. Combining the data from all doves (with and without ingested pellets), frequencies of elevated liver (≥ 6 ppm dry weight) and bone (≥ 20 ppm dry weight) lead concentrations were 8.3% and 26.8%, respectively, on areas where lead shot was allowed and 2.0% and 11.1%, respectively, on the area requiring nontoxic shot.

In doves without ingested lead pellets, lead concentrations in liver differed by sex (females >males, $P = 0.0053$), but not age ($P = 0.0666$), and concentrations in bone differed by age (AHY >HY, P = <0.0001), but not sex ($P = 0.1917$) (Tables 6, 7). Lead concentrations in wing bones of Mourning Doves without ingested lead pellets were greater in birds from areas where lead shot was allowed than from the area where nontoxic shot was required ($P < 0.0001$) (Table 8), but concentrations of lead in liver did not differ between lead and nontoxic shot areas ($P = 0.2198$).

Table 1. Number of hunter-harvested Mourning Doves collected from areas where the use of lead shot was permitted and the number with ingested lead pellets, 1998–2000.

State/Place name	Location	No. doves collected	No. with ingested pellets (%)
Arizona			
Buckeye Granary	33°22'N, 112°35'W	10	0
Curtis Road	32°36'N, 111°34'W	100	4 (4.0)
Gila Bend (14 mi SE)	33°09'N, 112°44'W	24	0
Gila River	32°43'N, 114°31'W	6	0
Gila Valley	32°46'N, 114°31'W	221	44 (19.9)
Hog Canyon (Unit 43-A)	31°40'N, 111°43'W	8	0
Milligan Road	32°44'N, 111°29'W	141	1 (0.7)
Robbins/Powers Butte	33°19'N, 112°38'W	128	0
Wilcox (12 mi SE)	32°07'N, 109°59'W	10	0
Yuma Mesa	32°41'N, 114°36'W	30	2 (6.7)
Yuma South	32°35'N, 114°38'W	20	0
Yuma Valley	32°40'N, 114°43'W	83	11 (13.3)
Arizona Total		**781**	**62 (7.9)**
Georgia			
Di-Lane	32°57'N, 82°04'W	90	0
Rum Creek	33°04'N, 83°52'W	204	0
Georgia Total		**294**	**0**
Missouri			
James Reed	38°53'N, 94°20'W	574	2 (0.3)
Missouri Total		**574**	**2[a] (0.3)**
Oklahoma			
Beaver	36°49'N, 100°31'W	91	1 (1.1)
Blue River	34°19'N, 96°35'W	96	0
Council Hill (1 mi NE)	35°38'N, 95°38'W	53	1 (1.9)
Harper Co. Ranch	36°41'N, 99°41'W	198	0
Keefeton (2 mi SW)	35°38'N, 95°21'W	16	0
Love Co.	33°58'N, 97°11'W	111	0
Packsaddle	36°20'N, 96°15'W	103	1 (1.0)
Skiatook	35°56'N, 99°44'W	80	1 (1.3)
Oklahoma Total		**748**	**4 (0.5)**
Pennsylvania			
Bedminster	40°26'N, 75°11'W	36	0
Girard (2 mi NE)	41°59'N, 80°14'W	19	0
Lancaster Co.	40°02'N, 76°15'W	144	3 (2.1)
Lebanon Co.	40°22'N, 76°28'W	31	0
Lehigh Co.	40°37'N, 75°35'W	25	0
Northampton Co.	40°45'N, 75°18'W	59	1[b] (1.7)
Oakville	40°07'N, 77°27'W	9	0
Pennsylvania Total		**323**	**4 (1.2)**
South Carolina			
Lake Wallace	34°39'N, 79°41'W	498	9[c] (1.8)
Oakland Hunt Club	33°25'N, 80°06'W	348	17 (4.9)
Westvaco-Walworth	33°21'N, 80°16'W	359	9 (2.5)
South Carolina Total		**1205**	**35 (2.9)**
Tennessee			
Hermitage Field	36°14'N, 86°36'W	4	0
Larry Kent Field	36°11'N, 86°32'W	264	0
Percy Priest	36°01'N, 86°31'W	36	1 (2.8)
Tennessee Total		**304**	**1 (0.3)**
Total		**4229**	**108[d] (2.6)**
Total with ingested lead pellets			**106 (2.5)**

[a]See Schulz et al. (2002). [b]This dove had two ingested steel shot, no ingested lead shot.
[c]One of the nine doves had one ingested steel shot, no ingested lead shot.
[d]Includes two doves with ingested steel shot only.

Table 2. Number of hunter-harvested Mourning Doves collected from areas where the use of nontoxic shot was required and the number with ingested pellets, 1998–2000.

State/Place name	Location	No. doves collected	No. with ingested pellets (%)
Missouri			
EBCA	38°50'N, 92°30'W	310	15[a,b] (4.8)
Oklahoma			
Hackberry Flat	34°17'N, 98°58'W	345	1[c] (0.3)
Totals		655	16 (2.4)

[a]See Schulz et al. (2002).
[b]One of the 15 doves had 17 ingested steel shot and 4 ingested lead shot.
[c]This dove had 18 ingested steel shot and 2 ingested lead shot.

Table 3. Frequency of one or more ingested pellets in Mourning Doves collected on areas where lead shot was permitted and where nontoxic shot was required, 1998–2000.

Lead shot permitted				Nontoxic shot required	
No. lead pellets	No. doves (%)	No. lead pellets	No. doves (%)	No. steel pellets	No. doves (%)
1	45 (42.45)	10	1 (0.94)	1	5 (31.25)
2	20 (18.87)	11	2 (1.89)	2	2 (12.50)
3	8 (7.55)	13	2 (1.89)	3	2 (12.50)
4	11 (10.38)	14	1 (0.94)	7	2 (12.50)
5	1 (0.94)	15	1 (0.94)	8	1 (6.25)
6	4 (3.77)	17	1 (0.94)	20	1 (6.25)
8	4 (3.77)	23	1 (0.94)	21	2 (12.50)
9	3 (2.83)	43	1 (0.94)	23	1 (6.25)

Table 4. Frequency of ingested pellets (%) in Mourning Doves on areas where lead shot was allowed, and on nontoxic shot areas according to age, sex, and time of carcass collection, 1998–2000.

	Lead shot permitted[a]	Nontoxic shot required[b]
HY[c] (hatch year)	3.5 (105/2982)	2.9 (12/409)
AHY (after hatch year)	0.08 (1/1218)	0 (0/149)
HY male[d]	3.4 (52/1519)	2.8 (6/217)
HY female	3.0 (37/1244)	3.5 (6/172)
September 1-7	2.3[e] (64/2832)	2.8 (16/580)
September 8–December 24	3.0 (42/1397)	0 (0/75)

[a]Two doves collected on lead shot areas, each with one ingested steel pellet (but no ingested lead pellets) were excluded from these summaries. Sex was not determined for 16 doves with ingested pellets.
[b]Two birds collected on nontoxic areas had ingested both lead and steel pellets. Because ingested steel was present, they were included in the summaries. Sex was not determined for four doves with ingested pellets.
[c]The frequency of HY (hatch year) doves with ingested shot was significantly greater than the frequency for AHY (after hatch year) on lead shot areas (Fisher's exact test, $P < 0.0001$) and nontoxic shot areas (Fisher's exact test, $P = 0.0424$).
[d]For HY doves, the frequency of ingested pellets did not differ by sex on lead shot areas (Fisher's exact test, $P = 0.5182$) or nontoxic shot areas (Fisher's exact test, $P = 0.7716$).
[e]Not significantly different (Fisher's exact test, $P = 0.1446$) than the frequency of ingested lead pellets in doves that were collected between September 8 and December 24.

Table 5. Lead concentrations (ppm dry weight) in livers and wing bones of Mourning Doves with ingested lead or steel pellets (all hatch year birds, except one after hatch year dove from Arizona), 1998–2000.

		Liver			Wing bones			
State	n	Median	Q1[a]	Q3[b]	N	Median	Q1	Q3
Arizona	62	45.64	15.42	67.44	60	66.69	29.56	195.60
Missouri	2	0.22	0.11	0.32	2	70.21	48.30	92.11
Oklahoma	5	20.23	14.06	106.85	4	24.97	12.23	62.68
Pennsylvania	3	19.05	13.60	22.85	3	236.89	27.71	457.39
South Carolina	34	36.79	19.38	73.80	33	187.94	63.49	405.76
Tennessee	1[c]				1[d]			
Total	107[e]	36.89	14.22	72.03	103[f]	89.33	33.04	236.89

[a]Quartile 1, 25th percentile.
[b]Quartile 3, 75th percentile.
[c]Liver lead concentration = 11.96 ppm dw.
[d]Bone lead concentration = 27.29 ppm dw.
[e]Includes livers from 106 doves with ingested lead shot from areas where lead shot was allowed for hunting doves, and one dove with ingested lead shot from a nontoxic shot area (Hackberry Flat, OK).
[f]Wing bones from four doves were not analyzed because they were damaged by pellets.

Table 6. Lead concentrations (ppm dry weight) in livers of Mourning Doves without ingested lead or steel pellets, by sex (excludes 37 doves of undetermined sex), 1998–2000.

		Male			Female			
State	n	Median	Q1[a]	Q3[b]	n	Median	Q1	Q3
Arizona	228	0.26	0.13	0.59	115	0.29	0.16	0.60
Georgia	103	0.17	0.08	0.38	85	0.25	0.10	0.71
Missouri	74	0.33	0.14	0.98	65	0.30	0.11	0.52
Oklahoma[c]	110	0.21	0.08	0.47	90	0.22	0.09	1.44
Pennsylvania	141	0.28	0.16	0.75	149	0.33	0.19	0.65
South Carolina	283	0.34	0.16	0.73	319	0.44	0.20	0.86
Tennessee	52	0.24	0.14	0.43	49	0.31	0.16	0.61
Total	991	0.27[d]	0.13	0.62	872	0.34	0.15	0.73

[a]Quartile 1, 25th percentile.
[b]Quartile 3, 75th percentile.
[c]Includes samples from 100 Mourning Doves (55 males and 45 females) collected at an area where nontoxic shot was required.
[d]Significantly different than females ($P = 0.0053$).

Table 7. Lead concentrations (ppm dry weight) in wing bones of Mourning Doves without ingested lead or steel pellets, by age (excludes one dove of undetermined age), 1998–2000.

		HY (hatch year)			AHY (after hatch year)			
State	n	Median	Q1[a]	Q3[b]	n	Median	Q1	Q3
Arizona	34	2.70	1.00	53.03	27	2.09	1.33	5.44
Georgia	61	1.31	0.74	2.96	50	2.61	1.64	4.88
Missouri	66	1.94	1.14	5.26	63	2.92	1.78	6.22
Oklahoma[c]	80	0.57	0.28	1.32	36	1.30	0.78	2.86
Pennsylvania	65	5.30	2.73	21.34	15	11.33	6.42	43.52
South Carolina	62	1.88	0.87	11.75	58	2.91	1.44	14.38
Tennessee	52	1.16	0.68	2.06	18	2.04	1.35	3.21
Total	420	1.56[d]	0.75	6.14	267	2.67	1.44	9.65

[a]Quartile 1, 25th percentile.
[b]Quartile 3, 75th percentile.
[c]Includes samples from 62 Mourning Doves (52 HY and 10 AHY) collected at an area where nontoxic shot was required.
[d]Significantly different than AHY ($P < 0.0001$).

Table 8. Lead concentrations (ppm dry weight) in wing bones of Mourning Doves without ingested lead pellets by age and area (excludes one dove of undetermined age)[a], 1998–2000.

	Lead shot permitted				Nontoxic shot required			
	n	Median	Q1[b]	Q3[c]	n	Median	Q1	Q3
HY (hatch year)	368	1.74	0.89	6.88	52	0.56	0.24	1.12
AHY (after hatch year)	257	2.74	1.51	9.65	10	1.05	0.61	1.55

[a]Significantly different by age and area (factorial ANOVA, $P < 0.0001$).
[b]Quartile 1, 25th percentile.
[c]Quartile 3, 75th percentile.

DISCUSSION

In our study, the combined frequency of ingested lead and steel pellets (2.5%) was the same as that found in an Indiana study (Castrale 1991) and the frequency of ingested lead pellets found in our study (also 2.5%) is within the range (<1% to 6.4%) previously reported in studies of lead exposure in Mourning Doves that were more restricted in geographic scope (see Kendall et al. 1996). However, we found considerable variation among locations. Although our study was not designed to compare managed versus unmanaged dove hunting fields, it is interesting to note that the greatest frequencies of ingested lead pellets (13.3% and 19.9%) were on two agricultural areas (Gila Valley and Yuma Valley in Arizona) not specifically managed for dove hunting and that frequencies of ingested pellets on managed areas varied considerably. For example, no ingested pellets were found in doves from two managed areas in Georgia (Di-Lane and Rum Creek), although lead poisoning has been reported in a Northern Bobwhite (*Colinus virginianus*) found dead at Di-Lane (Lewis and Schweitzer 2000). At another managed area, Oakland Hunt Club in South Carolina, 4.9% of doves had ingested lead pellets. Variation in the types of agricultural and management practices, particularly the frequency and timing of cultivation in relation to the hunting season, are among the factors expected to influence the availability of shot for foraging doves (Kendall et al. 1996).

Some previous studies have indicated that the progression of the hunting season was accompanied by increased densities of spent shot and greater frequencies of ingested lead pellets in Mourning Doves (see Kendall et al. 1996). On areas where the use of lead shot was allowed, we found no difference in the frequency of ingested pellets in doves collected during the first week of the hunting season (September 1 through 7) compared with the frequency in doves collected in the latter part of the season (September 8 through December 24) (Table 4). Although we sampled large numbers of doves early (n = 2,832) and late (n = 1,397) in the hunting season, few areas were sampled during both periods. Thus, variation in frequencies of ingested pellets among locations may have masked variation within locations through time.

Of the Mourning Doves for which age was determined, 69.9% were HY and 30.1% were AHY birds, but 105 of 106 doves with ingested lead pellets and 12 of 12 doves with ingested steel pellets were HY birds (Table 4). Previously published studies of ingested shot in Mourning Doves, where age was determined, include reports of ingested pellets in immature doves only (Best et al. 1992, Locke and Bagley 1967) and a report in which adults and immatures had ingested pellets, but with no significant difference between the age classes (Castrale 1991). It is unknown why HY doves in our study had a higher frequency of ingested pellets than AHY birds, but it is possible that a collection bias occurred because HY doves with ingested pellets, of either lead or steel, were more likely to be harvested than AHY birds with ingested pellets. A similar bias could occur if adult doves were more susceptible to the effects of lead exposure than HY birds and thus became incapacitated more quickly and were not available for harvest. If that were the case, we would expect to see different patterns of age-related pellet ingestion on areas where lead shot was allowed vs. nontoxic shot areas. Thus, of these two hypotheses, our limited results from nontoxic shot areas support the former, because we found no AHY doves with ingested steel shot. It is also possible that HY doves simply ingest pellets more frequently than older doves.

The liver and bone lead concentrations (Table 5) that we found in Mourning Doves with ingested lead shot were generally lower than the lead concentrations reported in experimental dosing studies and in field cases of lead poisoning in Mourning Doves, where liver and bone lead concentrations have ranged from about 80 to >200 ppm dry weight and 115 to >400 ppm dry weight, respectively (Locke and Bagley 1967, Kendall et al. 1983, Buerger et al. 1986, Schulz et al. 2006a). It is not surprising that the hunter-harvested doves with ingested lead pellets in our sample had lower concentrations of lead in their tissues than birds that died of lead poisoning, because the severity of poisoning would not have progressed to a near-terminal stage in birds still able to take flight. Although sample sizes were small, we found that bone lead concentrations in HY and AHY doves without ingested lead pellets were lower on the area where nontoxic shot was required than on the areas where lead shot was allowed (Table 8). This finding suggests a lower level of lead exposure may occur in doves on areas where nontoxic shot is required, at least based on lead concentrations that we found in bones from a hunter-harvested sample of doves.

In two studies where doves were dosed with multiple lead shot, mortality started to occur after two and six days (McConnell 1967, Buerger et al. 1986). In Mourning Doves receiving 2 to 24 lead pellets, each additional pellet increased the hazard of death by 18% and the 19 to 21 day survival estimate for doves with 13 to 19 pellets was 8.3% (Schulz et al. 2006a). Even in doves with ≤2 ingested pellets, survival estimates were reduced to 57% (Schulz et al. 2006a). The results of these reports suggest that the doves in our study with lead pellets in their gizzards had recently ingested them. However, 92.5% had elevated liver lead concentrations (≥6 ppm dry weight) and 85.4% had elevated bone lead concentrations (≥20 ppm dry weight), and it is likely that they were experiencing physiological effects of lead exposure. Possible adverse effects include changes in the hematopoietic system, including increased heterophil/lymphocyte ratios and reductions in heme synthesis and packed cell volume (Pain 1996, Schulz et al. 2006a). Based on results of the dosing studies and the likelihood of physiological effects, we suspect that many of the doves with ingested lead pellets in our study would soon have succumbed to lead poisoning, or causes related to lead poisoning morbidity (such as predation), had they not been harvested.

The frequencies of elevated liver (≥6 ppm dry weight) and bone (≥20 ppm dry weight) lead concentrations (8.3% and 26.8%, respectively) in doves from areas where lead shot was permitted for hunting were greater than the frequency of ingested pellets (2.5%). Concentrations of lead in the blood, liver, and other soft tissues are somewhat mobile and reflect relatively recent exposure. Lead also moves quickly from the bloodstream into bone, but it tends to remain there and to accumulate in bone over time (Pain 1996). Experimental studies have shown that a portion of Mourning Doves may survive exposure to lead shot, and that some of the lead pellets will be passed in the feces (Buerger et al. 1986, Marn et al. 1988, Schulz et al. 2006a, Schulz et al. 2007). Thus, it is possible that some wild doves could have elevated concentrations of lead in their tissues caused by ingested lead shot which had been voided from the gizzard by the time the birds were collected. Because shot may be voided, it is to be expected that estimates of lead exposure based on frequency of ingested pellets will be lower than estimates based on tissue concentrations of lead. Previous work with waterfowl indicates that lead surveys based only on the prevalence of ingested lead shot will underestimate the extent of lead exposure when compared to other testing criteria, such as analysis of blood samples for lead (Anderson and Havera 1985), and our results suggest that a similar situation exists for Mourning Doves.

Our results include findings that frequencies of ingested shot in Mourning Doves were highly variable among locations, hunter-harvested HY doves were more likely to have ingested pellets than AHY birds, and that concentrations of lead in bone of doves without ingested lead pellets were lower on an area where nontoxic shot was required than on areas where lead shot was permitted. A number of questions remain to be addressed, however, for a better understanding of the full significance of lead shot ingestion by doves. First, what is the true frequency of ingested lead pellets in Mourning Doves? The answer to this question requires investigation of possible biases associated with hunter-harvested samples of birds. The fact that studies have shown that even small numbers of lead shot can kill doves raises the question of the role that Mourning Doves play in secondary lead poisoning of scavengers and predators that consume doves dead or dying of lead poisoning. A variety of raptor species has been reported to have died of lead shot poisoning, presumably from the consumption of shot in prey items (see Fisher et al. 2006). The poisoning of these and other avian and mammalian scavengers may be the result of consuming lead pellets ingested by doves or embedded in muscle. Finally, how many doves actually die of lead poisoning annually? If we assume a frequency in the range of 3% lead shot ingestion, as proposed by Kendal et al. (1996) for upland game birds, and supported by this study for Mourning Doves, what is the impact on dove numbers? Based on the toxicity of lead shot for Mourning Doves and reported frequencies of ingested lead pellets, Schulz et al. (2006b) suggested that annual losses due to lead poisoning may approach annual harvest estimates.

ACKNOWLEDGMENTS

Major funding support was provided by the US Fish and Wildlife Service, Webless Migratory Game Bird Research Program. We thank personnel of the following agencies for their assistance in obtaining dove carcasses: Arizona Dept. of Game and Fish, Georgia Dept. of Natural Resources, Missouri Dept. of Conservation, Oklahoma Dept. of Wildlife Conservation, Pennsylvania Game Commission, South Carolina Cooperative Fish and Wildlife Research Unit, South Carolina Dept. of Natural Resources, Tennessee Wildlife Resources Agency. We also thank the following individuals for their assistance: M. Baughman, J. Berdeen, L. Blewett, R. Bredesen, B. Carmichael, T. Creekmore, D. Dolton, B. Dukes, J. Dunn, R. Engel-Wilson, D. Finely, K. Fitts, D. Forster, F. Granitz, M. Gudlin, J. Hann, J. Hanna, S. Hansen, J. Heffelfinger, R. Henry, T. Hollbrock, A. Hollmén, D. Jochimsen, S. Kasten, L. Kent, W. Mahan, K. Miller, K. Morris, C. Muckenfuss, M. O'Meilia, D. Otis, A. Schrader, P. Smith, and S. Stokes, Jr. P. Flint and L. Locke provided helpful comments on the manuscript.

LITERATURE CITED

ANDERSON, W. L. AND S. P. HAVERA. 1985. Blood lead, protoporphyrin, and ingested shot for detecting lead poisoning in waterfowl. Wildlife Society Bulletin 13:26–31.

BEST, T. L., T. E. GARRISON, AND C. G. SCHMITT. 1992. Availability and ingestion of lead shot by Mourning Doves (*Zenaida macroura*) in southeastern New Mexico. The Southwestern Naturalist 37:287–292.

BUERGER, T. T., R. E. MIRARCHI, AND M. E. LISANO. 1986. Effects of lead shot ingestion on captive Mourning Dove survivability and reproduction. Journal of Wildlife Management 50:1–8.

CASTRALE, J. S. 1991. Spent shot ingestion by Mourning Doves in Indiana. Proceedings of the Indiana Academy of Science 100:197–202.

FISHER, I. J., D. J. PAIN, AND V. G. THOMAS. 2006. A review of lead poisoning from ammunition sources in terrestrial birds. Biological Conservation 131:421–432.

FRANSON, J. C. 1996. Interpretation of tissue lead residues in birds other than waterfowl. Pages 265–279 *in* W. N. Beyer, G. H. Heinz, and A. W. Redmon-Norwood (Eds.). Environmental Contaminants in Wildlife: Interpreting Tissue Concentrations. Lewis Publishers, Boca Raton, Florida, USA.

FRANSON, J. C. AND M. R. SMITH. 1999. Poisoning of wild birds from exposure to anticholinesterase compounds and lead: diagnostic methods and selected cases. Seminars in Avian and Exotic Pet Medicine 8:3–11.

HELSEL, D. R. 2005. Nondetects and Data Analysis. John Wiley & Sons, Hoboken, New Jersey, USA.

IHNAT, M. 1999. Metals and other elements at trace levels in foods, method 9.2.20A. Pages 20–20B *in* P. Cunniff (Ed.). Official Methods of Analysis of AOAC International. AOAC International, Gaithersburg, Maryland, USA.

KENDALL, R. J. AND P .F. SCANLON. 1979. Lead concentrations in Mourning Doves collected from middle Atlantic game management areas. Proceedings of the Annual Conference of the Southeastern Association of Fish and Wildlife Agencies 33:165–172.

KENDALL, R. J. AND P. F. SCANLON. 1982. Tissue lead concentrations and blood characteristics of Mourning Doves from southwestern Virginia. Archives of Environmental Contamination and Toxicology 11:269–272.

KENDALL, R. J., P. F. SCANLON, AND H. P. VEIT. 1983. Histologic and ultrastructural lesions of Mourning Doves (*Zenaida macroura*) poisoned by lead shot. Poultry Science 62:952–956.

KENDALL, R. J., T. E. LACHER, JR., C. BUNCK, B. DANIEL, C. DRIVER, C. E. GRUE, F. LEIGHTON, W. STANSLEY, P. G. WATANABE, AND M. WHITWORTH. 1996. An ecological risk assessment of lead shot exposure in non-waterfowl avian species: upland game birds and raptors. Environmental Toxicology and Chemistry 15:4–20.

LEWIS, J. C. AND E. LEGLER, JR. 1968. Lead shot ingestion by Mourning Doves and incidence in soil. Journal of Wildlife Management 32:476–482.

LEWIS, L. A. AND S. H. SCHWEITZER. 2000. Lead poisoning in a Northern Bobwhite in Georgia. Journal of Wildlife Diseases 36:180–183.

LOCKE, L. N. AND G. E. BAGLEY. 1967. Lead poisoning in a sample of Maryland Mourning Doves. Journal of Wildlife Management 31:515–518.

MARN, C. M., R. E. MIRARCHI, AND M. E. LISANO. 1988. Effects of diet and cold exposure on captive female Mourning Doves dosed with lead shot. Archives of Environmental Contamination and Toxicology 17:589–594.

MCCONNELL, C. A. 1967. Experimental lead poisoning of Bobwhite Quail and Mourning Doves. Proceedings of the Annual Conference of the Southeastern Association of Game and Fish Commissioners 21:208–219.

MIRARCHI, R. E. 1993. Aging, sexing and miscellaneous research techniques. Pages 399–408 *in* T. S. Baskett, M. W. Sayre, R. E. Tomlinson, and R. E. Mirarchi (Eds.). Ecology and Management of the Mourning Dove. Stackpole Books, Harrisburg, Pennsylvania, USA.

MIRARCHI, R. E., AND T. S. BASKETT. 1994. Mourning Dove (*Zenaida macroura*). *In* A. Poole and F. Gill (Eds.). The Birds of North America, no. 117. Academy of Natural Sciences, Philadelphia, and American Ornithologists' Union, Washington, DC, USA.

PAIN, D. J. 1996. Lead in waterfowl. Pages 251–264 *in* W. N. Beyer, G. H. Heinz, and A. W. Redmon-Norwood (Eds.). Environmental Contaminants in Wildlife: Interpreting Tissue Concentrations. Lewis Publishers, Boca Raton, Florida, USA.

SANDERSON, G. C. AND F. C. BELLROSE. 1986. A Review of the Problem of Lead Poisoning in Waterfowl (ed 2). Illinois Natural History Survey, Champaign, Illinois, USA.

SAS INSTITUTE, INC. 1996. SAS/STAT User's Guide, Version 6.12. Cary, NC, USA.

SCHEUHAMMER, A. M. AND S. L. NORRIS. 1996. The ecotoxicology of lead shot and lead fishing weights. Ecotoxicology 5:279–295.

SCHULZ, J. H., S. L. SHERIFF, Z. HE, C. E. BRAUN, R. D. DROBNEY, R. E. TOMLINSON, D. D. DOLTON, AND R. A. MONTGOMERY. 1995. Accuracy of techniques used to assign Mourning Dove age and gender. Journal of Wildlife Management 59:759–765.

SCHULZ, J. H., J. J. MILLSPAUGH, B. E. WASHBURN, G. R. WESTER, J. T. LANIGAN III, AND J. C. FRANSON. 2002. Spent shot availability and ingestion on areas managed for Mourning Doves. Wildlife Society Bulletin 30:112–120.

SCHULZ, J. H., J. J. MILLSPAUGH, A. J. BERMUDEZ, X. GAO, T. W. BONNOT, L. G. BRITT, AND M. PAINE. 2006a. Acute lead toxicosis in Mourning Doves. Journal of Wildlife Management 70:413–421.

SCHULZ, J. H., P. I. PADDING, AND J. J. MILLSPAUGH. 2006b. Will Mourning Dove crippling rates increase with nontoxic-shot regulations? Wildlife Society Bulletin 34:861–865.

SCHULZ, J. H., X. GAO, J. J. MILLSPAUGH, AND A. J. BERMUDEZ. 2007. Experimental lead pellet ingestion in Mourning Doves (*Zenaida macroura*). American Midland Naturalist 158:177–190.

SHENK, T. M. AND A. B. FRANKLIN. 2001. Models in natural resource management: an introduction. Pages 1–8 *in* T. M. Shenk and A. B. Franklin (Eds). Modeling in Natural Resource Management: Development, Interpretation, and Application. Island Press, Washington, DC, USA.

TOMLINSON, R. E., D. D. DOLTON, R. R. GEORGE, AND R. E. MIRARCHI. 1994. Mourning Dove. Pages 5–26 *in* T. C. Tacha and C. E. Braun (Eds.). Migratory Shore and Upland Game Bird Management in North America. International Association of Fish and Wildlife Agencies, Washington, DC, USA.

US FISH AND WILDLIFE SERVICE. 2005. Mourning Dove National Strategic Harvest Management Plan. US Department of the Interior, Washington, DC, USA.

ACUTE LEAD TOXICOSIS AND EXPERIMENTAL LEAD PELLET INGESTION IN MOURNING DOVES

JOHN H. SCHULZ[1], XIAOMING GAO[1], JOSHUA J. MILLSPAUGH[2], AND ALEX J. BERMUDEZ[3]

[1]*Missouri Department of Conservation, Resource Science Center, 1110 South College Avenue, Columbia, MO 65201, USA.*

[2]*Department of Fisheries and Wildlife Sciences, University of Missouri, Columbia, MO 65211, USA.*

[3]*University of Missouri, Veterinary Medicine Diagnostic Laboratory, Columbia, MO 65211, USA.*

EXTENDED ABSTRACT.—Mourning Dove (*Zenaida macroura*) hunting is becoming increasingly popular, especially hunting over managed shooting fields. Given the possible increase in lead shot availability on these areas, our original objective was to estimate the availability and ingestion of spent shot at the Eagle Bluffs Conservation Area (EBCA; hunted with non-toxic shot) and the James A. Reed Memorial Wildlife Area (JARWA; hunted with lead shot) in Missouri (Schulz et al. 2002). During 1998, we collected soil samples one or two weeks prior to the hunting season (pre-hunt) and after four days of dove hunting (post-hunt). We also collected information on the number of doves harvested, number of shots fired, shotgun gauge, and shotshell size used. Dove carcasses were collected on both areas during 1998–99. At EBCA, 60 hunters deposited an estimated 64,775 pellets/ha of non-toxic shot on or around the managed field. At JARWA, approximately 1,086,275 pellets/ha of lead shot were deposited by 728 hunters. Our post-hunt estimates of spent shot availability from soil sampling were 0 pellets/ha for EBCA and 6,342 pellets/ha for JARWA. Our findings suggest that existing soil sampling protocols may not provide accurate estimates of spent shot availability in managed dove shooting fields. During 1998–1999, 15 of 310 (4.8%) Mourning Doves collected from EBCA had ingested non-toxic shot. For doves that ingested shot, 6 (40.0%) contained ≥7 shot pellets. In comparison, only 2 of 574 (0.3%) doves collected from JARWA had ingested lead shot. Because a greater proportion of doves were found to have ingested multiple steel pellets compared to lead pellets, we suggested that doves feeding in fields hunted with lead shot may succumb to acute lead toxicosis and thus become unavailable to harvest, resulting in an underestimate of lead shot ingestion rates. These findings may partially explain why previous studies have shown relatively few doves with ingested lead shot despite feeding on areas with high lead shot availability (Kendall et al. 1996, Mirarchi and Baskett 1994).

Our next objective was to test the acute lead toxicosis hypothesis, that free-ranging Mourning Doves may ingest spent lead pellets, succumb to lead toxicosis, and die in a relatively short time (Schulz et al. 2006) . We tested this hypothesis by administering 157 captive Mourning Doves 2–24 lead pellets, monitoring pellet retention and short-term survival, and measuring related physiological characteristics. During the 19- to 21-day post-treatment period, 104 doves that received lead pellets died (deceased doves) and 53 survived (survivors); all 22 birds in a control group survived. Within 24-h of treatment, blood lead levels increased almost twice as fast for deceased doves compared to survivors ($F_{1,208}$ = 55.49; P <0.001). During the first week, heterophil:lymphocyte (H:L) ratios increased twice as fast for deceased doves than with survivors ($F_{1,198}$ = 23.14, P <0.001). Post-treatment survival differed

(χ^2 = 37.4, P = 0.001) among the five groups of doves that retained different numbers of pellets, and survival ranged from 0.57 (95% CI: 0.44–0.74) for doves that retained ≤2 lead pellets 2-days post-treatment compared to 0.08 (95% CI: 0.022–0.31) for those doves that retained 13–19 lead pellets on 2-days post-treatment; significant differences existed among the five groups. After controlling for dove pretreatment body mass, each additional lead pellet increased the hazard of death by 18.0% (95% CI: 1.132–1.230, P <0.001) and 25.7% (95% CI: 1.175–1.345, P <0.001) for males and females, respectively. For each 1-g increase in pretreatment body mass, the hazard of death decreased 2.5% (P = 0.04) for males and 3.8% (P = 0.02) for females. Deceased doves had the highest lead levels in liver (49.20 ± 3.23 ppm) and kidney (258.16 ± 21.85 ppm) tissues, whereas controls showed the lowest levels (liver, 0.08 ± 0.041 ppm; kidney, 0.17 ± 0.10 ppm). For doves dosed with pellets, we observed simultaneous increases in blood lead levels and H:L ratios, whereas packed-cell volume (PCV) values declined. Our results therefore support an acute lead toxicosis hypothesis.

Next, we conducted an experiment to determine if doves held in captivity freely ingest lead shotgun pellets, investigate the relationship between pellet density and ingestion, and monitor physiological impacts of doves ingesting pellets (Schulz et al. 2007). We conducted two trials of the experiment with 60 doves per trial. We randomly assigned 10 doves to one of six groups per trial; 10, 25, 50, 100, 200 pellets mixed with food and a control group with no pellets. We monitored ingestion by examining x-rays of doves 1-day post-treatment and monitored the effects of lead ingestion by measuring H:L ratios, PCV, blood lead, liver lead and kidney lead. Pooled data from both trials showed 6 of 117 (5.1%) doves ingested lead pellets. Two Mourning Doves ingested multiple lead pellets in each of the treatments containing a mixture of 25, 100 and 200 lead pellets and food. Doves ingesting lead pellets had higher blood lead levels than before treatment (P = 0.031). Post-treatment H:L ratios, however, were not different compared to pre-treatment values (P = 0.109). Although post-treatment PCV decreased for four of six doves ingesting lead pellets, overall they were not lower than their pre-treatment values (P = 0.344). Liver (P <0.0001) and kidney (P = 0.0012) lead levels for doves ingesting pellets were higher than doves without ingested pellets. Our lead pellet ingestion rates were similar to previously reported ingestion rates from hunter-killed doves (Kendall et al. 1996, Otis et al. 2008), and our physiological measurements confirm earlier reports of a rapid and acute lead toxicosis (Schulz et al. 2006). Similar to previous field research (Lewis and Legler 1968, Castrale 1991, Best et al. 1992), we did not observe a relationship between pellet density in the food and ad libitum pellet ingestion.

We recommend that management agencies initiate development of a long-term strategic plan aimed at implementing a nontoxic shot regulation for Mourning Dove hunting. Although one approach would be to ban lead shot for Mourning Dove hunting on managed public hunting areas, we believe it is vitally important to ensure that policy development and implementation have a consensus among stakeholders. *Received 30 April 2008, accepted 8 August 2008.*

SCHULZ, J.H., X. GAO, J.J. MILLSPAUGH, AND A.J. BERMUDEZ. 2009. Acute lead toxicosis and experimental lead pellet ingestion in Mourning Doves. Extended abstract *in* R.T. Watson, M. Fuller, M. Pokras, and W.G. Hunt (Eds.). Ingestion of Lead from Spent Ammunition: Implications for Wildlife and Humans. The Peregrine Fund, Boise, Idaho, USA. DOI 10.4080/ilsa.2009.0203

Key words: Lead, Mourning Dove, pellet, shot, toxicosis.

LITERATURE CITED

BEST, T. L., T. E. GARRISON, AND C. G. SCHMITT. 1992. Availability and ingestion of lead shot by Mourning Doves (*Zenaida macroura*) in southeastern New Mexico. The Southwestern Naturalist 37:287–292.

CASTRALE, J. S. 1991. Spent shot ingestion by Mourning Doves in Indiana. Proceedings of the Indiana Academy of Science 100:197–202.

KENDALL, R. J., T. E. LACHER, JR., C. BUNCK, B. DANIEL, C. DRIVER, C. E. GRUE, F. LEIGHTON, W. STANSLEY, P. G. WATANABE, AND M. WHITWORTH. 1996. An ecological risk assessment of lead shot exposure in non-waterfowl avian species: upland game birds and raptors. Environmental Toxicology and Chemistry 15:4–20.

LEWIS, J. C., AND E. LEGLER, JR. 1968. Lead shot ingestion by Mourning Doves and incidence in soil. Journal of Wildlife Management 32:476–482.

MIRARCHI, R. E., AND T. S. BASKETT. 1994. Mourning Dove (*Zenaida macroura*). *In* A. Poole and F. Gill (Eds.), The Birds of North America, No. 117. Academy of Natural Sciences, Philadelphia, and American Ornithologists' Union, Washington, D.C., USA.

OTIS, D. L., J. H. SCHULZ, D. A. MILLER, R. E. MIRARCHI, AND T. S. BASKETT. 2008. Mourning Dove (*Zenaida macroura*). *In* A. Poole (Ed.). The Birds of North America Online. Ithaca: Cornell Lab of Ornithology. Retrieved from the Birds of North America Online Database: http://bna.birds.cornell.edu/bna/species/117.

SCHULZ, J. H., J. J. MILLSPAUGH, B. E. WASHBURN, G. R. WESTER, J. T. LANIGAN III, AND J. C. FRANSON. 2002. Assessing spent shot availability on areas managed for Mourning Doves. Wildlife Society Bulletin 30(1):112–120.

SCHULZ, J.H., J.J. MILLSPAUGH, A.J. BERMUDEZ, XIAOMING GAO, T.W. BONNOT, L.G. BRITT, AND M. PAINE. 2006. Acute lead toxicosis in Mourning Doves. Journal of Wildlife Management 70(2):413–421.

SCHULZ, J.H., XIAOMING GAO, J.J. MILLSPAUGH, AND A.J. BERMUDEZ. 2007. Experimental lead pellet ingestion in Mourning Doves (Zenaida macroura). American Midland Naturalist 158(1):177–190.

CAUSES AND CONSEQUENCES OF INGESTED LEAD PELLETS IN CHUKARS

R. JUSTIN BINGHAM[1], RANDY T. LARSEN[1], JOHN A. BISSONETTE[1], AND JERRAN T. FLINDERS[2]

[1]*US Geological Survey - Utah Cooperative Fish and Wildlife Research Unit, Department of Wildland Resources, Utah State University, Logan, UT 84322-5200, USA.*
E-mail: r.j.bing@aggiemail.usu.edu

[2]*Department of Plant and Wildlife Sciences, Brigham Young University, 275 WIDB, Provo, UT 84602-5253, USA.*

EXTENDED ABSTRACT.—Lead-pellet ingestion and the resulting toxicosis are well-documented in waterfowl, raptors, and Mourning Doves (*Zenaida macroura*) (Kendall et al. 1996). Ingestion of lead shot by other avian taxa is less well-understood, but a growing body of literature suggests it does occur and can be a significant source of mortality (Keymer and Stebbings 1987, Lewis and Schweitzer 2000, Vyas et al. 2000, Walter and Reese 2003, Butler 2005). We are currently investigating the ingestion of lead pellets by Chukars (*Alectoris chukar)* in Utah.

We have carefully processed hunter-harvested Chukars by removing the gizzards, inspecting them, and excluding any gizzards with penetration wounds. We have documented ingestion of lead pellets by Chukars throughout four counties in western Utah. We have found ingested lead-pellets in 8.74% of gizzards from our sample (n = 286). We used Inductively-Coupled Plasma/Mass Spectroscopy (ICP/MS) to analyze Chukar livers for lead residues. Toxicology results show elevated concentrations of lead (>0.5 ppm) in 14% (n = 50) of livers from our sample. Elevated concentrations of lead ranged from 0.7 to 42.6 ppm (wet weight). We consider lead concentrations from our sample that are greater than 0.5 ppm (ww) to be elevated because: 1) 43/50 (86%) of analyzed livers from our sample population contained less than 0.5 ppm lead and 2) the frequency of individuals in our sample between the categories of >0.1<0.5 ppm (n = 10) and >0.5<1.0 ppm (n = 1) decreased by an order of magnitude. Table 1 shows the categories for all our sample of analyzed livers. Regarding liver tissue, lead concentrations greater than 2 ppm (ww) are considered indicative of chronic exposure, whereas values greater than 6 ppm (ww) denote acute exposure (Pain et al. 1993). The discovery that multiple Chukars from independent populations have ingested lead pellets warrants additional investigation into the causes and consequences of lead-pellet ingestion by Chukars.

Ingestion of lead pellets by Chukars is likely related to 1) the arid, rocky, and alkaline nature of Chukar habitat, which reduces pellet settlement and dissolution (Shranck and Dollahon 1975, Walter and Reese 2003), 2) similarities in appearance between lead pellets and grit and food sources used by Chukars, and 3) use of natural and man-made water sources (Best et al. 1992). The inert nature of lead allows it to persist in acidic soils for up to 300 years (Jorgensen and Willems 1987) and presumably longer in basic soils such as those characteristic of Chukar habitat. We found that nearly a third of examined grit from our sample of Chukars, and all lead shot sizes that are generally used for upland game (#'s 4–8) are intercepted by a soil sieve with mesh having 2 mm diameter openings. Chukars from our sample commonly contained Indian Ricegrass (*Stipa hymenoides*) seeds in their crops and gizzards. These seeds have a strong resemblance in size, shape, and color to lead pellets.

Table 1. Lead concentrations and their corresponding frequencies and percents for 50 Chukar livers analyzed using ICP/MS.

Category	Frequency	Percent of Sample
<0.05[a]	20	40
>0.05 <0.1	13	26
>0.1 <0.5	10	20
>0.5 <1	1	2
>1 <2	4	8
>2 <6[b]	1	2
>6[c]	1	2

[a] All values are reported in ppm (ww).

[b] Values >2 ppm (ww) are consistent with chronic exposure to lead (Pain et al. 1993).

[c] Values >6 ppm (ww) are consistent with acute exposure to lead (Pain et al. 1993).

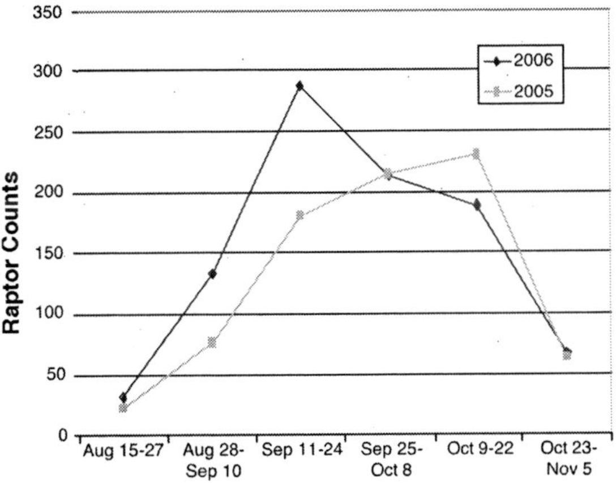

Figure 1. Raptor counts conducted by Hawkwatch International during the fall migrations of 2005 and 2006 in the Goshute Mountains of Eastern Nevada directly west of our Utah study area.

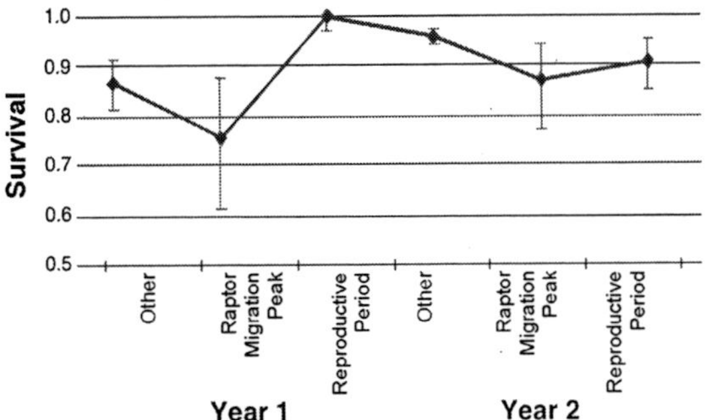

Figure 2. Two-week survival for radio-marked Chukars in our study area during 2005 and 2006. Survival estimates were less during peak raptor migration in both years (P = 0.06 in 2005) and (P <0.05 in 2006).

To monitor and assess the survival and probable causes of mortality of Chukars, we trapped Chukars on water sources using funnel traps and subsequently fitted these Chukars with nine and 14 g backpack-style radios from Advanced Telemetry Systems. We employed known fate models in Program MARK 4.1 (White and Burnham 1999) to estimate seasonal rates of survival and used model selection (Burnham and Anderson 2002) to evaluate hypotheses concerning seasonal differences in survival. Results from our radio-telemetry research showed that Chukar mortalities in our study area were due primarily to raptors and shooting. We obtained estimates of raptor migration for 2005 and 2006 using counts from the nearby Goshute Mountains of Eastern Nevada conducted by Hawkwatch International (Smith and Neal 2005, 2006). Figure 1 portrays the complete data from the 2005 and 2006 counts. During 2005 and 2006, nearly three-fourths (74%) of identified fatalities (n = 42) were attributed to raptors. Additionally, Chukars showed markedly-decreased survival rates during the fall raptor migrations of 2005 (P = 0.06) and 2006 (P <0.05). During the peak of raptor migration for 2005 and 2006, two-week rates of survival for radio-marked Chukars were lower than during any other two-week interval of the year, including the vulnerable period concomitant with reproductive behavior (see Figure 2).

Nearly half of Chukar mortalities that were attributed to raptors occurred during the fall migration of raptors through our study area. Also, fall migration of raptors coincides with the hunting season for all species of upland game in Utah including Chukars. Research concerning the relation between the hunting season for waterfowl and concentrations of blood lead in Marsh Harriers (*Circus aeruginosus*) has shown that amounts of blood lead rise significantly during the hunting season (Pain et al. 1997). Similar risks may be apparent to both migrating and resident raptors when feeding on upland game, particularly Chukars, in Utah.

Consuming Chukars may expose humans and wildlife to unhealthy concentrations of lead (Scheuhammer et al. 1998). Additional research is needed to clarify this risk. We are currently conducting further work to understand and assess the magnitude of this challenge. However, our research is largely incomplete at this time. We are addressing such questions as: 1) why do Chukars eat lead pellets, 2) what are the toxicological consequences that accompany such ingestion, 3) do natural and man-made water sources contribute to elevated concentrations of lead pellets in soil, and 4) is Chukar breast unhealthy for consumption if it has been harvested with lead pellets? *Received 9 June 2008, accepted 4 September 2008.*

BINGHAM, R. J., R. T. LARSEN, J. A. BISSONETTE, AND J. T. FLINDERS. 2009. Causes and consequences of ingested lead pellets in Chukars. Extended abstract *in* R. T. Watson, M. Fuller, M. Pokras, and W. G. Hunt (Eds.). Ingestion of Lead from Spent Ammunition: Implications for Wildlife and Humans. The Peregrine Fund, Boise, Idaho, USA. DOI 10.4080/ilsa.2009.0204

Key Words: *Alectoris chukar,* Chukar, ingested lead pellets, lead concentration, raptor, survival, predation, Utah.

LITERATURE CITED

BEST, T. L., T. E. GARRISON, AND C. G. SCHMITT. 1992. Availability and ingestion of lead shot by Mourning Doves (*Zenaida macroura*) in Southeastern New Mexico. The Southwestern Naturalist 37:287–292.

BURNHAM, K. P., AND D. R. ANDERSON. 2002. Model Selection and Multimodel Inference: Practical Information-Theoretic Approach, 2nd ed. Springer-Verlag, New York, USA.

BUTLER, D. A. 2005. Incidence of lead shot ingestion in Red-legged Partridges (*Alectoris rufa*) in Great Britain. Veterinary Record 157:661–662.

JORGENSEN, S. S., AND M. WILLEMS. 1987. The fate of lead in soils:The transformation of lead pellets of shooting range soils. Ambio 16:11-15.

KENDALL, R. J., T. E. LACHER, JR., C. BUNCK, B. DANIEL, C. DRIVER, C. E. GRUE, F. LEIGHTON, W. STANSLEY, P. G. WATANABE, AND M. WHITWORTH. 1996. An ecological risk assessment of lead shot exposure in non-waterfowl avian species: Upland game birds and raptors. Environmental Toxicology and Chemistry 15:4–20.

KEYMER, I. F., AND R. ST. J. STEBBINGS. 1987. Lead poisoning in a Partridge (*Perdix perdix*) after ingestion of gunshot. Veterinary Record 120:276–277.

LEWIS, L. A., AND S. H. SCHWEITZER. 2000. Lead poisoning in a Northern Bobwhite in Georgia. Journal of Wildlife Diseases 36:180–183.

PAIN, D. J., C. AMIARD-TRIQUET, C. BABOUX, G. BURNELEAU, L. EON, AND P. NICOLAU-GILLAUMET. 1993. Lead poisoning in wild populations of Marsh Harrier (*Circus aeruginosus*) from Charente-Maritime, France: Relationship with the hunting season. Biological Conservation 81:1–7.

PAIN, D. J., C. BAVOUX, AND G. BURNELEAU. 1997. Seasonal blood lead concentrations in Marsh Harriers (*Circus aeruginosus*) from Charente-Maritime, France: Relationship with the hunting season. Biological Conservation 81:1–7.

SCHEUHAMMER, A. M., J. A. PERRAULT, E. ROUTHIER, B. M. BRAUNE, AND G. D. CAMPBELL. 1998. Elevated lead concentrations in edible portions of game birds harvested with lead shot. Environmental Pollution 102:251–257.

SHRANCK, B. W., AND G. R. DOLLAHON. 1975. Lead shot incidence on a New Mexico public hunting area. Wildlife Society Bulletin 3:157–161.

SMITH, J. P., AND M. C. NEAL. 2005. Fall 2005 raptor migration studies in the Goshute Mountains of northeastern Nevada. Technical Report. Hawk Watch International, Salt Lake City, Utah, USA.

SMITH, J. P., AND M. C. NEAL. 2006. Fall 2006 raptor migration studies in the Goshute Mountains of northeastern Nevada. Technical Report. Hawk Watch International, Salt Lake City, Utah, USA.

VYAS, N. B., J. W. SPANN, G. H. HEINZ, W. N. BEYER, J. A. JAQUETTE, AND J. M. MENGELKOCH. 2000. Lead poisoning of passerines at a trap and skeet range. Environmental Pollution 107:159–166.

WALTER, H., AND K. P. REESE. 2003. Fall diet of Chukars (*Alectoris chukar*) in Eastern Oregon and discovery of ingested lead pellets. Western North American Naturalist 63:402–405.

WHITE, G. C., AND K. P. BURNHAM. 1999. Program MARK: Survival estimation from populations of marked animals. Bird Study 46 (Supplement):120–138.

LEAD EXPOSURE IN WISCONSIN BIRDS

SEAN M. STROM, JULIE A. LANGENBERG, NANCY K. BUSINGA, AND JASMINE K. BATTEN

Wisconsin Department of Natural Resources, Bureau of Wildlife Management, 101 S. Webster St., P.O. Box 7921, Madison, WI 53707, USA.
E-mail: Sean.Strom@Wisconsin.gov

ABSTRACT.—The Wildlife Health Program of the Wisconsin Department of Natural Resources has monitored lead (Pb) exposure in numerous avian species including Bald Eagles (*Haliaeetus leucocephalus*), Trumpeter Swans (*Cygnus buccinator*), Common Loons (*Gavia immer*), and American Woodcock (*Scolopax minor*). A comprehensive review of Trumpeter Swan health data indicated approximately 25% of Trumpeter Swan fatalities were attributed to lead toxicity. Similarly, approximately 15% of live-sampled Trumpeter Swans had blood lead levels above background concentrations (20 µg/dL). A similar review of necropsy data for Bald Eagles revealed approximately 15% of all Bald Eagle deaths in Wisconsin were attributed to lead toxicity. A noticeable increase in the percent of fatalities attributed to lead toxicity began in October and peaked in December. This pattern overlapped with the hunting seasons in Wisconsin suggesting lead ammunition could be a major source of lead exposure in eagles. A surveillance program examining lead toxicity as a factor in mortality of Common Loons was initiated in 2006. To date, approximately 30% of the dead loons submitted for necropsy were found to be lead poisoned. Lead fishing gear was recovered from the GI tracts of loons in all cases where lead toxicity was a major contributor to the cause of death. A comprehensive study investigating lead levels in woodcock from Wisconsin was completed in 2002. The results of the study indicated American Woodcock were exposed to lead on their breeding grounds in Wisconsin, resulting in high accumulations of lead in bone tissue. Bone lead concentrations considered to be toxic in waterfowl were observed in all age classes of woodcock. Stable isotope analysis was conducted on a subset of bone samples from young-of-year birds in order to identify the source of the lead. The results were inconclusive but did not rule out anthropogenic sources, and although the pathway of lead exposure was not identified, the data suggest a local and dietary source. It is clear that numerous species of Wisconsin wildlife are being exposed to potentially harmful levels of lead, and lead poisoning remains a significant mortality factor for many of these species. The prevalence of lead poisoning cases in Wisconsin is unlikely to decrease until the amount of lead discharged into the Wisconsin environment is reduced. *Received 9 July 2008, accepted 3 September 2008.*

STROM, S. M., J. A. LANGENBERG, N. K. BUSINGA, AND J. K. BATTEN. 2009. Lead exposure in Wisconsin birds. *In* R. T. Watson, M. Fuller, M. Pokras, and W. G. Hunt (Eds.). Ingestion of Lead from Spent Ammunition: Implications for Wildlife and Humans. The Peregrine Fund, Boise, Idaho, USA. DOI 10.4080/ilsa.2009.0205

Key words: American Woodcock, Bald Eagle, Common Loon, lead, Trumpeter Swan, Wisconsin.

THE DELETERIOUS IMPACTS on waterbirds of ingesting spent lead (Pb) shot have been recognized for over 100 years (Grinnell 1894). Ingestion of spent shot has been implicated in substantial die-offs of numerous waterfowl species (Bellrose 1959, Sanderson and Bellrose 1986) and as a result, non-toxic shot has been required to hunt waterfowl in the United States since 1991 (Anderson 1992). Even though the federal ban on lead shot for waterfowl hunting has been in effect for nearly 20 years, lead poisoning continues to be a significant mortality factor for several species of birds in Wisconsin. Lead is still regularly deposited in the environment as a result of lead shot used for upland game hunting, lead in rifle bullets, and lead fishing weights.

Lead is one of the most toxic metals known, with adverse impacts ranging from slight alterations of biochemical and physiological systems to serious damage in organs leading to death of the individual. Lead modifies the function and structure of the kidney, bone, the central nervous system, and the hematopoietic system and also has adverse biochemical, histopathological, fetotoxic, teratogenic and reproductive effects (Eisler 1988). Mortality associated with exposure of birds to lead has been shown to result from direct consumption of spent lead shot, consumption of lead shot or bullet fragments embedded in food items, and from the ingestion of lead fishing weights (Franson 1996).

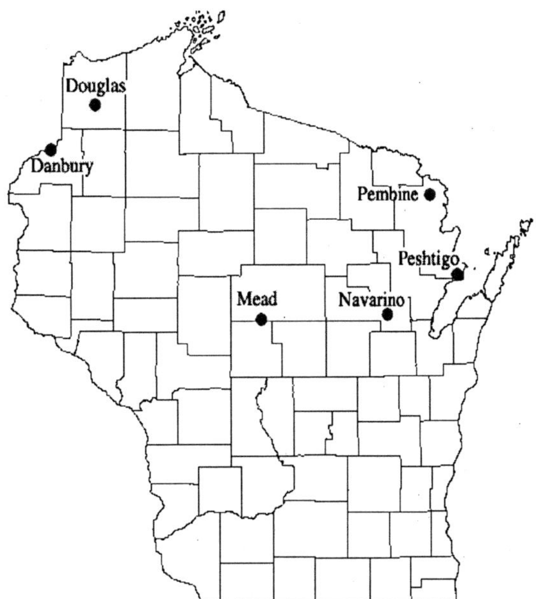

Figure 1: American Woodcock collection sites 1999–2001.

In Wisconsin, lead exposure has been monitored in numerous avian species including Bald Eagles (*Haliaeetus leucocephalus*), Trumpeter Swans (*Cygnus buccinator*), Common Loons (*Gavia immer*), and American Woodcock (*Scolopax minor*). In this paper, we report on lead as a mortality factor for Bald Eagles, Trumpeter Swans, and Common Loons. In addition, we report on concentrations of lead in bone of American Woodcock of different ages.

METHODS

American Woodcock.—Detailed methods of woodcock sample collection and analysis can be found in Strom et al. (2005). Briefly, woodcock were harvested between 1999 and 2001 at selected locations in Wisconsin using steel shot (Figure 1). Birds were collected a minimum of two weeks prior to the regular hunting season (late September through early November) to increase the probability that only locally exposed birds (young of year) would be collected. Woodcock age and sex were determined via plumage characteristics as described by Sepik (1994). Pointing dogs were used to assist in finding woodcock nests and broods. When a brood was found, one chick was randomly selected and euthanized via cervical dislocation for tissue analysis.

For sample collection, the humerus and radius/ulna were excised and excess tissue and cartilage removed. This process was followed for all woodcock collected, regardless of age or year collected. The GI tracts of all woodcock were radiographed and the radiographs were evaluated for any objects of comparable radiodensity with metal. The GI tracts were also dissected and visually inspected for lead pellets or metallic particles. The sex of the animal was confirmed at the time of dissection. All lead analyses were conducted at the Wisconsin Veterinary Diagnostic Laboratory, Madison, Wisconsin.

Bone samples were forced-air dried overnight at 100ºC to remove water, finely crushed in a mortar, and extracted with diethyl ether to remove fat prior to weighing for lead analysis. All tissue samples (bone, liver and blood) were digested in Teflon microwave vessels, under pressure with nitric acid in a microwave. After the samples cooled to room temperature, the contents were transferred to a volumetric flask and brought to volume with purified water. Samples were analyzed on a Perkin-Elmer

Table 1: Bone lead concentrations (μg/g dry weight) in Wisconsin American Woodcock.

Collection Site		n	Mean	Median	SD	Range
Navarino						
	Chick	7	30.1	27.9A	22.5	9.6 – 72.0
	Young of Year (YOY)	12	18.4A	11.25	18.0	<3.0–48.0
	Adult	14	22.0A	15.9	12.8	8.1–48.0
Mead						
	Chick	3	76.0	87.0B	24.4	48.0–93.0
	YOY	13	39.2A	29.1	50.0	4.5–198.0
	Adult	10	63.8B	33.3	56.0	18.9–171.0
Peshtigo						
	YOY	6	67.5A	40.5	80.7	6.9–222.0
	Adult	9	25.7AB	18.0	23.7	5.1–81.0
Pembine						
	YOY	6	18.45A	19.7	11.3	<3.0–33.0
	Adult	11	18.4A	15.6	12.0	5.1–45.0
Danbury						
	YOY	5	12.5A	5.7	14.8	4.8–39.0
	Adult	5	32.6AB	22.8	22.2	8.7–63.0
Douglas						
	YOY	4	10.5A	7.4	8.9	3.9–23.4
	Adult	3	6.4AB	6.6	0.62	5.7–6.9

For each age class, means or medians sharing the same letter are not significantly different.

Table 2: Mortality factors of Trumpeter Swans, Bald Eagles and Common Loons submitted to the Wisconsin Department of Natural Resources for necropsy.

		Diagnosis (%)						
Species	N	Lead Poisoning	Trauma	Undetermined	Gunshot	Electrocution	Drowning	Other
Trumpeter Swan	143	25	25	19	12	10	1	7
Bald Eagle	583	16	48	16	0	12	0	9
Common Loon	26	29	31	4	7	0	4	25

5100ZL graphite furnace atomic absorption spectrometer. A tissue and/or blood control sample was tested with each analytical run to ensure accuracy.

Trumpeter Swan, Bald Eagle and Common Loon.—Trumpeter Swans, Bald eagles, and Common Loons were submitted to the Wisconsin Department of Natural Resource's Wildlife Health Program for necropsy, as part of statewide mortality monitoring programs for these species. At the time of necropsy, all carcasses were evaluated for metal foreign bodies using radiography. If radio-dense objects were observed, attempts were made to recover the object from the carcass. Liver samples were collected from all individuals and submitted for laboratory lead analysis as described above. For Trumpeter Swans, data accumulated between the years 1991–2007 were reviewed and causes of mortality were summarized. A similar review was performed for Bald Eagle data collected between the years 2000–2007. A program investigating causes of Common Loon mortality was initiated in 2006.

Data Analysis.—For statistical analyses, samples with metal concentrations below the detection limit were assigned a value equal to one-half the detection limit. All statistical analyses were carried out using Excel and SYSTAT software packages. Nonparametric methods (Kruskal-Wallis tests) were utilized to determine differences between tissue

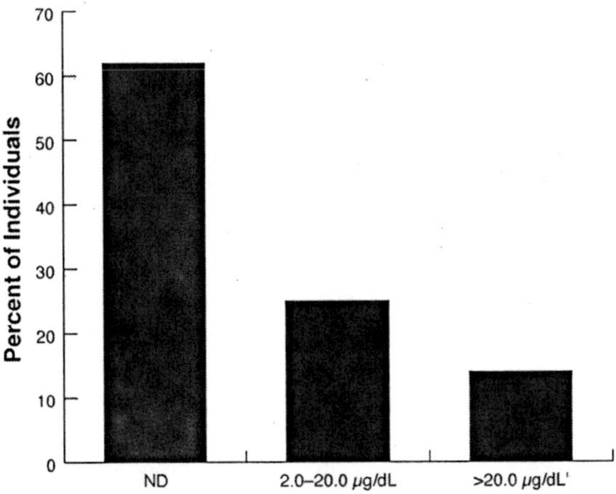

Figure 2: Blood lead concentrations (µg/dL) of Trumpeter Swans from Wisconsin (n = 1172 individuals). ND = below the limit of detection (detection limit range 2.0–5.0 µg/dL).

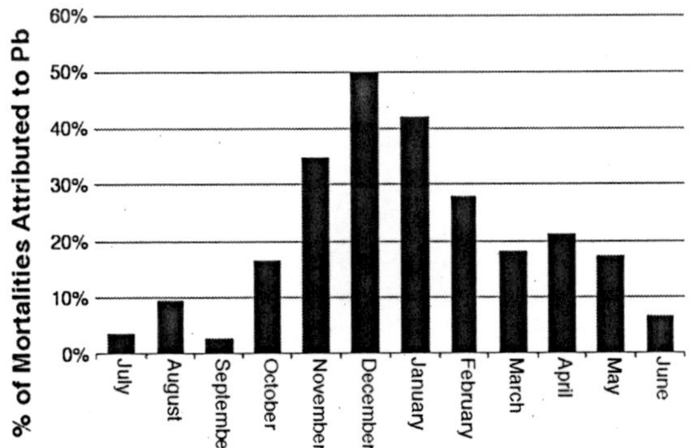

Figure 3: Percent of Bald Eagle fatalities attributed to lead toxicity according to month.

lead concentrations from woodcock collected at various sites across the state. The level of statistical significance was set at $\alpha = 0.05$.

RESULTS

American Woodcock.—The mean bone lead concentrations from chicks, young-of-year, and adult woodcock collected at various sites throughout Wisconsin are listed in Table 1. No significant differences were observed between bone lead concentrations in young-of-year woodcock from any of the areas sampled (p = 0.174). However, a significant difference was observed with the adult data (p = 0.006), with bone lead concentrations of Navarino adults being significantly different than those of Mead adults. Furthermore, bone lead concentrations of Pembine adults were significantly different than those of Mead adults. The concentration of lead in the wing bones from Mead chicks was significantly different than those from Navarino (p = 0.03). Radiography of the GI tracts of the woodcock did not reveal any lead pellets or metallic particles. Furthermore, no lead pellets or metallic objects were observed during the visual inspection of the GI tracts.

Trumpeter Swans.—Between 1991 and 2007, 143 swan fatalities were investigated. Approximately 25% of the fatalities were attributed to lead toxicity. The diagnosis of lead toxicity as a cause of mortality was often based only on tissue lead concentrations, as clinical history and histologic assessment of tissues were not available. Collisions with power lines as well as shootings also appear to be significant mortality factors.

During this same period, 1,172 blood samples were collected from swans and analyzed for lead (Figure 2). Lead was detected in 38% of the blood samples, with a geometric mean of 16 µg/dL. Approximately 15% of all sampled swans had blood lead concentrations above 20µg/dL. The overwhelming majority (80%) of the blood samples were collected from juvenile swans. Therefore, due to the age bias in sampling, age-specific differences in lead accumulation could not be not examined.

Bald Eagles.—Between 2000 and 2007, nearly 600 Bald Eagles were submitted to Wildlife Health for necropsy. Lead toxicity was diagnosed as the cause of death for 16% of the eagles during this period. A unique temporal pattern in eagle mortality was observed, with a dramatic increase in the percent of fatalities attributed to lead toxicity beginning in October, peaking in December, and then decreasing through the winter months (Figure 3). This pattern overlaps with the hunting seasons in Wisconsin which suggests lead ammunition could be a major source of lead exposure in eagles.

Common Loons.—Between 2006 and 2008, 26 Common Loons were submitted for necropsy. Approximately 30% of the dead loons submitted for necropsy were judged to have died from lead poisoning. Remnants of lead fishing tackle were recovered from the GI tracts of loons in all cases where lead toxicity was a major contributor to the cause of death.

DISCUSSION

Our data indicate lead exposure is a significant health problem for several bird species in Wisconsin, accounting for 15–30% of fatalities in Trumpeter Swans, Bald Eagles, and Common Loons. For our study, liver lead levels of >6.0 µg/g (wet weight) were considered indicative of lead toxicosis and bone lead levels >20 µg/g dry weight were considered significantly elevated (Franson 1996, Pain 1996). Blood lead levels >20 µg/dL were considered above background (Pain 1996).

Scheuhammer et al. (1999) documented a high incidence of elevated lead levels (>20 µg/g dry weight) in wing bones from adult (52%) and young-of-year woodcock (29%) from eastern Canada. Strom et al. (2005) found elevated lead levels in wing bones of woodcock from Wisconsin in 43% of young-of-year woodcock and 64% of woodcock chicks sampled.

Although the prevalence of lead exposure in waterfowl has declined following implementation of non-toxic shot for waterfowl hunting (Samuel and Bowers 2000), exposure to lead continues to be a significant factor in Trumpeter Swan mortality. The incidence of lead related mortality in Trumpeter Swans in Wisconsin (26%) is slightly higher than the 20% lead-related mortality observed in Trumpeter Swans from the tri-state area of Idaho, Montana, and Wyoming (Blus et al. 1989), but much lower than that observed in Trumpeter and Tundra Swans (81%) in Washington State (Degernes et al. 2006).

The geometric mean of blood lead levels observed in Trumpeter Swans from Wisconsin (16 µg/dL) was greater than the geometric mean reported in blood from Trumpeter Swans in the western U.S. (Blus et al. 1989). A blood lead concentration below 20 µg/dL has been suggested as a background level in waterfowl (Pain 1996, Beyer et al. 2000). Considering over 60% of the blood samples analyzed in the present study were below the detection limit (detection limit range 2.0–5.0 µg/dL) it is plausible the true background blood lead concentration for apparently healthy Trumpeter Swans is below the 20 µg/dL threshold.

The occurrence of lead poisoning in Bald Eagles from Wisconsin (15%) is higher than that reported in other studies where prevalence ranged from 0% to 5% (Coon et al. 1970, Deem 2003, Harris and Sleeman 2007, Morishita et al. 1998, Wendell et al. 2002). However, our findings are consistent with those of Elliot et al. (1992) who observed that 14% of Bald Eagles found dead in British Columbia, Canada, between 1988 and 1991 died as a result of lead poisoning and Wayland et al. (2003) who observed 14% of 546 Bald Eagle deaths in western Canada were attributable to lead toxicosis.

The proportion of lead poisoning among loon fatalities in Wisconsin is comparable to that observed in Canada (26%–30%) (Daoust et al. 1998, Scheuhammer and Norris 1996) but lower than that of breeding loons in the New England states (44%–52%) (Pokras and Chafel 1992, Sidor et al. 2003). The contrast of our results to the New England studies may be due to the much smaller sample size in the present study (n = 26), and may also reflect differences in regional landscape, human population, or land and wetland usage (Sidor et al. 2003). It nevertheless appears that lead exposure is a major mortality factor for loons in Wisconsin.

The percentages of lead poisoning cases in Wisconsin birds from this study are likely underestimated because, in many cases where cause of death was undetermined, liver lead analysis was not performed. Moreover, human-related mortality agents tend to occur in areas frequented by humans, whereas lead-poisoned fatalities are more randomly distributed and thus less likely to be discovered.

Sources of Lead.—Scheuhammer et al. (2003) found no evidence of lead shot ingestion in woodcock from Canada with elevated bone-lead concentrations. Likewise, we did not detect any ingested lead shot in any of the woodcock radiographed for the present study. However, ingestion of lead shot has been observed in other species related to woodcock, such as godwits, dowitchers, and snipe (Hall and Fisher 1985, Pain 1990). Scheuhammer et al. (2003) suggest that because woodcock do not have a large muscular gizzard, it is possible that lead pellets ingested by woodcock are quickly voided, and thus not observed in studies of woodcock gizzard and stomach contents.

Stable lead isotope analysis on a small number of bone samples from Wisconsin young-of-year woodcock produced inconclusive results. The observed lead isotope ratios in Wisconsin woodcock overlapped with ratios of both lead shotgun pellets and Precambrian lead, i.e. soil-associated lead (Strom et al. unpub. data). Similarly, Scheuhammer et al. (2003) observed lead isotope ratios in wing bones of woodcock that were consistent with, though not proof of, lead shot ingestion because the range of ratios for wing bones overlapped with ratios of lead shot pellets. In any case, ingestion of lead shot used for upland game bird hunting cannot be ruled out as a primary source of high bone-lead accumulation in woodcock from Wisconsin.

The feeding habits of Trumpeter Swans and Bald Eagles may predispose these species to lead exposure. Trumpeter Swans typically forage by digging up large amounts of sediments in lakes and streams (Banko 1960). Ingestion of these large volumes of sediments increases the likelihood of birds ingesting lead shot or sinkers (Blus et al. 1989). Of the lead poisoning cases observed in Trumpeter Swans in this study, 39% had lead artifacts in their ventriculus at the time of necropsy.

The increased prevalence of lead poisoning cases in Bald Eagles during October to January overlaps with the hunting seasons in Wisconsin, suggesting lead ammunition is likely the primary source of exposure in Bald Eagles. In addition to lead shot, lead bullet fragments may be a significant source of lead exposure in eagles. This pattern of increased lead exposure during hunting seasons has also been observed in Golden Eagles (*Aquila chrysaetos*) (Bloom et al. 1989), California Condors (*Gymnogyps californianus*) (Cade 2007, Hall et al. 2007, Hunt et al. 2007, Sorenson and Burnett 2007), and Common Ravens (*Corvus corax*) (Craighead and Bedrosian 2008). Hunt et al. (2006) observed that as lead bullets pass through deer, the bullets often fragment, leaving numerous lead fragments which are distributed throughout the carcass or the gut pile left on the landscape after the animal is field dressed. It is likely eagles feed on either unretrieved animals or scavenge the gut piles, thus exposing themselves to harmful levels of lead.

Conclusions.—Multiple species of Wisconsin wildlife are being exposed to potentially harmful levels of lead, and lead poisoning remains a significant mortality factor for some of these species. The association of lead-related mortality in Bald Eagles with the major hunting seasons in Wisconsin implies that exposure to spent lead ammunition should be considered as a potentially significant mortality factor in species which feed upon hunter-killed animals. It is unlikely the prevalence of lead poisoning cases will decrease until the amount of lead discharged into the Wisconsin environment is reduced.

LITERATURE CITED

ANDERSON, W. L. 1992. Legislation and lawsuits in the United States and their effects on non-toxic shot regulations. Pages 56-60 *in* D. J. Pain (Ed.). Lead Poisoning in Waterfowl. Proceedings of an IWRB Workshop, Brussels, Belgium, 13–15 June 1991. International Waterfowl and Wetlands Research Bureau Special Publication 16, Slimbridge, UK.

BANKO, W. E. 1960. The Trumpeter Swan. North American Fauna No. 63, US Fish and Wildlife Service, Washington, DC, USA.

BELLROSE, F. C. 1959. Lead poisoning as a mortality factor in waterfowl populations. Illinois Natural History Survey Bulletin 27:235-288.

BEYER, W. N., D. J. AUDET, G. H. HEINZ, D. J. HOFFMAN, AND D. DAY. 2000. Relation of waterfowl poisoning to sediment lead concentrations in the Coeur d'Alene River Basin. Ecotoxicology 9:207-218.

BLOOM, P. H., J. M. SCOTT, O. H. PATTEE, AND M. R. SMITH. 1989. Lead contamination of Golden Eagles (*Aquila chrysaetos*) within the range of the Californian Condor (*Gymnogyps californianus*). Pages 481–482 *in* B-U. Meyburgh and R. D. Chancellor (Eds.). Raptors in the Modern World. Proceedings of the 3rd World Conference on Birds of Prey and Owls, Eliat, Israel. World Working Group on Birds of Prey, Berlin.

BLUS, L. J., R. K. STROUD, B. REISWIG, AND T. MCENEANEY. 1989. Lead poisoning and other mortality factors in Trumpeter Swans. Environmental Toxicology and Chemistry 8:263–271.

CADE, T. J. 2007. Exposure of California Condors to lead from spent ammunition. Journal of Wildlife Management 71(7):2125–2133.

COON, N. C., L. N. LOCKE, E. CROMARTIE, AND W. L. REICHEL. 1970. Causes of Bald Eagle mortality, 1960–1965. Journal of Wildlife Diseases 6:72-76.

CRAIGHEAD, D., AND B. BEDROSIAN. 2008. Blood lead levels of Common Ravens with access to big-game offal. Journal of Wildlife Management 72(1):240-245.

DAOUST, P. Y., G. CONBOY, S. MCBURNEY, AND N. BURGESS. 1998. Interactive mortality factors in Common Loons from maritime Canada. Journal of Wildlife Diseases 34(3):524-531.

DEEM, S. L. 2003. Fungal diseases of birds of prey. Veterinary Clinics of North America: Exotic Animal Practice 6:363-376.

DEGERNES, L., S. HEILMAN, M. TROGDON, M. JORDAN, M. DAVISON, D. KRAEGE, M. CORREA, AND P. COWEN. 2006. Epidemiologic investigation of lead poisoning in Trumpeter Swans in Washington state, USA, 2000–2002. Journal of Wildlife Diseases 42(2):345-358.

EISLER, R. 1988. Lead hazards to fish, wildlife, and invertebrates: A synoptic review. US Fish & Wildlife Service, biol. report no. 85, Washington, DC, USA.

ELLIOT, J. E., K. M. LANGELIER, A. M. SCHEUHAMMER, P. H. SINCLAIR, AND P. E. WHITEHEAD. 1992. Incidence of lead poisoning in Bald Eagles and lead shot in waterfowl gizzards from British Columbia 1988–1991. Progress Note 200. Canadian Wildlife Service, Ottawa, Ontario, Canada.

FRANSON, J. C. 1996. Interpretation of tissue lead residues in birds other than waterfowl. Pages 265–279 *in* W. N. Beyer, G. H. Heinz, and A. W. Redwood-Norwood (Eds.). Environmental Contaminants in Wildlife: Interpreting Tissue Concentrations. CRC Press, Lewis Publishers, New York, USA.

GRINNELL, G. B. 1894. Lead poisoning. Forest and Stream 42(6):117–118.

HALL, S. L., AND F. M. FISHER, JR. 1985. Lead concentrations in tissues of marsh birds: Relationship of feeding habits and grit preference to spent shot ingestion. Bulletin of Environmental Contamination and Toxicology 35:1–8.

HALL, M., J. GRANTHAM, R. POSEY, AND A. MEE. 2007. Lead exposure among reintroduced California Condors in southern California. Pages 163–184 *in* A. Mee and L. S. Hall (Eds.). California Condors in the 21st century. Series in Ornithology no. 2. Nuttall Ornithological Club and American Ornithologists' Union, Cambridge, Massachusetts, USA.

HARRIS, M. C., AND J. M. SLEEMAN. 2007. Morbidity and mortality of Bald Eagles (*Haliaeetus leucocephalus*) and Peregrine Falcons (*Falco peregrinus*) admitted to The Wildlife Center of Virginia, 1993–2003. Journal of Zoo and Wildlife Medicine 38(1):62-66.

HUNT, W. G., W. BURNHAM, C. N. PARISH, K. K. BURNHAM, B. MUTCH, AND J. L. OAKS. 2006. Bullet fragments in deer remains: Implications for lead exposure in avian scavengers. Wildlife Society Bulletin 34:167–170.

HUNT, W. G., C. N. PARISH, S. C. FARRY, T. G. LORD, AND R. SIEG. 2007. Movements of introduced California Condors in Arizona in relation to lead exposure. Pages 79–96 *in* A. Mee and L. S. Hall (Eds.). California Condors in the 21st century. Series in Ornithology no. 2. Nuttall Ornithological Club and American Ornithologists' Union, Cambridge, Massachusetts, USA.

MORISHITA, T. Y., A. T. FULLERTON, L. J. LOWENSTINE, I. A. GARDNER, AND D. L. BROOKS. 1998. Morbidity and mortality in free-living raptorial birds of northern California: A retrospective study. Journal of Avian Medicine and Surgery 12:78-81.

PAIN, D. J. 1990. Lead poisoning of waterfowl: A review. Pages 172–181 *in* G. Matthews (Ed.). Managing Waterfowl Populations. IWRB, Slimbridge, UK.

PAIN, D. J. 1996. Lead in waterfowl. *In* W. N. Neyer, G. H. Heinz, and A. W. Redmon-Norwood (Eds.). Environmental Contaminants in Wildlife: Interpreting Tissue Concentrations. CRC Press, Inc., Boca Raton, Florida, USA.

POKRAS, M. A., AND R. M. CHAFEL. 1992. Lead toxicosis from ingested fishing sinkers in adult Common Loons (*Gavia immer*) in New England. Journal of Zoo and Wildlife Medicine 23:92–97.

SAMUEL, M. D., AND E. F. BOWERS. 2000. Lead exposure in American Black Ducks after implementation of non-toxic shot. Journal of Wildlife Management 64:947–953.

SANDERSON, G. C., AND F. C. BELLROSE. 1986. A review of the problem of lead poisoning in waterfowl. Illinois Natural History Survey Special Publication 4:1–34.

SCHEUHAMMER, A. M., AND S. L. NORRIS. 1996. The ecotoxicology of lead shot and lead fishing weights. Ecotoxicology 5:279–295.

SCHEUHAMMER, A. M., C. A. ROGERS, AND D. BOND. 1999. Elevated lead exposure in American Woodcock (*Scolopax minor*) in Eastern Canada. Archives of Environmental Contamination and Toxicology 36:334–340.

SCHEUHAMMER, A. M., D. E. BOND, N. M. BURGESS, AND J. RODRIGUE. 2003. Lead and stable lead isotope ratios in soil, earthworms, and bones of American Woodcock (*Scolopax minor*) from Eastern Canada. Environmental Toxicology and Chemistry 22:2585–2591.

SEPIK, G. F., 1994. A woodcock in the hand. Ruffed Grouse Society, Coraopolis, Pennsylvania, USA.

SIDOR, I. F., M. A. POKRAS, A. R. MAJOR, R. H. POPPENGA, K. M. TAYLOR, AND R. M. MICONI. 2003. Mortality of Common Loons in New England, 1987 to 2000. Journal of Wildlife Diseases 39(2):306–315.

SORENSON, K. J., AND J. L. BURNETT. 2007. Lead concentrations in the blood of Big Sur California Condors. Pages 185–195 *in* A. Mee and L. S. Hall (Eds.). California Condors in the 21st century. Series in Ornithology no. 2. Nuttall Ornithological Club and American Ornithologists' Union, Cambridge, Massachusetts, USA.

STROM, S. M., K. A. PATNODE, J. A. LANGENBERG, B. L. BODENSTEIN, AND A. M. SCHEUHAMMER. 2005. Lead contamination in American Woodcock (*Scolopax minor*) from Wisconsin. Archives of Environmental Contamination and Toxicology 49:396–402.

WAYLAND, M. L., K. WILSON, J. E. ELLIOT, M. J. R. MILLER, T. BOLLINGER, M. MCADIE, K. LANGELIER, J. KEATING, AND J. M. W. FROESE. 2003. Mortality, morbidity and lead poisoning of eagles in western Canada. Journal of Raptor Research 37:8–18.

WENDELL, M. D., J. M. SLEEMAN, AND G. KRATZ. 2002. Retrospective study of morbidity and mortality of raptors admitted to Colorado State University Veterinary Teaching Hospital during 1995 to 1998. Journal of Wildlife Diseases 38:101-106.

A RELATIONSHIP BETWEEN BLOOD LEAD LEVELS OF COMMON RAVENS AND THE HUNTING SEASON IN THE SOUTHERN YELLOWSTONE ECOSYSTEM

DEREK CRAIGHEAD AND BRYAN BEDROSIAN

Craighead Beringia South, P.O. Box 147, Kelly, WY 83011, USA.
E-mail:bryan@beringiasouth.org

ABSTRACT.—We recently found evidence to support the supposition that Common Ravens (*Corvus corax*) were ingesting lead from hunter-provided offal in the southern Yellowstone Ecosystem. Since those data were analyzed, we have collected an additional 237 samples from ravens in the same study area spanning an additional two hunting seasons. In total, we collected 153 individual blood samples during the hunting seasons of 2006/07 and 2007/08. Those new samples exhibited a median level of 10.0 µg/dL with a range of 2.7–51.7 µg/dL. We also collected 84 additional samples during the non-hunting season which exhibit a median blood lead level of 2.2 µg/dL with a range of 0.0–19.3 µg/dL. Comparatively, 50% of the hunting season sample exhibited blood lead levels >10µg/dL, while only 3% were greater than 10µg/dL during the non-hunt. We combine this new data with previous data collected to further understand the link between ingested lead and Common Ravens. *Received 24 June 2008, accepted 20 August 2008.*

CRAIGHEAD, D., AND B. BEDROSIAN. 2009. A relationship between blood lead levels of Common Ravens and the hunting season in the southern Yellowstone Ecosystem. *In* R. T. Watson, M. Fuller, M. Pokras, and W. G. Hunt (Eds.). Ingestion of Lead from Spent Ammunition: Implications for Wildlife and Humans. The Peregrine Fund, Boise, Idaho, USA. DOI 10.4080/ilsa.2009.0206

Key words: *Corvus corax*, heavy metal, lead poisoning, toxicosis.

AN INCREASING NUMBER OF STUDIES have implicated lead-core rifle bullets as a source for lead ingestion in bird species (e.g., Church et al. 2006, Craighead and Bedrosian 2008, Parish et al. 2007). Hunt et al. (2006) and Craighead and Bedrosian (2008) x-rayed offal piles left by hunters and found hundreds of lead fragments large enough to be seen in radiographs. Many scavenging species are long-lived, such as California Condors (*Gymnogyps californianus*), eagles, and Common Ravens (*Corvus corax*), and individuals within a population may be exposed to lead annually over their lifetimes. After lead is ingested it moves from the blood through soft tissues and is eventually sequestered in bone where it remains as a potential long-term source of toxicosis. There have been few long-term studies documenting the impact of lead on wild individuals and the possible influences on a population level.

We recently published data that implicated lead rifle bullet fragments in causing increased blood lead levels (BLL) of Common Ravens during the hunting seasons in the southern Greater Yellowstone Ecosystem (Craighead and Bedrosian 2008). Those BLL data were gathered over a 15-month period ending on 30 March 2006 that covered two hunting seasons and two non-hunting seasons. We continued this research (Craighead and Bedrosian 2008) for an additional 23 months and combined those data with previously published data to obtain a more robust dataset and analysis of BLL of Common Ravens.

METHODS

We captured Common Ravens within the Jackson Hole valley of northwestern Wyoming, which is a part of the Greater Yellowstone Ecosystem. An Elk (*Cervus canadensis*) and Bison (*Bison bison*) hunt occurs annually within and surrounding the study area, but very little hunting (recreational or other) occurs outside of the hunting season due to the protection afforded by Grand Teton National Park and the National Elk Refuge. This makes for an ideal comparative situation to test the hunting seasons and non-hunting seasons for blood lead levels. See Craighead and Bedrosian (2008) for more details regarding study area and capture methods.

We captured Common Ravens from 14 December 2004 through 7 March 2008, and gathered an additional 237 blood samples from ravens beyond those analyzed in Craighead and Bedrosian (2008), spanning an additional two hunting seasons. Of those, 153 samples were collected during the hunting seasons (109 and 44 during the 2006/07 and 2007/08 seasons, respectively) and 84 during the non-hunting seasons (59 and 25 in 2007 and 2008, respectively). In total, we collected 307 samples during four hunting seasons and 231 samples during four non-hunting seasons.

We collected blood samples intravenously from the brachial vein of captured ravens and stored the whole blood samples in lithium heparin Microtainer® blood tubes (Becton Dickinson, Franklin Lakes, NJ). Blood samples were analyzed for lead content (µg/dL) within 24 hr of collection using a Leadcare® portable blood lead analyzer (ESA Biosciences Inc., Chelmsford, MA).

We first tested for differences between different hunting and non-hunting seasons using a Kruskal-Wallis ANOVA test and obtained 95% confidence intervals using Wilcoxon signed-rank tests. We followed the Kruskal-Wallis with Mann-Whitney tests on individual years to determine annual differences. We then combined years (2004–2008) and tested for differences between hunting and non-hunting season using Kruskal-Wallis tests due to non-normality. We tested for differences between sexes during both the hunting season and non-hunting season using Mann-Whitney tests (Craighead and Bedrosian 2008). We attempted to correlate median BLL during the hunting season with the combined harvests from the National Elk Refuge and Grand Teton National Park using linear regression. Sex was assigned with two separate discriminant functions using footpad length and mass (Bedrosian et al. 2008).

RESULTS

We detected a difference among years for hunting season samples ($P = 0.005$, $H = 13.03$). Specifically, the 2004/05 and 2007/08 hunting seasons exhibited higher median BLLs than the two middle years (Figure 1). We found no differences between years of the non-hunting season samples ($P = 0.211$, $H = 4.52$). After pooling years, we found a median BLL of 10.0 µg/dL (SE = 0.528) for the hunting season (range = 1.0–55.5 µg/dL) and a median BLL of 2.0 µg/dL (SE = 0.151) for the non-hunt (range = 0.0–19.3). We detected a significant relationship between the annual median raven BLLs and the combined large-game harvest success from the National Elk Refuge and Grand Teton National Park ($P = 0.048$, Figure 2).

Using discriminant functions (Bedrosian et al. 2008), we were able to classify 413 individuals as male or female. We classified 125 individuals as male and 107 samples as female during the hunting seasons and detected no difference in median BLLs between sexes ($P = 0.220$, $W = 13937.5$). In non-hunt samples, we classified 69 males and 112 females and similarly found no difference between sexes ($P = 0.248$, $W = 6675.0$).

DISCUSSION

These additional data in combination with those reported in Craighead and Bedrosian (2008) provide more robust and defensible conclusions concerning lead ingestion in Common Ravens residing in the southern Greater Yellowstone Ecosystem. Notably, larger sample sizes negated our previous findings that females ingest greater amounts of lead during the hunting season. However, over 150 additional blood lead samples gathered during the two additional hunting seasons did not alter our original conclusions that raven BLLs increased significantly during the hunting seasons.

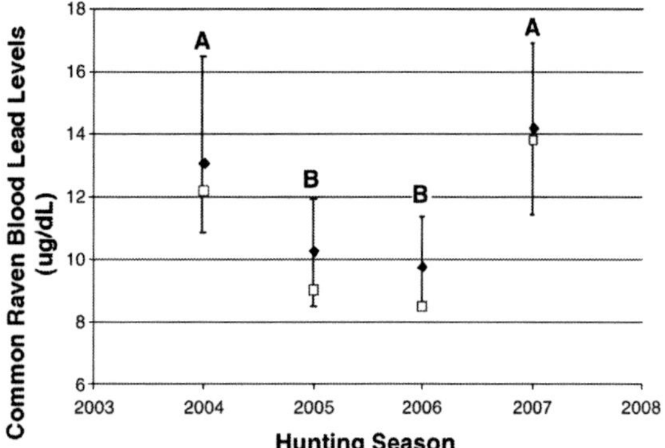

Figure 1. Median blood lead levels (open squares) of Common Ravens during the hunting seasons in Jackson Hole, Wyoming. Closed diamonds are Wilcoxon signed-rank estimated medians with 95% CI bars. A and B designate statistical similarities and differences tested with Kruskal-Wallis ANOVA (P = 0.005) and individual Mann-Whitney tests (α = 0.05).

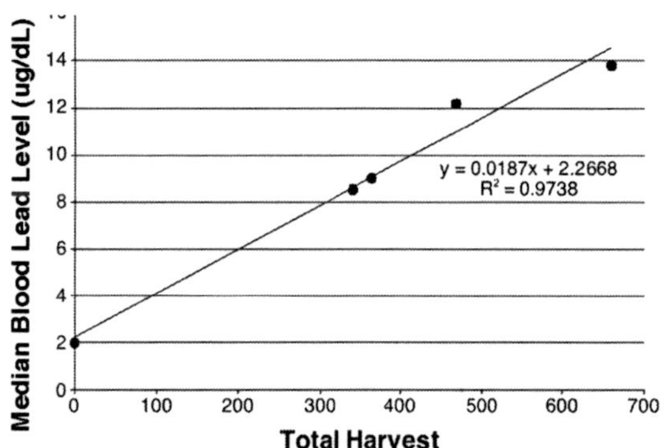

Figure 2. Regression analysis of median blood lead levels of Common Ravens during the hunting seasons in Jackson Hole, Wyoming and total large-game harvest rates (Elk and Bison) within Grand Teton National Park and the National Elk Refuge for the respective season (P = 0.048). Median blood lead level during the non-hunting season for all years combined was tested with an assumed harvest of zero.

With the additional data across years, we detected differences in hunting season median BLLs among years. We found that ravens exhibited more elevated BLLs in the 2004/05 and 2007/08 hunting seasons, when compared with the 2005/06 and 2006/07 seasons (Figure 1). These findings suggest that the median BLL in Common Ravens can be predicted by the large-game harvest rates on the National Elk Refuge and Grand Teton National Park (Figure 2). While there are many other animals harvested in the surrounding national forests and wilderness areas that ravens have access to, the harvest rates in the valley are likely indicative of harvest proportions in the surrounding areas.

Our data clearly indicate that ravens are being exposed to a lead source during the hunting season that is not present at other times of the year. Given the foraging behavior of ravens and their affinity for hunter-provided offal (Wilmers et al. 2003), gut piles are the likely source of ingested lead (Hunt et al. 2006, 2007). Given the annual deposition of lead from spent rifle bullets and the long lifespan of ravens (Boarman and Heinrich 1999), there are many negative implications for the cumulative, lifetime impacts on raven health and that of their populations in this, and other, regions. Similarly, other scavenging species such as Golden Eagles (*Aquila chrysaetos*) and Bald Eagles (*Haliaeetus leucocephalus*) are likely ingesting lead at similar, or higher, rates than ravens (Wilmers et al. 2003, Bedrosian and Craighead 2009, this volume). We suggest using alternative, non-lead rifle bullets to reduce the incidence of lead ingestion by scavenging wildlife.

LITERATURE CITED

BEDROSIAN, B., AND D. CRAIGHEAD. 2009. Blood lead levels of Bald and Golden Eagles sampled during and after hunting seasons in the Greater Yellowstone Ecosystem. *In* R. T. Watson, M. Fuller, M. Pokras, and W. G. Hunt (Eds.). Ingestion of Lead from Spent Ammunition: Implications for Wildlife and Humans. The Peregrine Fund, Boise, Idaho, USA. DOI 10.4080/ilsa.2009.0209

BEDROSIAN, B., J. LOUTSCH, AND D. CRAIGHEAD. 2008. Using morphometrics to determine the sex of Common Ravens. Northwestern Naturalist 89:46–52.

BOARMAN, W. I., AND B. HEINRICH. 1999. Common Raven (*Corvus corax*) *in* A. Poole and F. Gill (Eds.). The Birds of North America. The Academy of Natural Sciences, Philadelphia, and The American Ornithologists' Union, Washington, DC, USA.

CRAIGHEAD, D., AND B. BEDROSIAN. 2008. Blood lead levels of Common Ravens with access to big-game offal. Journal of Wildlife Management. 72:240–245.

CHURCH, M. E., R. GWIAZDA, R. W. RISEBROUGH, K. SORENSON, C. P. CHAMBERLAIN, S. FARRY, W. HEINRICH, B. A. RIDEOUT, AND D. R. SMITH. 2006. Ammunition is the principal source of lead accumulated by California Condors re-introduced to the wild. Environmental Science and Technology 40: 6143–6150.

HUNT, W. G., W. BURNHAM, C. PARISH, K. BURNHAM, B. MUTCH, AND J. L. OAKS. 2006. Bullet fragments in deer remains: Implications for lead exposure in avian scavengers. Wildlife Society Bulletin 33:167–170.

HUNT, W. G., C. N. PARISH, S. C. FARRY, R. SEIG, AND T. G. LORD. 2007. Movements of introduced California Condors in Arizona in relation to lead exposure. Pages 79–98 *in* A. Mee, L. S. Hall, and J. Grantham (Eds.). California Condors in the 21st Century. Special Publication of the American Ornithologists' Union and Nuttall Ornithological Club.

PARISH, C. N., W. R. HEINRICH, AND W. G. HUNT. 2007. Lead exposure, diagnosis, and treatment in California Condors released in Arizona. Pages 97–108 *in* A. Mee, L. S. Hall and J. Grantham (Eds.). California Condors in the 21st Century. Series in Ornithology, no. 2. American Ornithologists Union, Washington, DC, and Nuttall Ornithological Club, Cambridge, Massachusetts, USA.

WILMERS, C. C., D. R. STAHLER, R. L. CRABTREE, D. W. SMITH, AND W. M. GETZ. 2003. Resource dispersion and consumer dominance: scavenging at wolf- and hunter-killed carcasses in Greater Yellowstone, USA. Ecology Letters 6:996–1003.

LEAD INGESTION BY SCAVENGING MAMMALIAN CARNIVORES IN THE YELLOWSTONE ECOSYSTEM

TOM ROGERS[1,2], BRYAN BEDROSIAN[1], DEREK CRAIGHEAD[1], HOWARD QUIGLEY[1], AND KERRY FORESMAN[2]

[1]*Craighead Beringia South, P.O. Box 147, Kelly, WY 83011, USA.*
E-mail: thomasalanrogers@gmail.com

[2] *University of Montana, Division of Biological Sciences, 32 Campus Drive, Missoula, MT 59812, USA.*

EXTENDED ABSTRACT.—Ingestion of lead from spent ammunition is a potential challenge to the conservation of large carnivores and scavengers. Evidence suggests large carnivores such as Black Bears (*Ursus americanus*), Grizzly Bears (*U. arctos*), Wolves (*Canis lupis*), and Coyotes (*C. latrans*) scavenge to varying degrees on ungulate offal piles abandoned by hunters (Wilmers et al. 2003). Other top carnivores, such as Cougars (*Puma concolor*), may be less attracted to offal piles and thus less dependent on them, but may periodically still be exposed to lead at biologically significant levels because of the tendency to occasionally scavenge. Grizzly Bears alter their movement patterns outside of Yellowstone National Park during the fall hunting season to take advantage of unrecovered offal and wounded Elk (*Cervus canadensis*) left by hunters (Ruth et al. 2003, Haroldson et al. 2004). The Southern Yellowstone Ecosystem is host to one of the densest Elk populations in North America as well as a complete large carnivore guild. An annual big game hunt results in an abundant harvest and provides an ideal situation to test for the occurrence among predators and scavengers of lead ingestion from spent rifle bullets.

We have begun collecting samples of liver, hair, blood, and feces from Black and Grizzly Bears, Wolves, Coyotes, and Cougars, and tested samples for the presence of lead using inductively coupled plasma mass spectroscopy to determine if there is a seasonal correlation of lead ingestion during the hunting season. We also hope to determine if carnivores, such as Grizzly Bears, that scavenge to a greater extent on offal piles or on the unretrieved carcasses of animals mortally wounded by hunters, exhibit higher levels of lead ingestion than do species with lesser tendency to scavenge, such as Wolves and Cougars.

In a pilot study, blood samples from live captured Grizzly Bears were the most abundant sample type we were able to obtain, though limited samples of other material were also obtained. During the non-hunting season (March–August), no Grizzly Bear blood samples (n = 11) exhibited lead exposure (>10 μg/dL). However, during the hunting season (September–November), 46% of 13 samples showed exposure with blood lead levels >10 μg/dL. Of six liver samples collected from Wolves during the non-hunting season, none have shown signs of lead exposure. This preliminary evidence suggests mammalian carnivores in areas of high hunting density may exhibit the same temporal pattern of lead exposure from ingestion of rifle bullet fragments during the hunting season as avian scavengers (Cade 2007, Craighead and Bedrosian 2008, Parish et al. 2009, this volume). This study will continue as a master's thesis by Tom Rogers in the fall of 2008 at the University of Montana. *Received 12 June 2008, accepted 24 July 2008.*

ROGERS, T., B. BEDROSIAN, D. CRAIGHEAD, H. QUIGLEY, AND K. FORESMAN. 2009. Lead ingestion by scavenging mammalian carnivores in the Yellowstone ecosystem. Extended abstract *in* R.T. Watson, M. Fuller, M. Pokras, and W.G. Hunt (Eds.). Ingestion of Lead from Spent Ammunition: Implications for Wildlife and Humans. The Peregrine Fund, Boise, Idaho, USA. DOI 10.4080/ilsa.2009.0121

Key words: Ammunition, carnivore, ingestion, lead, mammal, scavenger, Yellowstone.

LITERATURE CITED

CADE, T. J. 2007. Exposure of California Condors to lead from spent ammunition. Journal of Wildlife Management 71:2125–2133.

CRAIGHEAD, D., AND B. BEDROSIAN. 2008. Blood lead levels of Common Ravens with access to big game offal. Journal of Wildlife Management 72(1): 240–245.

HAROLDSON, M., C. SCHWARTZ, S. CHERRY, AND D. MOODY. 2004. Possible effects of elk harvest on Grizzly Bears in the Greater Yellowstone Ecosystem. Journal of Wildlife Management 68:129–137.

PARISH, C. N., W. G. HUNT, E. FELTES, R. SIEG, AND K. ORR. 2009. Lead exposure among a re-introduced population of California Condors in northern Arizona and southern Utah. *In* R. T. Watson, M. Fuller, M. Pokras, and W. G. Hunt (Eds.). Ingestion of Lead from Spent Ammunition: Implications for Wildlife and Humans. The Peregrine Fund, Boise, Idaho, USA. DOI 10.4080/ilsa.2009.0217

RUTH, T., D. SMITH, M. HAROLDSON, P. BUOTTE, C. SCHWARTZ, H. QUIGLEY, S. CHERRY, K. MURPHY, D. WYERS, AND K. FREY. 2003. Large-carnivore response to recreational big-game hunting along the Yellowstone National Park and Absaroka-Beartooth Wilderness boundary. Wildlife Society Bulletin 31(4):1–12.

WILMERS, C., D. STAHLER, R. CRABTREE, D. SMITH, AND W. GETZ. 2003. Resource dispersion and consumer dominance: Scavenging at wolf- and hunter-killed carcasses in Greater Yellowstone, USA. Ecology Letters 6:996–1003.

POTENTIAL SOURCES OF LEAD EXPOSURE FOR BALD EAGLES: A RETROSPECTIVE STUDY

PATRICK T. REDIG[1], DONALD R. SMITH[2] AND LUIS CRUZ-MARTINEZ[1]

[1]*The Raptor Center, University of Minnesota, 1920 Fitch Avenue, St. Paul, MN 55108, USA.*
E-mail: redig001@umn.edu

[2]*Physical Sciences Building, Room 442, University of California, Santa Cruz, CA 95064, USA.*

EXTENDED ABSTRACT.—A 12-year (1996–2008) retrospective study of lead poisoning in Bald Eagles (*Haliaeetus leucocephalus*) was conducted at The Raptor Center at the University of Minnesota. The objectives of this study were to 1) Investigate trends in lead poisoning as a cause of morbidity and mortality in Bald Eagles admitted to The Raptor Center and 2) Examine evidence to determine the potential source(s) of the lead and more specifically to investigate incidence with reference to hunting of White-tailed Deer.

For the former objective, the incidence of lead poisoning from this 12-year period was compared to the incidence of lead poisoning reported on for the preceding 16 years (Kramer and Redig 1997). The latter objective was investigated based on the analysis of four epidemiological parameters: a) temporal/seasonal prevalence and relationship with deer hunting season start dates in Minnesota, Wisconsin and Iowa; b) spatial/geographical data (correlation of the animal recovery location with deer hunting zones); c) lead isotope ratio analysis of blood and metal fragments (found in the gastrointestinal tract) of lead-poisoned Bald Eagles; and, d) comparison of kidney copper concentration from lead-exposed vs. not-exposed eagles.

Our results showed a continuing trend on the incidence of lead poisoning in Bald Eagles admitted at The Raptor Center. No significant difference was seen in the number of cases admitted per year between the current 12-year period and the preceding 16-year period. Similarly, the mean blood lead concentration remained unchanged with a low level chronic exposure predominating.

A temporal-spatial association was found between deer hunting season onset and incidence of eagle poisoning. The majority of cases occurred during late fall and early winter. A significantly higher number of poisoned Bald Eagles were recovered from the deer hunting rifle zone, suggesting a greater bioavailability of lead fragments when compared to the shotgun zone.

The lead isotope ratio analysis yielded the following results. First, most of the paired blood-metal fragments samples have a closely matched isotopic signature and secondly, the isotope ratio of the majority of the blood samples and stomach contents samples (from lead exposed eagles) were within the isotope ratio from ammunition samples reported by Church et al. (2006).

The kidney copper concentration was significantly higher in lead exposed eagles. This implies that copper fragments (from copper-jacketed rifle bullets) are being ingested by eagles along with lead fragments.

Lead poisoning continues to be a cause of morbidity and mortality to Bald Eagles in the Upper Midwest (Minnesota, Wisconsin and Iowa). We conclude that none of the four epidemiological parameters examined here can be used as standalone evidence as to the source of the lead; however, taken together, they significantly reduce the validity of any other possible explanations other than ammunition of lead exposure for Bald Eagles. *Received 31 May 2008, accepted 25 September 2008.*

REDIG, P. T., D. R. SMITH, AND L. CRUZ-MARTINEZ. 2009. Potential sources of lead exposure for Bald Eagles: A retrospective study. Extended abstract *in* R. T. Watson, M. Fuller, M. Pokras, and W. G. Hunt (Eds.). Ingestion of Lead from Spent Ammunition: Implications for Wildlife and Humans. The Peregrine Fund, Boise, Idaho, USA. DOI 10.4080/ilsa.2009.0208

Key words: Ammunition, Bald Eagle, isotope, lead, poisoning, spatial, temporal.

LITERATURE CITED

CHURCH, M. E., R. GWIAZDA, R. W. RISEBROUGH, K. SORENSON, C. P. CHAMBERLAIN, S. FARRY, W. HEINRICH, B. A. RIDEOUT, AND D. R. SMITH. 2006. Ammunition is the principal source of lead accumulated by California Condors re-introduced to the wild. Environmental Science and Technology 40:6143–6150.

KRAMER, J. L., AND P. T. REDIG. 1997. Sixteen years of lead poisoning in eagles, 1980–95: An epizootiologic view. Journal of Raptor Research 31(4):327–332.

BALD EAGLE LEAD POISONING IN WINTER

KAY NEUMANN

Saving Our Avian Resources, 25494 320th St., Dedham, IA 51440, USA.
E-mail: diversityfarms@iowatelecom.net

ABSTRACT.—Wildlife rehabilitators across the state of Iowa began gathering lead poisoning information on Bald Eagles (*Haliaeetus leucocephalus*) in January 2004 for this ongoing project. Blood, liver, or bone samples were analyzed for lead levels from 62 of the 82 eagles currently in the database. Thirty-nine eagles showed lead levels in their blood above 0.2 ppm or lead levels in their liver above 6 ppm, which could be lethal poisoning without chelation treatment. Seven eagles showed exposure levels of lead (between 0.1 ppm and 0.2 ppm in blood samples, between 1 ppm and 6 ppm in liver samples, and between 10 ppm and 20 ppm in bone). Several of the eagles admitted with traumatic injuries showed underlying lead exposure or poisoning. Over fifty percent of the eagles being admitted to Iowa wildlife rehabilitators have ingested lead. Behavioral observations, time-of-year data analysis, and x-ray information point to lead shrapnel left in slug-shot White-tailed Deer (*Odocoileus virginianus*) carcasses to be a source of this ingested lead. With thousands of Bald Eagles spending the winter in Iowa (up to one fifth of the lower 48 states population), this poisoning mortality could be significant and is preventable. Educational efforts are being directed at encouraging deer hunters to switch from lead to non-toxic (copper) slugs and bullets. *Received 30 May 2008, accepted 4 September 2008.*

NEUMANN, K. 2009. Bald Eagle lead poisoning in winter. *In* R. T. Watson, M. Fuller, M. Pokras, and W. G. Hunt (Eds.). Ingestion of Lead from Spent Ammunition: Implications for Wildlife and Humans. The Peregrine Fund, Boise, Idaho, USA. DOI 10.4080/ilsa.2009.0119

Key Words: Bald Eagles, Iowa, lead poisoning, White-tailed Deer, lead slugs.

LEAD POISONING IN BALD EAGLES has been well documented in eagles brought in for rehabilitation (Kramer and Redig 1997, Franson et al. 2002) and in wild populations (Hermata and Restani 1995, Miller et al. 1998, Wayland and Bollinger 1999). In a review of published data, Sanborn (2002) found the sources of lead poisoning in wildlife overwhelmingly to be spent ammunition and fishing tackle. A unique convergence of events seems to be occurring in Iowa that could have potentially significant consequences for Bald Eagles. Iowa has become a major wintering area for eagles, documented by Christmas Bird Count (National Audubon Society 2002) and Iowa's Midwinter Bald Eagle Survey (Iowa Department of Natural Resources 2007). This influx of eagles to the state occurs during the major hunting seasons. As reducing and controlling the White-tailed Deer numbers in the state has become a priority for the Iowa Department of Natural Resources (IDNR), the number of deer harvested has increased along with new and lengthened seasons for hunting with guns. The increased harvest results in an increased availability of deer carcasses (Nixon et al. 2001). The added seasons for deer harvest with guns provides for carcasses to be available over more of the eagle's wintering cycle then when deer were hunted during a limited part of December.

In the fall of 2004 and winter of 2005 wildlife rehabilitators in Iowa noticed an increase in the number of Bald Eagles admitted for treatment, and most of the increase in cases was associated with lead poisoning or exposure. Wildlife rehabilitators are an excellent source of information and monitoring for wildlife mortality factors (e.g., West Nile virus, avian flu, illegal activities, secondary poisonings). Their work provides a sample of what is happening to wild animals that would cause death without human intervention. The data gathered by Iowa wildlife rehabilitators show an alarming rate of lead poisoning and exposure in Bald Eagles during their wintering cycle in Iowa.

METHODS

Saving Our Avian Resources (SOAR) began collecting Bald Eagle data from other Iowa rehabilitators in 2004. SOAR, MacBride Raptor Project, Orphaned and Injured Wildlife, Inc., Wildlife Care Clinic, Inc., and Blackhawk Wildlife Rehabilitation Project are the main facilities in Iowa working with Bald Eagles. Data recorded included: date admitted or found dead, county where found, eagle's age, facility doing treatment, lead levels (in blood, liver, or bone), full body x-ray results, and end results (released, died, euthanized, permanent captive). Lead levels were determined from liver samples in the cases when the eagle died before blood could be obtained. Data include lead levels from nine liver samples and one bone sample from eagle carcasses found and retrieved from the field by the public or Conservation Officers.

Blood lead levels were determined by laboratory analysis (Antech Diagnostics, Southaven, MS) or on-site analysis using ESA, Inc., LeadCare® system. Blood lead levels higher than 0.1 ppm but lower than 0.2 ppm were considered lead exposure cases. Exposure being defined as the animal has a higher than background level of lead in its system, indicating that it ingested an anthropogenic source of lead. Any blood lead level >0.1 ppm was considered abnormal (P. Redig pers. comm.). Exposure levels are below that at which chelation treatment is begun. Blood lead levels higher than 0.2 ppm were considered poisoning cases and chelation treatment was started if clinical symptoms warranted (Antech Diagnostics). These interpretations of blood lead levels agree with the thresholds for humans (AAP 2005 and CDC 2008). Liver and bone samples were analyzed for lead levels by Iowa State University Diagnostic Laboratory. Lead levels in liver higher than 1 ppm but lower than 6 ppm were considered lead exposure cases. Lead levels in liver greater than 6 ppm were considered lethal poisoning cases. Any liver level ≥1 ppm was considered abnormal. Lead levels in bone between 10 ppm and 20 ppm were considered exposure levels; greater than 20 ppm were considered poisoning levels. Any lead level in bone ≥10 ppm was considered abnormal (Puls 1994). Exposure levels may not cause death by themselves. Poisoning levels can be lethal without treatment. All animals admitted to a rehabilitator are considered a wild fatality; the animal would have died without human intervention. Data were analyzed on a by-month basis to determine timing of most lead poisoning or exposure cases.

Full body x-rays were examined for any remaining evidence of solid lead in eagles' digestive tracts. Trailcams were placed near salvaged, relocated, road-killed deer carcasses from early December to late March to assess use by wildlife. Deer harvest and eagle wintering population data were examined for associations to the number of abnormal lead level cases. Attendees at Iowa's Deer Classic were surveyed for attitudes regarding the impacts of lead on wildlife and human health, and their use of and satisfaction with non-lead ammunition.

RESULTS

Data were gathered on 82 Bald Eagles admitted to Iowa wildlife rehabilitators from 1 January 2004 to 30 April 2008 (Table 1 and Figure 1). Sixty-two of these eagles were tested for lead levels. Thirty-nine of these tests showed poisoning (potentially lethal) levels of lead in blood or liver (≥0.2 ppm in blood and ≥6 ppm in liver). Seven of those tested showed exposure levels of lead (between 0.1 and 0.2 ppm in blood or between 1 ppm and 6 ppm in liver or between 10 ppm and 20 ppm in bone). Abnormal lead levels (any level ≥0.1 ppm in blood or ≥1 ppm in liver or ≥10 ppm in bone) ranged from 37.5% of all eagles admitted (with 62.5% being tested) in 2004 to 70.0% of all eagles admitted (with 90.0% being tested) in 2005. Over the study period, 56% of all eagles admitted to Iowa wildlife rehabilitators had abnormal lead levels. Of the 46 eagles from Iowa with abnormal lead levels, 38 died (82.6%), four are permanent education birds due to secondary

Table 1. Bald Eagles admitted to Iowa wildlife rehabilitators.

Year	Total no. of eagles admitted to Iowa wildlife rehabilitators	No. tested for lead (% of total)	No. showing lead exposure	No. showing lead poisoning (lethal without treatment)	% of total showing abnormal lead levels
2004	8	5 (62.5%)	0	3	37.5%
2005	20	18 (90.0%)	1	13	70.0%
2006	20	11 (55.0%)	2	7	45.0%
2007	23	17 (73.9%)	2	11	56.5%
1 January through 30 April, 2008	11	11 (100.0%)	2	5	63.6%
TOTAL	82	62 (75.6%)	7	39	56.0%

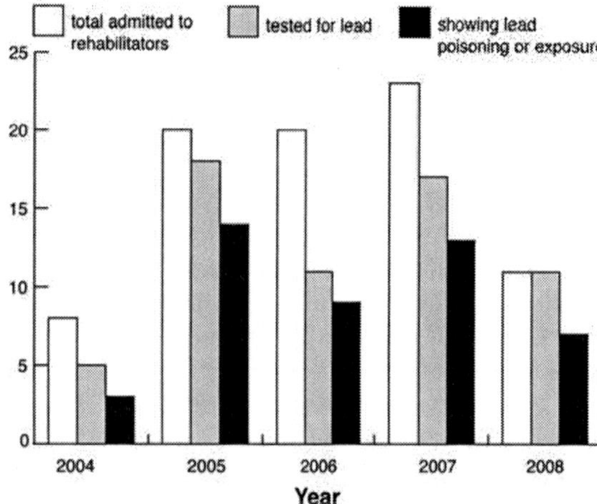

Figure 1. Incidence of abnormal lead levels in Iowa wintering Bald Eagles.

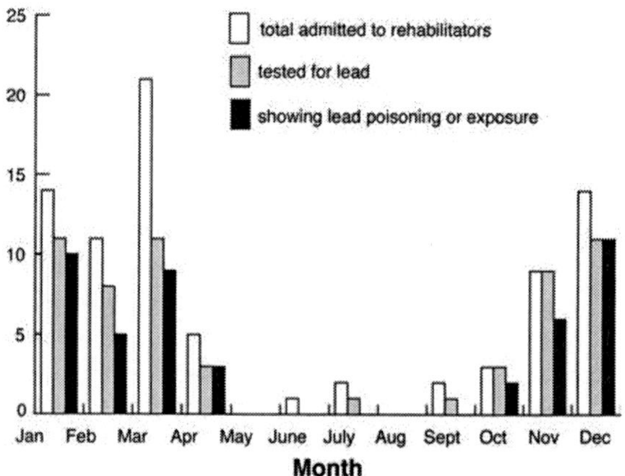

Figure 2. Seasonality of Bald Eagles admitted to Iowa wildlife rehabilitators.

trauma injuries (8.7%), and four were released to the wild (8.7%).

Time-of-year data (Figure 2) revealed a strong relationship of poisoning cases with both the increase in the numbers of eagles as they move into the state during the fall and the start of hunting seasons. Cases increased through the winter (December and January) peak deer harvest months. Poisoning cases showed another spike during early spring (March), as eagles traveled across country away from water sources, during their early spring movement north. This March movement was documented by a peak in the number of eagles observed feeding from carcasses placed near a trailcam. It is likely that more scavenging occurs during this northward migration period and more carcasses would be available to visual scavengers after snow melt. The number of eagles admitted with abnormal lead levels drops to zero in May and no poisoning cases are seen again until October.

Eagles readily feed on deer carcasses, in groups, for several days. A marked increase in eagle use of carcasses was noted the first week in March (Figure 3).

One White-tailed Deer shot with two lead slugs was x-rayed for shrapnel content. There were many remaining lead fragments in the carcass large enough to be lethal to a Bald Eagle (Figure 4).

Whole body x-rays of eagles showed lead shrapnel (Figs. 5a and 5b), shot (Figure 5c), or larger pieces of lead (Figure 5d) in eagle digestive tracts. Fifty-nine eagles were x-rayed, with seven eagles showing lead remaining in digestive tracts. It is very difficult to find the source of lead poisoning remaining

Figure 3. Bald Eagles feeding on deer carcasses in March.

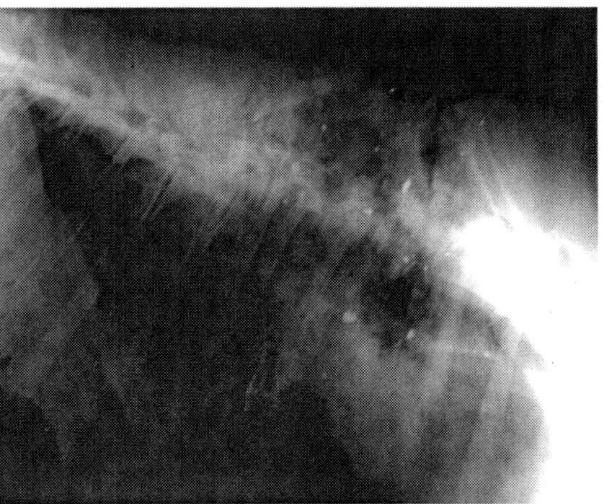

Figure 4. Lead shrapnel in deer shot with two lead slugs.

in the digestive tract of the eagle due to the eagle's very efficient digestive system and the time until the bird is debilitated enough to be caught in the field.

Of the 151 respondents to a Hunter Ammunition Survey, 62.9 % knew that lead fragments in scavenged meals could harm wildlife and 74.2% knew that lead fragments found in game meat could pose a human health risk. Forty-three percent of the respondents had used solid copper slugs or bullets to hunt big game. Of these, 87.7% liked the performance of the copper. A high percentage of respondents, 58.3%, had used non-lead/non-toxic shot for upland game hunting. Of these, 93.2% were satisfied with the results using the lead-free ammunition. When asked "For wildlife and human health reasons, do you think that lead ammunition should be phased out in favor of non-lead/non-toxic alternatives?" 83.4% said yes, 12% said no, and 4.6% said they were unsure.

DISCUSSION

An average of 56% of all Bald Eagles admitted to Iowa wildlife rehabilitators had abnormal lead levels. This is a much higher rate than that reported from other state rehabilitation efforts. The Raptor Center, University of Minnesota averages 25% of eagle patients with abnormal lead levels (The Raptor Center 2008). Raptor Recovery Nebraska averages 20% of eagle patients with abnormal lead levels (B. Finch Raptor Recovery Nebraska, pers. comm.). The amount of effort (percentage of total cases tested for lead) may explain the higher rate of poisoning cases in Iowa as compared to other rehabilitation facilities. These facilities may not test individuals that are admitted with obvious trauma and attribute the cause for admission as the trauma, with the underlying lead levels going undetected. Ten (21.7%) of the eagles with abnormal lead levels in Iowa's data set also suffered from trauma (five gunshot, five impact trauma). The trauma may have occurred due to the lead in the bird's system, which may make it less wary, increase reaction time, and affect vision and nervous system function.

Another explanation for the lower rate of eagle lead exposure or poisoning in other states is that Iowa has become a significant wintering area for a large number of Bald Eagles. There may be fewer eagles in Minnesota and Nebraska during the winter. This is supported by the Iowa Mid-Winter Bald Eagle Survey (IDNR 2007) (Figure 6a) and Christmas Bird Count data (National Audubon Society 2002) (Figure 6b). Therefore, the comparatively high incidence of lead contamination in wintering Bald Eagles from Iowa is a matter of concern.

Eagles are attracted to Iowa by open water in river systems and below reservoir dams. Hundreds of eagles congregate in these fishing spots throughout the winter months. Iowa also has ample scavenging opportunities, as the hunting seasons are ongoing during the time thousands of eagles are wintering in the state. Deer carcasses may be the most available

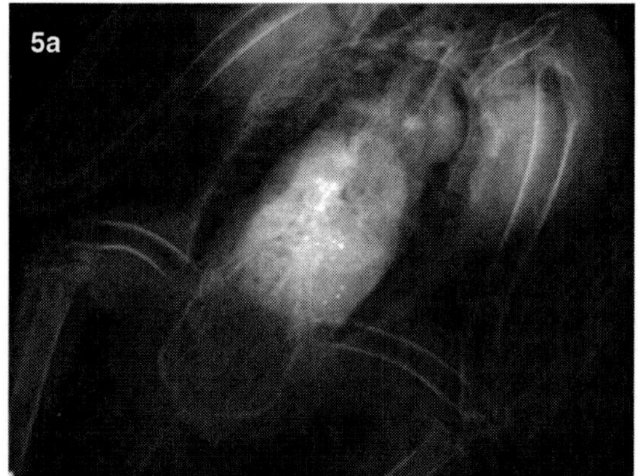

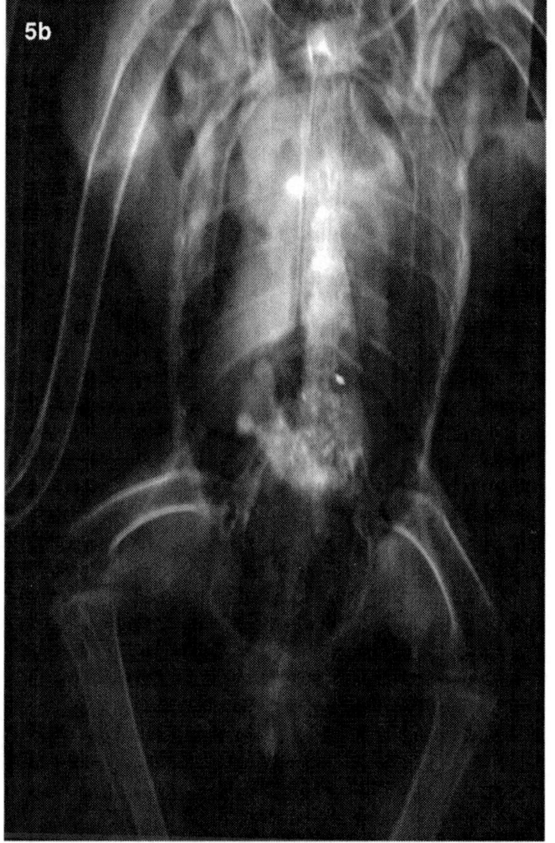

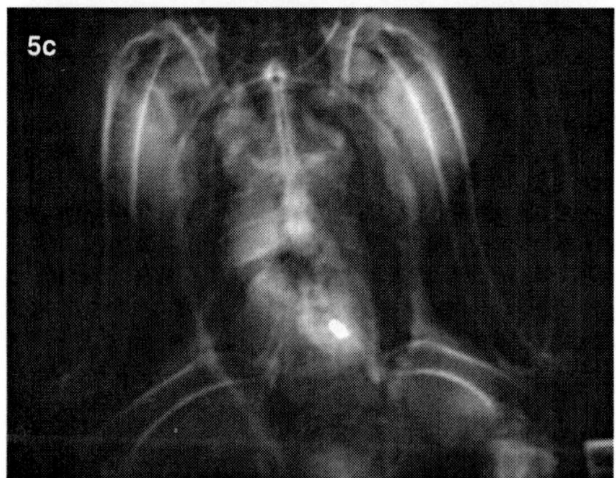

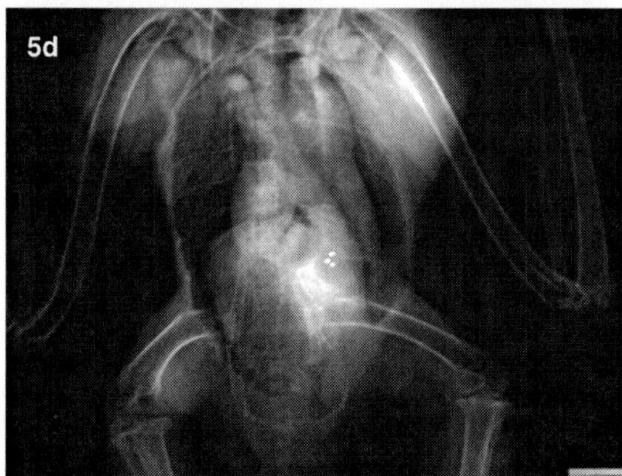

Figure 5a. Lead shrapnel in Bald Eagle digestive tract.

Figure 5b. Lead shrapnel in Bald Eagle digestive tract.

Figure 5c. Lead shot in Bald Eagle digestive tract.

Figure 5d. Large piece of lead in Bald Eagle digestive tract.

and preferred item for scavenging. A youth deer season is held in mid-September, a special doe season in November, and the main shotgun seasons are in December, with an extended doe season in January. Cases of eagle poisonings peak in December and January. These expanded deer hunting opportunities could also make carcasses available for a longer timeframe during the eagle's wintering cycle. Nixon et al. (2001) estimated that the number of animals wounded but not retrieved could be 10% or more of total gun harvest. From 12,000 to 18,000 deer carcasses could have been available to eagles annually throughout the data collection timeframe (Figure 7). This does not account for deer poached and left in the field.

X-ray evidence of eagles and deer carcasses support the theory that lead shrapnel in lead slug wounded but not retrieved deer carcasses is a source of lead poisoning for Iowa wintering Bald Eagles. Hunt et al. (2006) found lead shrapnel readily available to avian scavengers in deer shot with lead rifle bullets, also. In clinical trials, less than 200 mg of ingested lead was enough to kill an eagle (Hoffman et al. 1981, Pattee et al. 1981). Thus a very small piece of shrapnel could be lethal to an eagle. Other x-ray evidence implicates lead shot used in upland game hunting. Timing of poisoning cases however seems more closely linked to deer harvest than upland game harvest. Eighty percent of pheasant harvest occurs within the first three weeks of the season, beginning the third weekend in October (IDNR 2007). Pheasant or rabbit carcasses would be more hidden than a large deer carcass and would be less likely to be available into February and March. It is illegal to hunt waterfowl with lead ammunition, so it is hoped that waterfowl would no longer be a source of secondary lead for eagles. The timing, however, of the waterfowl harvest closely coincides with pheasant harvest and most hunting is done early in the season and most waterfowl have moved south by the end of November. Wounded deer may not die immediately, they may die later in the winter or early spring as they are stressed by lower food availability and spring snows. Thus, fresh deer carcasses may become available as the winter progresses. Deer and, to some extent, upland game are thus likely sources of spent lead from the fall to early spring.

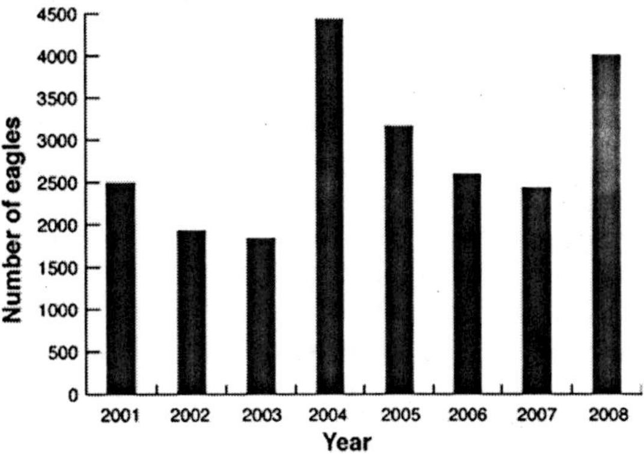

Figure 6a. Iowa mid-winter Bald Eagle survey.

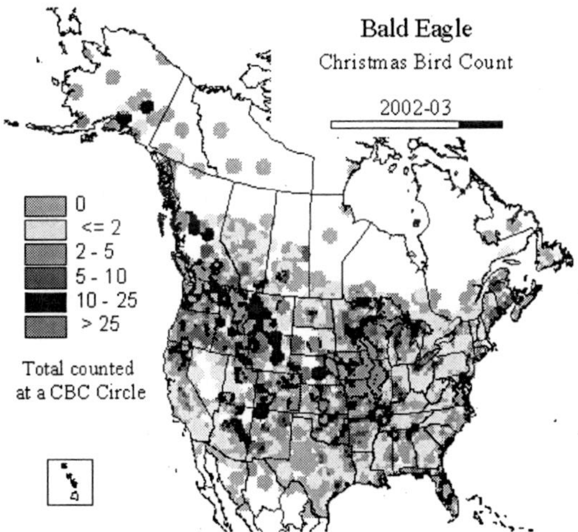

Figure 6b. Bald Eagle Christmas Bird Count Data.

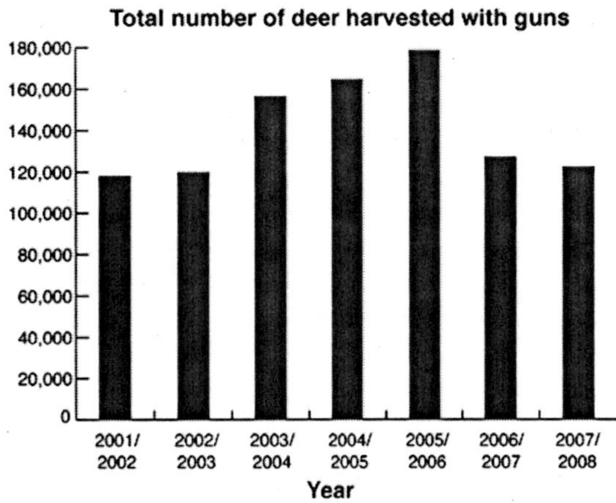

Figure 7. Deer Carcass Availability.

How much is too much? Preventable poisoning of protected wildlife species should not be an acceptable form of mortality. Lead poisoning deaths could be a population risk factor in a species that is at the top of its food web, is slow to reproduce, and should have a 20 to 25 year life span. Iowa wildlife rehabilitators have documented ten to twelve eagles dying each year in Iowa with abnormal lead levels. Not all eagles were able to be tested for lead, so this mortality rate due to lead may be an underestimation. This number also is only some fraction of the number of eagle deaths each year due to lead ingestion, as not all sick and dead eagles are found and brought in for testing. With over half of Bald Eagles admitted due to lead ingestion there is reason for alarm, concern, and action to prevent further mortality due to this factor. For comparison, more random events seem to occur at a much lower percentage of the total number of eagles admitted. Gunshot wounds, for example, were recorded in ten of the 82 eagles in this database (12.2%). The data do not indicate that the increasing number of Bald Eagles being admitted by Iowa wildlife rehabilitators is simply a function of the increasing numbers of eagles in wild populations. If this were the case, it would be expected that a variety of causes would be seen for admittance (miscellaneous trauma, fractures, starvation, disease, etc.), at percentages relative to that seen for gunshot wounds. This has been the case for other species. As Coopers Hawk (*Accipter cooperii*) populations have increased, rehabilitators have admitted an increasing number of them with a wide variety of problems (Cancilla pers. comm.). This has not been the case with Bald Eagles. As the numbers of Bald Eagles has increased, Iowa wildlife rehabilitators have admitted more of them but not with a variety of injuries, instead with one overwhelming cause; lead ingestion.

Thousands of eagles (up to one fifth of the lower 48 states' population) spend the winter in Iowa (IDNR 2007). With IDNR trying to reduce the deer herd with lengthened seasons and increased harvest, deer carcasses are readily available throughout the fall, winter, and early spring. These converging factors could cause a non- sustainable number of bald eagle deaths from lead ingestion annually. With the United States Fish and Wildlife Service (USFWS 2008) population estimates for Bald Eagles nearing 10,000 pair in the lower 48 states and the eagle's very recent removal from the endangered species list, there may not be an excess of individuals to compensate for this annual mortality on a major wintering ground. With delisting, Bald Eagle nesting and wintering habitat will no longer have mandatory protections. There is concern that loss of habitat may cause Bald Eagle populations to decrease. The Southwest Bald Eagle population (Arizona, New Mexico, and Mexico) has not shown a clear long-term increase in numbers and has a low productivity rate and a high adult mortality rate (RRF 2006). Sub-lethal effects of lead may effect reproductive success and shorten life span. There was a documented 10 year lag time with the decline and recovery of Bald Eagles from the effects of DDT, another poison (Bednarz et al. 1990). By the time a detectable change is seen in the number of breeding pairs, it may be too late to reverse those declines. Accurate breeding pair numbers, productivity, and winter survey data needs to be gathered to monitor Bald Eagle population health. Adult survival and turnover rate at historic nesting territories would be other good indicators of population health; these numbers are not available for Iowa nesters. It is estimated that there were 500,000 Bald Eagles living in what is now the lower 48 states at pre-settlement time. It is estimated there were more than 3,000 pairs on the Chesapeake Bay alone (Buelher 2000). They have endured bounties, chemical contamination, and habitat loss. With complete protection, they have returned to less than 10 % of their historic numbers, while other species of birds of prey have recovered completely from adversity. Peregrine Falcon (*Falco peregrinus*) numbers have returned to their pre-DDT levels with low adult turnover at nest sites (White et al. 2002). In many parts of the United States, Peregrine Falcons occur in higher numbers than ever known historically (GRIN 2008). Bald Eagles, a large, adaptable predator, have been unable to recover to numbers even near their historic levels (Buelher 2000). Perhaps the lethal and sub-lethal effects of lead are preventing a more robust return in Bald Eagle numbers across all of their range.

Educational efforts are underway to urge hunters to use solid copper deer slugs instead of lead. Deer hunters are reporting good success using the solid copper slugs. Many hunters have switched to copper slugs and/or muzzleloader bullets for better performance. They also have the added satisfaction of knowing that they will not contribute to secondary poisoning of wildlife. With a majority of hunters in our survey responding positively to the phase out of lead ammunition for wildlife and human health, it does not appear that hunters' attitudes will be a barrier to getting lead out of recreational activities.

ACKNOWLEDGMENTS

Iowa Wildlife Rehabilitators Association; Drs. Rexanne Struve and Gary Riordan of Veterinary Associates, Manning, Iowa; Dr. Ross Dirks, Dickinson County Small Animal Clinic, Spirit Lake, Iowa; Dr. William Clark, Iowa State University; Kristene Lake and Jodeanne Cancilla, MacBride Raptor Project; Linda Hinshaw, Orphaned and Injured Wildlife, Inc.; Wildlife Care Clinic, Inc; Blackhawk Wildlife Rehabilitation Project; Iowa Department of Natural Resources Conservation Officers; Liz Garst, Rachel Garst, Darwin Pierce, and Dean Jackson; Whiterock Conservancy; Marla Mertz, Marion County Conservation Board; and Jon Judson, Diversity Farms, Inc.

LITERATURE CITED

AMERICAN ACADEMY OF PEDIATRICS (AAP). 1998. Screening for elevated blood lead levels–Committee on Environmental Health. Pediatrics. 101(6):1072–1078. Policy updated. 2005. http://aappolicy.aappublications.org/cgi/content/full/pediatrics;101/6/1072

BEDNARZ, J. C., D. KLEM, JR., L. J. GOODRICH, AND S. E. SENNER. 1990. Migration counts of raptors at Hawk Mountain, Pennsylvania, as indicators of population trends, 1934–1986. Auk 107:96–109.

BUEHLER, D. A. 2000. Bald Eagle. *In* A. Poole (Ed.). The Birds of North America Online (A. Poole, Ed.). Ithaca: Cornell Laboratory of Ornithology; Retrieved from the Birds of North America Online database: http://bna.birds.cornell.edu/bna/species/506 doi:10.2173/bna.506

CENTERS FOR DISEASE CONTROL (CDC). 2007. Interpreting and managing blood lead levels <10 µg/dL in children and reducing childhood exposures to lead: Recommendations of CDC's Advisory Committee on Childhood Lead Poisoning Prevention. Recommendations and Reports 56(RR08):1–14, 16.

CANCILLA, J. 2008. Director, MacBride Raptor Project. Cedar Rapids, Iowa, USA.

FRANSON, J. C., L. SILEO, AND N. J. THOMAS. 2002. Causes of eagle deaths. National Biological Survey, National Wildlife Health Center, Madison, Wisconsin, USA.

GRIN. GLOBAL RAPTOR INFORMATION NETWORK. 2008. Species account: Peregrine Falcon *Falco peregrinus*. [Online.] Available at http://www.globalraptors.org. Accessed 26 August 2008.

HARMATA, A. R., AND M. RESTANI. 1995. Environmental contaminants and cholinesterase in blood of vernal migrant Bald and Golden Eagles in Montana. Intermountain Journal of Sciences 1:1–15.

HOFFMAN, D. J., O. H. PATTEE, S. N. WIEMEYER, AND B. MULHERN. 1981. Effects of lead shot ingestion on g-aminolevulinic acid dehydratase activity, hemoglobin concentration, and serum chemistry in Bald Eagles. Journal of Wildlife Distribution 17:423–431.

IOWA DEPARTMENT OF NATURAL RESOURCES. 2007. Trends in Iowa Wildlife Populations and Harvest. Iowa Department of Natural Resources—Wildlife Bureau, Des Moines, Iowa, USA. www.iowadnr.gov

KRAMER, J. L., AND P. T. REDIG. 1997. Sixteen years of lead poisoning in eagles, 1980–95: An epizootiologic view. Journal of Raptor Research 31:327–332.

MILLER, M. J. R., M. RESTANI, A. R. HARMATA, G. R. BORTOLLI, AND M. E. WAYLAND. 1998. A comparison of blood lead levels in Bald Eagles from two regions of the great plains of North America. Journal of Wildlife Diseases 34(4):704–714.

NATIONAL AUDUBON SOCIETY. 2002. The Christmas Bird Count Historical Results. [Online.] Available at http://www.audubon.org/bird/cbc. Accessed May 2008.

NIXON, C. M., L. P. HANSEN, P. A. BREWER, J. E. CHELSVIG, T. L. ESKER, D. ETTER, J. B. SULLIVAN, R. G. KOERKENMEIER, AND P. C. MANKIN. 2001. Survival of White-tailed Deer in intensively farmed areas of Illinois. Canadian Journal of Zoology 79:581–588.

PATTEE, O. H., S. N. WIEMEYER, B. M. MULHERN, L. SILEO, AND J. W. CARPENTER. 1981. Experimental lead-shot poisoning in Bald Eagles. Journal of Wildlife Management 45:806–810.

PULS, R. 1994. Mineral Levels in Animal Health. Second Edition. Sherpa International, Aldergrove, British Columbia, Canada.

RRF. RAPTOR RESEARCH FOUNDATION, INC. 2006. Raptor Research Foundation comments on Bald Eagle delisting documents. Olympia, Washington, USA.

REDIG, P. 2008. Director, The Raptor Center, University of Minnesota, St. Paul, Minnesota, USA.

SANBORN, W. 2002. Lead poisoning of North American wildlife from lead shot and lead fishing tackle. HawkWatch International, Salt Lake City, Utah, USA.

THE RAPTOR CENTER. 2008. University of Minnesota, St. Paul, Minnesota, USA. http://www.raptor.cvm.umn.edu/raptor/news/healthtopics/leadpoisoning/home.html

WAYLAND, M., AND T. BOLLINGER. 1999. Lead exposure and poisoning in Bald Eagles and Golden Eagles in the Canadian prairie provinces. Environmental Pollution 104:341–350.

WHITE, C. M., N. J. CLUM, T. J. CADE AND W. G. HUNT. 2002. Peregrine Falcon. *In* A. Poole (Ed.). The Birds of North America Online (A. Poole, Ed.). Ithaca: Cornell Laboratory of Ornithology; Retrieved from the Birds of North America Online database: http://bna.birds.cornell.edu/bna/species/660 doi:10.2173/bna.660

US FISH AND WILDLIFE SERVICE. 2008. http://www.fws.gov/midwest/eagle/population/index.html

BLOOD LEAD LEVELS OF BALD AND GOLDEN EAGLES SAMPLED DURING AND AFTER HUNTING SEASONS IN THE GREATER YELLOWSTONE ECOSYSTEM

BRYAN BEDROSIAN AND DEREK CRAIGHEAD

Craighead Beringia South, P.O. Box 147, Kelly, WY 83011, USA.
E-mail: bryan@beringiasouth.org

EXTENDED ABSTRACT.—Recently, we discovered a significant amount of lead ingestion in Common Ravens (*Corvus corax*) from the southern Yellowstone Ecosystem during the large-game hunting seasons (Craighead and Bedrosian 2008). Our results provided further evidence that hunter discarded viscera of large-game animals is a source of lead in the ecosystem. However, there are many species that feed on hunter provided offal (Wilmers et al. 2003, Hunt et al. 2006) and are thus potentially exposed to lead throughout the duration of the hunting season (mid-September through December). We expanded the scope of our research to include both Bald Eagles (*Haliaeetus leucocephalus*) and Golden Eagles (*Aquila chrysaetos*). We measured the blood lead levels of both species during and after large-game hunts for two years. We tested 63 eagles (47 Bald Eagles and 16 Golden Eagles) and found a median blood lead level of 41.0 µg/dL (range = 3.2–523 µg/dL), 74.9% of all birds tested exhibited elevated lead levels (>20 µg/dL), and 14.3% exhibited levels associated with clinical poisoning (>100 µg/dL). We found no difference between Bald and Golden Eagles during the non-hunting season (median = 29.9 and 21.9 µg/dL, respectively; $P = 0.792$). We could not separate species during the hunting season due to inadequate sample size for Golden Eagles in this period (n = 3). The median blood lead levels for eagles during the hunting season was significantly higher than the non-hunting season (56.0 vs. 27.7 µg/dL, respectively; $P = 0.01$). We found no difference in blood lead levels for different age groups and found no evidence to suggest that increased blood lead levels decreased body condition, as measured by an index of regression residuals of bill depth and mass (Craighead and Bedrosian 2008). We were also able to collect data from a sample of nestling Bald Eagles (n = 9) and Golden Eagles (n = 1) to begin understanding baseline lead levels for these species. We found a median blood lead level of 0.3 µg/dL for nine nestlings from both species (range = 0.0 – 0.8 µg/dL). These results confirmed that both Bald and Golden Eagles are ingesting large amounts of lead during the hunting season in the southern Yellowstone Ecosystem. Further, the magnitude of lead in the blood of many eagles is extremely high and likely results in the death of some individuals (Pattee et al. 1981). While it is clear that eagles are ingesting large amounts of lead during the hunting season, the long-term, cumulative impacts of annual exposure are uninvestigated. *Received 24 June 2008, accepted 12 August 2008.*

BEDROSIAN, B., AND D. CRAIGHEAD. 2009. Blood lead levels of Bald and Golden Eagles sampled during and after hunting seasons in the Greater Yellowstone Ecosystem. Extended abstract *in* R. T. Watson, M. Fuller, M. Pokras, and W. G. Hunt (Eds.). Ingestion of Lead from Spent Ammunition: Implications for Wildlife and Humans. The Peregrine Fund, Boise, Idaho, USA. DOI 10.4080/ilsa.2009.0209

Key words: Bald Eagle, blood, Common Raven, game hunting, Golden Eagle, lead, season.

LITERATURE CITED

CRAIGHEAD, D., AND B. BEDROSIAN. 2008. Blood lead levels of Common Ravens with access to big-game offal. Journal of Wildlife Management. 72:240–245.

HUNT, W. G., W. BURNHAM, C. PARISH, K. BURNHAM, B. MUTCH, AND J. L. OAKS. 2006. Bullet fragments in deer remains: implications for lead exposure in avian scavengers. Wildlife Society Bulletin 33:167–170.

PATTEE, O. H., S. N. WIENMEYER, B. M. MULHERN, L. SILEO, AND J. W. CARPENTER. 1981. Experimental lead–shot poisoning in Bald Eagles. Journal of Wildlife Management 45:806–810.

WILMERS, C. C., D. R. STAHLER, R. L. CRABTREE, D. W. SMITH, AND W. M. GETZ. 2003. Resource dispersion and consumer dominance: scavenging at wolf- and hunter-killed carcasses in Greater Yellowstone, USA. Ecology Letters 6:996–1003.

BLOOD-LEAD LEVELS OF FALL MIGRANT GOLDEN EAGLES IN WEST-CENTRAL MONTANA

ROBERT DOMENECH[1] AND HEIKO LANGNER[2]

[1]*Raptor View Research Institute, P.O. Box 4323, Missoula, MT 59806, USA.*
E-mail: rob.domenech@raptorview.org

[2]*University of Montana, Geosciences Department, Missoula, MT 59812, USA.*

EXTENDED ABSTRACT.—Lead has long been documented as a serious environmental hazard to eagles and other predatory and scavenging avian species (Redig et al. 1984, Kramer and Redig 1997). The use of lead shotgun pellets for waterfowl hunting on federal and state lands was banned in 1991 due to lead poisoning in Bald Eagles (*Haliaeetus leucocephalus*), Golden Eagles (*Aquila chrysaetos*) and numerous waterfowl species. Spring migrating eagles sampled in west-central Montana between 1983 and 1985 showed elevated blood-lead levels in 85% of 86 Golden Eagles and 97% of 37 Bald Eagles (Harmata and Restani 1995). The authors suggested shot from waterfowl hunting and fragmented lead-core rifle bullets in ground squirrels (*Spermophilus spp.*) as a possible lead source. More recently, lead poisoning from spent ammunition has been identified as the leading cause of death in California Condors (*Gymnogyps californianus*), prompting the recent ban of lead ammunition within the "California Condor Recovery Zone" (Hunt et al. 2006, Cade 2007). Another study on Common Ravens (*Corvus corax*) in Wyoming has shown a direct correlation between elevated blood-lead levels and the onset of rifle hunting season (Craighead and Bedrosian pers. comm.)

We sampled blood from 42 Golden Eagles captured on migration during the fall of 2006 and 2007 to quantify a suite of possible heavy metal contaminants, with an emphasis on lead. Eagles were trapped using traditional ridgeline trapping techniques with bow nets, and using a harnessed Rock Dove (*Columbia livia*) for a lure (Bloom 1987). Small blood samples (2 ml) were taken from the brachial vein, and whole blood samples (1 ml) were frozen for later analysis using Inductively Coupled Plasma-Mass Spectrometry (ICP-MS) following hot acid digestion. We employed a portable field instrument (ESA LeadCare II®, Biosciences Inc.) to analyze lead in a subset of 20 samples. Our data suggest that both methods produce equivalent results.

Total lead concentrations ranged from <0.5 to 481 µg/dL with a median value of 13.6 µg/dL. When separated into four exposure stages (Redig 1984), our results were as follows: 18 eagles contained *background* levels of 0–10 µg/dL, 19 eagles were considered *sub-clinically exposed* at 10–60 µg/dL, two birds were *clinically exposed* (60–100 µg/dL), and three exhibited *acute exposure* of >100 µg/dL. In all, we found that 58% of the 42 fall migrant Golden Eagles sampled had elevated blood-lead levels.

We speculate that the five birds (12%) showing at least clinical exposure levels (≥60 µg/dL) had recently ingested lead-tainted carcasses and/or offal piles, likely during migration. Eagles with lower, but detectable blood lead levels may have had earlier exposure with the majority of the lead already deposited in other organs and bone. We surmise the use of lead-core ammunition for hunting is the major source for lead exposure in Golden Eagles, though we cannot identify a particular source species or region, in part because of the overlapping timing of hunting seasons for various game species in different regions of the Rocky Mountains and the very large area visited by Golden Eagles during migration season.

We are uncertain whether our preliminary numbers represent the northern migratory population of Golden Eagles as a whole, but a serious threat to the welfare of the species on a landscape level appears plausible. We believe an intensive educational outreach campaign and a switch away from lead-containing hunting ammunition to alternative, less toxic materials are appropriate ways to protect these and other scavenging species, as well as human consumers of gun-killed animals. *Received 5 September 2008, accepted 17 November 2008.*

DOMENECH, R., AND H. LANGNER. 2009. Blood-lead levels of fall migrant Golden Eagles in west-central Montana. Extended abstract *in* R. T. Watson, M. Fuller, M. Pokras, and W. G. Hunt (Eds.). Ingestion of Lead from Spent Ammunition: Implications for Wildlife and Humans. The Peregrine Fund, Boise, Idaho, USA. DOI 10.4080/ilsa.2009.0210

Key words: Fall migration, Golden Eagle, lead poisoning, Rocky Mountains.

LITERATURE CITED

BLOOM, P. H. 1987. Capturing and Handling Raptors. Pages 99–123 *in* B. A. Millsap, K. W. Cline, B. G. Pendleton, and D. A. Bird (Eds.). Raptor Management Techniques Manual. National Wildlife Federation, Washington, DC, USA.

CADE, T. J. 2007. Exposure of California Condors to lead from spent ammunition. Journal of Wildlife Management 71(7):2125–2133.

HARMATA, A. R., AND M. RESTANI. 1995. Environmental contaminants and cholinesterase in blood of vernal migrant Bald and Golden Eagles in Montana. Intermountain Journal of Sciences 1:1–15.

HUNT, W. G., W. BURNHAM, C. N. PARISH, K. K. BURNHAM, B. MUTCH, AND J. L. OAKS. 2006. Bullet fragments in deer remains: implications for lead exposure in avian scavengers. Wildlife Society Bulletin 34:167–170.

KRAMER, J. L., AND P. T. REDIG. 1997. Sixteen years of lead poisoning in eagles, 1980–95: an epizootiologic view. Journal of Raptor Research 31:327–332.

REDIG, P. T. 1984. An investigation into the effects of lead poisoning on Bald Eagles and other raptors: final report. Minnesota Endangered Species Program Study 100A-100B, University of Minnesota, St. Paul, Minnesota, USA.

SURVEY OF LEAD TOXICOSIS IN FREE-RANGING RAPTORS FROM CENTRAL ARGENTINA

MIGUEL D. SAGGESE[1], AGUSTÍN QUAGLIA[2,7], SERGIO A. LAMBERTUCCI[3], MARÍA S. BO[4], JOSÉ H. SARASOLA[5], ROBERTO PEREYRA- LOBOS[6], AND JUAN J. MACEDA[7]

[1]*College of Veterinary Medicine, Western University of Health Sciences, Pomona, CA, USA.*
E-mail: msaggese@westernu.edu

[2]*Facultad de Ciencias Veterinarias, Universidad de Buenos Aires, Buenos Aires, Argentina.*

[3]*Laboratorio Ecotono, Departamento de Ecología, Centro Universitario Bariloche, Universidad Nacional del Comahue-INIBIOMA/CONICET, Río Negro, Argentina.*

[4]*Facultad de Ciencias Exactas y Naturales, Universidad Nacional de Mar del Plata, Buenos Aires, Argentina.*

[5]*CECARA—Facultad de Ciencias Exactas y Naturales, Universidad de La Pampa, La Pampa, Argentina.*

[6]*Programa para la Conservación del Aguila Coronada, Dirección de Recursos Naturales, Mendoza, Argentina.*

[7]*Fundación Historia Natural Félix de Azara, Buenos Aires, Argentina.*

ABSTRACT.—Lead toxicosis is a problem recognized worldwide in raptors that has seriously impacted the recovery efforts of several endangered species. Vultures, eagles and kites are commonly affected because, as scavengers, they ingest lead ammunition residues when feeding on the remains of gun-killed animals. In South America, lead toxicosis in birds has been scarcely investigated. Raptors have been occasionally reported with the presumptive diagnosis of lead poisoning in hunting areas of central Argentina, although no systematic surveys have been conducted. Given the current understanding and knowledge we have on lead toxicosis in raptors in other parts of the world, the intense year-round wildlife hunting activities occurring in central Argentina, and the clinical diagnosis of lead poisoning in some birds presented to rehabilitation centers, we hypothesize that scavenging birds of prey in central Argentina could be systematically exposed to this heavy metal. Recently, we detected lead in the blood and bones of Argentine Solitary Crowned Eagles (*Harpyhaliaetus coronatus*), a severely endangered species from central and northern Argentina. Together with recent findings of lead in Andean Condors (*Vultur gryphus*) and in waterfowl, these studies are a first step to determine the extent of this problem in Argentine birds. Systematic studies of lead toxicity in wild birds in Argentina are needed in order to educate hunters toward a switch to non-lead substitutes. *Received 4 August 2008, accepted 22 October 2008.*

SAGGESE, M .D., A. QUAGLIA, S. A. LAMBERTUCCI, M. S. BO, J. H. SARASOLA, R. PEREYRA- LOBOS, AND J. J. MACEDA. 2009. Survey of lead toxicosis in free-ranging raptors from central Argentina. *In* R. T. Watson, M. Fuller, M. Pokras, and W. G. Hunt (Eds.). Ingestion of Lead from Spent Ammunition: Implications for Wildlife and Humans. The Peregrine Fund, Boise, Idaho, USA. DOI 10.4080/ilsa.2009.0211

Key words: Ammunition, Argentina, lead toxicity, raptors, wildlife.

LEAD TOXICOSIS caused by the ingestion of contaminated tissues of animals shot with lead ammunition is a phenomena recognized worldwide and an emerging health problem for wildlife and humans (De Francisco et al. 2003, Fischer et al. 2006, Stroud and Hunt 2009, Friend et al. 2009, Matz and Flint 2009, Beintema 2001, Mateo 2009). Among numerous reports and reviews on the toxicological effects, clinical manifestations, diagnosis, and treatment of spent lead ammunition in humans and wildlife are Samour and Naldo (2002), Grandjean and White (2002), Murata et al. (2003), Parish et al. (2006), Wynne and Stringfield (2007), and Kosnett (2009). The risk of lead exposure from spent ammunition in humans is higher for those populations that rely on game meat as a main source of animal protein (Johansen et al. 2004, Mateo et al. 2007, Cornatzer et al. 2009, Verbrugge et al. 2009, Hunt et al. 2009).

Raptors, both scavengers and predators, are exposed to lead when feeding on the remains of shot animals or when hunting prey injured with lead shot or bullets (Church et al. 2006, Cade 2007). Particularly susceptible to lead toxicosis are scavenger species (Platt et al. 1999, García-Fernandez et al. 1997, Hunt et al. 2006) like the California Condor (*Gymnogyps californianus*), a severely endangered species (Snyder and Snyder 2000). Spent lead ammunition appears to be not only the main cause of its decline in the wild during the 20th Century, but continues to undermine current recovery efforts aimed to establish self-sustaining populations in California, Utah and Arizona (Sorenson et al. 2000, Parish et al. 2006, 2009, Woods et al. 2006, Cade 2007). Furthermore, lead toxicosis has been reported as an important contributing factor in the population decline of other endangered species, including Steller's Sea Eagles (*Haliaetus pelagicus*), White-tailed Eagles (*Haliaetus albicilla*), Red Kites (*Milvus milvus*), Spanish Imperial Eagles (*Aquila adalberti*), and Egyptian Vultures (*Neophron percnopterus*) (Kurosawa 2005, Pain et al. 2005, 2007, Saito 2009, Krone et al. 2009, Gangoso et al. 2009, Mateo 2009).

Given the accumulating body of evidence about spent lead ammunition and its relationship with toxicosis in raptors worldwide, it is likely that the same occurs in Argentine birds of prey in the presence of comparable ecological and epidemiological conditions. Moreover, given their top position in the food chain, raptors may become valuable sentinels to detect the presence of lead toxicosis caused by spent ammunition in other species of wildlife. However, the investigation of lead exposure in Argentine birds has thus far received little attention. Therefore, the main goal of this communication is to review and summarize the limited available information about spent lead ammunition in Argentine raptors and other wildlife and to discuss future research needs.

MATERIALS AND METHODS

We conducted a detailed search of the literature about lead toxicosis and spent lead ammunition in raptors and other wildlife in Argentina. We searched BioOne, Raptor Information System, Pubmed, Global Raptor Information Network, SORA and Scielo databases, as well as various technical and local reports. Given the relative lack of official information about the extent of hunting activities in most Argentine provinces, we also consulted Argentine colleagues and hunters. Additional information about hunting activities in Argentina was obtained from web pages publicizing hunting activities and from provincial wildlife agency web pages. We also present limited unpublished data recently gained by the authors on lead levels in selected species of raptors.

RESULTS AND DISCUSSION

Lead Toxicosis in Argentina.—When compared with the information available for North America and Europe, knowledge about the effects of spent lead ammunition in raptors and other species in Argentina is limited. Recognized sources for humans include foundries, mines, car painting shops, batteries, cemeteries, plumbing, toys, waste from electronic and other industries, gas, soldering, prints, and pottery (Lacasaña et al. 1996, Hansen et al. 1999). Lead is a significant health problem in humans in Argentina, especially in children (Hansen et al. 1999, Martinez et al. 2003). Recent studies estimated that 10–40% of children less than 15 years old living in Buenos Aires and Córdoba have blood lead levels over 10 μg/dL (Garcia and Mercer 2003). As elsewhere, spent ammunition has not been reported as a source of lead for humans in Argentina or in other Latin-American countries (Lacasaña et al. 1996). Whether the lack of such reports reflects a lack of exposure to this source or an uninvestigated problem is unknown.

Prior to 2006, studies on lead and other heavy metals in Argentine wildlife focused mainly on aquatic invertebrates (Amin et al. 1996, 1998, Ferrer et al. 2000, Perez et al. 2005 and references therein) and mammals (Marcovecchio et al. 1990, 1994). Two scientific communications reported low levels of this element in a small number of marine birds (Gil et al. 1997, 2006). In 2007, two different groups of researchers found lead and other heavy metals in blood samples of Olrog´s Gull (*Larus atlanticus*) (L. La Sala pers. comm.) and in feathers of Common Terns (*Sterna hirundo*) (L. Mauco et al. unpubl. data). However, the source of lead for these birds may have originated in invertebrate prey consumed by these birds and not from ammunition (L. LaSala pers. comm., L. Mauco et al. unpubl. data).

Evidence for spent lead ammunition exposure and toxicity in Argentine birds is recent and comes from ongoing studies conducted in waterfowl and raptors in central Argentina. In 2007, lead shotgun pellets were found in two species of ducks in Santa Fe province (H. Ferreyra pers. comm.). In the same year, a blood lead concentration of 0.27 μg/dL was detected in a severely anemic and dehydrated nestling Crowned Eagle (*Harpyhaliaetus coronatus*) (A. Quaglia et al. unpubl. data.). An opportunistic investigation of lead in a Turkey Vulture (*Cathartes aura*) found dead in the same area revealed a bone lead concentration of 3.5 μg/g (A. Quaglia et al. unpubl. data). To the author´s knowledge, this is the first time that lead exposure has been confirmed in waterfowl and raptors in this country.

Additional evidence of exposure to lead came from Andean Condors (*Vultur gryphus*), a species very sensitive to this metal (Pattee et al. 2006). High blood lead levels were recently confirmed by laboratory investigation in Andean Condors admitted for rehabilitation at Fundación and Zoo Temaikén in Buenos Aires province (G. Gachen pers. comm.). Further evidence of lead toxicosis in this species came from another ongoing study in northern Patagonia; lead ammunition fragments observed in the ventriculus of a radiographed bird and corresponding high lead concentrations in wing feathers were detected there in Andean Condors (S. Lambertucci et al. unpubl. data).

Taken together, these limited data confirm that Argentine birds of prey are being exposed to lead in several and distant geographic areas of Argentina, although the quantitative extent of this exposure is unknown. These findings are of concern, considering that Crowned Eagles and Andean Condors are endangered species in the Neotropical region (BirdLife 2008).

Sources of Spent Lead Ammunition for Argentine Raptors.—Given its large area and diversity of habitats, Argentina is a country rich in large and small game species. Both native and exotic mammals and birds are hunted for sport and also for subsistence throughout the country. Commonly hunted species include tinamous (Tinamiformes), geese and ducks (Anseriformes), doves and pigeons (Columbiformes), quail (Galliformes), foxes and wildcats (Carnivora), hares and rabbits (Lagomorpha), wild boars, guanacos, cervids, and antelope (Artyodactyla), among others. Alien species, such as introduced cervids (e.g. *Axis axis*, *Cervus elaphus*, *Dama dama*) and bovids (*Antilope cervicapra*) are bred and kept in semi-captivity as well as in wild conditions for hunting on numerous private ranches in Patagonia and central provinces of Argentina (Novillo and Ojeda 2008, and references therein).

Another important source of spent lead ammunition for raptors and other wildlife in Argentina appears to be hunted waterfowl. Despite the banning of lead ammunition for waterfowl hunting in North America and Europe, it is still legally used in Argentina. With the convenience of travel, thousands of hunters visit wetlands and rice fields in Buenos Aires, Santa Fé, Corrientes, and Entre Rios provinces for waterfowl hunting, thus contributing to the seeding of spent lead ammunition in these water bodies and providing lead shot prey for raptors.

In Argentina, sport and recreational hunting is regulated through national and provincial laws. Hunting seasons, species, and daily and seasonal bag limits, are determined by provincial wildlife agencies. However, enforcement of these regulations is poorly accomplished due to the paucity of trained personnel and limited economic resources. As a result, large areas of the country lack effective police control. Thus, sport and subsistence hunting of large and small game remains in practice poorly regulated and difficult to control in most provinces. Moreover, subsistence hunting of waterfowl and other small game occurs year round in most Argentine provinces and has dramatically increased in recent decades as the result of economic crisis, unemployment, and rise in poverty levels (Anonymous 1997).

Of special concern is the sport hunting of pigeons and doves in central Argentina (Gordillo 2008). This activity has increased considerably in recent years in provinces like Córdoba, where several species of pigeons and Eared Doves (*Zenaida auriculata)* are considered overpopulated (Mourton et al. 1974). Hunting of these doves is allowed year round, without any limits regarding the numbers of birds killed daily. Given the increasing interest of hunters around the world in these game species, a large number of ranches have transformed completely or partially into private hunting grounds in recent years. In the last 15 years, it has been estimated that more than 10,000 hunters, coming mainly from North America and Europe, visit Córdoba every year attracted by the low rates and unlimited number of birds they can shoot each day (Gordillo 2008, M.L. Pignata pers. comm.). Recent estimations suggest that each hunter shoots more than 1000 cartridges a day (Gordillo 2008, M.L. Pignata pers. comm.). The total amount of lead ammunition spent in northwestern Córdoba, where dove hunting activities predominate, has been calculated to be approximately 1,600 tons of lead per year (Gordillo 2008). Thousands of doves and pigeons are killed, most of these birds remaining unrecovered in the fields (Gordillo 2008) where scavenging raptors are attracted to the easy food (M.L. Pignata pers. comm., S. Seipke pers.comm.). The carcasses are also given as "payment" to local children that assist the hunters and are used as food.

Species of Argentine Raptors at Risk of Lead Toxicosis.—Given the extension and diversity of habitats in Argentina, ranging from mountain forests and high Andean plateaus to prairie lands and patagonian steppes, approximately 1000 species of birds have been recorded in the country (Narosky and Yzurieta 2003). Among them, 65 (6.5%) species are diurnal birds of prey. Nine (56%) species of Argentine raptors are considered obligate scavengers (Andean Condor, King Vulture *Sarcorhampus papa,* Turkey Vulture, Lesser Yellow-Headed Vulture *Cathartes burrovianus,* Greater Yellow-Headed Vulture *Cathartes melambrotus,* Black Vulture *Coragyps atratus,* Mountain Caracara *Phalcoboenus megalopterus,* White-throated Caracara *Phalcoboenus albogularis,* and Striated Caracara *Phalcoboenus australis),* while seven (44%) species are considered facultative scavengers (Crowned Eagle, Solitary Eagle *Harpyhaliaetus solitarius,* Great Black-Hawk *Buteogallus urubitinga,* Black-Chested Buzzard-Eagle *Geranoaetus melanoleucus,* Southern Caracara *Caracara plancus,* Chimango Caracara *Milvago chimango,* Yellow-headed Caracara *Milvago chimachima*). Thus, a significant percentage (25%) of Argentine diurnal birds of prey are potentially exposed to spent lead ammunition, including three species of particular concern (Andean Condor, Crowned Eagle and Striated Caracara) given their conservation status in Argentina and South America. In addition, Peregrine Falcons (*Falco peregrinus*), Aplomado Falcons (*F. femoralis*), and other bird-eating falcons and accipiters may ingest lead shot from waterfowl, columbiforms, and others carrying lead shot in their tissues.

Current Studies on Lead Toxicosis in Argentine Raptors.—Studies on the population impact of environmental pollutants have rarely been conducted in Argentina (Saggese 2007). Considering the increasing body of evidence on the effects of spent lead ammunition in raptors worldwide and the extensive and widely distributed hunting activities occurring in Argentina, it is expected that in some areas scavenging birds of prey may be systematically exposed to this heavy metal. Whereas the overall impact that lead may be having on Argentine raptor populations is unknown, significant effects upon species with severely reduced and localized populations like those of Crowned Eagles and Andean Condors appear highly probable. Scientific documentation of exposure to spent lead ammunition and its effects on these and other species in Argentina is therefore required if there is to be education of hunters and lead's eventual replacement with non-lead substitutes.

Currently, five different research groups are systematically trapping wild birds of prey or collecting samples for biomedical studies in Argentina. Two groups are located in La Pampa province and the others are in Mendoza, Buenos Aires and Northern Patagonia provinces, all areas of high hunting pressure. Species being studied are Crowned Eagles, Chimango Caracaras, Andean Condors, and others. Given that blood from these wild birds is regularly collected for nutritional and physiological studies, simultaneous investigations of blood lead levels are feasible. Raptors admitted to rehabilitation centers are another potentially useful source to investigate lead exposure and toxicity. Therefore, a survey aimed to investigate the prevalence of lead exposure and toxicosis in Argentine raptors will begin in 2009. Blood from free-ranging birds, as well those admitted to several rehabilitation centers in Mendoza and Buenos Aires provinces, will be collected and analyzed using an automatic electrochemical system (LeadCare® Blood Lead Testing System, ESA Inc, Chelmsford, MA, USA). Results of this study are expected to demonstrate and quantify lead exposure in both facultative and obligate scavengers, and to contribute to the design of further studies identifying lead sources in wild raptors.

Further Recommendations and Conclusions.—Spent lead ammunition affects not only wildlife, but it may affect humans and even domestic animals as well. Understanding its social, medical, and biological consequences requires a multidisciplinary team (Pokras and Kneeland 2009). A conservation medicine approach based on a collaborative working relationship between those interested in human and animal health has therefore been recommended (Saggese 2007, Pokras and Kneeland 2009). Needed evidence about exposure and the effects of spent lead ammunition in Argentine wildlife and humans include: (1) the investigation of stomach contents in hunted waterfowl and upland game bird carcasses (Degernes et al. 2006), (2) radiological investigation of raptor pellets and carcasses (Martin et al. 2008), (3) investigation of lead concentrations in feathers, liver, bone, bone marrow and kidneys of scavenger birds of prey, waterfowl, and ground-feeding birds (Martin et al. 2008, Mateo 2009), (4) imaging examination (x-ray, computed tomography) of hunted animal remains abandoned in the field (Hunt et al. 2009), (5) investigation of blood lead levels in hunters and their families that consume game meat (Verbrugge et al. 2009), and (6) the use of lead isotope patterns to identify the source of this element (Matz and Flint 2009).

The effects of spent lead ammunition on Argentine raptor populations are currently unknown. Biomedical studies, including the investigation of other heavy metals and environmental pollutants, should be added to the study of factors that may cause demographic changes in Argentine birds of prey. Given the current understanding we have about lead toxicosis in other parts of the world, together with the intense year-round wildlife hunting activities detected in central Argentina, we hypothesize that scavenging birds of prey may be systematically exposed to lead, particularly in central Argentina and Patagonia where most hunting activities seem to concentrate. Results of future, rapid, short-term assessments of this problem, combined with the now widely available scientific background information about the perniciousness of lead, may support a move toward nontoxic substitutes for lead bullets and shotgun pellets.

ACKNOWLEDGEMENTS

We thank the organizers and sponsors of this conference for convening such an important scientific event that allowed us to present our work with Argentine birds of prey as well as to learn more about the effects of spent lead ammunition in wildlife and humans. The following institutions and people supported our studies on Argentine birds of prey and contributed to this presentation: The Schubot Exotic Bird Health Center—Texas A&M University, College of Veterinary Medicine—Western University of Health Sciences, Dirección de Recursos Naturales de la Provincia de Mendoza, Dirección de Fauna Silvestre Argentina, Direcciones de fauna provinciales, Park Rangers at Telteca Wildlife Refuge, CECARA—Facultad de Ciencias Exactas y Naturales—Universidad Nacional de La Pampa,

Universidad de Mar del Plata, Universidad Nacional del Comahue, PCRAR—Fundación Bioandina and Zoo Buenos Aires, The Peregrine Fund, Fundación Temaikén, R. Watson, R. Mateo, K. Saito, G. Wiemeyer, A. Capdevielle, H. Ferreyra, L. La Sala, M.L. Pignata, O. Krone, S. Seipke, L. Mauco, L. Biondi, R. Rodriguez, M. Uhart, M. Romano, G. Gachen, D. Avery, P. Redig, R. Jensen, L. Rodriguez, H. Ibañez, M. Rivolta, G. Hunt and R. Aguilar. Special thanks to Dr. I. Tizard, The Schubot Exotic Bird Health Center, S. Kasielke and Los Angeles Zoo and Botanical Gardens for supporting our past and future studies on lead toxicosis in Argentine birds of prey. We dedicate this communication to the memory of Jane Goggin, The Raptor Center-University of Minnesota longtime rehabilitation coordinator, for her relentless wildlife conservation efforts, her sincere humanity, and for her dedicated care of thousands of Bald Eagles, California Condors, and other species affected by lead poisoning during the last 20 years.

LITERATURE CITED

AMIN, O., J. ANDRADE, J. MARCOVECCHIO, AND L. COMOGLIO. 1996. Heavy metal concentrations in the mussel Mytilus edulis chilensis from the coast near Ushuaia city (Tierra del Fuego, Argentina). Pages 335–339 *in* J. Marcovecchio (Ed.). International Conference on Pollution Processes in Coastal Environments.

AMIN, O., E. RODRÍGUEZ, M. HERNANDO, L. COMOGLIO, L. LOPEZ, AND D. MEDESANI. 1998. Effects of lead and cadmium on hatching of the Southern King Crab Lithodes santolla (Decapoda, Anomura). Invertebrate Reproduction and Development 33:81–85.

ANONYMOUS. 1997. Cómo viven los argentinos que tienen que cazar para comer. Diario Clarín. [Online.] Available at http//:www.clarin.com/diario/1997/09/21/e-05201d.htm (accessed May 07 2008).

BEINTEMA, N. 2001. Lead poisoning in waterbirds. International Report 2000. Wetlands International, Wageningen, The Netherlands.

BIRDLIFE INTERNATIONAL. 2008. Species factsheet: Harpyhaliaetus coronatus. [Online.] Available at Http://www.birdlife.org. (accessed May 12 2008).

CADE, T. J. 2007. Exposure of California Condors to lead from spent ammunition. Journal of Wildlife Management 71:2125–2133.

CHURCH, M. E., R. GWIAZDA, R. W. RISEBROUGH, K. SORENSON, C. P. CHAMBERLAIN, S. FARRY, W. HEINRICH, B. A. RIDEOUT, AND D. R. SMITH. 2006. Ammunition is the principal source of lead accumulated by California Condors reintroduced to the wild. Environmental Sciences and Technology 40:6143–6150.

CORNATZER, W. E., E. F. FOGARTY, AND E. W. CORNATZER. 2009. Qualitative and quantitative detection of lead bullet fragments in random venison packages donated to the Community Action Food Centers of North Dakota, 2007. *In* R. T. Watson, M. Fuller, M. Pokras, and W. G. Hunt (Eds.). Ingestion of Lead from Spent Ammunition: Implications for Wildlife and Humans. The Peregrine Fund, Boise, Idaho, USA. DOI 10.4080/ilsa.2009.0111

DE FRANCISCO, N., J. D. RUIZ-TROYA, AND E. I. AGUERA. 2003. Lead and lead toxicity in domestic and free living birds. Avian Pathology 32:3–13.

DEGERNES, L., S. HEILMAN, M. TROGDON, M. JORDAN, M. DAVISON, D. KRAEGE, M. CORREA, AND P. COWEN. 2006. Epidemiologic investigation of lead poisoning in Trumpeter and Tundra swans in Washington State, USA, 2000–2002. Journal of Wildlife Diseases 42:345–358.

FERRER, L., E. CONTARDI, S. ANDRADE, R. ASTEASUAIN, A. PUCCI, AND J. MARCOVECCHIO. 2000. Environmental cadmium and lead concentrations in the Bahia Blanca Estuary (Argentina). Potential toxic effects of Cd and Pb on crab larvae. Oceanología 43:493–504.

FISHER, I. J., D. J. PAIN, AND V. G. THOMAS. 2006. A review of lead poisoning from ammunition sources in terrestrial birds. Biological Conservation 131:421–432.

FRIEND, M., J.C. FRANSON, AND W.L. ANDERSON. 2009. Biological and societal dimensions of lead poisoning in birds in the USA. *In* Watson, R. T., M. Fuller, M. Pokras, and W. G. Hunt, (Eds.). Ingestion of Lead from Spent Ammunition: Implications for Wildlife and Humans. The Peregrine Fund, Boise, Idaho, USA. DOI 10.4080/ilsa.2009.0104

GANGOSO, L., P. ÁLVAREZ-LLORET, A. RODRÍGUEZ-NAVARRO, R. MATEO, F. HIRALDO, AND J. ANTONIO DONÁZAR. 2009. Long-Term Effects of Lead Poisoning on Bone Mineralization in Egyptian Vulture *Neophron percnopterus.* Abstract *in* R.T. Watson, M. Fuller, M. Pokras, and W.G. Hunt, (Eds.). Ingestion of Lead from Spent Ammunition: Implications for Wildlife and Humans. The Peregrine Fund, Boise, Idaho, USA. DOI 10.4080/ilsa.2009.0214

GARCÍA, S. I., AND R. MERCER. 2003. Salud infantil y plomo en Argentina. Salud Pública de México 45 Suplemento 2:s252–s255.

GARCÍA-FERNÁNDEZ, A. J., M. MOTAS-GUZMÁN, I. NAVAS, P. MARÍA-MOJICA, A. LUNA, AND J. A. SÁNCHEZ-GARCÍA. 1997. Environmental exposure and distribution of lead in four species of raptors in southeastern Spain. Archives of Environmental Contamination and Toxicology 33:76–82.

GIL, M., M. HARVEY, H. BELDOMÉNICO, S. GARCÍA, M. COMMENDATORE, P. GANDINI, E. FRERE, P. DORIO, E. CRESPO, AND J. L. ESTEVES. 1997. Contaminación por metales y plaguicidas organoclorados en organismos marinos de la zona costera patagónica. Informes Técnicos del Plan de Manejo Integrado de la Zona Costera Patagónica-Fundacion Patagonia Natural 32:1–28.

GIL, M. N., A. TORRES, M. HARVEY, AND J. L. ESTEVES. 2006. Metales pesados en organismos marinos de la zona costera de la Patagonia argentina continental. Revista de Biología Marina y Oceanografia 41:167–176.

GORDILLO, S. 2008. Las rutas silenciosas del plomo en el norte Cordobés. [Online.] Available at www..ecoportal.net/content/view/full/76508 (accessed May 15 2008).

GRANDJEAN, P., AND R. WHITE. 2002. Neurodevelopmental disorders. Pages 66–78 *in* Children's Health and Environment: a Review of the Evidence. World Health Organization and European Environmental Agency, Rome, Italy.

HANSEN, C., R. RUTELER, E. PROCOPOVICH, G. PAGAN, B. DIAZ, N. GAIT, M. MEDICINA, M. MEZZANO, S. BRITOS, AND S. FULGINITI. 1999. Niveles de plomo en sangre en niños de la ciudad de Córdoba. Medicina 59:167–170.

HUNT, W. G., W. BURNHAM, C. N. PARISH, K. K. BURNHAM, B. MUTCH, AND J. L. OAKS. 2006. Bullet fragments in deer remains: implications for lead exposure in avian scavengers. Wildlife Society Bulletin 34:167–170.

HUNT, W. G., R. T. WATSON, J. L. OAKS, C. N. PARISH, K. K. BURNHAM, R. L. TUCKER, J. R. BELTHOFF, AND G. HART. 2009. Lead bullet fragments in venison from rifle-killed deer: Potential for human dietary exposure. *In* R. T. Watson, M. Fuller, M. Pokras, and W. G. Hunt (Eds.). Ingestion of Lead from Spent Ammunition: Implications for Wildlife and Humans. The Peregrine Fund, Boise, Idaho, USA. DOI 10.4080/ilsa.2009.0112

JOHANSEN, P., G. ASMUND, AND F. RIGET. 2004. High human exposure to lead through consumption of birds hunted with lead shot. Environmental Pollution 127:125–129.

KOSNETT, M. J. 2009. Health effects of low dose lead exposure in adults and children, and preventable risk posed by the consumption of game meat harvested with lead ammunition. *In* R. T. Watson, M. Fuller, M. Pokras, and W. G. Hunt, (Eds.). Ingestion of Lead from Spent Ammunition: Implications for Wildlife and Humans. The Peregrine Fund, Boise, Idaho, USA. DOI 10.4080/ilsa.2009.0103

KRONE, O., N. KENNTNER, A. TRINOGGA, M. NADJAFZADEH, F. SCHOLZ, J. SULAWA, K. TOTSCHEK, P. SCHUCK-WERSIG, AND R. ZIESCHANK. 2009. Lead poisoning in White-tailed Sea Eagles: Causes and approaches to solutions in Germany. *In* R. T. Watson, M. Fuller, M. Pokras, and W. G. Hunt, (Eds.). Ingestion of Lead from Spent Ammunition: Implications for Wildlife and Humans. The Peregrine Fund, Boise, Idaho, USA. DOI 10.4080/ilsa.2009.0207

KUROSAWA, N. 2005. Lead poisoning in Steller's Sea Eagles and White-tailed Sea Eagles. Pages 107–109 *in* M. Ueta and M. J. McGrady (Eds.). First Symposium on Steller´s and White-tailed Sea Eagles in East Asia.

LACASAÑA, M., I. ROMIEU, R. MCCONNELL, Y GRUPO DE TRABAJO SOBRE PLOMO DE LA OPS. 1996. El problema de exposición al plomo en América Latina y el caribe. Organización Panamericana de la Salud. Centro Panamericano de Ecología Humana y Salud, Metepec, Estado de México.

MARCOVECCHIO, J., V. MORENO, R. BASTIDA, M. GERPE, AND D. RODRÍGUEZ. 1990. Tissue distribution of heavy metals in small cetaceans from the southwestern Atlantic Ocean. Marine Pollution Bulletin 21:299–304.

MARCOVECCHIO, J., M. GERPE, R. BASTIDA, D. RODRÍGUEZ, AND G. MORON. 1994. Environmental contamination and marine mammals in coastal waters from Argentina: an overview. Science of the Total Environment 154:141–151.

MARTIN, P. A., D. CAMPBELL, K. HUGHES, AND T. MCDANIEL. 2008. Lead in the tissues of terrestrial raptors in southern Ontario, Canada, 1995–2001. Science of the Total Environment 391:96–103.

MARTINEZ RIERA, N., N. SORIA, G. FELDMAN, AND N. RIERA. 2003. Niveles de plombemia y otros marcadores en niños expuestos a una fundición de plomo en Lastenia, Tucumán, Argentina. Revista de Toxicología en Linea. [Online.] Available at www.sertox.com.ar/retel/default.htm (accessed May 15 2008).

MATEO, R., M. RODRÍGUEZ-DE LA CRUZ, D. VIDAL, M. REGLERO, AND P. CAMARERO. 2007. Transfer of lead from shot pellets to game meat during cooking. The Science of the Total Environment 372:480–485.

MATEO, R. 2009. Lead poisoning in wild birds in Europe and the regulations adopted by different countries. *In* R.T. Watson, M. Fuller, M. Pokras, and W.G. Hunt, (Eds.). Ingestion of Lead from Spent Ammunition: Implications for Wildlife and Humans. The Peregrine Fund, Boise, Idaho, USA. DOI 10.4080/ilsa.2009.0107

MATZ, A., AND P. FLINT. 2009. Lead isotopes indicate lead shot exposure in Alaska-breeding waterfowl. Abstract *In* R.T. Watson, M. Fuller, M. Pokras, and W.G. Hunt, (Eds.). Ingestion of Lead from Spent Ammunition: Implications for Wildlife and Humans. The Peregrine Fund, Boise, Idaho, USA. DOI 10.4080/ilsa.2009.0113

MOURTON, R. K., E. H. BUCHER, M. NORES, E. GOMEZ, AND J. REARTES. 1974. The ecology of the Eared Dove (Zenaida auriculata) in Argentina. Condor 76:80–88.

MURATA, K., T. SAKAI, Y. MORITA, K. YWATA, AND M. DAKEISHI. 2003. Critical dose of lead affecting δ-aminolevulinic acid levels. Journal of Occupational Health 45:209–214.

NAROSKY, T., AND D. YZURIETA. 2003. Guía para la identificación de las Aves de Argentina y Uruguay. Asociación Ornitológica del Plata. Vazquez Mazzini Ed. Buenos Aires.

NOVILLO, A., AND R. A. OJEDA. 2008. The exotic mammals of Argentina. Biological Invasions. DOI 10.1007/s10530-007-9208-8

PAIN, D. J., I. CARTER, A. W. SAINSBURY, R. F. SHORE, P. EDEN, M. A, TAGGART, S. KONSTANTINOS, L. A. WALKER, A. MEHARG, A. RAAB. 2007. Lead contamination and associated disease in captive and reintroduced Red Kites *Milvus milvus* in England. The Science of the Total Environment 376:116–127.

PAIN, D. J., A. A. MEHARG, M. FERRER, M. TAGGART, AND V. PENTERIANI. 2005. Lead concentrations in bones and feathers of the globally threatened Spanish Imperial Eagle. Biological Conservation 121:603–610.

PARISH, C. N., W. R. HEINRICH, AND W. G. HUNT. 2006. Lead exposure, diagnosis, and treatment in California Condors released in Arizona. *In* A. Mee, L. S. Hall, and J. Grantham (Eds.). California Condors in the 21st Century. Series in Ornithology, no. 2. American Ornithologists' Union, Washington, DC, and Nuttall Ornithological Club, Cambridge, Massachusetts, USA.

PARISH, C. N., W. G. HUNT, E. FELTES, R. SIEG, AND K. ORR. 2009. Lead exposure among a reintroduced population of California Condors in northern Arizona and southern Utah. *In* R.T. Watson, M. Fuller, M. Pokras, and W.G. Hunt, (Eds.). Ingestion of Lead from Spent Ammunition: Implications for Wildlife and Humans. The Peregrine Fund, Boise, Idaho, USA. DOI 10.4080/ilsa.2009.0217

PATTEE, O. H., J. W. CARPENTER, S. H. FRITTS, B. A. RATTNER, S. N. WIEMEYER, J. A. ROYLE, AND M. R. SMITH. 2006. Lead poisoning in captive Andean Condors (*Vultur gryphus*). Journal of Wildlife Diseases 42:772–779.

PEREZ, A.A., M. A. FAJARDO, A. M. STROBL, L. B. PEREZ, A. PIÑEYRO, AND C. M. LOPEZ. 2005. Contenido de plomo, cromo y cadmio en moluscos comestibles del golfo San Jorge (Argentina). Acta Toxicológica Argentina 13:20–25.

PLATT, S. R., K. E. HELMICK, J. GRAHAM, R. A. BENNETT, L. PHILLIPS, C. L. CHRISMAN, AND P. E. GINN. 1999. Peripheral neuropathy in a Turkey Vulture with lead toxicosis. Journal of American Veterinary Medical Association 214:1218–1220.

POKRAS, M. A., AND M. R. KNEELAND. Understanding lead uptake and effects across species lines: A conservation medicine approach. *In* R.T. Watson, M. Fuller, M. Pokras, and W.G. Hunt, (Eds.). Ingestion of Lead from Spent Ammunition: Implications for Wildlife and Humans. The Peregrine Fund, Boise, Idaho, USA. DOI 10.4080/ilsa.2009.0101

SAITO, K. 2009. Lead poisoning of Steller's Sea-Eagle (*Haliaeetus pelagicus*) and White-tailed Eagle (*Haliaeetus albicilla*) caused by the ingestion of lead bullets and slugs, in Hokkaido Japan. *In* R.T. Watson, M. Fuller, M. Pokras, and W.G. Hunt, (Eds.). Ingestion of Lead from Spent Ammunition: Implications for Wildlife and Humans. The Peregrine Fund, Boise, Idaho, USA. DOI 10.4080/ilsa.2009.0304

SAGGESE, M. D. 2007. Medicina de la conservación, enfermedades y aves rapaces. El Hornero 22(2):117-130.

SAMOUR, J. H., AND J. N. NALDO. 2002. Diagnosis and therapeutic management of lead toxicosis in falcons in Saudi Arabia. Journal of Avian Medicine and Surgery 16:16–20.

SNYDER, N., AND H. SNYDER. 2000. The California Condor: A Saga of Natural History and Conservation. Princeton University Press, Princeton, New Jersey, USA.

SORENSON, K. J., L. J. BURNETT, AND J. R. DAVIS. 2000. Status of the California Condor and mortality factors affecting recovery. Endangered Species Update 18:120–123.

STROUD, R. K., AND W. G. HUNT. 2009. Gunshot wounds: A source of lead in the environment. *In* R.T. Watson, M. Fuller, M. Pokras, and W.G. Hunt, (Eds.). Ingestion of Lead from Spent Ammunition: Implications for Wildlife and Humans. The Peregrine Fund, Boise, Idaho, USA. DOI 10.4080/ilsa.2009.0109

VERBRUGGE, L. A., S. G. WENZEL, J. E. BERNER, AND A. C. MATZ. 2009. Human exposure to lead from ammunition in the circumpolar north. *In* R.T. Watson, M. Fuller, M. Pokras, and W.G. Hunt, (Eds.). Ingestion of Lead from Spent Ammunition: Implications for Wildlife and Humans. The Peregrine Fund, Boise, Idaho, USA. DOI 10.4080/ilsa.2009.0110

WOODS, C. P., W. R. HEINRICH, AND S. C. FARRY. 2006. Survival and reproduction of California Condors released in Arizona. *In* A. Mee, L. S. Hall, and J. Grantham (Eds.). California Condors in the 21st Century. Series in Ornithology, no. 2. American Ornithologists' Union, Washington, DC, and Nuttall Ornithological Club, Cambridge, Massachusetts, USA.

WYNNE, J., AND C. STRINGFIELD. 2007. Treatment of lead toxicity and crop stasis in a California Condor (*Gymnogyps californianus*). Journal of Zoo and Wildlife Medicine 38:588–590.

RISK ASSESSMENT OF LEAD POISONING IN RAPTORS CAUSED BY RECREATIONAL SHOOTING OF PRAIRIE DOGS

ROBERT M. STEPHENS[1,3], ARAN S. JOHNSON[1], REGAN E. PLUMB[1], KIMBERLY DICKERSON[2], MARK C. MCKINSTRY[1], AND STANLEY H. ANDERSON[1,4]

[1]*US Geological Survey, Wyoming Cooperative Fish and Wildlife Research Unit, Department 3166, 1000 East University Avenue, Laramie, WY 82071, USA.* E-mail: Kimberly_dickerson@fws.gov

[2]*US Fish & Wildlife Service, 5353 Yellowstone Road, Suite 308A, Cheyenne, WY 82009, USA.*

EXTENDED ABSTRACT.—Recreational shooting of Black-tailed Prairie Dogs (Cynomys ludovicianus) is a common activity at Thunder Basin National Grassland (TBNG), Wyoming where annual use by prairie dog shooters can be as high as 8,500 shooter-use-days. The prairie dog carcasses left in the area are scavenged by raptors and other animals susceptible to lead (Pb) poisoning if they consume Pb bullet fragments or Pb shot. In 2000, a local rehabilitator noted an increase of Pb poisoning cases in raptors from the area. We collected 22 shooter-killed prairie dog carcasses from TBNG in 2001 to determine if Pb fragments remained embedded in the tissue that potentially would be consumed by raptors. Radiographs of 19 of the 22 prairie dog carcasses showed fragments consistent with Pb.

In 2002, we conducted a more in-depth study to determine if Pb poisoning was occurring in raptors at TBNG by documenting the number of raptors on prairie dogs at colonies where shooting occurred, assaying bullet fragments in shot prairie dogs to determine Pb content, and analyzing blood and feather samples of Ferruginous Hawk (Buteo regalis) and Golden Eagle (Aquila chrysaetos) nestlings and feathers from Burrowing Owls (Athene cunicularia) for clinical signs of Pb poisoning. We observed raptors foraging at prairie dog colonies and collected data on the number of shooters at prairie dog colonies. Shooter intensity did not predict raptor use when compared to a site near Rawlins, Wyoming where shooting did not occur. We detected metal fragments in 4 of 10 shot prairie dog carcasses collected. The total weight of the fragments found in each carcass ranged from 10 – 146 mg. Copper was the primary metal detected in three of four carcasses; but, significant amounts of Pb were found in the three carcasses. These fragments contained an average of 11.5 mg Pb and weighed an average of 24.7 mg with copper presumably making up the difference. Fragments <25 mg are considered small enough to be ingested and pose a risk of Pb absorption (Pauli and Buskirk 2007). Blood Pb concentrations in Ferruginous Hawk nestlings were below sub-clinical levels at TBNG and the control site. Analysis of red blood cell delta-aminolevulinic acid dehydratase activity, hemoglobin levels, and protoporphyrin levels also did not indicate Pb poisoning in Ferruginous Hawk nestlings. Additionally, blood and feather samples from Golden Eagle nestlings and feather samples from Burrowing Owls (juveniles and adults) at TBNG did not indicate Pb poisoning.

There are several possible reasons why Pb levels we found in Ferruginous Hawks and Golden Eagles (and possibly Burrowing Owls) that scavenge on the carcasses of shot prairie dogs were low. First, a sylvatic plague (Yersinia pestis) epizootic drastically reduced prairie dog numbers at many of the colonies in TBNG during 2001–2002. Second, 13% of TBNG was closed in 2001 to prairie dog shooting in an effort to reintroduce Black-footed Ferrets (Mustela nigripes). New shooting regulations and a

[3]Idaho Fish and Game, 3316 16th Street, Lewiston, ID 83501, USA. E-mail: rstephens@idfg.idaho.gov
[4]Deceased.

dwindling prairie dog population reduced the number of shooters that visited TBNG during the course of our study relative to previous years. Finally, long-term surveys in this area indicate that lagomorphs were abundant during our study, reducing the likelihood of raptors scavenging shot prairie dogs. Further study is needed to determine if the occurrence of Pb poisoning in TBNG relates to prairie dog and raptor abundance, availability of alternate food sources, and regulations on shooting.

Received 30 May 2008, accepted 22 October 2008.

STEPHENS, R. M., A. S. JOHNSON, R. E. PLUMB, K. DICKERSON, M. C. MCKINSTRY, AND S. H. ANDERSON. 2009. Risk assessment of lead poisoning in raptors caused by recreational shooting of prairie dogs. Extended abstract in R.T. Watson, M. Fuller, M. Pokras, and W.G. Hunt (Eds.). Ingestion of Lead from Spent Ammunition: Implications for Wildlife and Humans. The Peregrine Fund, Boise, Idaho, USA. DOI 10.4080/ilsa.2009.0212

Key words: Lead, raptors, prairie dogs, recreational shooting, scavenging.

This paper published in full as: STEPHENS, R. M., A. S. JOHNSON, R. E. PLUMB, K. DICKERSON, M. C. MCKINSTRY, AND S. H. ANDERSON. 2008. Risk assessment of lead poisoning in raptors caused by recreational shooting of prairie dogs. Intermountain Journal of Science 13(4):116-123. Also available as a US Fish and Wildlife Report: STEPHENS, R.M.; A.S. JOHNSON; R. PLUMB; K.K. DICKERSON; M.C. MCKINSTRY; AND S.H. ANDERSON. 2006. Secondary lead poisoning in Golden Eagle and Ferruginous Hawk chicks consuming shot Black-tailed Prairie Dogs, Thunder Basin National Grassland, Wyoming. US Fish and Wildlife Service Contamination Report Number R6/720C/05. Cheyenne, Wyoming.
http://www.fws.gov/mountain%2Dprairie/contaminants/papers/r6ecpubs_wy.htm

LITERATURE CITED

BECHARD, M. J., AND J. K. SCHMUTZ. 1995. Ferruginous Hawk (*Buteo regalis*). *In* A. Poole and F. Gill (Eds.). The Birds of North America, No. 172. The Academy of Natural Sciences, Philadelphia, and The American Ornithologists' Union, Washington, D.C., USA.

BELLROSE, F. C. 1959. Lead poisoning as a mortality factor in waterfowl populations. Illinois Natural History Bulletin 27:235–288.

BURCH, H. B., AND A. L. SIEGEL. 1971. Improved method for measurement of delta-aminolevulinic acid dehydratase activity of human erythrocytes. Clinical Chemistry 17:1038–1041.

BURGER, J., AND M. GOCHFELD. 2000. Metals in Laysan Albatrosses from Midway Atoll. Arch. Environmental Contamination and Toxicology 38:254–259.

CRAIG, T. H., J. W. CONNELLY, E. H. CRAIG, AND T. L. PARKER. 1990. Lead concentrations in Golden and Bald Eagles. Wilson Bulletin 102:130–133.

EISLER, R. 1998. Copper hazards to fish, wildlife, and invertebrates: a synoptic review. US Geological Survey, Biological Resources Division, Biological Science Report USGS/BRD/BSR--1998-0002.

FERNANDEZ, F. J., AND D. HILLIGOSS. 1982. Improved graphite furnace method for determination of lead in blood using matrix modification and the L'vov platform. Journal of Analytical Atomic Spectroscopy. 3:130–131.

FORBES, R. M., AND G. C. SANDERSON. 1978. Lead toxicity in domestic animals and wildlife. Pages 225–277 *in* J. O. Nriagu (Ed.). The Biogeochemistry of Lead in the Environment. Part B: Biological Effects. Elsevier/ North Holland Biomedical Press, Amsterdam, New York, and Oxford.

FRANSON, J. 1996. Interpretation of tissue lead residues in birds other than waterfowl. Pages 265–279 *in* W. N. Beyer, G. H. Heinz, and A. W. Redmon-Norwood (Eds.). Environmental Contaminants in Wildlife: Interpreting Tissue Concentrations. Lewis Publishers, Boca Raton, Florida, USA.

FRANSON, J. C., W. L. HOHMAN, J. L. MOORE, AND M. R. SMITH. 1996. Efficacy of protoporphyrin as a predictive marker for lead exposure in Canvasback Ducks: Effect of sample storage time. Environmental Monitoring Assessment 43:181–188.

HARMATA, A. R., AND M. RESTANI. 1995. Environmental contaminants and cholinesterase in blood of vernal migrant Bald and Golden Eagles in Montana. Intermountain Journal of Sciences 1:1–15.

HENNY, C. J., L. J. BLUS, D. J. HOFFMAN, L. SILEO, D. J. AUDET, AND M. R. SNYDER. 2000. Field evaluation of lead effects on Canada Geese and Mallards in the Coeur d'Alene River Basin, Idaho. Archives of Environmental Contamination and Toxicology 39:97–112.

HOFFMAN, D. J., J. C. FRANSON, O. H. PATTEE, C. M. BUNCK, AND H. C. MURRAY. 1985. Biochemical and hematological effects of lead ingestion in nestling American Kestrels (*Falco sparverius*). Comparative Biochemistry and Physiology 80:431–439.

HOFFMAN, D. J., B. A. RATTNER, G. A. BURTON JR., AND J. CAIRNS JR. 1995. Handbook of Ecotoxicology. Lewis Publishers, Ann Arbor, Michigan, USA.

KOCHERT, M. N., K. STEENHOF, C. L. MCINTYRE, AND E. H. CRAIG. 2002. Golden Eagle (*Aquila chrysaetos*). *In* A. Poole, and F. Gill (Eds.). The Birds of North America, No. 684. The Birds of North America, Inc., Philadelphia, Pennsylvania, USA.

MCANNIS, D. M. 1990. Home range, activity budgets, and habitat use of Ferruginous Hawks (*Buteo regalis*) breeding in southwest Idaho. Master's thesis, Boise State University, Boise, Idaho, USA.

MILLER, M. J. R., M. E. WAYLAND, E. H. DZUS, AND G. R. BORTOLOTTI. 2000. Availability and ingestion of lead shotshell pellets by migrant Bald Eagles in Saskatchewan. Journal of Raptor Research 34:167–174.

PATTEE, O. H., AND S. K. HENNES. 1983. Bald Eagles and waterfowl: the lead shot connection. Transactions of the North American Wildlife and Natural Resources Conference 48:230–237.

PAULI, J. N., AND S. W. BUSKIRK. 2007. Recreational shooting of Prairie Dogs: A portal for lead entering wildlife food chains. Journal of Wildlife Management 71:103–108.

RAWSON, R. E., G. D. DELGIUDICE, H. E. DZIUK, AND L. D. MECH. 1992. Energy metabolism and hematology of White-tailed Deer fawns. Journal of Wildlife Diseases 28:91–94.

SARI, M., S. DE PEE, E. MARTINI, S. HERMAN, SUGIATMI, W. M. BLOEM, AND R. YIP. 2001. Estimating the prevalence of anemia: a comparison of three methods. Bulletin of the World Health Organization 79:506–511.

WAYLAND, M., AND T. BOLLINGER. 1999. Lead exposure and poisoning in Bald Eagles and Golden Eagles in the Canadian prairie provinces. Environmental Pollution 104:341–350.

WYOMING STATE VETERINARY LAB. 2001. Toxicology Section, Lead Standard Operating Procedure, Laramie, Wyoming, USA.

LEAD IN GRIFFON AND CINEREOUS VULTURES IN CENTRAL SPAIN: CORRELATIONS BETWEEN CLINICAL SIGNS AND BLOOD LEAD LEVELS

JULIA RODRIGUEZ-RAMOS[1], VALERIA GUTIERREZ[2], URSULA HÖFLE[1,2], RAFAEL MATEO[2], LIDIA MONSALVE[2], ELENA CRESPO[3], AND JUAN MANUEL BLANCO[1]

[1]*Centro de Estudios de Rapaces Ibéricas CERI, Sevilleja de la Jara, Spain.*

[2]*Instituto de Investigación en recursos Cinegéticos IREC, Ciudad Real, Spain.*

[3]*CRFS, El Chaparillo, Ciudad Real, Spain.*

EXTENDED ABSTRACT.—As in other regions, vultures in Spain are often exposed to spent lead ammunition from carcasses of small and large game. This exposure may have increased after the ban on abandoning carcasses of domestic ruminants in the field due to the bovine spongiform encephalitis (BSE) crisis, both because the vultures consume hunting bag residues more frequently and because malnutrition may lead to mobilisation of lead stores (Iñigo and Atienza 2007). Although cases of clinical intoxication have been reported in numerous species including Griffon Vultures (*Gyps fulvus*) and Cinereous Vultures (*Aegypius monachus*) (Mateo et al. 1997, Mateo et al. 2003, Hernandez and Margalida 2008), little information on the potential correlation of blood lead levels and clinical signs and on potential subclinical effects of lead in vultures is available. Although other sources of lead may exist, in a study of live-trapped Griffon Vultures in southern Spain in which high blood lead levels were detected, the authors concluded that the ingestion of spent lead ammunition alone was responsible for the exposure (Garcia-Fernandez et al. 2005). An experimental study in Turkey Vultures (*Cathartes aura*) showed that there was great individual variation in susceptibility to lead, and that weakness and lack of coordination were present in most of the intoxicated birds, while only very high levels of lead produced lead toxicosis (Carpenter et al. 2003).

In this study, we analysed samples from vultures admitted to rehabilitation centers in South-central Spain, comparing blood lead levels, clinical signs, and hematological data in order to determine any clinical/subclinical effect of exposure to high lead levels and the degree of exposure to which it might be related. In addition, our aim was to compare the blood lead levels obtained with the LeadCare® blood lead testing system (ESA Biosciences, Inc. Chelmsford, MA, USA), the system used in our rehabilitation center network for the rapid confirmation of clinical lead intoxications, with results obtained by standard laboratory methods.

Blood samples were taken from 56 Griffon, 13 Cinereous, and one Egyptian Vulture (*Neophron percnopterus*) upon admission to four different rehabilitation centers in central Spain throughout the years 2006 and 2007. Blood lead levels were measured with the LeadCare® device that uses anodic stripping voltammetry (ASV), and standard laboratory methods (ICP-inductively coupled plasma atomic emission spectrometry). Clinical signs suggestive of potential lead toxicosis included disorientation, ataxia and impaired landing, posterior paresis, and hematology ranged from a slightly increased polychromatic index to a severe hypochromic anemia.

A good correlation was observed between the results for blood lead levels obtained with the LeadCare® device and standard laboratory methods (Figure 1). One Griffon Vulture and the Egyptian Vulture were confirmed to have clinical lead intoxications.

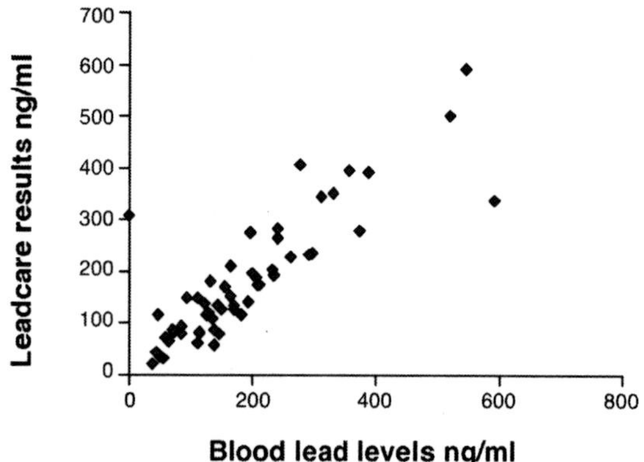

Figure 1. Correlation of blood lead levels obtained using the LeadCare® device and standard laboratory (ICP) methods.

While the Griffon Vulture also had severe traumatic lesions and was euthanized, the Egyptian Vulture was treated with oral and parenteral Calcium-EDTA, and recovered. Mean blood lead levels among the other birds were 144.41 ng/ml, with a range from 0.37 to 591.98 ng/ml in Eurasian Griffon Vultures, and mean of 70.07 ng/ml with a range from 0.75 to 512.16 ng/ml in Cinereous Vultures. Approximately 35% of the Griffon Vultures and 23% of the Cinereous Vultures tested had blood lead levels above 200 ng/ml. Using nonparametric statistics we were not able to detect any correlation between clinical signs, hematological values, body condition, and blood lead levels in either of the species. Griffon Vultures appeared to be somewhat more exposed to lead than Cinereous Vultures, although this trend was not significant. In general, lead levels appeared to be higher in individuals admitted to the rehabilitation centers between mid-August and mid-February, coinciding with the large and small game hunting seasons, suggesting that spent lead ammunition alone is responsible for the uptake. To the best of the author's knowledge the lead intoxication in the Egyptian Vulture is the first description of lead intoxication in this species. *Received 19 June 2008, accepted 27 October 2008.*

RODRIGUEZ-RAMOS, J., V. GUTIERREZ, U. HÖFLE, R. MATEO, L. MONSALVE, E. CRESPO, AND J. M. BLANCO. 2009. Lead in Griffon and Cinereous Vultures in Central Spain: Correlations between clinical signs and blood lead levels. Extended abstract *in* R. T. Watson, M. Fuller, M. Pokras, and W. G. Hunt (Eds.). Ingestion of Lead from Spent Ammunition: Implications for Wildlife and Humans. The Peregrine Fund, Boise, Idaho, USA. DOI 10.4080/ilsa.2009.0213

Key words: Ammunition, blood lead, clinical signs, hunting, lead, toxicosis, vultures.

LITERATURE CITED

CARPENTER, J. W., O. H. PATTEE, S. H. FRITTS, B. A. RATTNER, S. N. WIEMEYER, J. A. ROYLE, AND M. N. SMITH. 2003. Experimental lead poisoning in Turkey Vultures. (*Cathartes aura*). Journal of Wildlife Diseases 39:96–104.

GARCIA-FERNANDEZ, A. J., E. MARTINEZ-LOPEZ, D. ROMERO, P. MARIA-MOJICA, A. GODINO, AND P. JIMENEZ. 2005. High levels of blood lead in Griffon Vultures (*Gyps fulvus*) from Cazorla Natural Park (southern Spain). Environmental Toxicology 20:459-463.

HERNÁNDEZ, M., AND A. MARGALIDA. 2008. Pesticide abuse in Europe: effects on the Cinereous Vulture (*Aegypius monachus*) population in Spain. Ecotoxicology 17:264–272.

IÑIGO, A., AND J. C. ATIENZA. 2007. Efectos del Reglamento 1774/2002 y las decisions adoptadas por la Comisión Europea en 2003 y 2005 sobre las aves necrófagas en la peninsula Ibérica y sus posibles soluciones. Official report to the European Comisión. SEO/BirdLife 15-06-2007.

MATEO, R., R. MOLINA, J. GRÍFOLS, AND R. GUITART. 1997. Lead poisoning in a free ranging Griffon Vulture (*Gyps fulvus*). Veterinary Record 140:47-48.

MATEO, R., M. TAGGART, AND A. A. MEHARG. 2003. Lead and arsenic in bones of birds of prey from Spain. Environmental Pollution 126:107-14.

LONG-TERM EFFECTS OF LEAD POISONING ON BONE MINERALIZATION IN EGYPTIAN VULTURE *NEOPHRON PERCNOPTERUS*

LAURA GANGOSO[1], PEDRO ÁLVAREZ-LLORET[2], ALEJANDRO RODRÍGUEZ-NAVARRO[2], RAFAEL MATEO[3], FERNANDO HIRALDO[1], AND JOSÉ ANTONIO DONÁZAR[1]

[1]*Dept. of Conservation Biology, Estación Biológica de Doñana, C.S.I.C., Avda Mª Luisa s/n, 41013 Sevilla, Spain.* E-mail: laurag@ebd.csic.es

[2]*Dept. of Mineralogy and Petrology, University of Granada, Avda Fuentenueva s/n, 18002 Granada, Spain.*

[3]*Instituto de Investigación en Recursos Cinegéticos, IREC (CSIC, UCLM, JCCM), Ronda de Toledo s/n, 13005 Ciudad Real, Spain.*

ABSTRACT.—Poisoning from lead shot has been well documented globally. However, despite its recognized importance as a threat factor for populations of vertebrates of conservation concern, very little is still known about its hidden long-term effects. Long-lived species are particularly susceptible to bioaccumulation of lead in bone tissues. In this paper we gain insights into the sub-lethal effects of lead contamination on Egyptian Vultures (*Neophron percnopterus*), a globally threatened species. We compared two populations (Canary Islands and Iberian Peninsula) differing in exposure to the ingestion of lead ammunition. The island population, being sedentary, has a greater exposure to the ingestion of lead hunting shot during the winter hunting season. To determine the sub-lethal effects of lead, we analyzed the consequences of the accumulation of this contaminant in bone tissue.

Blood lead levels were higher in the island population showing clear seasonal trends, being highest during the hunting season. Moreover, males were more susceptible to lead accumulation than females. Bone lead concentration increased with age, reflecting a bioaccumulation effect. The comparison of quantitative measurements obtained from deconvoluted FTIR spectra showed that the bone composition was significantly altered by this contaminant and, in particular, the degree of mineralization decreased as lead concentration levels increased. These results demonstrate the existence of long-term effects of lead poisoning which may be of importance in the declines of threatened populations of long-lived species exposed to this contaminant.

GANGOSO, L., P. ÁLVAREZ-LLORET, A. RODRÍGUEZ-NAVARRO, R. MATEO, F. HIRALDO, AND J. ANTONIO DONÁZAR. 2009. Long-term effects of lead poisoning on bone mineralization in Egyptian Vulture *Neophron percnopterus.* Abstract *in* R. T. Watson, M. Fuller, M. Pokras, and W. G. Hunt (Eds.). Ingestion of Lead from Spent Ammunition: Implications for Wildlife and Humans. The Peregrine Fund, Boise, Idaho, USA. DOI 10.4080/ilsa.2009.0214

Key words: Avian, bone, effects, island, lead, long-lived, scavenger, vulture.

This paper published in full as: GANGOSO, L., P. ALVAREZ-LLORET, A.A.B. RODRIGUEZ-NAVARRO, R. MATEO, F. HIRALDO, AND J. A. DONAZAR. 2008. Long-term effects of lead poisoning on bone mineralization in vultures exposed to ammunition sources. Environmental Pollution, DOI 10.1016/j.envpol.2008.09.015.

BLOOD-LEAD CONCENTRATIONS IN CALIFORNIA CONDORS RELEASED AT PINNACLES NATIONAL MONUMENT, CALIFORNIA

JAMES R. PETTERSON[1], KELLY J. SORENSON[2], COURT VANTASSELL[1], JOSEPH BURNETT[2], SCOTT SCHERBINSKI[1], ALACIA WELCH[1], AND SAYRE FLANNAGAN[2]

[1]*U.S. National Park Service, Pinnacles National Monument, 5000 Hwy. 146, Paicines, CA 95043, USA.* E-mail: jim_petterson@nps.gov

[2]*Ventana Wildlife Society, Species Recovery Program, HC 67 Box 99 Monterey, CA 93940, USA.*

ABSTRACT.—Twenty-four California Condors (*Gymnogyps californianus*) were released at Pinnacles National Monument, California between December 2003 and August 2007 as part of a larger program to recover the endangered species to the wild. We collected 63 independent, post-release blood samples from 20 individuals and analyzed their lead concentration patterns. Of the 20 individuals, we collected pre-release samples on nine to compare blood lead values before and after release. Of the 63 post-release samples, 24 (38.1%) were above background (20–59 µg/dL), two (3.2%) were clinically affected (60–99 µg/dL), and two more (3.2%) were indicative of acute toxicity (≥100 µg/dL). Fifteen (75.0%) of individuals sampled were exposed at least once and eight (40.0%) were exposed on two or more occasions. We found a significant difference comparing samples collected before release and within one year after release from the same individuals, revealing that even young, inexperienced condors in this area are vulnerable to lead exposure. We show a lack of seasonal and annual trends of lead exposure to condors in this area and discuss possible explanations. Observations were made of free-flying condors feeding on ground squirrels, feral pigs, cattle, and marine mammals, suggesting it is plausible that elevated blood lead levels resulted, in part, due to inadvertent ingestion of spent lead ammunition. Convinced that lead ammunition is the primary source of exposure to condors, California lawmakers and the California Fish and Game Commission adopted a lead ammunition ban within the range of the condor in California beginning on July 1, 2008. Reducing lead exposure in the long-run will largely depend on hunters' willingness to switch to non-lead alternatives and studies such as this one should be continued into the foreseeable future to monitor the extent and effect of hunter compliance.

PETTERSON, J. R., K. J. SORENSON, C. VANTASSELL, J. BURNETT, S. SCHERBINSKI, A. WELCH, AND S. FLANNAGAN. 2009. Blood-lead concentrations in California Condors released at Pinnacles National Monument, California. Abstract *in* R. T. Watson, M. Fuller, M. Pokras, and W. G. Hunt (Eds.). Ingestion of Lead from Spent Ammunition: Implications for Wildlife and Humans. The Peregrine Fund, Boise, Idaho, USA. DOI 10.4080/ilsa.2009.0215

Key words: Ammunition, blood, California, condor, endangered, lead, hunter, toxic.

BLOOD CHEMISTRY VALUES OF CALIFORNIA CONDORS EXPOSED TO LEAD

MOLLY CHURCH[1], KAREN ROSENTHAL[1], DONALD R. SMITH[2], KATHRYN PARMENTIER[2], KEN ARON[3], AND DALE HOAG[3]

[1]*University of Pennsylvania, School of Veterinary Medicine, 3800 Spruce Street, Philadelphia, PA 19104, USA.*

[2]*University of California, Santa Cruz, Environmental Toxicology Department, 430 Physical Sciences Building, Santa Cruz, CA 95064, USA.*

[3]*ABAXIS, 3240 Whipple Road, Union City, CA 94587, USA.*

ABSTRACT.—Clinical pathology provides a noninvasive means to elucidate health status in wildlife species. The California Condor (*Gymnogyps californianus*) is an endangered species that has been reintroduced into the wild due to successful captive breeding. Condors are periodically recaptured from the wild in order to evaluate battery function on wing-tag radio transmitters and to collect blood samples to assess blood lead levels. It has been shown previously that blood lead levels in condors increase with time spent in the wild. Lead is a well-known neurotoxin and nephrotoxin; however, there are no specific blood chemistry markers for lead-induced nerve tissue damage or renal disease. The effect of lead on blood chemistry values in wild bird species is unknown. It is thought that lead interferes with second messenger receptors in neurons and induces Schwann cell degeneration. Lead also decreases renal glomerular filtration rates, which would then lead to an increase in serum concentrations of molecules normally filtered by the kidney. This project analyzed 12 plasma samples from condors collected between April and November of 2007. Four of the condors had not yet been released into the wild (prerelease), and eight had been released (released) for at least one month prior to the time of blood collection. Samples were run in duplicate on an Abaxis® VetScan VS2® instruments utilizing Avian/Reptilian Profile Plus (ALB, AST, BA, Ca, CK, GLOB, GLU, K^+, Na^+, PHOS, TP, UA) and Mammalian Liver Profile (ALB, ALP, ALT, BA, BUN, CHOL, GGT, TBIL) rotors. Increases in phosphorous and in uric acid are used as indicators of renal disease in birds, however, in this study, no correlation was found between phosphorus or uric acid and blood lead levels. In the two condors with the highest blood lead levels (64 and 20 μg/dL), CK, and K were at levels above the currently published reference ranges for these blood chemistry values. CK is an enzyme found in skeletal muscle, cardiac muscle and brain tissue, and increases in CK activity have been associated with lead toxicity, perhaps due to neuronal damage. K is an electrolyte that may be increased due to kidney dysfunction caused by lead-protein complex formation in renal tubular cells. This study served as an initial survey of blood chemistry values in condors with established blood lead levels, and will be expanded in the future.

CHURCH, M., K. ROSENTHAL, D. R. SMITH, K. PARMENTIER, K. ARON, AND D. HOAG. 2009. Blood chemistry values of California Condors exposed to lead. Abstract *in* R. T. Watson, M. Fuller, M. Pokras, and W. G. Hunt (Eds.). Ingestion of Lead from Spent Ammunition: Implications for Wildlife and Humans. The Peregrine Fund, Boise, Idaho, USA. DOI 10.4080/ilsa.2009.0118

Key words: Blood, chemistry, California Condor, lead.

EFFECTIVENESS OF ACTION TO REDUCE EXPOSURE OF FREE-RANGING CALIFORNIA CONDORS IN ARIZONA AND UTAH TO LEAD FROM SPENT AMMUNITION

RHYS E. GREEN[1,2], W. GRAINGER HUNT[3], CHRISTOPHER N. PARISH[3], AND IAN NEWTON[4]

[1]*Conservation Science Group, University of Cambridge, Department of Zoology, Downing Street, Cambridge CB2 3EJ, UK.*

[2]*Royal Society for the Protection of Birds, The Lodge, Sandy, Bedfordshire SG19 2DL, UK.*

[3]*The Peregrine Fund, 5668 Flying Hawk Lane, Boise ID 83709, USA.*

[4]*Centre for Ecology and Hydrology, Monks Wood Experimental Station, Abbots Ripton, Huntingdon PE28 2LS, UK.*

ABSTRACT.—California Condors (*Gymnogyps californianus*) released into the wild in Arizona ranged widely in Arizona and Utah. Previous studies have shown that the blood lead concentrations of many of the birds rise because of ingestion of spent lead ammunition. Condors were routinely recaptured and treated to reduce their lead levels as necessary but, even so, several died from lead poisoning. We used tracking data from VHF and satellite tags, together with the results of routine testing of blood lead concentrations, to estimate daily changes in blood lead level in relation to the location of each bird. The mean daily increment in blood lead concentration depended upon both the location of the bird and the time of year. Birds that spent time during the deer hunting season in two areas in which deer were shot with lead ammunition (Kaibab Plateau (Arizona) and Zion (Utah)) were especially likely to have high blood lead levels. The influence upon blood lead level of presence in a particular area declined with time elapsed since the bird was last there. We estimated the daily blood lead level for each bird and its influence upon daily mortality rate from lead poisoning. Condors with high blood lead over a protracted period were much more likely to die than birds with low blood lead or short-term elevation. We simulated the effect of ending the existing lead exposure reduction measures at Kaibab Plateau, which encourage the voluntary use of non-lead ammunition and removal of gut piles of deer and elk killed using lead ammunition. The estimated mortality rate due to lead in the absence of this program was sufficiently high that the condor population would be expected to decline rapidly. The extension of the existing lead reduction program to cover Zion (Utah), as well as the Kaibab plateau, would be expected to reduce mortality caused by lead substantially and allow the condor population to increase. *Reproduced with permission from PloS ONE 3(12).*[5]

GREEN, R. E., W. G. HUNT, C. N. PARISH, AND I. NEWTON. 2009. Effectiveness of action to reduce exposure of free-ranging California Condors in Arizona and Utah to lead from spent ammunition. Reproduced *in* R. T. Watson, M. Fuller, M. Pokras, and W. G. Hunt (Eds.). Ingestion of Lead from Spent Ammunition: Implications for Wildlife and Humans. The Peregrine Fund, Boise, Idaho, USA. DOI 10.4080/ilsa.2009.0218

---

[5] Reproduced in accordance with the Creative Commons Attribution License with permission of the authors from: Green R. E., W. G. Hunt, C. N. Parish, and I. Newton. 2008. Effectiveness of action to reduce exposure of free-ranging California Condors in Arizona and Utah to lead from spent ammunition. PLoS ONE 3(12):e4022. doi:10.1371/journal.pone.0004022

Key words: Ammunition, blood lead, condor, foraging movements, lead absorption model, lead ammunition reduction, lead exposure, hunting season, hunting zone, model, mortality

THE CALIFORNIA CONDOR (*Gymnogyps californianus*) became extinct in the wild in the 1987 when the last wild individual was captured and added to the captive flock, which then consisted of 27 birds. Since 1992, releases of these birds and their captive-bred progeny have re-established wild populations of condors in California, Mexico and around the Grand Canyon in Arizona and Utah. Individual condors in these populations have suffered from lead poisoning caused by ingested ammunition, which is the most frequently diagnosed cause of death among Grand Canyon condors. This holds despite intensive efforts to monitor blood concentrations of lead and to treat birds with high levels using chelating agents (Parish et al. 2007). The condors in the Grand Canyon population range widely in Arizona and Utah and feed on carrion, a proportion of which comes from the carcasses of game animals shot by hunters using lead ammunition. Ingestion of shotgun pellets and fragments of bullets in flesh from such carcasses is the route by which lead poisoning occurs. Condors are located as frequently as possible using satellite tags and VHF radio tags and those that cease to move are recovered. Birds are also captured routinely and their blood lead concentrations measured. Any individuals with high levels are held for treatment to reduce the burden of lead in the body before release. Action is also taken on the Kaibab Plateau, Arizona to reduce exposure of condors to lead by encouraging hunters to use non-lead bullets and to remove potentially contaminated gut piles. The level of condor mortality caused by lead that would occur in the absence of chelation therapy and lead exposure reduction is of interest because it might not always be practical to locate birds daily and trap all condors routinely once or twice per year for blood lead monitoring, and implementation of lead exposure reduction schemes requires resources (Sullivan et al. 2007). Could the reintroduced population persist if the lead exposure reduction and treatment programs ceased or were reduced in scope? What would be the effect of reducing exposure to spent lead ammunition throughout the range of this population? As a step towards addressing these questions, we report here a statistical model of blood lead levels in free-ranging condors, which extends previous analyses (Hunt et al. 2007). We took advantage of the unusually complete radio-tracking data, which allow the influence on blood lead of the location of condors within their geographical range to be assessed. Our objectives were to model the distribution of blood lead levels throughout the year in the absence of treatment, and then to estimate the mortality rates that would prevail. Finally, we used the model to explore the possible effects on condor mortality of withdrawing or increasing measures to reduce exposure of condors to spent lead ammunition.

MATERIALS AND METHODS

Field Studies.—We used data for 2005, 2006 and 2007 derived from the monitoring of movements and blood lead levels of free-ranging condors (Parish et al. 2007). The dependent variable in our analyses was the concentration of lead in the blood of a condor determined within five days after capture. Blood lead levels were determined using a portable field tester (LeadCare Blood Lead Testing System). Some blood samples were also analyzed by atomic absorption spectroscopy at the Louisiana State University Diagnostics Laboratory using a Perkin Elmer Analyst 800. Levels of lead in the same blood sample measured using the field tester and in the laboratory were strongly correlated, but laboratory measurements gave significantly higher values (see Figure 2 of Parish et al. 2007). Using 99 cases in which the lead concentration in the same blood sample had been determined by both methods, we found that the mean concentration of lead measured in the laboratory was larger than that from the field tester by a factor of 1.914. In all analyses we therefore used a laboratory determination whenever one was available and otherwise adjusted the field tester measurement using this correction factor.

We modeled the blood lead level in each free-ranging condor in relation to the locations it had used before it was recaptured for testing. During the study period, roost locations of condors marked with VHF or satellite tags were determined on the majority of days for all tagged condors, and attributed to one of the following five zones; Paria (Vermilion Cliffs), Colorado River Corridor, Kaibab Plateau, South Zone and North Zone (Utah). A location was taken to be a roost location if it was obtained later than16.00h. local time. Condors are known to range widely, even within a day (Hunt et al. 2007), so the ideal analysis would take into account the bird's location at several times during each day. However, only the data for satellite tagged birds would permit this. Roost locations were recorded for as many days as possible during the period beginning with the initial release of each bird, or its release after capture for blood lead monitoring and ending with another capture at which blood lead concentration was determined. For days on which the roost location was not recorded, we interpolated the roost zone used by assuming that it was the same as that on the nearest day with data available. Overall, it was necessary to interpolate the roost zone on 27.2% of days, with the range of this proportion for individual birds being 11.1% to 59.9%. We had eligible data derived from 60 individual condors consisting of 322 pairs of blood lead measurements preceded by periods comprising, in total, 41,230 bird-days with known or interpolated roost locations.

Numbers of deer, elk and buffalo reported as killed by hunters in each zone in 2005–2007 were obtained from the Arizona Game and Fish Department and the Utah Department of Natural Resources. We estimated the number of carcasses and gut piles potentially contaminated with lead and left in the field for scavengers by using information collected on the proportion of kills made with lead ammunition and the number of lead-killed animals from which gut piles were brought in by hunters for safe disposal. We also assumed that in addition to the number of animals reported as killed with lead bullets, an additional 10% of that number were wounded and died unrecovered soon after, thereby becoming available to condors.

Analysis and Statistical Modeling of Blood Lead Data.—We assumed that, with no further ingestion, the relationship between blood lead concentration and time after ingestion of fragments of metallic lead could be described by a simple three compartment model, with one-way movement of lead between successive pairs of compartments. Although this model is a simplification, it has the advantage of requiring the estimation of only two parameters and seems likely to capture the main features of real changes in blood lead. We assumed that a constant proportion of the ingested lead enters the blood from the gut per unit time and that fragments are not expelled from the gut within the period that significant absorption is occurring. Hence, the proportion of the lead ingested that remains in the gut at time t (in days) since ingestion is given by $\exp(-k_1 t)$, where k_1 is a constant, and $(1-\exp(-k_1 t))$ is the proportion of lead ingested that has moved from the gut to the blood by that time. We also assumed that a constant proportion per unit time of the lead present in the blood was lost to another compartment, such that the amount in the blood would decline by a proportion $(1-\exp(-k_2))$ per day in the absence of absorption. The quantity of lead in the blood, as a proportion of that ingested, is then given by the function

$$g(t) = (k_1/(k_1 - k_2))\,(\exp(-k_2 t) - \exp(-k_1 t)) \qquad \text{Eq.(1).}$$

Assuming that blood volume is constant, blood lead concentration is proportional to $g(t)$. Note that this expression approximates to $g(t) = \exp(-k_2 t)$ when k_1 is much larger than k_2. That is, when absorption from the gut is very rapid, blood lead concentration declines exponentially with time since ingestion. The model is illustrated for a single value of k_2 and three values of k_1 in Figure 1.

We next used the function $g(t)$ to explore how the concentration of lead in the blood of an average condor would be expected to change over time, given the possibility of ingestion of lead on more than one day. We assumed that the condor spends some time in areas where there is a high risk each day of ingesting lead and some time in low risk areas. We imagined a large number of condors, all showing the same movement pattern. On each successive day, the average quantity of lead ingested by the birds would, if it was all absorbed immediately, increase the average concentration of lead in the blood by an amount m, which we call the mean daily blood lead increment. In fact, we would expect the component of the concentration of lead in the blood derived from the lead ingested on a given day to be given by $m\,g(t)$, where t is the time elapsed since that day. How the lead ingested on successive days would influence the total concentration of lead in the blood can be visualized with

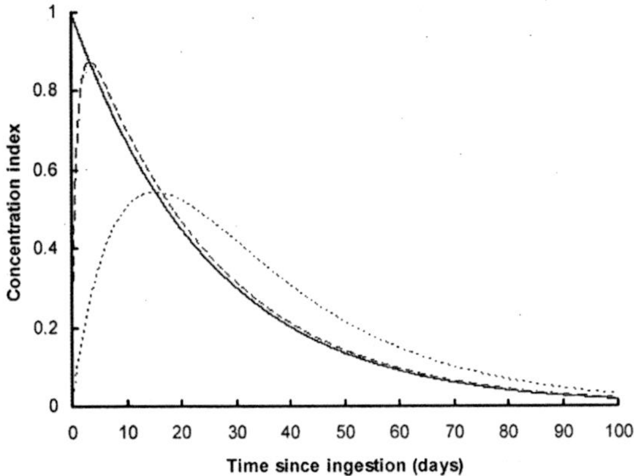

Figure 1. Models of the relationship between blood lead concentration and time since ingestion of metallic lead in California Condors. The schematic diagram shows the family of models (see text) assumed to describe changes in blood lead concentration in untreated condors in relation to time since ingestion, in the absence of any further ingestion. The value of model parameter k_2 is 0.04 for all three curves, but k_1 is very large for the solid line, k_1 = 1 for the dashed line, and k_1 = 0.1 for the dotted line.

the aid of the schematic diagrams shown in Figure 2. For simplicity, we use g(t) = exp($-k_2t$) in this illustration, but the equivalent for the function given by Eq.(1) can easily be envisaged. Consider first a sample of birds that remain in an area with a low risk of ingesting lead (area A). If the average amount of lead ingested per day is small relative to the rate at which it is eliminated from the blood, we would expect that the average blood lead concentration would decline over time (Figure 2(a)). Hence, we would expect that the average of measurements of blood lead concentration at the end of a period in which the condors had remained in a low risk area would be lower than the average of measurements made at the beginning of the period (shown in the diagram by circles). However, if the condors instead spend part of the period in a high risk area (area B) and the rest in A, we would expect that the average blood lead concentration would be higher at the second measurement than at the first (Figure 2(b)). Furthermore, the timing of the birds' visit to the high risk area would be expected to influence the change in blood lead between the two measurements. In this case, a visit to area A late in the interval between two blood lead measurements (Figure 2(c)) is expected to result in a larger increase in blood lead than a visit of the same duration earlier in the interval (Figure 2(b)). Note that this effect of the timing of the visit to area A might be different if g(t) was of the form that the blood lead concentration from a given day's ingested lead first increased and then decreased (as in Figure 1). The pattern of variation in changes over time in blood lead shown by these illustrations suggests

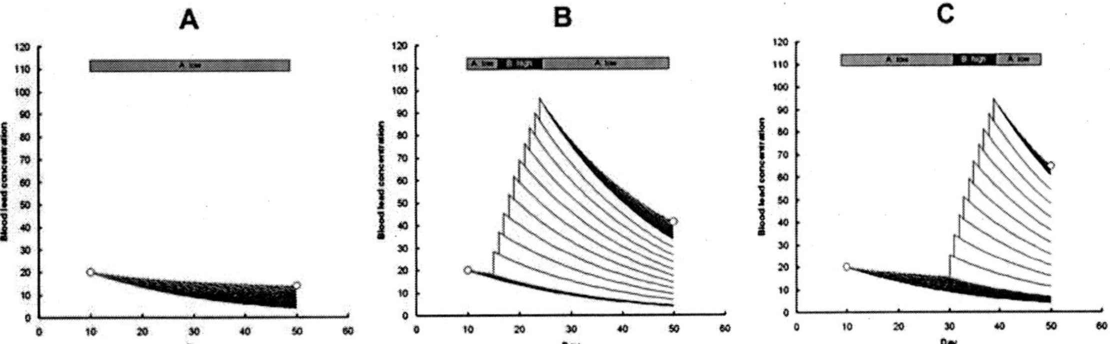

Figure 2. Hypothetical changes in blood lead concentration over time for California Condors moving between zones with high and low daily risk of ingesting lead. The schematic diagram shows average hypothetical changes in blood lead concentration (µg dL^{-1}) for condors moving between zones with high and low daily risk of ingesting lead. Curves show the time course of components of blood lead derived from each day's intake. Birds remaining in the low risk zone (A) throughout would be expected to show a decline in blood lead over the time between a pair of measurements (open circles). Birds visiting the high risk zone early in the interval between measurements (B) would be expected to show an increase in blood lead concentration. Birds visiting the high risk zone for the same number of days, but later in the interval (C), would be expected to show the highest increase in blood lead concentration.

that, given sufficient pairs of observations of blood lead concentration for condors whose location was known during the intervening period, it would be possible to estimate the parameters of g(t) and the mean daily blood lead increment for different zones.

It is evident from Figure 2 that the average blood lead measurement v_s at time t_s is given by

$$v_s = v_f\, g(t_s - t_f) + \sum_{t_i = t_f + 1}^{t_i = t_s} m_i g(t_s - t_i) \qquad \text{Eq. (2),}$$

where v_f is the previous blood lead measurement at time t_f, and m_i is the mean blood lead increment for the location where the bird was on day t_i of the period intervening between t_f and t_s. Applying this model to real data on pairs of blood lead measurements from condors whose location was known during the intervening period, we used Eq.(2) to calculate E(v_s), the expected value of v_s , from the observed value v_f and provisional starting values of the parameters k_1, k_2 and m_i. The values of the m_i were assumed to be specific to each zone used by the condors and to be different between the hunting season and outside the hunting season. Hence, there were 10 values of m_i, two for each of the five zones. For each observation, we then calculated the log-likelihood of observing blood lead level v_s, given the expected value E(v_s), assuming that observed values were distributed log-normally around the expected mean value with variance s^2. We summed these log-likelihoods to give the total log-likelihood of the data under the model. We then used a simplex procedure to find the values of the parameters k_1, k_2, m_i. and s which maximized the log-likelihood. Because the logic underlying our model does not permit the mean daily blood lead increment to be negative, estimated values of m_i were constrained to exceed zero using log transformation. Some v_s values measured only with the field tester were known to exceed its upper limit of quantification, but the actual values were unknown. These were treated as right-censored observations exceeding the upper limit of quantification in the likelihood calculation (Kalbfleisch 1985). Confidence limits for parameter estimates were obtained by bootstrap sampling, with replacement, from the actual data, with individual condors as bootstrap sampling units. Sets of parameter estimates were obtained by the maximum-likelihood method described above for each of 1,000 bootstrap samples and the central 950 of the estimates of each parameter were used to define its 95% confidence interval.

Analysis of Mortality Caused by Lead Intoxication in Relation to Movements.—We wished to estimate the daily probability of a condor dying from lead poisoning as a function of its recent history of movement among zones in a way that would be likely to reflect its exposure to lead. It was not practical to model the daily probability of death from lead poisoning as a function of recent presence in all of the five zones within and outside the hunting season. This is because only eight deaths from lead poisoning were observed in 2005–2007 and, of these, two deaths occurred early in 2005 with inadequate data on their prior movement history, leaving only six birds with sufficient information. A model with many parameters cannot be supported by this small sample size. Therefore, we undertook the modeling of mortality in two steps. First, we used the zone- and season-specific estimates of mean daily blood lead increments m from the analysis described above and the observed movements of each bird to reconstruct expected values E(v) of blood lead concentration from Eq.(2) for every day on which each bird was free-ranging. Next, we took a weighted mean of the E(v) on days up to and including each focal day on which the bird was free-ranging. We considered it necessary to use in the model the weighted mean of E(v) on the focal day and a set of previous days, rather than just E(v) on the focal day itself, because it seemed biologically realistic for the probability of death to be determined by a bird's recent history of blood lead concentration. We used a weighting function, so that the weight for a given day t_i on or previous to the focal day t^* was $\exp(-(t^* - t_i)^2/q^2)$, where q is a constant. We used logistic regression to fit the relationship between the daily probability of death from lead poisoning and the weighted mean of E(v) on and prior to each focal day. We used a bisection search to determine the value of the parameter q of the weighting function that maximized the log-likelihood of the data. The data to which the model was fitted comprised all bird-days in 2005–2007 on which condors were free-ranging, including the two bird-days on which deaths from lead poisoning of free-living birds occurred and four bird-days on which birds were taken into captivity with high blood lead levels from which they subsequently died. We obtained confidence limits of the parameters of the logistic regression and of q by a bootstrap procedure. We drew a bootstrap sample, with replacement, from the data contributed by the 60 individual condors, with individual birds as bootstrap units. From this sample we estimated the m and k values as described above, reconstructed the $E(v)$, and then estimated q and the logistic regression parameters. We took 1,000 such bootstrap samples and took the central 950 estimates for each parameter as its 95% confidence interval.

Analysis of Movements Among Zones.—We wished to use information on movement patterns to simulate condor mortality from lead intoxication in the absence of intervention to remove condors from the wild and to treat those with high blood lead levels. Clearly, we could not use the observed movement pattern directly for this because runs of days in particular zones that would otherwise have occurred were disrupted by capture and removal from the wild. We therefore made a statistical model, based upon the observed sequences of movement among zones, to describe the probability that a condor present in a given zone on one day (the origin zone) would move to another specified zone (the destination zone) on the next day, rather than remaining in the origin zone or moving elsewhere. Inspection of the data on observed and interpolated roost locations for 2005–2007 indicated that there was an annual cycle in the use made of the different zones. Hence, we fitted a model in which the logit of the daily probability of a condor moving between two specified zones was a sinusoidal function of time of year z (the date expressed as a proportion of the calendar year). We also wished to allow the probability of movement to be free to vary systematically with time across the whole three year period. We therefore also included a quadratic function relating the logit probability of movement to the time, in years y elapsed since 31 December 2004. The expression used was

$$\text{logit(prob)} = f_0 + 0.5 f_1(1+\sin(2\pi(f_2 + z))) + f_3 y + f_4 y^2 \qquad \text{Eq.(3)},$$

where the f are constants. We estimated the parameters of this logistic regression model using a maximum-likelihood method.

Estimating Daily Blood Lead Concentrations.—We used a Monte Carlo process, together with the statistical model of movements among zones described above, to simulate movements of condors among zones for the three-year period 2005–2007. For each of 10,000 simulated condors, we generated a random number on each successive day. We used it and the probabilities for that day of movement from the origin zone to each of the four destination zones, obtained from Eq.(3), to determine which zone each bird moved to, or whether it remained within the origin zone. The origin zone on the first simulated day was selected by generating a random number and using the observed proportions of known locations of birds on 1 January 2005 to allocate the bird to a starting zone. We then ran the model for one year using 2005 dates as a burn-in procedure and then continued to run the model for a further three years, starting again from 1 January 2005; only this latter period being used further. We summarized the simulated movements by calculating the proportion of simulated condors in each zone on each day of the three-year period.

For each simulated condor, we also estimated its expected blood lead concentration on each day by allowing the lead present on the previous day to change according to the function $g(t)$, with maximum-likelihood estimates of its parameters, and by adding the mean daily blood lead increment expected for the zone and season, again using the maximum-likelihood estimates. We calculated the geometric mean and the variance of the simulated blood lead concentrations on each day. The variance was obtained by adding together the variance of the modeled blood lead levels calculated across simulated individuals and fitted value of the residual variance s^2 from the model of blood lead concentration in relation to zone (see above). Proportions of simulated condors in different categories of blood lead concentration on each day were then calculated from the geometric mean and variance. Categories were defined according to Franson (1996) and Fisher et al. (2006).

Estimating Mortality Caused by Lead Intoxication.—We also used the Monte Carlo model of condor movements and blood lead levels, described above, to estimate the mortality rate caused by lead intoxication. On each successive day of the simulated sequence, we calculated the weighted mean expected blood lead level, using values of E(v) for the focal day and previous days and the weighting function, with the maximum-likelihood value of q. We then used the maximum-likelihood values of the parameters of the logistic regression relating the daily probability of death to weighted blood lead level to calculate the expected probability of death. A random number between zero and one was generated and, if it was less than the expected value of the probability of death, the condor was simulated to have died from lead poisoning. The simulation for that bird was then terminated. The procedure was repeated for all 10,000 simulated condors. From the set of simulations, we calculated the proportion of the cohort of birds present on 1 January 2005 that had not yet died from lead poisoning on each successive day until the end of 2007. This gave a simulated survivorship curve, ignoring mortality from other causes. We also obtained the proportion of birds simulated as not having died from lead intoxication at the end of the three years. We raised this proportion to the power 1/3 and then subtracted the result from one to give the simulated average annual mortality rate from lead poisoning.

We ran the Monte Carlo simulations for each of the bootstrap sets of parameter values described in the analysis of mortality and took the central 95% of bootstrap estimates of annual mortality caused by lead poisoning to be its 95% confidence limits.

RESULTS

Relationship of Blood Lead Level to Time Since Ingestion.—We fitted the model summarized in Eq.(2) to the 322 pairs of blood lead measurements derived from 60 condors, as described above. We used both the one-parameter and two-parameter forms of the function g(t) that relates blood lead concentration to time since ingestion. The one-parameter version of g(t) gave a higher log-likelihood and we therefore selected it for reasons of parsimony. The maximum-likelihood value of the parameter k_2 was 0.0408 (95% confidence limits 0.0286–0.0581). This function describes an exponential decline in blood lead concentration with a half-life of 17.0 days (95% confidence limits 11.9–24.2 days).

Mean Daily Blood Lead Increment in Relation to Zone and Season.—The maximum-likelihood estimates of mean daily blood lead increment m from the fitted model summarized in Eq.(2), with the one-parameter version of g(t), are shown in Table 1. The fitted model performed well in accounting for variation in blood lead concentration. There was a high correlation between log-transformed observed and modeled blood lead values (r = 0.708) and deviations from the expected values were approximately uniform across the range of expected values (Figure 3).

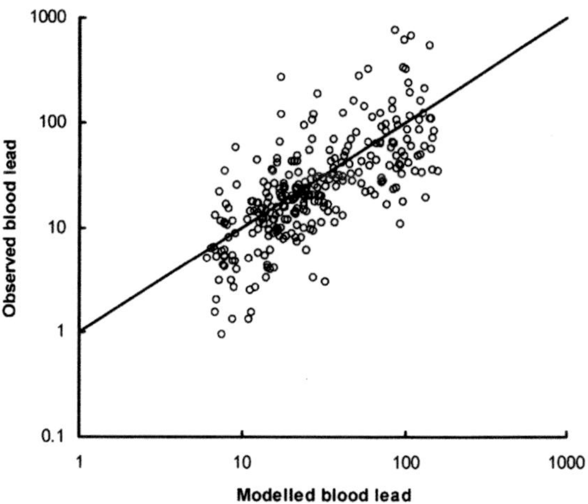

Figure 3. Relationship between observed and modeled blood lead concentration (µg dL^{-1}) for 296 measurements on free-living California Condors 2005-2007. The model is that described in the text with maximum-likelihood parameter estimates given in Table 1. Measurements that exceeded the limit of quantification of the field tester and were not duplicated by a laboratory measurement were excluded. The line shows the expected relationship if the observed and modeled measurements were identical.

Table 1. Maximum-likelihood estimates of mean daily increment of blood lead concentration (µg dL^{-1} d^{-1}) for free-ranging California Condors, with 95% bootstrap confidence limits, outside and within the hunting seasons of 2005–2007, for five zones. Also shown are the reported numbers of deer, elk and buffalo killed per day within and outside the deer and elk hunting seasons in 2005–2007, and the number of these per day (including additional wounding losses of deer and elk) left as lead contaminated carcasses or gut piles. Small numbers of buffalo killed in Kaibab in March–August were included in the non-hunting season total.

	Mean daily increment	95% Confidence limits			Reported kills per day	Lead contaminated remains per day
Outside hunting season						
Colorado River Corridor	1.21	0.76	-	1.93	0.00	0.00
Kaibab Plateau	0.79	0.40	-	1.56	0.04	0.02
Paria	0.27	0.16	-	0.46	0.00	0.00
South Zone	1.29	0.71	-	2.32	0.00	0.00
North Zone	1.26	0.77	-	2.05	0.00	0.00
Hunting season						
Colorado River Corridor	0.00	0.00	-	0.92	0.35	0.16
Kaibab Plateau	15.36	8.00	-	29.51	36.41	14.76
Paria	0.24	0.05	-	1.19	0.08	0.04
South Zone	2.13	0.00	-	8.27	0.00	0.00
North Zone	14.15	5.75	-	34.81	40.11	41.23

The estimates of mean daily blood lead increment were low (<3 µg dL^{-1} d^{-1}) for most zones and seasons, but strikingly higher (>14 µg dL^{-1} d^{-1}) in the Kaibab and North zones during the hunting season. The daily blood lead increment was much lower outside the hunting season than within it for both the Kaibab and North zones, with the difference between hunting and non-hunting seasons being smaller and inconsistent in direction for the other zones (Table 1).

The statistics collected on hunting showed that the average number of large game animals killed per day of the season was much higher during the hunting seasons in Kaibab and North zones than outside the hunting season in these zones, and also higher than for all other zones, both within and outside the hunting season (Table 1). The estimated number of lead-contaminated carcasses and gut piles produced per day of the season, after allowing for the effects of the program to encourage the use of copper bullets and to remove contaminated gut piles, was also higher in Kaibab and North zones during the hunting season than in other zones and seasons. However, the lead reduction program in Kaibab resulted in estimated lead exposure being lower during the hunting season in Kaibab than in North zone, despite the numbers of kills per day being similar in both zones.

Mortality Caused by Lead Intoxication in Relation to Blood Lead Concentration Reconstructed Using Movement History.—Although only six condors with sufficient data for analysis died from lead intoxication during our study period, there was a marked difference between the blood lead concentration history of these birds, reconstructed using information on their movements, and the equivalent results for birds that did not die from lead intoxication. Simulated blood lead concentration was higher on the day of death or final capture of the birds that died, and for a substantial period beforehand, compared with days upon which death from lead intoxication did not occur. This difference was greatest when a value of 125.3 days was chosen for the parameter q, which is used in calculating the weighted mean concentration over the days prior to the focal day (Figure 4). Other parameter values of the fitted logistic regression model that relates the daily probability of death to the weighted mean reconstructed blood lead concentration are given in Table 2.

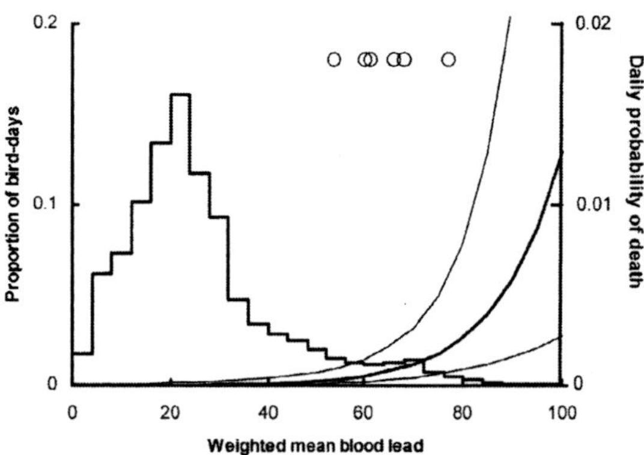

Figure 4. Daily mortality rate of California Condors from lead poisoning in relation to the weighted mean of reconstructed blood lead concentration (µg dL^{-1}) prior to the focal day. The histogram (left-hand scale) shows the distribution of weighted mean blood lead values on days when monitored condors did not die from lead poisoning. Circles show values on the day of death or last capture for the six birds that did die from lead poisoning. The thick curve shows the logistic regression model, fitted to these data, relating daily probability of death (right-hand scale) to the weighted mean of modeled blood lead concentration. Thin curves show 95% bootstrap confidence limits.

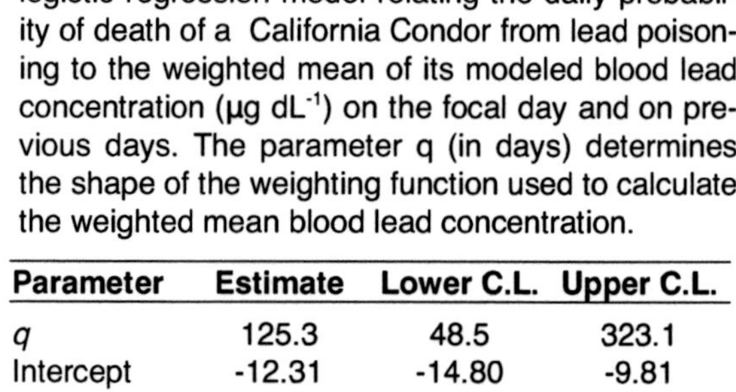

Table 2. Maximum-likelihood estimates, with 95% bootstrap confidence limits, for the parameters of a logistic regression model relating the daily probability of death of a California Condor from lead poisoning to the weighted mean of its modeled blood lead concentration (µg dL^{-1}) on the focal day and on previous days. The parameter q (in days) determines the shape of the weighting function used to calculate the weighted mean blood lead concentration.

Parameter	Estimate	Lower C.L.	Upper C.L.
q	125.3	48.5	323.1
Intercept	-12.31	-14.80	-9.81
Slope	0.07971	0.04460	0.1148

Movements Among Zones.—The parameter estimates of the logistic regression models relating the log-transformed probabilities of movements between each pair of origin and destination zones to year and time of year are shown in Table 3. The proportions of birds simulated as present in the five zones are shown in relation to time of year, with results for the three simulated years pooled, in Figure 5(b). The pattern of change through the year in the proportions of birds in each zone resembles that in the raw roost location data (Figure 5(a)). The Kaibab Plateau and North zones were most used from July to November. The Paria and Colorado River zones were most used from December to April. The South zone was most used from February to May.

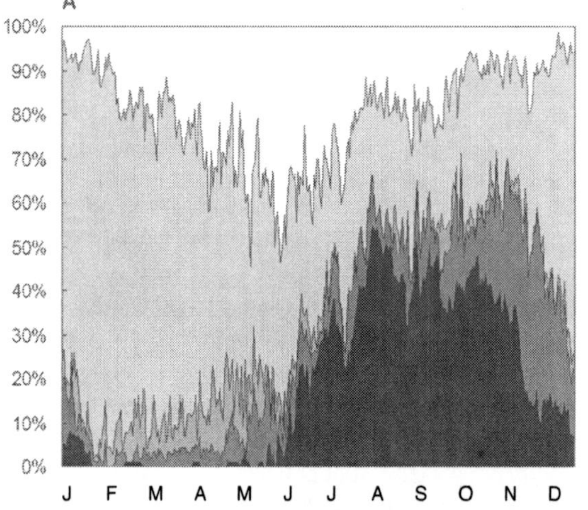

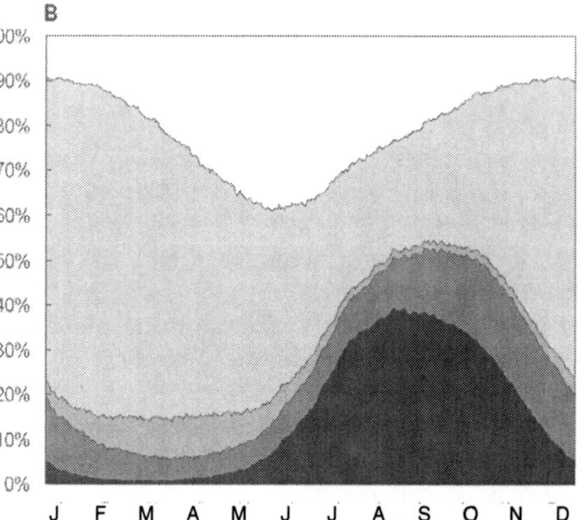

Figure 5. Movements of California Condors among zones. (A) Observed proportions of free-ranging condors in each of five zones (black = North Zone, dark gray = Kaibab Plateau, mid-gray = Colorado River Corridor, light gray = Paria, white = South Zone) in relation to time of year for observations pooled over the three-year period 2005-2007. (B) Modeled proportions of condors in the same zones from the Markov chain model described in the text.

Table 3. Fitted values of parameters of a Markov chain model (see text) relating the daily probability of a tagged California Condor moving from a zone in which it was initially present (origin) to a given other zone (destination) to season and year.

Origin zone	Destination zone	f_0	f_1	f_2	f_3	f_4
Colorado River	Kaibab Plateau	-2.250	-2.284	0.212	-2.714	0.986
Colorado River	Paria	-0.732	-0.905	0.408	1.356	-0.593
Colorado River	South Zone	-1.178	-3.254	0.237	-1.417	0.461
Colorado River	North Zone	-25.324	-6.807	0.146	22.813	-5.309
Kaibab Plateau	Colorado River	-2.267	-2.706	0.347	-3.693	1.291
Kaibab Plateau	Paria	-0.408	-1.060	0.377	-0.628	0.094
Kaibab Plateau	South Zone	-1.110	-1.719	0.221	-0.577	0.137
Kaibab Plateau	North Zone	-2.126	-2.674	0.177	-0.323	0.147
Paria	Colorado River	-3.348	-2.204	0.422	1.412	-0.478
Paria	Kaibab Plateau	-3.031	-1.882	0.118	0.747	-0.186
Paria	South Zone	-1.311	-2.966	0.286	-0.072	-0.071
Paria	North Zone	-4.107	-3.967	0.108	1.047	-0.168
South Zone	Colorado River	-5.322	-0.770	0.363	0.390	-0.109
South Zone	Kaibab Plateau	-4.269	1.721	0.463	0.857	-0.313
South Zone	Paria	-0.914	-1.288	0.351	0.131	-0.157
South Zone	North Zone	-4.215	-4.224	0.101	0.692	-0.032
North Zone	Colorado River	-15.185	-0.130	0.041	9.266	-2.042
North Zone	Kaibab Plateau	-4.278	2.494	0.049	-0.203	0.061
North Zone	Paria	-1.678	-1.676	0.487	-0.625	0.118
North Zone	South Zone	-4.763	-1.566	0.382	1.060	-0.190

Estimated Daily Blood Lead Concentration.—The reconstructions of blood lead concentrations for the period 2005–2007 showed large peaks in geometric mean concentrations in November-December of each year (Figure 6). Measured over the whole three-year period, blood lead concentrations exceeded the upper bound of the normal range on about half of the simulated condor-days. There was a peak in the proportion of birds simulated as having lethal lead levels in November of each year, which reached 5.1, 6.3 and 8.5% of condors in 2005, 2006 and 2007 respectively (Figure 7).

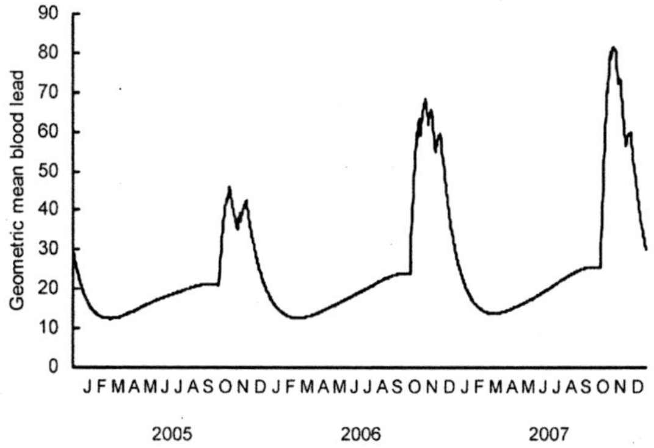

Figure 6. Modeled geometric mean blood lead concentration (µg dL^{-1}) of free-ranging California Condors for the three years 2005 to 2007.

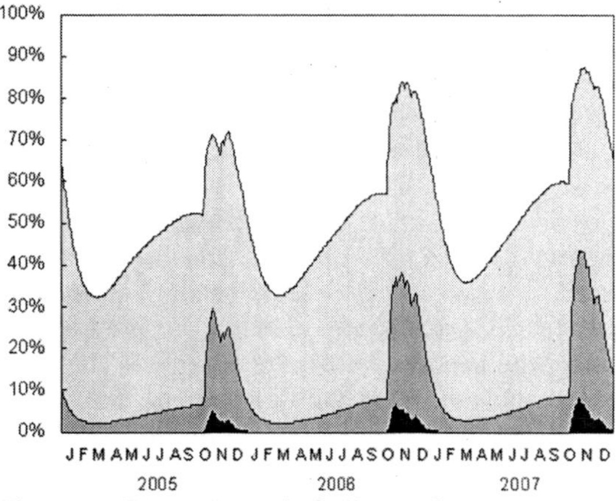

Figure 7. Proportions of California Condors with lethal (black, >500 µg dL^{-1}), toxic (dark gray, >100 µg dL^{-1}), subclinical (light gray, >20 µg dL^{-1}) and normal (white, <20 µg dL^{-1}) blood lead concentration, as reconstructed from Markov chain simulated movement patterns for the three-year period 2005–2007.

Table 4. Modeled proportions of California Condors dying from lead poisoning during the period 2005–2007, and equivalent annual death rates, under various scenarios for the reduction of exposure to spent lead ammunition. The scenarios represent the 2005–2007 pattern of exposure ("As now"), the recent pattern, but with the effects of the program to reduce lead exposure at Kaibab Plateau (Arizona–AZ) removed ("No reduction program"), and with the average proportion of reduction of lead exposure as achieved at Kaibab plateau in 2005–2007 applied to both Kaibab Plateau and North zones (Utah–UT)- this scenario is called "UT reduction as in AZ now").

Lead exposure scenario	Three-year rate	Annual rate	95% confidence limits of annual rate		
As now	0.1450	0.0509	0.0146	-	0.1519
No reduction program	0.5002	0.2064	0.0428	-	0.5029
UT reduction as in AZ now	0.0458	0.0155	0.0048	-	0.0479

Mortality Caused by Lead Intoxication in the Absence of Chelation Therapy.—The estimated proportion of a cohort of condors present on 1 January 2005 which had yet to die from lead intoxication showed three periods of rapid decline in November-January in each of the three years (Figure 8). Over the whole three-year period, the simulations showed that 14.5% of condors present at the start of the period had died from lead intoxication by the end of it. This is equivalent to an annual probability of dying from lead intoxication of 5.1% (Table 4).

Effect Upon Mortality Caused by Lead Intoxication of Ending the Program to Reduce Exposure to Lead or Increasing the Area Covered by the Program.—We simulated the effect upon mortality of the existing program to reduce the exposure of condors to lead by changing the parameter value for *m*, which specifies the mean daily blood lead increment for each zone and season. Over the three hunting seasons 2005–2007, it was estimated that the program to reduce exposure to lead in the Kaibab zone diminished the number of potentially lead-contaminated carcasses and gut piles remaining for scavengers by 63%, compared with what would otherwise have been present. Hence, we multiplied the estimated value of *m* for the Kaibab zone during the hunting season in Table 1 by a factor of 2.7 (= 1/1–0.63) to simulate what would have occurred without the lead reduction program. Simulations of mortality caused by lead in the absence of the Kaibab reduction program indicate a much higher mortality; a 50% rate in three years, which is equivalent to an annual mortality rate of 20.6% (Table 4).

We also simulated the effect of implementing a lead reduction program in the North zone (Utah) with similar effectiveness to the existing program in the Kaibab zone. To do this, we used the value of *m* for the Kaibab zone in the hunting season shown in Table 1 (i.e. the value obtained with the existing program in place), but we divided the value of *m* in Table 1 for the North zone in the hunting season by 2.7. Simulations of mortality caused by lead with lead reduction programs in both Kaibab and North zones indicate lower mortality; a 4.6% mortality rate in three years, which is equivalent to an annual rate of 1.6% (Table 4).

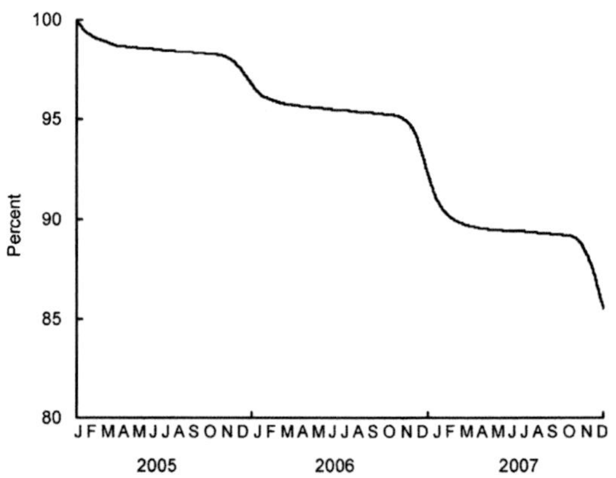

Figure 8. Modeled percentage of a hypothetical cohort of California Condors alive on 1 January 2005 which had not yet died from lead poisoning on each day of the ensuing three-year period. It was assumed that birds were only subject to mortality caused by spent lead ammunition under the current exposure scenario, and did not receive chelation therapy.

DISCUSSION

Our analysis of changes in blood lead concentration between release and recapture reveals a striking tendency for condors to run a high risk of acquiring elevated blood lead concentrations when they visited the Kaibab and North zones during the hunting seasons for deer and elk in 2005 - 2007. These two zones are those in which the largest numbers of these quarry species were hunted. A previous analysis (Hunt et al. 2007), using data for July 2001 to June 2005, also showed that condors that visited the Kaibab zone were more likely to acquire high blood lead levels upon recapture than those that did not. Our analysis differs in also finding a high risk when condors visited the North zone in Utah. The difference probably arises because our present analysis includes more recent data (2005–2007) and condors greatly increased their use of the North zone from 2004. Taken together with evidence of shotgun pellets and bullet fragments in recaptured condors with high blood lead levels (Parish et al. 2007) and observations of condors at carcasses of hunter-killed deer and elk (Hunt et al. 2007), our results indicate that ingestion of contaminated tissue from carcasses and gut piles of game animals killed with lead ammunition in the Kaibab and North zones is the largest source of lead for condors in the Grand Canyon population.

Changes in blood lead levels were affected by the location of the condors over a considerable period prior to sampling. The results were consistent with ingestion of lead being followed, in the absence of further ingestion, by a progressive exponential diminution in its concentration in the blood, with a half-life of 17 days. This pattern is broadly similar to that found in a previous study of captive condors with initially high levels of lead in the blood (Fry and Maurer 2003).

The daily death rate from lead intoxication was highest if blood lead levels, simulated using observed movements and the model of lead acquisition and depletion, remained high over a period exceeding one hundred days. Short periods with similarly high blood lead resulted in a lower death rate per day. We think it likely that this reflects a cumulative effect of protracted high blood lead levels on organ function. Condors that visited the Kaibab and North zones less frequently during the hunting season, and which therefore were not simulated as acquiring such high concentrations for such a long period, were much less likely to die from lead intoxication. The Kaibab and North zones were probably especially attractive to condors during the hunting season because of the large quantity of remains of hunted deer and elk available as food there, compared with other parts of the study area.

We used a previously published population model (Meretsky et al. 2000) to assess likely long-term trends in the numbers of condors in the absence of further releases and without chelation and other treatment of birds with elevated blood lead concentrations. According to this model, the condor population would tend to decline under present conditions unless natural adult mortality was at the lower end of the likely range or reproduction was at the "maximum conceivable" level (Table 5). Since the assumptions of the "maximum conceivable" scenario are extremely unlikely to apply to any real population of condors, this indicates that the Grand Canyon condor population is unlikely to be self-sustaining at current levels of exposure to lead.

Our simulations indicated a large effect of the existing program to reduce exposure of condors to lead from carcasses and gut piles of deer and elk killed using lead ammunition. At present, this program only covers the Kaibab zone. A simulation of the likely mortality rate due to lead if this program was not in place indicates a very high death rate. A previously published population model (Meretsky et al. 2000) indicates that, without releases and chelation therapy, the condor population would decline rapidly under these circumstances, regardless of whether the "most likely" or "maximum conceivable" levels of reproduction are assumed (Table 5). This implies that continuation of the existing lead reduction program in the Kaibab zone, which is conducted by the Arizona Game and Fish Department, is important for the future persistence of the condor population.

Simulation of the extension of the existing lead reduction program to cover the North zone (Utah), as well as the Kaibab Plateau, indicates that this would reduce mortality caused by lead intoxication substantially. With this level of mortality due to lead, the population model indicates that the condor

Table 5. Annual percentage rates of increase or decrease of a stable age structure model California Condor population under various scenarios for the reduction of exposure to spent lead ammunition. Details of the scenarios are given in the text and the legend to Table 4. For each scenario, the rate of increase is shown for three plausible annual survival rates for adults in the absence of mortality caused by lead (0.90, 0.95 and 0.99) and for the annual non-lead mortality rate of immatures in the absence of lead being equal to that of adults or twice that of adults. Deaths caused by lead poisoning are assumed to occur in addition to those from other causes. Results are shown for the "Most likely" and "Maximum conceivable" reproductive scenarios (Meretsky et al. 2000).

	Non-lead mortality: immature = adult			Non-lead mortality: immature = 2 x adult		
	0.90	0.95	0.99	0.90	0.95	0.99
Most likely						
No Pb mortality	-0.7	+4.8	+9.2	-4.8	+2.6	+8.8
As now	-5.8	-0.5	+3.7	-9.6	-2.6	+3.2
No reduction	-21.2	-16.8	-13.3	-24.4	-18.6	-13.7
UT reduction as in AZ now	-2.2	+3.2	+7.5	-6.2	+1.0	+7.1
Maximum conceivable						
No Pb mortality	+4.4	+10.2	+14.8	-0.8	+7.4	+14.2
As now	-0.9	+4.6	+9.0	-5.9	+2.0	+8.4
No reduction	-17.2	-12.6	-8.9	-21.3	-14.7	-9.3
UT reduction as in AZ now	+2.8	+8.5	+13.0	-2.4	+5.8	+12.5

population would increase unless reproduction was at the "most likely" level and adult mortality not caused by lead was at the worst-case (upper) end of the likely range (Table 5). Hence, we recommend that lead reduction programs are implemented as effectively as possible in both the Kaibab and North zones. Only then is a self-sustaining population of free-living California Condors likely to persist within the Grand Canyon area currently occupied without continued veterinary management and the regular addition of captive-bred birds.

Poisoning caused by ingestion of spent lead ammunition is a widespread hazard to many species of wild birds (Fisher et al. 2006), but few previous studies have estimated the effect on mortality rates of differential exposure to lead ammunition of free-ranging individuals (Tavecchia et al. 2001, Grand et al. 1998). Aside from their application to the problem of conserving California Condors, we believe that the methods developed for our study may be more widely applicable to other species with large home ranges, within which localized sources of contamination occur.

ACKNOWLEDGMENTS

The data were collected as part of The Peregrine Fund's California Condor restoration project, which is supported by the US Fish and Wildlife Service, Arizona Game and Fish Department, Bureau of Land Management, The Charles Engelhard Foundation, Liz Claiborne and Art Ortenberg Foundation, Nina Mason Pulliam Charitable Trust, Grand Canyon Conservation Fund, National Fish and Wildlife Foundation, Jane Smith Turner Foundation, and other important donors. We thank R. Watson, T. Cade, F. Sergio and an anonymous reviewer for advice and comments. R. Sieg and K. Sullivan (Arizona Game and Fish Department) and J. Parrish (Utah Department of Natural Resources) gave access to data on hunting and lead exposure reduction. E. Feltes, T. Lord, R. Benefield, E. Buechley, M. Dominguez, B. Fairchild, V. Frary, R. Gay, T. Hauck, M. Jenkins, J. Jones, M. Maglione, D. McGraw, F. Nebenburgh, K. Parmentier, M. Podolsky, S. Putz, J. Wiedmaier, E. Weis, J. Wilmarth, and S. Wolf collected field data.

AUTHOR CONTRIBUTIONS

Organized and carried out data collection: WGH, CNP. Conceived and designed the analyses: REG, WGH, IN, CNP. Analyzed the data: REG. Wrote the paper: REG, WGH, IN, CNP.

LITERATURE CITED

FISHER, I. J., D. J. PAIN, V. G. REYNOLDS. 2006. A review of lead poisoning from ammunition sources in terrestrial birds. Biological Conservation 131:421–432.

FRANSON, J. C. 1996. Interpretation of tissue lead residues in birds other than waterfowl. Pages 265–279 *in* W. N. Beyer, G. H. Heinz, A. W. Redmon-Norwood (Eds.). Environmental contaminants in wildlife: interpreting tissue concentrations. SETAC CRC Lewis Publishers, Boca Raton, Florida, USA.

FRY, D. M., J. R. MAURER. 2003. Assessment of lead contamination sources exposing California Condors. California Department of Fish and Game, Sacramento, California, USA.

GRAND, J. B., P. L. FLINT, M. R. PETERSEN, C. L. MORAN. 1998. Effect of lead poisoning on spectacled eider survival rates. Journal of Wildlife Management 62:1103–1109.

HUNT, W. G., C. N. PARISH, S. C. FARRY, T. G. LORD, R. SIEG. 2007. Movements of introduced condors in Arizona in relation to lead exposure. Pages 79–96 *in* A. Mee, L. S. Hall and J. Grantham (Eds.). California Condors in the 21st Century. Series in Ornithology, no. 2. American Ornithologists Union, Washington, DC, and Nuttall Ornithological Club, Cambridge, Massachusetts, USA.

KALBFLEISCH, J. G. 1985. Probability and Statistical Inference, Volume 2: Statistical Inference. Springer-Verlag, New York, USA.

MERETSKY, V. J., N. F. R. SNYDER, S. R. BEISSINGER, D. A. CLENDENEN, J. W. WILEY. 2000. Demography of the California Condor: Implications for Re-establishment. Conservation Biology 14:957–967.

PARISH, C. N., W. R. HEINRICH, AND W. G. HUNT. 2007. Lead exposure, diagnosis, and treatment in California Condors released in Arizona. Pages 97–108 *in* A. Mee, L. S. Hall and J. Grantham (Eds.). California Condors in the 21st Century. Series in Ornithology, no. 2. American Ornithologists Union, Washington, DC, and Nuttall Ornithological Club, Cambridge, Massachusetts, USA.

SULLIVAN, K., R. SIEG, C. PARISH. 2007. Arizona's efforts to reduce lead exposure in California Condors. Pages 109-121 *in* A. Mee, L. S. Hall and J. Grantham (Eds.). California Condors in the 21st Century. Series in Ornithology, no. 2. American Ornithologists Union, Washington, DC, and Nuttall Ornithological Club, Cambridge, Massachusetts, USA.

TAVECCHIA, G., R. PRADEL, J.-D. LEBRETON, A. R. JOHNSON, J.-Y. MONDAIN-MONVAL. 2001. The effect of lead exposure on survival of adult mallards in the Camargue, southern France. Journal of Applied Ecology 38:1197–1207.

BULLET FRAGMENTS IN DEER REMAINS: IMPLICATIONS FOR LEAD EXPOSURE IN SCAVENGERS

GRAINGER HUNT[1], WILLIAM BURNHAM[1], CHRIS PARISH[1], KURT BURNHAM[1], BRIAN MUTCH[1], AND J. LINDSAY OAKS[2]

[1]*The Peregrine Fund, 5668 West Flying Hawk Lane, Boise, ID 83709, USA.*
E-mail: grainger@peregrinefund.org

[2]*Department of Veterinary Microbiology and Pathology, Washington State University, Pullman, WA 99164-7040, USA.*

ABSTRACT.—Bullet fragments in rifle-killed deer carrion have been implicated as agents of lead intoxication and death in Bald Eagles (*Haliaeetus leucocephalus*), Golden Eagles (*Aquila chrysaetos*), California Condors (*Gymnogyps californianus*), and other avian scavengers. Deer offal piles are present and available to scavengers in autumn, and the degree of exposure depends upon incidence, abundance, and distribution of fragments per offal pile and carcass lost to wounding. In radiographs of selected portions of the remains of 38 deer supplied by cooperating, licensed hunters in 2002–2004, we found metal fragments broadly distributed along wound channels. Ninety-four percent of samples of deer killed with lead-based bullets contained fragments, and 90% of 20 offal piles showed fragments: 5 with 0–9 fragments, 5 with 10–100, 5 with 100–199, and 5 showing >200 fragments. In contrast, we counted a total of only six fragments in 4 whole deer killed with copper expanding bullets. These findings suggest a high potential for scavenger exposure to lead. *Reproduced with permission from the Wildlife Society Bulletin 34(1):167-170, 2006.*

HUNT, G., W. BURNHAM, C. PARISH, K. BURNHAM, B. MUTCH, AND J. L. OAKS. 2009. Bullet fragments in deer remains: Implications for lead exposure in scavengers. *In* R. T. Watson, M. Fuller, M. Pokras, and W. G. Hunt (Eds.). Ingestion of Lead from Spent Ammunition: Implications for Wildlife and Humans. The Peregrine Fund, Boise, Idaho, USA. DOI 10.4080/ilsa.2009.0123

Key words: Bullet fragmentation, lead, lead poisoning, raptors, scavengers.

AVIAN PREDATORS AND SCAVENGERS are susceptible to lead poisoning when they ingest pellets or fragments in the tissues of animals wounded or killed by lead-based bullets (Franson 1996, Locke and Thomas 1996, Wayland and Bollinger 1999). Toxic effects of ingested lead include neural degeneration, modification of kidney structure and bone, and inhibition of blood formation and nerve transmission (Eisler 1988). Shotgun pellets experimentally fed to five Bald Eagles (*Haliaeetus leucocephalus*) killed four of them, and severe clinical signs prompted euthanization of the fifth (Hoffman et al. 1981, Pattee et al. 1981). Residual weights of recovered pellets showed that the five eagles dissolved (mobilized) totals of 19, 38, 42, 129, and 184 mg of lead, each less in mass than a single #4 pellet of 209 mg.

Harmata and Restani (1995) found lead in the blood of 97% of 37 Bald Eagles and 85% of 86 Golden Eagles (*Aquila chrysaetos*) captured as spring migrants in Montana during 1985–1993. Pattee et al. (1990) reported that 36% of 162 free-ranging Golden Eagles captured during 1985–86 in southern California had been exposed to lead, and 9% had blood lead levels >0.6 ppm. Six of nine dead or moribund eagles (*Haliaeetus* spp.) in Japan died of lead poisoning; five had lead bullet fragments in their stomachs, and evidence implicated hunter-killed deer as the primary vector (Iwata et al. 2000). Lead ingestion was a principal cause of recorded death in wild California Condors (*Gymnogyps californianus*) prior to the mid-1980s when the population was brought into captivity (Wiemeyer et al. 1988), and in subsequently reintroduced, captive-bred condors tracked with radio-telemetry in Arizona (Cade et al. 2004). Kramer and Redig (1997) found a reduction in blood lead concentrations in Bald and Golden Eagles after a 1987 ban on lead shot for waterfowl hunting in Minnesota and Wisconsin; however, they found no change in prevalence of lead poisoning, a finding the authors attributed in part to offal piles from hunter-killed deer.

The availability of ungulate offal piles can be high in some regions. For example, the ten-year mean (1992–2001) of 676,739 White-tailed Deer (*Odocoileus virginianus*) annually harvested by rifle hunters in Wisconsin would have produced an average density of about five offal piles per km^2 for the area of the entire state (Dhuey 2004). An unknown number of additional whole carcasses lost to wounding is present in the landscape during and after hunting seasons, possibly on the order of 10% or more (Nixon et al. 2001). The extent to which avian scavengers encounter lead in deer carrion is therefore not so much a question of carrion availability, but rather one of lead incidence, abundance, and distribution per offal pile or carcass. Our examination of these three factors using radiographic data strengthens the body of evidence that deer killed by rifle bullets are a potentially important pathway of lead contamination to scavenger food webs.

METHODS

We obtained whole or partial remains of 38 deer (*Odocoileus virginianus* and *O. hemionus*) killed with standard, center-fire, breach-loading rifles by participating, licensed hunters engaged in normal hunting practices in Wyoming and California during 2002–2004. Thirty-four (89%) of the deer were killed by single shots to the thorax as determined by carcass examination and hunter interviews. The samples consisted of 15 offal piles discarded by hunters in the field, 10 deer carcasses in which tissues and viscera anterior to the diaphragm were left in place (abdominal viscera removed), four eviscerated carcasses, and nine whole deer carcasses; the latter were eviscerated on polyethylene sheets to sequester offal for radiography.

Hunters chose rifles, bullets, and bullet weights. Hunters used seven standard deer rifle calibers, and the mean weight of 37 bullets was 145 grains (SD = 18, range 100–180). Thirty-four were standard copper-jacketed, lead core bullets, and four were monolithic copper expanding "X-bullets." Seventeen of the former were of lead-tipped configuration (five brands), 12 were polymer-tipped (five brands), two were hollow points (one brand), and three were of unrecorded structure. Shot distances varied from 37 to >200 m (mean of 12 ranged distances = 158 m, SD = 77).

Local veterinarians radiographed areas of bullet transition of all carcasses and offal dorsoventrally and laterally; and adjusted exposures to maximize contrast (e.g., 56–70 kvp, 100 mAS, 0.3 sec). We placed a 2.5-cm grid transparency on selected radiographs, and using a hand (reading) lens for clarity, we counted all unambiguous metal fragments (opaque to radiation) in each cell and summed the counts. We verified the presence of metal particles in one sample by dissection. We estimated the width of the fragment arrays (excluding outliers) in five samples by extrapolation from the width of a 9-mm-diameter carbon-fiber tube inserted through the wound channel and aligned perpendicular to the x-ray beam. We did not attempt to distinguish between copper and lead in fragment counts. Copper, which is less frangible than lead, accounted for 30% of the mass in one standard (.308 caliber, 150-grain) hunting bullet we analyzed.

RESULTS

Most radiographs showed a profusion of small (<2 mm) metal fragments broadly distributed along wound channels. In deer killed by lead-based bullets, radiographs showed fragments in 18 of 20 offal piles (range = 2–521 fragments, mean = 160,

SD = 157). Five showed 0–9 fragments, five had 10–99, five had 100–199, and five showed >200 fragments. We counted 416–783 fragments (mean = 551, SD = 139) in the five whole deer carcasses (Figure 1), and 25–472 (mean = 213, SD = 172) in 10 carcasses containing thoracic organs but no abdominal viscera. Nine eviscerated carcasses showed fragments (range = 38–544, mean = 181, SD = = 153). Fragment clusters in five samples radiated as far as 15 cm from wound channels; the average of 30 measurements of the most far-reaching clusters in 11 radiographs was 7 cm (SD = 3). Magnification of one sample of excised tissue showed that fragments ranged in size from a few of >5 mm to tiny ones beyond the limit of unaided vision, estimated to be about 0.5-mm. Copper bullets resisted fragmentation: we counted a total of only six fragments in four (whole) deer killed with these bullets, and only one in the offal piles (Table 1).

DISCUSSION

The surprisingly high incidence of metal retention in carcasses as a result of fragmentation, and the density and distribution of fragments within them, suggest a high potential exposure of scavengers to lead. All whole or eviscerated deer killed with lead-based bullets contained fragments, 74% of them showing >100 visible fragments. The high proportion (90%) of offal piles containing fragments is not surprising, given that gut piles contain the thoracic organs normally targeted by hunters. The minuteness of many fragments may explain why lead is often unseen in radiographs of lead-poisoned birds (Kramer and Redig 1977); small fragments may be overlooked or completely digested. Ingestion of very small particles of lead would explain the accumulation of sublethal levels in the blood of Golden Eagles during the hunting season (Wayland and Bollinger 1999).

Experiments on projectile toxicity have focused on shotgun pellets (Hoffman et al. 1981), and extrapolations from those experiments may underestimate the effects of rifle bullet fragments under natural conditions. The toxicity of ingested bullet fragments that are irregularly shaped must be greater than those of shotgun pellets of comparable mass because pellets, being spherical, have less surface area exposed to stomach acids. Moreover, the apparent high densities and small sizes of bullet particles likely contribute to their multiple ingestion by individual scavengers, and surface area within an aggregate of ingested particles would be greater than that of spherical pellets or intact bullets of comparable mass.

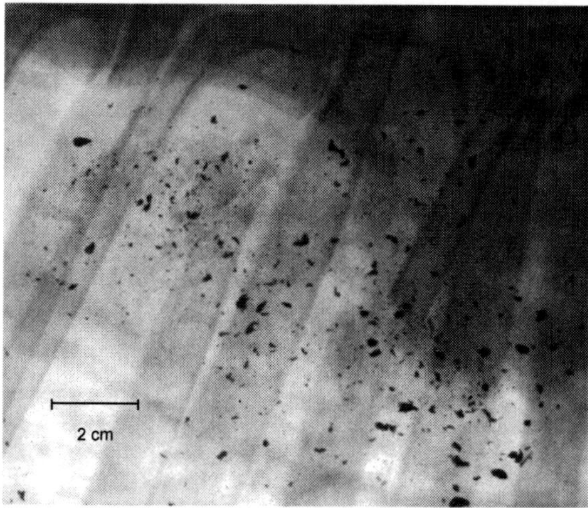

Figure 1. Lateral-view radiograph of the mid-thorax of an adult female White-tailed Deer killed by a standard copper-jacketed, lead-core, soft-point hunting bullet in northern Wyoming in 2004. The fragment array surrounding the bullet path was approximately 12 cm in diameter, excluding outliers.

Table 1. Metal fragments counted in radiographs of the remains of nine deer in which whole carcasses were available for study. Lead-based bullets (one brand) were of standard, copper-jacketed, soft point (lead-tipped) configuration. Expanding copper bullets (one brand) contained no lead. The deer were killed in northern Wyoming in fall 2004.

	Offal Pile	***Eviscerated Carcass***
Lead-based bullets	218	236
	450	214
	521	95
	67	224
	161	544
Copper bullets	0	0
	0	2
	0	0
	1	2

Based on these findings, we recommend further study on the frangibility of the various kinds of rifle bullets and continued use of carcass radiography to determine the incidence of bullet fragments in carcasses. Meanwhile, deer gut piles left in the field and whole deer carcasses lost to wounding should be considered as potentially poisonous to scavengers. Evidence of the perniciousness of ingested lead (Pattee et al. 1981) as manifested, for example, in the minute absorbed amounts that killed eagles under experimental conditions, give incentive for change to safer alternatives. Copper is less toxic than lead and less frangible. In reviewing a wide range of ballistics literature, we have encountered numerous test results and accounts supporting the efficacy of copper expanding bullets in hunting (see McMurchy 2002).

ACKNOWLEDGMENTS

We thank P. Widener, R. Berry, P. Pelissier, M. Murray, and the Mountain View Veterinary Hospital, Sheridan, Wyoming for special help with this project. Hunters providing deer for this study included P. Jenny, B. Widener, L. Widener, P. Widener, P. Hunt, and three of the authors. Additional assistance was provided by A. Brubaker, T. Cade, K. Evans, J. Fry, M. Gilbert, W. Heinrich, R. Jackman, S. Marrug, A. Matz, R. Mutch, B. Oakleaf, T. Hunt, R. Tucker (Washington State University), the Arizona Game and Fish Department, the Wyoming Game and Fish Department, and The Peregrine Fund Research Library. We thank M. Goldstein, R. Risebrough, V. Thomas, and two anonymous reviewers for helpful comments on the manuscript. Financial support was provided by The Peregrine Fund.

LITERATURE CITED

CADE, T. J., S. A. H. OSBORN, W. G. HUNT, AND C. P. WOODS. 2004. Commentary on released California Condors in Arizona. Pages 11–25 *in* R. D. Chancellor and B. U. Meyburg Eds.). Raptors Worldwide. World Working Group on Birds of Prey and Owls, Berlin and MME/Birdlife Hungary, Budapest, Hungary.

DHUEY, B. 2004. Wisconsin Big Game Hunting Summary. Wisconsin Department of Natural Resources Report Pub-WM-284 2004.

EISLER, R. 1988. Lead hazards to fish, wildlife, and invertebrates: a synoptic review. United States Fish and Wildlife Service, Biological Report 85 (1.14), Patuxent Wildlife Research Center, Laurel, Maryland, USA.

FRANSON, J. C. 1996. Interpretation of tissue lead residues in birds other than waterfowl. Pages 265–279 *in* W. N. Beyer, G. H. Heinz, and A. W. Redmon-Norwood (Eds.). Environmental Contaminants in Wildlife: Interpreting Tissue Concentrations. CRC Press, Boca Raton, Florida, USA.

HARMATA, A. R., AND M. RESTANI. 1995. Environmental contaminants and cholinesterase in blood of vernal migrant bald and golden eagles in Montana. Intermountain Journal of Sciences 1:1–15.

HOFFMAN, D. J., O. H. PATTEE, S. N. WIEMEYER, AND B. MULHERN. 1981. Effects of lead shot ingestion on g-aminolevulinic acid dehyratase activity, hemoglobin concentration, and serum chemistry in Bald Eagles. Journal of Wildlife Distribution 17: 423–431.

IWATA, H., M. WATANABE, E. Y. KIM, R. GOTOH, G. YASUNAGA, S. TANABE, Y. MASUDA, AND S. FUJITA. 2000. Contamination by chlorinated hydrocarbons and lead in Steller's Sea Eagle and White-tailed Sea Eagle from Hokkaido, Japan. Pages 91–106 *in* M. Ueta and M. J. McGrady (Eds.). First symposium on Steller's and White-tailed Sea Eagles in East Asia. Wild Bird Society of Japan, Tokyo, Japan.

KENDALL, R. J., T. E. LACHER, JR., C. BUNCK, B. DANIEL, C. DRIVER, C. E. GRUE, F. LEIGHTON, W. STANSLEY, P. G. WATANABE, AND M. WHITWORTH. 1996. An ecological risk assessment of lead shot exposure in non-waterfowl avian species: upland game birds and raptors. Environmental Toxicology and Chemistry 15:4-20.

KRAMER, J. L., AND P. T. REDIG. 1997. Sixteen years of lead poisoning in eagles, 1980-95: An epizootiologic view. Journal of Raptor Research 31:327–332.

LOCKE, L. N., AND N. J. THOMAS. 1996. Lead poisoning of waterfowl and raptors. Pages 108–117 *in* A. Fairbrother, L. N. Locke, and G. L. Huff (Eds.). Noninfectious Diseases of Wildlife 2nd ed. Iowa State University Press, Ames, Iowa, USA.

MCMURCHY, I. 2003. Barnes XLC bullets. American Hunter 31(1):70–71.

NIXON, C. M., L. P. HANSEN, P. A. BREWER, J. E. CHELSVIG, T. L. ESKER, D. ETTER, J. B. SULLIVAN, R. G. KOERKENMEIER, AND P. C. MANKIN. 2001. Survival of White-tailed Deer in intensively farmed areas of Illinois. Canadian Journal of Zoology 79:581–588.

PATTEE, O. H., P. H. BLOOM, J. M. SCOTT AND M. R. SMITH. 1990. Lead hazards within the range of the California Condor. Condor 92:931–937.

PATTEE, O. H., S. N. WIEMEYER, B. M. MULHERN, L. SILEO, AND J. W. CARPENTER. 1981. Experimental lead-shot poisoning in Bald Eagles. Journal of Wildlife Management 45:806–810.

WAYLAND, M. AND T. BOLLINGER. 1999. Lead exposure and poisoning in Bald eagles and Golden Eagles in the Canadian prairie provinces. Environmental Pollution 104:341–350.

WIEMEYER, S. N., J. M. SCOTT, M. P. ANDERSON, P. H. BLOOM AND C. J. STAFFORD. 1988. Environmental contaminants in California Condors. Journal of Wildlife Management 52:238–247.

LEAD EXPOSURE AMONG A REINTRODUCED POPULATION OF CALIFORNIA CONDORS IN NORTHERN ARIZONA AND SOUTHERN UTAH

CHRISTOPHER N. PARISH[1], W. GRAINGER HUNT[1], EDWARD FELTES[1], RON SIEG[2], AND KATHY ORR[3]

[1]*The Peregrine Fund, 5668 West Flying Hawk Lane, Boise, ID 83709, USA.*
E-mail: cparish@peregrinefund.org

[2]*Arizona Game and Fish Department, 3500 South Lake Mary Road, Flagstaff, AZ 86004, USA.*

[3]*The Phoenix Zoo, 455 North Galvin Parkway, Phoenix, AZ 85008, USA.*

ABSTRACT.—Lead poisoning remains the leading cause of death among free-ranging California Condors released by The Peregrine Fund in Arizona from 1996 to 2007 in an ongoing effort to establish a self-sustaining population. Daily monitoring of radio-tagged condors by means of VHF and GPS telemetry shows them ranging from the Grand Canyon National Park to the Zion region of southern Utah. Increased proficiency of condors at finding carrion in the wild corresponds with a greater incidence of lead exposure. Periodic testing reveals spikes in blood lead levels during November and December commensurate with the deer hunting seasons and condor movement to deer hunting areas. These data combined with information collected on food types supports the hypothesis that lead ammunition residues in rifle- and shotgun-killed animals are the principle source of lead contamination among these scavengers in northern Arizona and southern Utah. Sustaining the population requires an intensive management regime of testing and treatment for lead exposure. Reducing or eliminating the availability of lead is essential to reestablishment of condors in the wild. *Received 15 September 2008, accepted 31 October 2008.*

PARISH, C. N., W. G. HUNT, E. FELTES, R. SIEG, AND K. ORR. 2009. Lead exposure among a reintroduced population of California Condors in northern Arizona and southern Utah. *In* R. T. Watson, M. Fuller, M. Pokras, and W. G. Hunt (Eds.). Ingestion of Lead from Spent Ammunition: Implications for Wildlife and Humans. The Peregrine Fund, Boise, Idaho, USA. DOI 10.4080/ilsa.2009.0217

Key words: Bullet fragmentation, California Condor, chelation, lead exposure, lead poisoning, mortality.

THE CALIFORNIA CONDOR (*Gymnogyps californianus*) was among the first species listed on the Endangered Species Act of 1973 when only about 60 individuals remained in the wild (Snyder and Snyder 2000). The fossil record shows that condors once ranged throughout most of the southern United States, from California to Florida, but a drastic range reduction appears to have coincided with the extinction of Pleistocene megafauna about 10,000 years ago (Emslie 1987, Snyder and Snyder 2000). Evidence suggests that condors thereafter persisted along the Pacific Coast by scavenging on the remains of marine mammals and fish (Fox-Dobbs 2006), and then returned to the interior southwest in the 1700s with the introduction of livestock as a food base (Emslie 1987). But even with this probable increase in carrion availability, the species' naturally slow reproductive rate proved insufficient in offsetting human-related mortality, and the population declined. Mortality agents such as egg collecting, shooting, and poisoning, both intentional and inadvertent, were doubtless contributing factors (Koford 1953). In general, however, the difficulty of recovering dead condors for necropsy failed to reveal all the mortality agents involved and their relative contributions to the decline of condors.

Intensive monitoring by radiotelemetry in the 1980s marked the beginning of understanding, but by that time the population had dwindled to fewer than 30 birds. As extinction grew imminent, all remaining condors were removed from the wild to form a captive population. Fourteen founders among them eventually gave rise to captive-bred young in numbers sufficient to begin the reintroduction phase of a recovery effort (Grantham 2007). Condors were first reintroduced to their recent historic range in California in 1992 after a pilot experiment with Andean condors. Soon thereafter, in accordance with the federal recovery plan, efforts turned towards establishing a second, disjunct population. The U.S. Fish and Wildlife Service invited The Peregrine Fund to establish a captive breeding facility in Boise, Idaho, followed by a release program in northern Arizona. Releases began there in 1996 with an "experimental non-essential" design under Section 10(j) of the Endangered Species Act.

The area chosen for release and now occupied by free-ranging condors stretches from the Grand Canyon in northern Arizona north into the canyon lands and forests of southern Utah (Figure 1). Abrupt altitudinal differentiation, from deep red rock canyons (*ca.* 865 m msl) to uplifted mesas (3055 m msl) with immense cliffs, supports a variety of vegetational associations. Wind deflection and thermals create abundant updrafts upon which condors travel easily and rapidly. The condor's primary food consists of carcasses and partial remains of Mule Deer (*Odocoileus hemionus*), domestic cattle (*Bos taurus*), domestic sheep (*Ovis aries*), and Elk (*Cervus elaphus*). Supplementary food in the form of dairy calf carcasses is made available every three or four days at the Vermilion Cliffs Release Site (Figure 1) to which the condors periodically return.

Each condor wears a numbered patagial tag and one or two radio transmitters, either standard VHF for ground tracking and/or satellite-reporting with GPS capability. A team of biologists track the condor

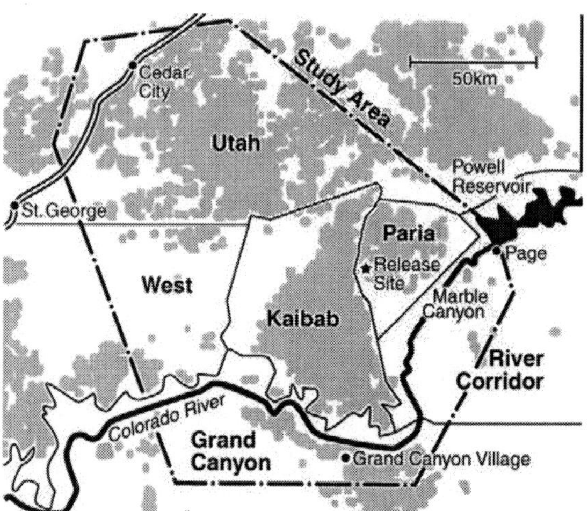

Figure 1. Study area as defined by condor movements in Arizona and Utah.

population on a daily basis with the goal of monitoring the movements and activity of as many individuals as possible. These locational data guide day-to-day management decisions and are used retrospectively to interpret behavior, foraging ecology, nesting activity, and potential encounters with contaminants. The current overall range (33,843 km^2) is divided into six zones based on movement patterns apparent during the first eight years of the program (Figure 1). Thus far, all condors return to the release site three or more times during the year, although the frequency of such visits for some individuals has diminished as they matured. Virtually all condors have been present at or near the release site during the coldest times of the year. Release site visits offer the opportunity to capture condors for examination, transmitter refitting, and blood sampling.

As of spring 2008, there have been 102 condors released, 40 fatalities, six birds returned to captivity, and nine wild young produced by reintroduced condors, leaving 64 condors in the wilds of southern Utah and northern Arizona. Details about the Arizona reintroduction program are given by Cade et al. 2004, Woods et al. 2007, Hunt et al. 2007, Parish et al. 2007, Sullivan et al. 2007, Osborne 2007, and Cade 2007. The present report focuses on testing, treatment, and mortality with emphasis on lead poisoning occurring since June 2005, the extent of our last published update (Parish et al. 2007).

Table 1. Information on population, reproduction, mortality, lead exposure, and treatment of California Condors in northern Arizona from 2000 through 2007.

Year	No. in Wild	No. Exposed to Lead	No. Tested for Pb	No. Treated for Pb	Blood-lead tests >15µg/dL	Blood-lead tests >65µg/dL	Deaths*	No. Birds of Breeding Age	No. Wild Young Fledged
2000	28	17 (61%)	25	9 (32%)	18	15	10(3)	7	0
2001	25	12 (48%)	25	1 (4%)	12	0	0	9	0
2002	31	23 (74%)	31	13 (42%)	29	11	4(1)	11	0
2003	40	30 (75%)	40	7 (18%)	43	7	1	14	0
2004	43	35 (81%)	43	18 (42%)	56	15	1	16	2
2005	56	29 (52%)	56	11 (20%)	40	8	6 (2)	22	2
2006	57	54 (95%)	57	40 (70%)	86	37	6 (3)	33	0
2007	61	50 (82%)	59	25 (41%)	52	18	4	40	2

*confirmed lead deaths in parentheses

LEAD POISONING

During the early years of the Arizona project, most fatalities resulted from predation and other mishaps associated with the inexperience of newly released condors (Woods et al. 2007). As the numbers of such fatalities diminished with the development of a wild flock and the application of adaptive management, lead poisoning emerged as the primary mortality factor. We first became aware of it in spring 2000 when 12 or more condors ingested shotgun pellets from an unknown source (Woods et al. 2007). This episode, which resulted in several deaths and emergency treatments, gave rise to a regular monitoring program of blood lead levels of condors periodically recaptured when they returned to the release site (Parish et al. 2007).

Condors prior to release and those subsequently feeding only upon proffered calf carcasses showed blood lead levels no higher than 12 µg/dL. Response to condors showing >30 µg/dL on a portable field analyzer (which underestimated laboratory values; see Parish et al. 2007, Green et al. 2009, this volume, Bedrosian et al. 2009, this volume) normally consisted of holding and retesting after a few days to determine if lead levels were increasing or decreasing. If increasing, or in cases of high exposure (>60 µg/dL), we administered chelation therapy, and where indicated we radiographed the condor to determine if radiodense objects existed in the digestive tract. If so, and they failed to soon pass from the stomach into the intestine, we administered phsyllium fiber or, when necessary, performed surgery to remove the object(s), the analysis of which invariably confirmed the diagnosis of lead poisoning (Sullivan et al. 2007, Chesley et al. 2009, this volume). We continued holding lead-exposed birds in captivity until a significant decrease in lead levels was apparent, at which point the birds were released (Parish et al. 2007).

These measures do not account for possible sublethal effects of lead on condors from repeated exposures, even low level exposures. In other organisms, lead is known to accumulate in soft tissue and bone where it is undetectable by blood testing (Mautino 1997). The result of that accumulation may conceivably impact adult fertility and reproduction, neural development of young, and other processes (see Gangoso et al. 2009, this volume).

We observed an abrupt increase in lead exposure in the fall of 2002, and the levels have been high each fall thereafter (Table 1). In examining the movements of condors in the weeks prior to the detection of lead exposure, we found that most had frequented the Kaibab Plateau, a popular deer hunting area nearby (Hunt et al. 2007). We hypothesized that the source might be bullet fragments in deer remains, so we studied the extent to which fragments are retained in deer gut piles and in deer lost to wounding. Hunt et al. (2006) reported that the gut piles of 18 of 20 deer killed with standard lead-based bullets contained bullet fragments, 10 showing over 100 fragments. Five whole deer carcasses contained 416–783 fragments. Other evidence, including observations of condors associated with deer remains, spikes in condor lead levels during and just after the Kaibab deer seasons (November and December), and continued correspondence between lead levels and condor occurrence on the Kaibab Plateau all supported the hypothesis of hunter-killed deer as the primary source of exposure (Hunt et al. 2007). We documented rifle-killed coyote carcasses as an additional source, but were unable to determine the source of shotgun pellet ingestion which recurs occasionally.

Several variables likely contributed to yearly differences in overall rates of exposure and treatment apparent in Table 1. The relatively low number of exposed condors in 2003 probably resulted from early capture and retention of condors at the release site during the hunting season. The timing and distribution of snowfall relative to hunting seasons concentrated deer, hunters, and condors in some years, presumably increasing the probability of finding and feeding on the remains of shot animals.

The relationships apparent in the 2002–2004 analyses (Parish et al. 2007, Hunt et al. 2007) showing fall spikes in lead exposure (Figure 2) and their association with deer hunting areas have continued to the present (Figure 3; see Green et al. 2009, this volume). From 2002, increasing numbers of condors have summered in the Zion region of southern Utah and have remained there through much of the fall. The environment in this higher, somewhat wetter zone differs from the Arizona portion of the range in having a greater area of private ranches and abundant domestic livestock which extend to public land as well. Seasonal herds of sheep and cattle in the Zion region form a more plentiful and regular food base for condors until the livestock are removed in fall with the arrival of cold weather. Deer and Elk provide a continued source of carrion, especially with the fall hunting seasons. Condors depart the area with the arrival of snow cover and the consequent loss of food accessibility. Condors leaving Utah usually travel to the Kaibab Plateau and ultimately to the release site where proffered carcasses are continuously available.

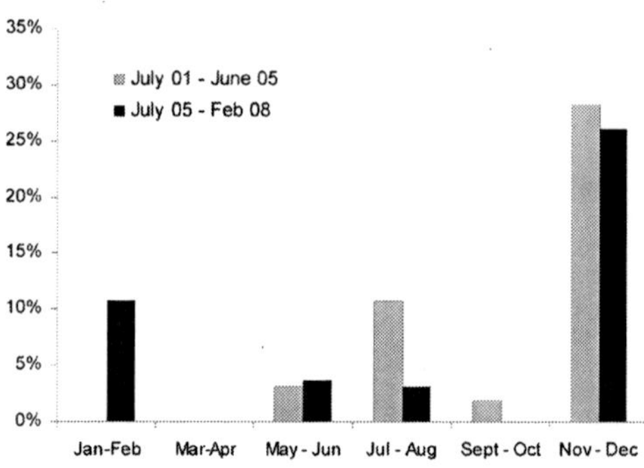

Figure 2. Annual pattern of toxic lead levels (>60 μg/dL) in condor blood from July 2001 through February 2008 in Arizona. The period July 01–June 05 was published in Parish et al. 2007. The period July 05–Feb 08 has not previously been published.

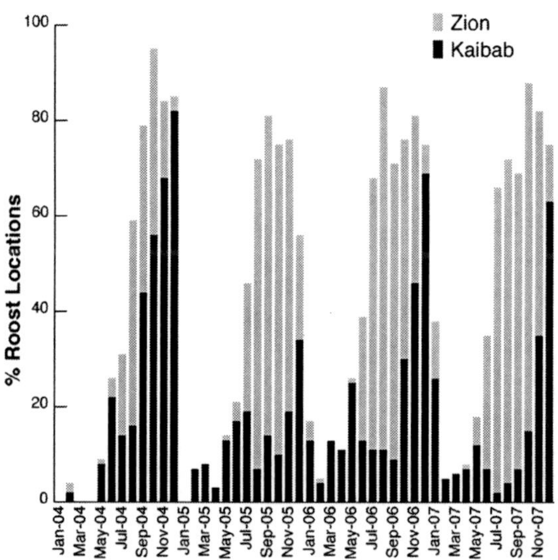

Figure 3. Percent of condor roost locations spent in the Zion and Kaibab zones.

DISCUSSION

Our findings suggest that the Kaibab Plateau and especially the Zion area are sufficient to sustain condors without reliance on food subsidies except in periods of heavy snow cover. This capability is enhanced by the ability of condors to forego eating for days at a time as they wait for favorable updrafts, and then range several hundred kilometers in a day to exploit seasonally changing food sources. One such opportunity emerges with the ungulate hunting seasons in the fall, offering sudden abundance of carrion at a time when the accessibility of food is declining with the onset of winter. The contribution of hunters in providing food during this period produces an important link in the annual cycle of food availability.

It appears on the basis of much published evidence that the occurrence of lead in gun-killed animal remains is the principal impediment to the establishment of a self-sustaining population in Arizona and Utah. The Arizona Game and Fish Department has made an effort, unprecedented in wildlife conservation, by implementing a hunter-education campaign and providing free, non-lead ammunition for hunters on the Kaibab and Paria plateaus beginning in 2005 (Sieg et al. 2009, this volume). Fortunately for condors, test reports of non-lead bullets available on today's market rate them comparable and even superior to their lead-based counterparts (see Jamison 2005). Response by hunters to Arizona's non-lead bullet program has been overwhelmingly favorable, with >80% participation in reducing the availability of ammunition lead to condors and other scavengers in fall 2007. Although lead exposure continues, its associated severity in 2007 was low, and no lead-related deaths occurred. The Utah Division of Wildlife is currently advancing a similar initiative in southern Utah, scheduled to begin in the fall 2009. According to Green et al. (2009, this volume), Utah's participation, if comparable to that in Arizona, will substantially strengthen the demography of condors and take them one step closer to establishment in the region. Meanwhile, funding permitting, The Peregrine Fund will continue monitoring condor movements and blood lead levels, and administering treatment when necessary.

ACKNOWLEDGMENTS

Funding for this project was provided by The Peregrine Fund, the U.S. Fish and Wildlife Service, the Arizona Game and Fish Department, and many other important donors. We would like to acknowledge our cooperators, and members of The Southwest Condor Working Group. We especially thank The Phoenix Zoo, Arizona Game and Fish Department, the National Park Service, and Liberty Wildlife for their help with transportation and medical attention to sick and injured condors. We thank our current team of dedicated biologists and those who have contributed in the past. Of special mention we thank Bill Burnham, Tom Cade, Bill Heinrich, Rick Watson, Lloyd Kiff, Travis Rosenberry, Kathy Sullivan, Tim Hauck, Jim Wilmarth, Eric Weis, Evan Beauchley, Mathew Podolsky, Maria Dominguez, Shaun Putz, Neil Paprocki, Marti Jenkins, Rob Gay, Thom Lord, Frank Nebenberg, Jonna Weidmaier, Sophie Osborn, Jesse Grantham, Bryan Bedrosian, and Maggie Sacher.

LITERATURE CITED

CADE, T. J. 2007. Exposure of California Condors to lead from spent ammunition. Journal of Wildlife Management 71(7): 2125–2133.

CADE, T. J., S. A. H. OSBORN, W. G. HUNT, AND C. WOODS. 2004. Commentary on released California Condors *Gymnogyps californianus* in Arizona. Pages 11–25 *in* R. D. Chancellor and B.-U. Meyburg (Eds.), Raptors Worldwide. Proceedings of the 6th World Conference on Birds of Prey and Owls. WWGBP/MME-Birdlife, Hungary.

CHESLEY, J., P. REINTHAL, C. N. PARISH, K. SULLIVAN, AND R. SIEG. 2009. Direct evidence for the source of lead contamination within the California Condor. Abstract *in* R. T. Watson, M. Fuller, M. Pokras, and W. G. Hunt (Eds.). Ingestion of Lead from Spent Ammunition: Implications for Wildlife and Humans. The Peregrine Fund, Boise, Idaho, USA. DOI 10.4080/ilsa.2009. 0219

EMSLIE, S. D. 1987. Age and diet of fossil California Condors in Grand Canyon, Arizona. Science 237:768–770.

FOX-DOBBS, K., T. A. STIDHAM, G. J BOWEN, AND S. D. EMSLIE. 2006. Dietary controls on extinction versus survival among avian megafauna in the late Pleistocene. Geology 34(8):685–688.

GANGOSO, L., P. ÁLVAREZ-LLORET, A. RODRÍGUEZ-NAVARRO, R. MATEO, F. HIRALDO, AND J. ANTONIO DONÁZAR. 2009. Long-Term Effects of Lead Poisoning on Bone Mineralization in Egyptian Vulture *Neophron percnopterus*. Abstract *in* R.T. Watson, M. Fuller, M. Pokras, and W.G. Hunt, (Eds.). Ingestion of Lead from Spent Ammunition: Implications for Wildlife and Humans. The Peregrine Fund, Boise, Idaho, USA. DOI 10.4080/ilsa.2009.0214

GRANTHAM, J. 2007. Reintroduction of California Condors into their historic range: The recovery program in California. Pages 123–138 *in* A. Mee, L. S. Hall and J. Grantham (Eds.). California Condors in the 21st Century. Series in Ornithology, no. 2. American Ornithologists Union, Washington, DC, and Nuttall Ornithological Club, Cambridge, Massachusetts, USA.

GREEN, R. E., W. G. HUNT, C. N. PARISH, AND I. NEWTON. 2009. Effectiveness of action to reduce exposure of free-ranging California Condors in Arizona and Utah to lead from spent ammunition. Reproduced *in* R. T. Watson, M. Fuller, M. Pokras, and W. G. Hunt (Eds.). Ingestion of Lead from Spent Ammunition: Implications for Wildlife and Humans. The Peregrine Fund, Boise, Idaho, USA. DOI 10.4080/ilsa.2009.0218

HUNT, W. G., W. BURNHAM, C. N. PARISH, K. BURNHAM, B. MUTCH, AND J. L. OAKS. 2006. Bullet fragments in deer remains: implications for lead exposure in scavengers. Wildlife Society Bulletin 34:168–171.

HUNT, W. G., C. N. PARISH, S. C. FARRY, R. SEIG, AND T. G. LORD. 2007. Movements of introduced California Condors in Arizona in relation to lead exposure. Pages 79–98 *in* A. Mee, L. S. Hall and J. Grantham (Eds.). California Condors in the 21st Century. Series in Ornithology, no. 2. American Ornithologists Union, Washington, DC, and Nuttall Ornithological Club, Cambridge, Massachusetts, USA.

JAMISON, R. 2005. Premium bullet shootout: a look at ten bullets and what they can do for you. Petersen's Hunting (August): 58–61.

KOFORD, C. B. 1953. The California Condor. National Audubon Research Report No. 4:1–154.

MAUTINO, M. 1997. Lead and zinc intoxication in zoological medicine: a review. Journal of Zoo and Wildlife Medicine 28(1):28–35.

OSBORN, S. A. H. 2007. Condors in Canyon Country. Grand Canyon Association, Grand Canyon, Arizona, USA.

PARISH, C. N., W. R. HEINRICH, AND W. G. HUNT. 2007. Lead exposure, diagnosis, and treatment in California Condors released in Arizona. Pages 97–108 *in* A. Mee, L. S. Hall and J. Grantham (Eds.). California Condors in the 21st Century. Series in Ornithology, no. 2. American Ornithologists Union, Washington, DC, and Nuttall Ornithological Club, Cambridge, Massachusetts, USA.

SNYDER, N., AND H. SNYDER. 2000. The California Condor: a saga of natural history and conservation. Academic Press, San Diego, California, USA.

SULLIVAN, K., R. SIEG, AND C. PARISH. 2007. Arizona's efforts to reduce lead exposure in California Condors. Pages 109–120 *in* A. Mee, L. S. Hall and J. Grantham (Eds.). California Condors in the 21st Century. Series in Ornithology, no. 2. American Ornithologists Union, Washington, DC, and Nuttall Ornithological Club, Cambridge, Massachusetts, USA.

WOODS, C. P., W. R. HEINRICH, S. C. FARRY, C. N. PARISH, S. A. H. OSBORN, AND T. J. CADE. 2007. Survival and reproduction of California Condors released in Arizona. Pages 57–78 *in* A. Mee, L. S. Hall and J. Grantham (Eds.). California Condors in the 21st Century. Series in Ornithology, no. 2. American Ornithologists Union, Washington, DC, and Nuttall Ornithological Club, Cambridge, Massachusetts, USA.

EVIDENCE FOR THE SOURCE OF LEAD CONTAMINATION WITHIN THE CALIFORNIA CONDOR

JOHN CHESLEY[1], PETER REINTHAL[1], CHRIS PARISH[2], KATHY SULLIVAN[3], AND RON SIEG[3]

[1]*The University of Arizona, Gould-Simpson Building #77, 1040 E 4th St, Tucson AZ, 85721, USA.*

[2]*The Peregrine Fund, 5668 West Flying Hawk Lane, Boise, ID 83709, USA.*

[3]*Arizona Game and Fish Department, 3500 S. Lake Mary Road, Flagstaff, AZ 86001, USA.*

ABSTRACT.—The California Condor (*Gymnogyps californianus*) is the largest bird species in North America. Prior to the 20th century these birds were abundant along the western coast of the U.S. However, losses of habitat, natural predation, shooting, and environmental contamination have all been thought to contribute to a precipitous population decline. Early studies suggested that the demise of the condor population was in part the result of incidental Pb poisoning from either direct ingestion of lead fragments from hunter-killed game or indirectly as the result of biologically incorporated Pb from the environment. A recent article for the National Rifle Association (Wright and Peddicord 2007) suggested that although condors are most likely adversely affected by elevated lead in their tissues and lead ammunition is used in condor range, there is little scientific evidence of actual ingestion of lead ammunition by condors, and there is little scientific evidence that the lead in the tissues of condors can be traced to ammunition.

Condors in Arizona were periodically captured and monitored for blood Pb concentrations; subsets of these blood samples were analyzed for Pb isotopic ratios. To date, Pb isotopic ratios have been measured in blood in 47 birds over 3 years. Multiple measurements have been undertaken on 18 birds, including metal fragments collected at the same time from two different birds. Birds with elevated blood Pb levels were isolated, x-rayed and the excrement monitored for metal fragments. Twelve fragments were collected from 6 different birds. Analyses of the metal fragments from these birds determined that the fragments were Pb, Cu, Fe-Cr alloy and Pb-Sn alloy.

We present Pb isotopic evidence that directly links ingested Pb fragments to Pb in the blood of condors. One condor was found to have metal fragments in both 2004 and 2007 and had differing blood Pb isotopic ratios, which were within analytical error of the fragments collected at the same time. In addition to identifying the possible source(s) of Pb in the blood of condors, lead isotopic measurements can be used to discern if the condor has undergone a significant poisoning event between blood collection periods and provide insight into the number of Pb toxicity events over the lifetime of a bird. These results support the hypothesis that bullet fragments are causing increased blood lead levels in condors.

CHESLEY, J., P. REINTHAL, C. PARISH, K. SULLIVAN, AND R. SIEG. 2009. Evidence for the source of lead contamination within the California Condor. Abstract *in* R.T. Watson, M. Fuller, M. Pokras, and W.G. Hunt (Eds.). Ingestion of Lead from Spent Ammunition: Implications for Wildlife and Humans. The Peregrine Fund, Boise, Idaho, USA. DOI 10.4080/ilsa.2009.0219

Key words: Ammunition, bullet, condor, fragments, isotope, lead, scavenger, wildlife.

LEAD INTOXICATION KINETICS IN CONDORS FROM CALIFORNIA

MICHAEL FRY[1], KELLY SORENSON[2], JESSE GRANTHAM[3], JOSEPH BURNETT[2], JOSEPH BRANDT[3], AND MICHAELA KOENIG[3]

[1]*American Bird Conservancy, 1731 Connecticut Ave, NW, Washington DC, 20009, USA.*

[2]*Ventana Wildlife Society, 19045 Portola Dr. Suite F, Salinas, CA 93908, USA.*

[3]*U.S. Fish and Wildlife Service, 2493 Portola Rd., Ventura, CA 93003, USA.*

ABSTRACT.—Lead intoxication is a major factor in morbidity and mortality of California Condors (*Gymnogyps californianus*) in the wild. Data were obtained on blood lead levels in condors in California from random blood testing in the wild provided by the US Fish and Wildlife Service and Ventana Wilderness Society. Since the year 2000 blood samples have been routinely collected from condors whenever birds were captured. An analysis of 469 blood samples taken from 95 different condors indicated that 79 of these condors have had at least one significant lead exposure incident. Some condors have been intoxicated many times: Condor 108 with 13 positive blood samples, Condor 98 with 12, three others each exposed at least 11 times. In total there have been 276 separate documented incidents of blood lead levels in excess of 10 µg/dL. 27 of these have been in excess of 50 µg/dL, and have required emergency clinical care to prevent permanent injury or death. The clinical consequences of recurrent lead poisoning is uncertain, but will likely result in long-term neurological injury.

Analysis of successive blood samples from individual condors allowed a determination of depuration kinetics and half-life of lead in blood. The relatively rapid half-life averaged 14 days (range 3–22 days), with uncertainty arising from the possibility of unobserved lead fragments remaining in the alimentary canal of exposed birds. The high incidence of exposure and rapid half-life indicate that most condors have been repeatedly exposed to lead in the wild. The behavioral consequences of lead exposure to birds will be discussed.

In 2007, the California State Legislature passed legislation requiring the use of "lead-free" bullets for big-game hunting within the condor range. The California Fish and Game Commission subsequently implemented regulation to allow hunting only with "lead-free" ammunition, defined as bullets containing not more than 1% lead. Toxicological modeling of this amount of lead impurity in bullet fragments indicates that even if condors consume major fragments of bullets, the dissolution of lead is unlikely to raise the blood lead levels above 1 µg/dL, a low level equivalent to the blood lead levels of condors being raised in Los Angeles or San Diego Zoos on a lead-free diet.

FRY, M., K. SORENSON, J. GRANTHAM, J. BURNETT, J. BRANDT, AND M. KOENIG. 2009. Lead intoxication kinetics in condors from California. Abstract *in* R.T. Watson, M. Fuller, M. Pokras, and W.G. Hunt (Eds.). Ingestion of Lead from Spent Ammunition: Implications for Wildlife and Humans. The Peregrine Fund, Boise, Idaho, USA. DOI 10.4080/ilsa.2009.0301

Key words: Ammunition, blood, bullet, condor, fragments, intoxication, lead.

FEATHER PB ISOTOPES REFLECT EXPOSURE HISTORY AND ALAD INHIBITION SHOWS SUB-CLINICAL TOXICITY IN CALIFORNIA CONDORS

KATHRYN PARMENTIER[1], ROBERTO GWIAZDA[1], JOSEPH BURNETT[2], KELLY SORENSON[2], SCOTT SCHERBINSKI[3], COURT VANTASSELL[3], ALACIA WELCH[3], MICHAELA KOENIG[4], JOSEPH BRANDT[4], JAMES PETTERSON[3], JESSE GRANTHAM[4], ROBERT RISEBROUGH[5], AND DONALD SMITH[1]

[1]*Department of Environmental Toxicology, University of California, Santa Cruz, CA 95064, USA.*

[2]*Ventana Wildlife Society, 19045 Portola Dr., Suite F-1, Salinas, CA 93924, USA.*

[3]*US National Park Service, Pinnacles National Monument, 5000 Hwy 146, Paicines, CA 95043, USA.*

[4]*US Fish and Wildlife Service, 2493 Portola Rd., Suite A, Ventura, CA 93003, USA.*

[5]*Bodega Bay Institute, 2711 Piedmont Ave., Berkeley, CA 94705, USA.*

ABSTRACT.—Environmental lead exposure continues to hinder the recovery of the California Condor (*Gymnogyps californianus*) in the wild, but the full scope of exposures and their effects on condors is not well-characterized, in part due to the challenges associated with conducting comprehensive exposure assessments and the limited availability of biological markers of sub-lethal effects. While cases of severe lead poisoning resulting in clinical morbidity and mortality in condors are well understood, no data exist showing negative health effects in asymptomatic wild condors. Here we (1) validate the utility of lead concentration and stable isotopic measurements in growing condor feathers to reconstruct a comprehensive exposure history over the preceding months, and (2) demonstrate that lead-exposed condors are suffering sub-clinical toxicity, based on δ-aminolevulinic acid dehydratase (ALAD) inhibition by lead. Growing flight feathers were sequentially sampled along the trailing margin and analyzed by ICP-MS for lead concentrations and stable isotopic compositions (n=5 condors). Two of these birds (# 306 and 318) were implicated in a well documented .22 caliber lead ammunition-exposure event from a discarded pig carcass. In addition, whole blood samples were collected from pre-release (n=4) and free-flying (n=5) condors and their lead concentrations, isotopic compositions, and ALAD activity determined. For condors 306 and 318, the lead concentration and isotopic composition of their feathers changed following the documented ammunition-exposure event, arriving at values that matched exactly the isotopic composition of recovered ammunition. Feather samples from the remaining condors analyzed to date (#'s 307, 336, 351) similarly evidence lead exposure events over the months prior to sampling. Preliminary results of nine condors with blood lead concentrations ranging between 1.9-64.0 µg/dL show a significant inverse relationship between blood lead concentration and ALAD activity, indicating significant inhibition of ALAD at blood lead levels below those where clinical chelation treatment is indicated. Our research demonstrates that sequential analyses of growing feathers is a valuable tool to comprehensively reconstruct exposure histories, and that sub-lethal exposure results in measurable toxicity in condors.

PARMENTIER, K., R. GWIAZDA, J. BURNETT, K. SORENSON, S. SCHERBINSKI, C. VANTASSELL, A. WELCH, M. KOENIG, J. BRANDT, J. PETTERSON, J. GRANTHAM, R. RISEBROUGH, AND D. SMITH. 2009. Feather Pb isotopes reflect exposure history and ALAD inhibition shows sub-clinical toxicity in California Condors.

Abstract *in* R. T. Watson, M. Fuller, M. Pokras, and W. G. Hunt (Eds.). Ingestion of Lead from Spent Ammunition: Implications for Wildlife and Humans. The Peregrine Fund, Boise, Idaho, USA. DOI 10.4080/ilsa.2009.0302

Key words: ALAD, δ-aminolevulinic acid dehydratase, blood, condor, feather, isotope, lead.

USE OF MACHINE LEARNING ALGORITHMS TO PREDICT THE INCIDENCE OF LEAD EXPOSURE IN GOLDEN EAGLES

ERICA H. CRAIG[1], TIM H. CRAIG[2], FALK HUETTMANN[3], MARK R. FULLER[4]

[1]*Aquila Environmental, P.O. Box 81291, Fairbanks, AK 99708, USA.*
E-mail: tecraig@acsalaska.net

[2]*Bureau of Land Management, Central Yukon Field Office, 1150 University Avenue, Fairbanks, AK 99709, USA.*

[3]*EWHALE lab, Biology and Wildlife Department, Institute of Arctic Biology, 419 IRVING I, University of Alaska-Fairbanks, Fairbanks, AK 99775-7000, USA.*

[4]*US Geological Survey, Forest and Rangeland Ecosystem Science Center, 970 Lusk St., Boise, ID 83706, USA.*

ABSTRACT.—Quantitative models can be used to predict the occurrence of wildlife relative to certain environmental conditions. Resolving the impacts of environmental contaminants on wildlife often involves complex data sets suitable for analysis with quantitative models; yet despite their potential, such models are not commonly used. In this paper, we use data collected from wintering Golden Eagles *(Aquila chrysaetos)* and GIS-based models to demonstrate the use of stochastic gradient boosting, a machine learning algorithm, to examine factors most likely to influence the incidence of elevated blood lead levels. This fast, data-mining algorithm is capable of constructing predictive but sensitive and generalized models from complex contaminants datasets, and preliminary results suggest it accurately identified patterns that clarified and extended results of analyses performed using traditional statistical techniques. The management implications of using these models are far-reaching in their potential for identifying members of a population most at risk to contaminants, factors most likely to influence the incidence of lead contamination in a population, and potential sources of lead in the landscape. *Received 1 July 2008, accepted 4 December 2008.*

CRAIG, E. H., T. H. CRAIG, F. HUETTMANN, AND M. R. FULLER. 2009. Use of machine learning algorithms to predict the incidence of lead exposure in Golden Eagles. *In* R. T. Watson, M. Fuller, M. Pokras, and W. G. Hunt (Eds.). Ingestion of Lead from Spent Ammunition: Implications for Wildlife and Humans. The Peregrine Fund, Boise, Idaho, USA. DOI 10.4080/ilsa.2009.0303

Key words: stochastic gradient boosting, machine learning algorithm, modeling, lead, Golden Eagles, contaminants, lead contamination, data mining, predictive modeling, TreeNet®, machine learning.

THE INVESTIGATION OF ENVIRONMENTAL CONTAMINANTS is challenging because the variables are many, complex, and difficult to interpret. The sheer complexity of such datasets can affect the ability of researchers to obtain accurate results using traditional statistical approaches and to find the best possible solutions for making sustainable decisions (Craig and Huettmann 2009). Such obstacles can result in under-utilization of available information because of the inability to identify meaningful biological patterns in data with multi-dimensional input variables.

Recently, the incidence of lead in the environment as a result of lead bullet fragments in hunter killed game (Hunt et al. 2006), and its potential effect on the population viability of raptors and scavenger species such as the California Condor (*Gymnogyps californianus*), has received considerable attention (Church et al. 2006, Cade 2007, Johnson et al. 2007, Craighead and Bedrosian 2008). The Golden Eagle (*Aquila chrysaetos*) is known to scavenge carrion, including the remains of gun-killed animals, and it is thought that elevated lead levels in the species are related to the ingestion of ammunition lead remaining in these animals (Craig et al. 1990, Pattee et al. 1990, Wayland and Bollinger 1999, Fisher et al. 2006). Reports from long-term studies of Golden Eagles in Idaho, Colorado, and California suggest declines in productivity (Leslie 1992, Steenhoff et al. 1997, Kochert and Steenhoff 2002, D. Bittner and J. Oakley in Kochert et al. 2002). Declines in counts of migrant raptors in some parts of the western United States (Hoffman and Smith 2003) are coincidental with ongoing environmental changes resulting from wildfires, invasive plants, oil and gas development, urban sprawl, and other factors (Leslie 1992, Kochert and Steenhoff 2002, Knick et al. 2003). Stress from sublethal lead contaminant loads has the potential to further affect survival and reproduction (Fisher et al. 2006, Craighead and Bedrosian 2008). However, the issue of environmental contaminants in wildlife populations and how they affect a species is complex. Sources of lead can vary spatially and temporally and at different scales, and may differentially impact segments of the population; numerous factors could potentially influence the occurrence of elevated lead in individuals and populations.

Traditional statistical approaches to such problems usually require assumptions about the distribution of variables, independence of variables, and linearity of the data, and typically restrict the number of predictor variables. Such methods are often labor intensive and thus, costly. They also require *a priori* knowledge and decisions about the data relevance and applicability that might introduce bias in the resultant models. For these reasons, researchers are increasingly using powerful and flexible data mining techniques for exploratory analysis of complex ecological questions (Hochachka et al. 2007). They are finding that such alternative approaches greatly outperform traditional modeling methodologies with non-linear and linear data (Olden and Jackson 2002, Prasad et al. 2006).

We demonstrate the advantages of using a data mining algorithm known as stochastic gradient boosting (SGB) to identify meaningful patterns and relationships (Friedman 2001, 2002) in the investigation of contaminants in wildlife. We use blood lead levels (BLL) from 323 wintering Golden Eagles in Idaho, and satellite telemetry locations from six of those eagles (Craig and Craig 1998) in a GIS environment in a sample application of this method.

METHODOLOGY AND DISCUSSION

Machine learning is a rapidly advancing field of artificial intelligence whereby computer programs strategize in response to diverse data input. Stochastic gradient boosting is a refinement of classification and regression tree analysis (Breiman et al. 1984). It is a hybrid "boosting and bagging" algorithm that randomly selects subsets of the sample data (without replacement) and uses them to fit a series of very simple decision trees in sequence. Each successive tree is built from the prediction residuals of the preceding tree(s) and the final model is a summation of the best partitioning of the data (Friedman 2002). Models are developed using the software known as TreeNet® (Salford Systems, Inc.). Model construction is fast, requires no *a priori* assumptions about the relationship between the response and predictor variables, does not limit the number of predictor variables, and is capable of uncovering the underlying structure in data that are non-additive or hierarchical in nature (Prasad et al. 2006). Resultant models are robust with a high degree of predictive accuracy (Hochachka et al. 2007). The user can define an unlimited number of predictor variables that might potentially influence the occurrence of the target variable (in the case of our example, blood lead levels in Golden Eagles).

These features are highly applicable to the field of applied ecology (Olden et al. 2008) and critical for evaluating the validity of models. With the rapidly changing environment and numerous threats to biological diversity, such accurate predictive capability is increasingly important. With SGB, evaluation can be accomplished by: 1) applying the model to an independent dataset; 2) withholding a randomly selected portion of the dataset and using it as "testing" data (we used this method for our example dataset, withholding 20% of the data to be used for independent testing); or 3) in the case of very small datasets, through k-fold cross-validation (Kohavi 1995). The software produces a confusion matrix of the percent of eagles (in the case of our example model) correctly predicted with the presence or absence of elevated BLL and a Receiver Operating Characteristic Curve (ROC) that describes the sensitivity and specificity of the model (Collinson 1998). To further aid in interpretation of the models, and the biological significance of the results, TreeNet® produces an index ranking the relative importance of the predictor variables. The top predictor variable is given a score of 100 and then all other predictors are ranked in descending order, based on their influence upon the target variable, in comparison with the top predictor. To further aid in the interpretation of the data, graphic displays of dependence plots for single variables and two way interactions among variables are produced (Friedman 2001, 2002). This information, combined with expert opinion about a species, can be a valuable tool in the development of models that have ecological validity.

We used six predictors to develop our models (age and sex of the eagle, year and month in which it was captured, location of capture, and time of day the bird was captured). We defined the number of nodes in each tree and the number of samples allowed in each terminal node; we used six-node trees (the default) and 10 samples in each terminal node. The final number of trees for best model performance was determined by the software, which indicated the point at which no further variation in the data was explained by the formation of additional trees. We used the binary logistic option for model development, with presence or absence of BLL above background ($\geq$0.20 ppm, Kramer and Redig 1997) as our binary response.

Our models indicated that the contaminants dataset for the wintering Golden Eagles was not simply linear in nature and further clarified results of analyses performed using traditional statistical techniques (Craig and Craig 1998). Similar to other studies, our preliminary model indicated that females were more likely to have elevated BLL than males (Craighead and Bedrosian 2008, Pain et al. 1993), and there existed a positive relationship between eagles exposed to lead and the month of December, when offal from hunter-killed game associated with gun hunting seasons was most often available (Pattee et al. 1990, Church et al. 2006, Craighead and Bedrosian 2008). Our models also allowed us to examine the interactions among variables, particularly in regard to differences observed in BLL among years and valleys. This information, in conjunction with our geospatial analysis, allowed us to examine the distribution of lead contaminated eagles relative to the location of historical lead mines within the study area and identified an area for additional research. We will further investigate the incidence and sources of lead for this wintering population by examining stable lead isotopes signatures among a sample of the wintering eagles.

CONCLUSION

Preliminary results suggest that SGB accurately and quickly produced a model that independently detected patterns of lead contamination in wintering Golden Eagles corroborative with other studies using more traditional analytical methods and those of our previous research on this population. Based on our model, it is likely that lead contamination in hunter-killed game carcasses is the principal source of lead exposure to the wintering population of Golden Eagles in this study, but differences among years, study sites, and timing of capture raise additional questions that need to be clarified. For example, we must examine the possible role of contamination from local historical lead mining sites.

The management implications of using predictive modeling to investigate the effects of contaminants in wildlife are potentially far-reaching. Our preliminary results suggest that machine learning algorithms, such as SGB, are particularly appropriate management tools for identifying patterns among data to generate hypotheses for further research. SGB also very quickly created structured predictive models as stand-alone products useful for identification of factors influencing the incidence of elevated lead levels in Golden Eagles, of possible sources of contamination, and of segments of the population most at risk. SGB is appropriate for approaching any similar problem, and we believe that linking such computational software with wildlife ecology and conservation management in an interdisciplinary framework is crucial for timely responses to critical ecological questions, and is relevant for species sustainability (Chernetsov and Huettmann 2005, Olden et al. 2008).

ACKNOWLEDGMENTS

Many individuals have provided technical support for this project. We would particularly like to acknowledge the contributions of R. Craig, J. Craig, and H. Craig. Funding was provided by the US Bureau of Land Management, US Geological Survey, University of Alaska Fairbanks, E-WHALE Lab, Western Ecological Studies Team, Idaho Department of Fish and Game, Golden Eagle Chapter of the Audubon Society and The Idaho Wildlife Society. We thank W. G. Hunt and C. J. Henny for reviewing our manuscript. Any use of trade, product, or firm names is for descriptive purposes only and does not imply endorsement by the US Government.

LITERATURE CITED

BREIMAN, L., J. H. FRIEDMAN, R. A. OLSHEN, AND C. J STONE. 1984. Classification and regression trees. CRC Press. Boca Raton, Florida, USA.

CADE, T. J. 2007. Exposure of California Condors to lead from spent ammunition. Journal of Wildlife Management 71:2125–2133.

CHERNETSOV, N., AND F. HUETTMANN. 2005. Linking global climate grid surfaces with local long-term migration monitoring data: Spatial computations for the Pied Flycatcher to assess climate-related population dynamics on a continental scale. Pages 133–142 *in* Lecture Notes in Computer Science (LNCS) 3482, International Conference on Computational Science and its Applications (ICCSA), Springer Berlin. Heidelberg, Germany.

CHURCH M. E., R. GWIAZDA, R. W. RISEBROUGH, K. SORENSON, C. P. CHAMBERLAIN, S. FARRY, W. HEINRICH, B. A. RIDEOUT, AND D. R. SMITH. 2006. Ammunition is the principal source of lead accumulated by California Condors reintroduced to the wild. Environmental Science and Technology 40:6143–50.

COLLINSON, P. 1998. Of bombers, radiologists, and cardiologists: time to ROC. Heart 80:215–217.

CRAIG, E., AND F. HUETTMANN. 2009. Using "blackbox" algorithms such as TreeNet and Random Forests for data-mining and for finding meaningful patterns, relationships and outliers in complex ecological data: An overview, an example using Golden Eagle satellite data and an outlook for a promising future. *In* Hsiao-fan Wang (Ed.). Intelligent Data Analysis: Developing New Methodologies through Pattern Discovery and Recovery. IGI Global, Hershey, Pennsylvania, USA.

CRAIG, E. H., AND T. H. CRAIG. 1998. Lead and mercury levels in Golden and Bald Eagles and annual movements of Golden Eagles wintering in east-central Idaho: 1990–1997. Idaho BLM Technical Bulletin No. 98–12.

CRAIG, T. H., J. W. CONNELLY, E. H. CRAIG, AND T. PARKER. 1990. Lead concentrations in Golden and Bald Eagles. Wilson Bulletin 102:130–133.

CRAIGHEAD, D., AND B. BEDROSIAN. 2008. Blood lead levels of Common Ravens with access to big-game offal. Journal of Wildlife Management 73:240–245.

FISHER, I. J., D. J. PAIN, AND V. G. THOMAS. 2006. A review of lead poisoning from ammunition sources in terrestrial birds. Biological Conservation 131:421–432.

FRIEDMAN, J. H. 2001. Greedy function approximation: A gradient boosting machine. Annals of Statistics 2:1189–1232.

FRIEDMAN, J. H. 2002. Stochastic gradient boosting: Nonlinear methods and data mining. Computational Statistics and Data Analysis 38:367–378.

HOCHACHKA, W. M., R. CARUANA, D. FINK, A. MUNSON, M. RIEDEWALD, D. SOROKINA, AND S. KELLING. 2007. Data-mining discovery of pattern and process in ecological systems. Journal of Wildlife Management 71:2427–2437.

HOFFMAN, S. W., AND J. P. SMITH. 2003. Population trends of migratory raptors in western North America, 1977–2001. Condor 105:397–419.

HUNT, W. G., W. BURNHAM, C. N. PARISH, K. K. BURNHAM, B. MUTCH, AND J. L. OAKS. 2006. Bullet fragments in deer remains: Implications for lead exposure in avian scavengers. Wildlife Society Bulletin 34:67–170.

JOHNSON, C. K., T. VODOVOZ, W. M. BOYCE, AND J. A. K. MAZET. 2007. Lead exposure in California Condors and sentinel species in California. Report prepared for the California Fish and Game Commission [Online.] Available at http://www.dfg.ca.gov/habcon/info/bm_research/bm_pdfrpts/WHCleadcondorsfinal.pdf

KNICK, S. T., D. S. DOBKIN, J. T. ROTENBERRY, M. A. SCHROEDER, W. M. VANDER HAEGEN, AND C. VAN RIPER III. 2003. Teetering on the edge or too late? Conservation and research issues for avifauna of sagebrush habitats. Condor 105:611–634.

KOCHERT, M. N., AND K. STEENHOF. 2002. Golden Eagles in the US and Canada: Status, trends, and conservation challenges. Journal of Raptor Research 36 (1 Suppl.): 32–40.

KOCHERT, M.N., K. STEENHOF, C.L. MCINTYRE, AND E.H. CRAIG. 2002. Golden Eagle (*Aquila chrysaetos*). *In* A. Poole and F. Gill (Eds.). The Birds of North America, No. 684. The Birds of North America, Inc., Philadelphia, Pennsylvania, USA.

KOHAVI, R. 1995. A study of cross-validation and bootstrap for accuracy estimation and model selection. (Pages 1137–1143). *In* Proceedings of the Fourteenth International Joint Conference on Artificial Intelligence. Morgan Kaufmann, San Mateo, California, USA.

KRAMER, J. L., AND P. T. REDIG. 1997. Sixteen years of lead poisoning in eagles, 1980–1995: An epizootiologic view. Journal of Raptor Research 31:327–332.

LESLIE, D. G. 1992. Population status, habitat and nest-site characteristics of a raptor community in eastern Colorado. Unpubl. Master's Thesis, Colorado State University, Fort Collins, Colorado, USA.

OLDEN, J. D., AND D. A. JACKSON. 2002. A comparison of statistical approaches for modeling fish species distributions. Freshwater Biology 47:1–20.

OLDEN, J. D., J. J. LAWLER, AND N. L. POFF. 2008. Machine learning without tears: A primer for ecologists. The Quarterly Review of Biology 83:171–193.

PAIN, D. J., C. AMIARD-TRIQUET, C. BAVOUX, G. BURNELEAU, L. EON, AND P. NICOLAU-GUILLAUMET. 1993. Lead poisoning in wild populations of Marsh Harriers *Circus aeruginosus* in the Camargue and Charente-Maritime, France. Ibis 135:379–386.

PATTEE, O. H., P. H. BLOOM, J. M. SCOTT, AND M. R. SMITH. 1990. Lead hazards within the range of the California Condor. Condor 92:931–937.

PRASAD, A. M., L. R. IVERSON, AND A. LIAW. 2006. Newer classification and regression tree techniques: Bagging and random forests for ecological prediction. Ecosystems 9:181–199.

STEENHOF, K., M. N. KOCHERT, AND T. L. MCDONALD. 1997. Interactive effects of prey and weather on Golden Eagle reproduction. Journal of Animal Ecology 66:350–362.

WAYLAND, M., AND T. BOLLINGER. 1999. Lead exposure and poisoning in Bald Eagles and Golden Eagles in the Canadian prairie provinces. Environmental Pollution 104:341–350.

LEAD SHOT POISONING IN SWANS: SOURCES OF PELLETS WITHIN WHATCOM COUNTY, WA, USA, AND SUMAS PRAIRIE, BC, CANADA

MICHAEL C. SMITH[1], MICHAEL A. DAVISON[2], CINDY M. SCHEXNIDER[3], LAURIE WILSON[4], JENNIFER BOHANNON[2], JAMES M. GRASSLEY[1], DONALD K. KRAEGE[2], W. SEAN BOYD[4], BARRY D. SMITH[4], MARTHA JORDAN[5], AND CHRISTIAN GRUE[1]

[1]*University of Washington, Washington Cooperative Fish and Wildlife Research Unit, School of Aquatic and Fishery Sciences, P.O. Box 355020, Seattle, WA 98195-5020, USA.* E-mail: mcsmith@u.washington.edu

[2]*Washington Department of Fish and Wildlife, 16018 Mill Creek Blvd., Mill Creek, WA 98012, USA.*

[3]*US Fish and Wildlife Service, 510 Desmond Dr. SE #102, Lacey, WA 98503, USA.*

[4]*Environment Canada, Pacific Wildlife Research Centre, 5421 Robertson Rd., RR#1 Delta, BC, V4K 3N2, Canada.*

[5]*The Trumpeter Swan Society, County Road 9 – Suite 100, Plymouth, MN 55441, USA.*

EXTENDED ABSTRACT.—Over the past nine years (1999–2008), swan populations in northwest Washington State and on Sumas Prairie, British Columbia (Figure 1) have declined by over 1,500 birds due to lead poisoning caused by ingestion of lead pellets. The large majority of mortalities involved Trumpeter Swans (*Cygnus buccinator*) and most have occurred in Whatcom County, Washington State. Swan mortalities in Skagit and Snohomish Counties (Figure 2) of Washington State have also been included in the study total as these individuals may have ingested lead pellets in Whatcom County and Sumas Prairie prior to moving south.

Lead shot use for waterfowl hunting has been banned in northwest Washington State since 1989 (WDFW 2001) and in the Sumas Prairie area of British Columbia since 1990 (Wilson et al. 1998). Lead shot is still permitted for upland bird hunting and target shooting in most of both areas. Swans are at risk from lead poisoning because of their method of feeding (Blus et al. 1989). Large amounts of plant and sediment material are consumed whole, along with small pebbles (grit) to aid in the grinding of food in the birds' gizzards. Because of their size, lead pellets may be unintentionally ingested when birds eat grit or seeds. The grinding action and acidic environment of the gizzard break down the pellets, allowing lead to enter the bloodstream (Shillinger et al. 1937, Bellrose 1975, WDFW 2001). Symptoms of lead poisoning in waterfowl (e.g., lethargy, muscular weakness) can appear as early as four days after ingestion of as few as two or three pellets, with death occurring in 17–21 days (USGS-NWHC, pers. comm.). If more pellets are ingested, less time may be required before mortality occurs (Bellrose 1975, Pain 1990).

In 2001, an international effort was initiated to locate the source(s) of the lead. Participants included the Washington Department of Fish and Wildlife, US Fish and Wildlife Service, Environment Can-

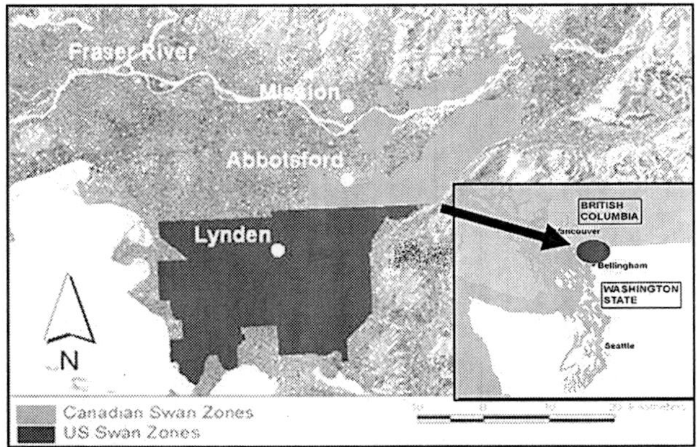

Figure 1. A map of the study area encompassing the lead related swan mortalities in Whatcom County, Washington State, and Sumas Prairie, British Columbia. The study area is ~100,000 ha, 58% in the United States and 42% in Canada.

Figure 2. A schematic of county boundaries in northwestern Washington State.

ada, The Trumpeter Swan Society, and the University of Washington (Washington Cooperative Fish and Wildlife Research Unit). Rocket propelled nets were used to capture a total of 249 Trumpeter and 42 Tundra Swans between 2001 and 2005. A blood sample was collected from each captured swan and analyzed for blood lead content (BC Ministry of Agriculture and Lands, Animal Health Centre, Abbotsford, BC.). Trumpeter Swans were fitted with either a VHF (n=243) or satellite (n=6) transmitter attached to a coded neck collar. Tundra Swans were fitted with a coded neck collar. Sick and dead swans were collected throughout the winter, and carcasses examined to determine cause of death, measure liver lead residues, and recover shot from gizzards. Results suggest that swans arrive on the wintering grounds with low blood lead levels, but some birds subsequently become exposed to lead after ingesting lead shot.

During the winter months (October-February) of 2001–2006, telemetry surveys were conducted each day and night to document locations of marked individuals. Population surveys were also conducted semi-weekly from November to January to obtain additional detail on population movements and to validate telemetry data. The locations of collared swans were used to identify forage areas and roost sites. Telemetry data for swans that died from lead poisoning after radio-collaring were used to identify and prioritize areas for lead shot density assessment (shot collected from soil/sediment sampling).

Lead shot has been found in fields and water bodies where hunting and target shooting have occurred. Relatively high densities of lead shot were found on the US side of Judson Lake, a ~100 acre lake spanning the US/Canada border. Waterfowl hunting still occurs on the US side of Judson Lake, but this activity has not occurred on the Canadian side for at least 30 years because of a landowner ban.

An adaptive management approach was initiated in October 2006 to test the hypothesis that Judson Lake is a primary source of lead shot. Swans were discouraged from using Judson Lake by both passive (windsocks, effigies) and active (noise makers, laser light, airboat) deterrent methods. Hazing activities ceased in late January 2007 and swans were allowed on the lake as the water depth at that time was believed to be sufficient to preclude swans from accessing sediments. The number of lead poisoned swans in 2006–2007 decreased by >50% compared to the average of the five previous years. However, population surveys showed that fewer swans foraged in an agricultural area near Judson Lake. Heavy snow and freezing temperatures forced most swans to leave the study area for up to two weeks during the predicted lead exposure period. Therefore, it was unclear if the decreased lead mortality was due to the exclusion of swans from Judson Lake, reduced use of the nearby agricultural area, or because swans departed the area during adverse weather.

Hazing at Judson Lake was repeated from October 2007 to January 2008. Preliminary results from visual inspections (liver lead tests to confirm cause of death are pending) indicate that swan lead mortality again decreased by >50% compared to the average of the five years previous to the hazing activity. During the winter of 2007–2008, there were no weather events forcing swans to depart the study area. However, as in 2006–2007, population surveys showed that fewer swans foraged in the agricultural area near Judson Lake. Lead shot density assessment of these fields is currently underway.

Lead related swan mortalities in Skagit and Snohomish Counties of Washington State have been included in this study. During years prior to hazing Judson Lake, nearly 78% of the lead mortalities have occurred in Whatcom County/Sumas Prairie. In both 2006–2007 and 2007–2008 there was a southward shift in mortalities with over 50% of the estimated lead mortalities occurring in Skagit and Snohomish counties. It is unclear if these individuals ingested lead pellets in Whatcom County and Sumas Prairie prior to moving south or accessed new source(s) in either Skagit or Snohomish County.

At this time conclusions are preliminary. Because mortalities have declined by 50% during the two years of hazing, we conclude that Judson Lake is a source of lead shot but it clearly is not the only source. Cooperators are continuing to analyze data to clarify the relative contribution of Judson Lake versus other possible sites in northwest Washington State and Sumas Prairie, British Columbia. Following this analysis, cooperators will determine the best course of action to reduce the availability of lead shot to swans during 2008–2009 and subsequent wintering periods. *Received 31 May 2008, accepted 28 July 2008.*

SMITH, M. C., M. A. DAVISON, C. M. SCHEXNIDER, L. WILSON, J. BOHANNON, J. M. GRASSLEY, D. K. KRAEGE, W. S. BOYD, B. D. SMITH, M. JORDAN, AND C. GRUE. 2009. Lead shot poisoning in swans: Sources of pellets within Whatcom County, WA, USA, and Sumas Prairie, BC, Canada. Extended abstract *in* R. T. Watson, M. Fuller, M. Pokras, and W. G. Hunt (Eds.). Ingestion of Lead from Spent Ammunition: Implications for Wildlife and Humans. The Peregrine Fund, Boise, Idaho, USA.
DOI 10.4080/ilsa.2009.0201

Key words: Lead poisoning, lead shot, lead shot sampling, swan, waterfowl, waterfowl hazing.

ACKNOWLEDGMENTS

We would like to thank all of the following contributors to this project, where their efforts and expertise and/or financial support made a project of this magnitude and duration possible: British Columbia Ministry of Agriculture and Lands Animal Health Centre (Victoria Bowes and all other reception and post-mortem room staff), British Columbia Ministry of Environment (Jack Evans), British Columbia Waterfowl Society, Ducks Unlimited Canada (Dan Buffet), Environment Canada – Canadian Wildlife Service (Garry Grigg, Sandi Lee, Rick McKelvey, Bob Elner, Saul Schneider, Oliver Busby, Gabriella Kardosi, Patti Dods, Jessica Beaubier, Robin Whittington, Judy Wheeler, Terry Sullivan, Stephen Symes, Tracy Sutherland, Courtney Albert), Monika's Wildlife Shelter (Monika Tolksdorf, Lorne Green, Gloria White), North Cascades Audubon Society (Tom Pratum, Sally Hansberry, Carl Decker, James Duemmel, Sally Hewitt, Anna Marie Bangs, Jeanette Opiela, Michelle Bodtke), Northwest Wildlife Rescue and Rehabilitation (Linda Williamson, Aimee Frazier, Alicia Hartley, Audrey Brocker, Carl Decker, Chris Smith, Cindy Hansen, Darla Arnason, Jerry Farrell, Julie Elsbree, Julie Straight, Kraig Hansen, Krista Unser, Megan Kink, Rachel Strachan, Terry Preston, Wayne Ginter), Pilchuck Valley Wildlife Rehabilitation Center (Sue Murphy, Mike Murphy, Sheena Murphy), Puget Sound Energy (Mel Walters), Sardis Wildlife Centers (Sharon Walters and assistants), The Central Valley Naturalists (Johanna Saaltink, Gerry Powers, Margaret & Richard Bunbury, Jacquie

Reznick, Maria Carmen Brackhaus, Annabelle & Walter Rempel, Glenn Ryder, Stan Olson, Henry Savard), The Trumpeter Swan Society (Ruth Shea, Gary Ivey, Paul Fischbach, Mike Schwitters, Connie Schwitters), University of Washington (Loveday Conquest, Dave Manuwal), United States Fish and Wildlife Service (Doug Zimmer, Brad Bortner, Paul Weyland, Bruce Conant), United States Geological Survey (John Takakawa, Alex Westhoff), Veterinarian (Laurel A. Degernes), Volunteers (Lila Emmer, Terry Sisson, Dan Loehr, Sammy Low, Adina Parsley, Andrea Warner, Roy Teo), Washington Cooperative Fish and Wildlife Research Unit (Verna Blackhurst, Morgan Sternberg, Rob Fisk, Megan Horne-Brine, Paul Debruyn, Amanda Winans, Tristen Frum, Don Kruse, Canon Leurkins), Washington Department of Fish and Wildlife (Lora Leschner, Ruth Milner, Doug Huddle, Maynard Axelson, Ron Friesz, Brad Otto, Russ Canniff, Briggs Hall, John Pierce, John Jacobsen, Shelly Snyder, Russ Wright, Tom Cyra, Jed Varney, Ed Argenio, Troy McCormick, Carrie Herziger, Paul Moorehead, Joel Cappello, Don Hubner, Lee Kantar, Kendall Fish Hatchery), Washington Waterfowl Association (Rodney Vandersypen, Ken Miller, Mike Vanderhey, Jill Vanderhey, Stephen Vanderhey), and Wolf Hollow Wildlife Rehabilitation Center. We also express a special thank you to the many landowners who allowed us access to their lands.

LITERATURE CITED

BELLROSE, F. 1975. Impact of ingested lead pellets on waterfowl. Pages 163–167 *in* Proceedings of the First International Waterfowl Symposium (MO). Ducks Unlimited, St. Louis, Missouri, USA.

BLUS, L. J., R. K. STROUD, B. REISWIG, AND T. MCENEANEY. 1989. Lead poisoning and other mortality factors in Trumpeter Swans. Environmental Toxicology and Chemistry 8:263–271.

PAIN, D. J. 1990. Lead shot ingestion by waterbirds in the Camargue, France: An investigation of levels and interspecific differences. Environmental Pollution 66:273–285.

SCHILLINGER, J. E., AND C. C. COTTOM. 1937. The importance of lead poisoning in waterfowl. Transcripts of the North American Wildlife Conference 2:398–403.

WDFW (WASHINGTON DEPARTMENT OF FISH AND WILDLIFE). 2001. Report to the Washington Fish and Wildlife Commission: The use of non-toxic shot for hunting in Washington. Washington Department of Fish and Wildlife Non-toxic Shot Working Group, Olympia, Washington, USA.

WILSON, L. K., J. E. ELLIOT, K. M. LANGELIER, A. M. SCHEUHAMMER, AND V. BOWES. 1998. Lead poisoning of Trumpeter Swans, *Cygnus buccinator*, in British Columbia, 1976–1994. Canadian Field-Naturalist 112:204–211.

LEAD POISONING OF TRUMPETER SWANS IN THE PACIFIC NORTHWEST: CAN RECOVERED SHOT PELLETS HELP TO ELUCIDATE THE SOURCE?

LAURIE K. WILSON[1], GARRY GRIGG[1], RANDY FORSYTH[2], MONIKA TOLKSDORF[3], VICTORIA BOWES[4], MICHAEL SMITH[5], AND ANTON SCHEUHAMMER[6]

[1]*Environment Canada – Canadian Wildlife Service, Pacific Wildlife Research Centre, 5421 Robertson Rd, RR#1, Delta, BC, V4K 3N2, Canada.*
E-mail: laurie.wilson@ec.gc.ca

[2]*Environment Canada – Wildlife Enforcement Division, Saskatoon, SA, Canada.*

[3]*Monika's Wildlife Shelter, Surrey, BC, Canada.*

[4]*British Columbia Ministry of Agriculture and Lands, Abbotsford, BC, Canada.*

[5]*Washington Department of Fish and Wildlife, Mill Creek, WA, USA.*

[6]*Environment Canada – Science & Technology, Ottawa, ON, Canada.*

EXTENDED ABSTRACT.—At least 2,577 Trumpeter (*Cygnus buccinator*) and Tundra Swans (*Cygnus columbianus*) have died in northwestern Washington and southwestern British Columbia over the past nine winters (1999–2008). An average of 283 swan carcasses were recovered annually, with annual variation ranging from 108 to 401 individuals. Most of the fatalities were attributed to ingestion of lead shot (80%; 1,376 of the 1,727 intact remains suitable for toxicological testing from 1999 to 2006). Degernes et al. (2006) previously reported on the approximate 400 swan fatalities that occurred in Washington during the winters of 2000–02. Lead related swan fatalities have been documented in Washington State in the past (Kendall and Driver 1982, Blus et al. 1989, Lagerquist et al. 1994), however, the only other large-scale mortality event occurred in 1992, involving approximately 100 swans (Wilson et al. 1998).

Swan lead fatalities from ingestion of spent lead shot have been occurring since at least 1925 (Munro 1925). Swans may intentionally ingest spent shot, mistaking it as grit or seeds or may inadvertently pick it up while foraging on associated vegetation (Blus 1994, Mateo and Guitart 2000). Once ingested, the shot is retained in the gizzard where the grinding action and acidic environment exacerbates the absorption of lead which is ultimately distributed to soft tissues and bones (Clemens et al. 1975). Ingestion of only two to three pellets may cause mortality in approximately three weeks (USGS-NWHC, pers. comm.).

The use of lead shot for waterfowl hunting was banned in Whatcom County, Washington in 1989 and Sumas Prairie, British Columbia in 1992. Lead shot continues to be permitted for upland hunting and target shooting and for the Migratory Bird Convention Act species including doves, Band-tailed Pigeons (*Columba fasciata*), and American Woodcock (*Scolopax minor*).

A working group comprised of Environment Canada, Washington Department of Fish and Wildlife, US Fish and Wildlife Service, Trumpeter Swan Society, University of Washington (Washington Cooperative Fish and Wildlife Research Unit), and other organizations made efforts to locate the source(s) of lead shot since 2001. To identity the sources of lead shot causing swan deaths, biologists have conducted a multi-faceted investigation (see Smith et al. 2009, this volume, for details). This extended abstract focuses on information obtained from the shot recovered from lead-poisoned swan gizzards and sediment/soil from suspected source areas (roost sites/forage fields). We attempt to identify specific roosts and forage fields responsible for swan poisonings by comparing the type and size of shot as well as the lead isotope ratios of shot collected from suspected source areas with shot recovered from the gizzards of lead-poisoned swans.

The results presented here are preliminary, as this investigation is ongoing. We plan to examine shot collected from more recently identified suspected source areas, determine the type and size of shot from a greater number of lead poisoned swans, and measure lead isotope ratios in lead poisoned and non-lead poisoned swans. Over the past two winters, an adaptive management action of hazing or preventing swans from using a major roost site (Judson Lake) was conducted to ascertain the role of the lake in the lead-related swan fatalities. The number of documented lead poisoned swans in these years declined by approximately 50% compared to the average of previous years. We still want to compare the type, size and lead isotope ratio of shot recovered from swan gizzards during these two discrete time periods.

Shot were recovered while conducting lead shot density assessment of suspected source areas (14 agricultural fields in Canada, eight permanent water-bodies (four in Canada, four in the USA), and six temporarily flooded agricultural fields (sheet-water) in the USA. Agricultural fields were sampled by collecting core samples in a spray pattern from hunting blinds (biased sampling design). Conversely, the entire permanent water-bodies and sheet-water areas were sampled in a grid pattern (non-biased sampling design). Cores were 6-inches in diameter with a depth of 6–12 inches. Shot, grit and vegetation were separated from individual cores through repeated washing using a series of different sized sieves. Shot was recovered from 1,727 dead swans collected either from major roosts or in response to calls from the public from 1999 to 2006. Individual swan carcasses were examined by a veterinarian to determine probable cause of death, and tissues were collected for additional testing. Shot, grit and vegetation were separated from individual gizzards though repeated washing, using a series of different sized sieves. Environment Canada Wildlife Enforcement Officers identified the size and type of shot through visual examination and testing with a magnet. To date, preliminary results on the size and type of shot recovered from the sediment/soil and swan gizzards includes only a portion of the total shot collected (6,052 shot from 108 lead poisoned swans were also analyzed for lead isotope ratios). Finally, lead isotope ratios were analyzed from shot collected in sediment/soil samples from suspected source areas (n=319), livers from 108 lead-poisoned swans (n=1 swan from a Canadian forage field and n=107 swans from permanent water-bodies), and a portion of the shot from the gizzards of the same 108 swans (n=1,078 shot; 10 shot from a Canadian field, 1,068 shot from permanent water-bodies). Lead isotope ratios were determined as described in Scheuhammer et al. (2003).

The location where a swan carcass is recovered does not necessarily reflect the location where shot were ingested. Swans which ingest lead shot may remain ill for a considerable time before dying. During this time it is probable that these individuals would seek shelter on a water-body which provides some protection from predators. Therefore, it was not surprising that, in our investigation, the majority of sick and dead swans were recovered from water-bodies (85%, 1,796/2,123) and the rest were collected from forage fields or along roadways.

Of the 1,270 gizzards examined from 1999 to 2006, an average of 22 lead and nine non-toxic shot were recovered per gizzard (total 38,695 shot). The majority of swans (72%) had <25 lead shot per gizzard, but 9% of swans had >50 lead shot per gizzard. The trend of swans ingesting more lead shot compared to non-toxic shot differs from the results of Anderson et al. (2000) which demonstrated that ducks ingested mostly non-toxic shot.

The specific size of lead and non-toxic shot from 108 swan gizzards (n=6,052 pellets) and from sediment/soil sampled at roosts/forage fields of suspected source areas (n=375 pellets) ranged from sizes typical of waterfowl hunting, to smaller sizes typical of upland game bird hunting and target shooting. The sediment/soil samples contained predominately lead shot sizes typically used for waterfowl hunting (#2–6) and a mix of non-toxic shot sizes typically used for waterfowl (#2–3) and upland game bird hunting (#4–6). Swan gizzards had a mix of lead shot sizes typically used for upland game bird hunting (#6) and target shooting (#7.5–8) and predominately non-toxic shot sizes typically used for upland game bird hunting (#4–6).

At least seven non-toxic shot alternatives are currently commercially available in Canada (steel, bismuth, tungsten iron, tungsten matrix, tungsten polymer, tin, and tungsten hevi, Forsyth 2005). Steel is the least expensive and most readily available non-toxic shot. Of the 197 shot recovered from sediment/soil which have been examined, 85% (n=168 pellets) was lead with comparatively fewer non-toxic shot. The majority of non-toxic shot was steel (79%, n=23 pellets), with only three tungsten-matrix and three bismuth pellets identified. The relative presence of shot types in the environment does not reflect current surveys on shot type used by hunters (Stevenson et al. 2005). Possible explanations for these differences could be continued presence of 'old' spent lead shot, continued use of lead shot for waterfowl hunting by non-compliant hunters, or continued use of lead shot for upland game bird hunting and backyard target shooting.

Lead isotope ratios can be used to help differentiate among sources of environmental lead. There are four stable isotopes of lead, three of which are products of radioactive decay (^{206}Pb, ^{207}Pb, ^{208}Pb; but not ^{204}Pb). The lead isotope ratios vary in different ores depending on the relative amount of radiogenic lead. Any combination of ratios can help identify geological origin of lead samples but ^{206}Pb:^{207}Pb is most commonly used. The ^{206}Pb:^{207}Pb ratio reflects the time at which uranium mixed into the ore. Ratios of 0.93–1.08 are associated with Canadian Precambrian Shield lead ores; ratios of 1.15–1.22 are associated with most non-Precambrian lead ores in North America; ratios of 1.27–1.37 are associated with Mississippi Valley lead ores (Scheuhammer and Templeton 1998). The ^{206}Pb:^{207}Pb ratio also can reflect differences in sources of atmospheric lead. Ratios of 1.13–1.16 are associated with leaded gasoline use in Canada (Sturges and Barrie 1987); ratios of 1.18–1.20 are associated with leaded gasoline used in the US (Flegal et al. 1989). The ^{206}Pb:^{207}Pb ratios measured in lead shot pellets from various manufacturers range widely (mean 1.18 ± 0.05) but generally do not overlap with Precambrian lead isotope ratios, thus reflecting predominant use of USA, and Central and South American lead sources in their manufacture (Scheuhammer and Templeton 1998).

Patterns of ^{206}Pb:^{207}Pb ratios in shot from suspected source areas, in shot from gizzards of lead-poisoned swans, and in liver of lead poisoned swans were compared. Only 56% of the ^{206}Pb:^{207}Pb ratios measured in the shot collected from sediment/soil fell within the range found for shot collected from the swan gizzards. Therefore, it appears that swans are not consuming the whole range of shot sizes recovered from agricultural fields and water-bodies. The reason for this phenomenon is currently unclear; however the distribution of shot sizes from the gizzards was also different from that found in sediments/soils.

To summarize, preliminary results do not indicate a specific identifiable source of lead shot responsible for the swan fatalities. There is no clear explanation for some preliminary results, but additional analyses are planned which may provide further insight and understanding. Should the preliminary results suggesting higher prevalence of shot from upland game bird hunting in gizzards of lead-poisoned swans continue to hold true, then this lead-poisoning investigation of the fatality of more than 2,500 swans over the past nine years may implicate lead shot for trap or skeet practice in areas frequented by waterfowl. *Received 30 May 2008, accepted 24 July 2008.*

WILSON, L. K., G. GRIGG, R. FORSYTH, M. TOLKSDORF, V. BOWES, M. SMITH, AND A. SCHEUHAMMER. 2009. Lead poisoning of Trumpeter Swans in the Pacific Northwest – Can recovered shot pellets help to elucidate the source? Extended abstract *in* R. T. Watson, M. Fuller, M. Pokras, and W. G. Hunt (Eds.). Ingestion of Lead from Spent Ammunition: Implications for Wildlife and Humans. The Peregrine Fund, Boise, Idaho, USA. DOI 10.4080/ilsa.2009.0120

Key Words: Lead isotope, lead poisoning, lead shot, Pacific Northwest, swan, waterfowl.

ACKNOWLEDGMENTS

The authors thank Sandi Lee, Oliver Busby, Lorne Green, Gabriella Kardosi, Patti Dodds, Mike Davison and Rus Caniff for their excellent technical help in collecting and processing soil/sediment samples; Sandi Lee, Judy Wheeler, Robin Whittington, Sue Murphy, Martha Jordan for recovering sick and dead swans; Sue Murphy for her swan collections and wildlife rehabilitation services; Gloria White and Lorne Green for shot recovery from cores and gizzards; and Judy Wheeler for coordination of sample processing. We also thank Canadian and US landowners for permitting access to their property to conduct this study. The Swan Lead Technical Working Group (Sean Boyd, Barry Smith, Jennifer Bohannon, Mike Davison, Don Kraege, Cindy Schexnider, Doug Zimmer, Martha Jordan) are to be thanked for their ongoing contribution to resolve this issue. The study was funded by the Environment Canada's Wildlife Toxicology Program and the Canadian Wildlife Service Migratory Birds Program with US field activities funded by the Washington Department of Fish and Wildlife, the US Fish & Wildlife Service and the Trumpeter Swan Society.

LITERATURE CITED

ANDERSON, W. L., S. P. HAVERA, AND B. W. ZERCHER. 2000. Ingestion of lead and nontoxic shotgun pellets by ducks in the Mississippi flyway. Journal of Wildlife Management 64:848–857.

BLUS, L. J. 1994. A review of lead poisoning in swans. Comparative Biochemistry and Physiology C: Pharmacology, Toxicology and Endocrinology 108C:250–267.

BLUS, L. J., R. K. STROUD, B. REISWIG, AND T. MCENEANEY. 1989. Lead poisoning and other mortality factors in Trumpeter Swans. Environmental Toxicology and Chemistry 8:263–271.

CLEMENS, E. T., L. KROOK, AND A. L. ARONSON. 1975. Pathogenesis of lead shot poisoning in the Mallard duck. Cornell Veterinarian 65:248–285.

DEGERNES, L., S. HEILMAN, M. TROGDON, M. JORDAN, M. DAVISON, D. KRAEGE, M. CORREA, AND P. COWEN. 2006. Epidemiologic investigation of lead poisoning in Trumpeter and Tundra Swans in Washington State, USA, 2000–2002. Journal of Wildlife Diseases 42(2):345–358.

FLEGAL, A. R., O. J. NRIAGU, S. NIEMEYER, AND K. H. COALE. 1989. Isotopic tracers of lead contamination in the Great Lakes. Nature 339:455–458.

FORSYTH, R. 2005. Understanding crippling rates and non-toxic shot in relation to hunting of waterfowl in North America. Environment Canada—Canadian Wildlife Service, Saskatoon, Saskatchewan, Canada.

KENDALL, R. J., AND C. J. DRIVER. 1982. Lead poisoning in swans in Washington state. Journal of Wildlife Diseases 18:385–387.

LAGERQUIST, J. E., M. DAVISON, AND W. J. FOREYT. 1994. Lead poisoning and other causes of mortality in Trumpeter (*Cygnus buccinator*) and Tundra (*C. columbianus*) Swans in western Washington. Journal of Wildlife Diseases 30(1):60–64.

MATEO, R., AND R. GUITART. 2000. The effects of grit supplementation and feed type on steel-shot ingestion in Mallards. Preventive Veterinary Medicine 44:221–229.

MUNRO, J. A. 1925. Lead poisoning in Trumpeter Swans. Canadian Field-Naturalist 39:160–162.

SCHEUHAMMER, A. M., AND D. M. TEMPLETON. 1998. Use of stable isotope ratios to distinguish sources of lead exposure in wild birds. Ecotoxicology 7:37–42.

SCHEUHAMMER, A. M., D. E. BOND, N. M. BURGESS, AND J. RODRIGUE. 2003. Lead and stable lead isotope ratios in soil, earthworms, and bones of American Woodcock (*Scolopax minor*) from Eastern Canada. Environmental Toxicology and Chemistry 22:2585–2591.

SMITH, M. C., M. A. DAVISON, C. M. SCHEXNIDER, L. WILSON, J. BOHANNON, J. M. GRASSLEY, D. K. KRAEGE, W. S. BOYD, B. D. SMITH, M. JORDAN, AND C. GRUE. 2009. Lead shot poisoning in swans: Sources of pellets within Whatcom County, WA, USA, and Sumas Prairie, BC, Canada. Extended abstract *in* R. T. Watson, M. Fuller, M. Pokras, and W. G. Hunt (Eds.). Ingestion of Lead from Spent Ammunition: Implications for Wildlife and Humans. The Peregrine Fund, Boise, Idaho, USA. DOI 10.4080/ilsa.2009.0201

STEVENSON, A. L., A. M. SCHEUHAMMER, AND H. M. CHAN. 2005. Effects of nontoxic shot regulations on lead accumulation in ducks and American Woodcock in Canada. Archives of Environmental Contamination and Toxicology 48:405–413.

STURGES, W. T., AND L. A. BARRIE. 1987. Lead 206/207 isotope ratios in the atmosphere of North America as tracers of US and Canadian emissions. Nature 329:144–146.

WILSON, L. K., J. E. ELLIOT, K. M. LANGELIER, A. M. SCHEUHAMMER, AND V. BOWES. 1998. Lead poisoning of Trumpeter Swans, *Cygnus buccinator*, in British Columbia, 1976–1994. Canadian Field-Naturalist 112:204–211.

LEAD OBJECTS INGESTED BY COMMON LOONS IN NEW ENGLAND

MARK A. POKRAS[1], MICHELLE R. KNEELAND[1], ANDREW MAJOR[2], ROSE MICONI[1], AND ROBERT H. POPPENGA[3]

[1]*Tufts Cummings School of Veterinary Medicine, 200 Westboro Road, North Grafton, MA 10536, USA.*
E-mail: mark.pokras@tufts.edu

[2]*United States Fish and Wildlife Service, 22 Bridge Street, Suite 400, Concord, NH 03301-4901, USA.*

[3]*California Animal Health & Food Safety Laboratory, University of California, Davis, CA 95616, USA.*

EXTENDED ABSTRACT.—Lead poisoning from ingested fishing gear has regularly been reported in four avian species: Common Loon (*Gavia immer*, Locke 1982), Mute Swan (*Cygnus olor*, Sears et al. 1989), Trumpeter Swan (*Cygnus buccinator*, Blus et al.1989), and Sandhill Crane (*Grus canadensis*, Windingstad et al. 1984). Aquatic birds may ingest lead objects while collecting gizzard stones or by preying on live bait or escaped fish with attached fishing gear. Evidence gathered from necropsies conducted at the Wildlife Clinic at Tufts Cummings School of Veterinary medicine suggests that ingestion of lead weights is the number one killer of breeding adult Common Loons in New England (Sidor et al. 2003). The current study quantifies the size, mass, and types of lead fishing gear ingested by Common Loons.

Between 1987 and 2000, 522 Common Loon carcasses were collected from the six New England states and submitted to the Tufts Wildlife Clinic for necropsy. Ingested lead objects were visually classified into the following six categories: Sinker, Jighead, Split Shot, Ammunition, Other (lead gear used for fishing that could not otherwise be classified, such as lead wires or tapes), and Unknown (original use could not be determined because of wear, fragmentation, or deformation). The recovered objects were weighed to the nearest 0.1 g (Pesola® spring scale), and a mechanical caliper (Fischer Scientific® type 6911) was used to measure length and width to the nearest 0.05 mm. Length was defined as the longest axis measurable, and width as the largest diameter perpendicular to the longest axis. Objects were then tested for lead using a commercial, buffered rhodizonate dye swab test (LeadCheck® Swabs, Hybrivet Systems, Inc). Toxicological analyses of loon body tissues were performed as described in Sidor et al. (2003).

Of the 522 loon carcasses examined, 118 (22.6%) had ingested lead objects (Figure 1). Of these 118 loons, 73 had more than one object in their gizzard, for a total of 222 lead objects recorded. The type of object ingested most frequently were sinkers at 48% of the total objects, followed by jigheads, split shot, and ammunition at 19%, 12%, and 11% respectively. The ammunition category consisted primarily of shotgun pellets, but also included one .22 caliber bullet and one .44 –.45 caliber bullet. Fifty percent of loons with ingested shotgun pellets had either two or three such projectiles present. About 36% of loons with ingested lead had other fishing-related objects (mostly hooks, swivels and monofilament line) present in the gastro-intestinal (GI) tracts. In this sample, all loons ingesting lead objects also had elevated liver lead levels consistent with lead poisoning.

Length, width, and mass varied among sinkers, jigs and split shot (Table 1). Of the 222 lead objects ingested, the largest weighed 25 g but 94% of them weighed less than 10 g. Over 94% were less than 25.4 mm in length, and 44% had a length of less than 10 mm (Figure 2). No jigheads recovered from the GI tracts had hooks remaining attached, thus length measurements of jigs includes only the lead portion.

Because of the grinding action of the gizzard and the presence of small stones against which the fishing gear is abraded, we suspect that measured sizes are somewhat smaller at necropsy than at the time they were first ingested by loons. Most ingested lead was similar in size to ingested stones, indicating loons may deliberately select lead objects because they fit whatever criteria birds have evolved to choose stones (Figure 2). While lead objects as small as 1 mm were found, no stones smaller than 6 mm were encountered. This may be explained by the fact that small stones can be passed out of the gizzard through the pylorus, while even the smallest lead objects would have inhibited gastrointestinal peristalsis and are more likely to be retained.

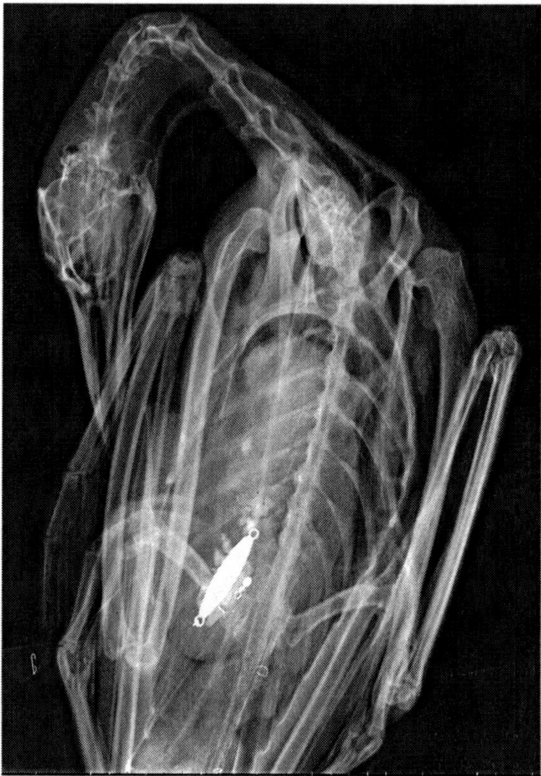

Figure 1. Radiograph of an adult Common Loon with an ingested lead sinker and jighead present in the gizzard.

The sizes of lead fishing gear encountered in the current study shows very close correlation with the sizes found by Franson et al. (2003), even though the frequency of lead ingestion by loons in New England in that sample population was much lower (7.5%). One possible reason for this difference is that all but two of the New England loons sampled by Franson et al. (2003) were live birds, while the present study reports solely on dead loons submitted for necropsy. It would be expected to find a higher frequency of lead ingestion in a sample of dead and moribund birds than in a sample of live, apparently healthy birds.

One can make the argument that, given what we know about the toxicity of lead to loons, humans, and a wide variety of other species, every effort should be made to utilize non-toxic alternatives and minimize the introduction of lead into the environment. Barring the complete elimination of lead for such sporting uses, a clear understanding of the sizes and types of gear that pose the greatest threat to loons and other fish-eating species will allow us to formulate rational policies for the protection and management of these species. *Received 12 June 2008, accepted 20 August 2008.*

Table 1. Dimensions of sinkers, jigheads and split shot recovered from 522 Common Loon carcasses collected from 1987 to 2000.

Lead Object	Sample Size, n	Length (mm)			Width (mm)			Mass (g)		
		mean	median	range	mean	median	range	mean	median	range
Sinkers	107	14.03 (±6.47)	12.7	4.0–40.2	5.89 (±2.55)	5.5	0.3–15.2	4.04 (±4.63)	2.4	0.3–25.0
Jighead	41	16.53 (±6.58)	15.3	5.2–33.6	6.55 (±2.33)	6.0	3.0–13.9	3.89 (±3.64)	3.18	0.3–18.1
Split Shot	26	5.79 (±2.33)	6.10	1.2–9.9	4.85 (±1.68)	4.5	1.5–8.13	1.60 (±1.04)	1.4	0.3–5.7

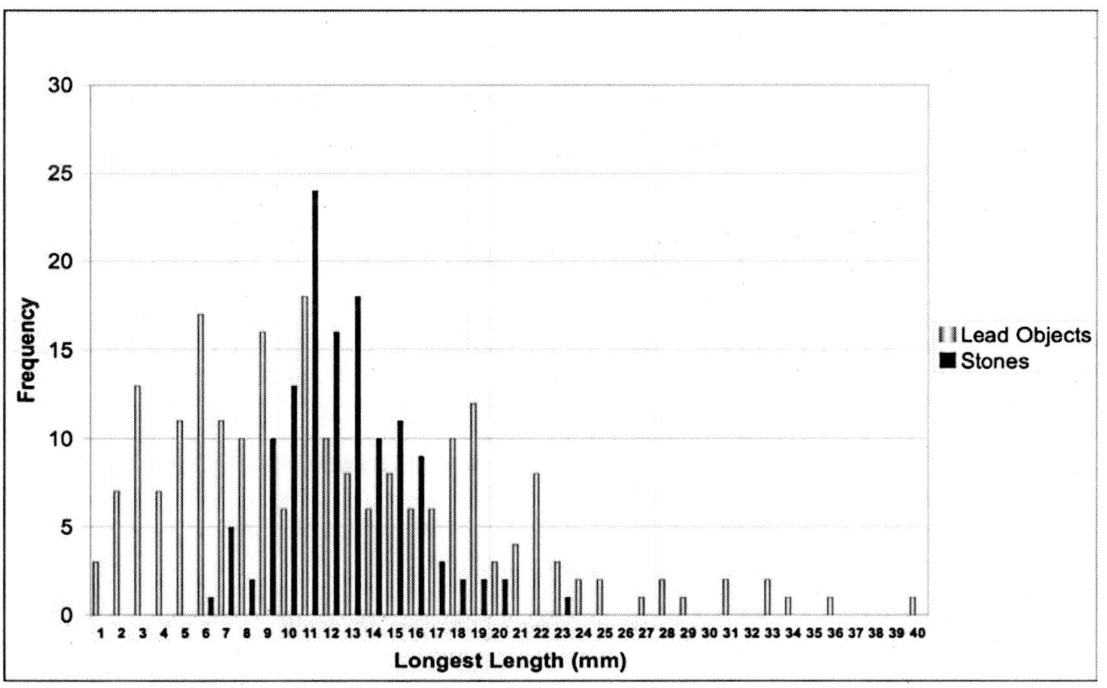

Figure 2. Lengths of lead objects and stones recovered from the ventriculus of 522 Common Loon carcasses.

POKRAS, M. A., M. R. KNEELAND, A. MAJOR, R. MICONI, AND R. H. POPPENGA. 2009. Lead objects ingested by Common Loons in New England. Extended abstract *in* R. T. Watson, M. Fuller, M. Pokras, and W. G. Hunt (Eds.). Ingestion of Lead from Spent Ammunition: Implications for Wildlife and Humans. The Peregrine Fund, Boise, Idaho, USA. DOI 10.4080/ilsa.2009.0116

Key words: Fishing, ingestion, lead, loon, sinkers.

ACKNOWLEDGMENTS

Thanks to the following people who have made these studies possible: Mr. Rawson Wood, Dr. J. McIntyre of Utica College, Dr. P. Spitzer of the Center for Northern Studies, and Ms. C. Perry of the USFWS, Dr. J. C. Franson of the National Wildlife Health Research Center (USGS) and Dr. R. Haebler of the USEPA, and the many veterinary students and volunteers who have helped with this project over the years. We gratefully acknowledge financial support from the USEPA, the USFWS, the Wharton Trust, Massachusetts Environmental Trust, the New Hampshire Charitable Foundation, the North American Loon Fund, and the Wood Family Trust. Participation of the following organizations has been critical to our study: Loon Preservation Committee of New Hampshire; Audubon Societies of Maine, New Hampshire, and Massachusetts; and the Fish and Wildlife Agencies of Maine, New Hampshire, Vermont, and Massachusetts.

Portions of this presentation have been published in the journal *Northeastern Naturalist* for which we thank the editors of that journal.

LITERATURE CITED

BLUS, L. J., R. K. STROUD, B. REISWIG, AND T. MCENEANEY. 1989. Lead poisoning and other mortality factors in Trumpeter Swans. Environmental Toxicology and Chemistry 8:263-271.

FRANSON, J. C., S. P. HANSEN, T. E. CREEKMORE, C. J. BRAND, D. C. EVERS, A. E. DUERR, AND S. DESTEFANO. 2003. Lead fishing weights and other fishing tackle in selected waterbirds. Waterbirds 26(3):345-352.

LOCKE, L. N., S. M. KERR, AND D. ZOROMSKI. 1982. Case report— Lead poisoning in Common Loons (*Gavia immer*). Avian Diseases 26(2):392-396.

SEARS, J., S. W. COOKE, Z. R. COOKE, AND T. J. HERON. 1989. A method for the treatment of lead poisoning in the Mute Swan (*Cygnus olor*) and its long-term success. British Veterinary Journal 145:586-595.

SIDOR, I. F., M. A. POKRAS, A. R. MAJOR, R. H. POPPENGA, K. M. TAYLOR, AND R. M. MICONI. 2003. Mortality of Common Loons in New England 1987 to 2000. Journal of Wildlife Disease 39(2):306-315.

WINDINGSTAD, R. M., KERR, S. M., AND L. N. LOCKE. 1984. Lead poisoning of Sandhill Cranes. (*Grus canadensis*). Prairie Naturalist 16(1):21-24.

DIFFERENCE BETWEEN BLOOD LEAD LEVEL DETECTION TECHNIQUES: ANALYSIS WITHIN AND AMONG THREE TECHNIQUES AND FOUR AVIAN SPECIES

BRYAN BEDROSIAN[1], CHRIS N. PARISH[2], AND DEREK CRAIGHEAD[1]

[1]*Craighead Beringia South, P.O. Box 147, Kelly, WY 83011, USA.* E-mail: bryan@beringiasouth.org

[2] *The Peregrine Fund, 5668 West Flying Hawk Lane, Boise, ID 83709, USA.*

EXTENDED ABSTRACT.—It can often be difficult to compare and interpret results among studies measuring lead levels in varying tissues because of different tissue uptake rates. Further, it can be difficult to directly compare data among studies that sample the same tissue if different methods are used to determine lead levels (e.g., graphite furnace atomic absorption vs. inductively coupled mass spectrometry) without adjusting for discrepancies between techniques. Several recent studies have utilized a relatively new technology (ESA LeadCare® portable field tester) to assess blood lead levels in birds (e.g., Parish et al. 2007, Craighead and Bedrosian 2008). For human blood samples, Pineau et al. (2002) found the LeadCare® system (LCS) tended to underestimate blood lead levels when compared with graphite furnace atomic absorption spectrometry (GFAAS), and others have called into question the validity of this technique (Bossarte et al. 2007). Similarly, Parish et al. (2007) found that the LCS underestimated blood lead levels of California Condors (*Gymnogyps californianus*) when compared with laboratory techniques.

We investigated potential discrepancies of blood lead level detection techniques among four avian species and three techniques. We sampled blood from California Condors, Common Ravens (*Corvus corax*), Bald Eagles (*Haliaeetus leucocephalus*), and Golden Eagles (*Aquila chrysaetos*) and compared inductively coupled plasma mass spectrometry (ICPMS), graphite furnace atomic absorption spectrometry, and the LCS. In general, we found that ICPMS and GFAAS are directly comparable for all four species, but ESA LeadCare® systematically underestimates the blood lead levels of all species, as compared to ICPMS and GFAAS. Further, we found that the differences between techniques may vary among species. We conclude that while the LCS may underestimate the lead levels of individuals, it is a useful technique to gather data inexpensively and rapidly if a species-specific technique differential can be determined. If precise data are required, or to adequately compare LCS results to GFASS and/or ICPMS, then a species-specific calibration between techniques must be designed by analyzing a sub-set of samples with different techniques. *Received 5 September 2008, accepted 22 October 2008.*

BEDROSIAN, B., C. N. PARISH, AND D. CRAIGHEAD. 2009. Differences between blood lead level detection techniques: Analysis within and among three techniques and four avian species. Extended abstract *in* R. T. Watson, M. Fuller, M. Pokras, and W. G. Hunt (Eds.). Ingestion of Lead from Spent Ammunition: Implications for Wildlife and Humans. The Peregrine Fund, Boise, Idaho, USA. DOI 10.4080/ilsa.2009.0122

Key words: Blood, detection, lead, method, species, technique.

LITERATURE CITED

BOSSARTE, R. M., M. J. BROWN AND R. L. JONES. 2007. Blood lead misclassification due to defective LeadCare® blood lead testing equipment. Clinical Chemistry. 53:994–995.

CRAIGHEAD, D., AND B. BEDROSIAN. 2008. Blood lead levels of Common Ravens with access to big-game offal. Journal of Wildlife Management. 72:240–245.

PARISH, C. N., W. R. HEINRICH, AND W. G. HUNT. 2007. Lead exposure, diagnosis, and treatment in California Condors released in Arizona. Pages 97–108 *in* A. Mee, L. S. Hall, and J. Grantham (Eds.). California Condors in the 21st Century. Series in Ornithology, no. 2. American Ornithologists' Union, Washington, DC, and Nuttall Ornithological Club, Cambridge, Massachusetts, USA.

PINEAU, A., B. FAUCONNEAU, M. RAFAEL, A. VIALLEFONT, AND O. GUILLARD. 2002. Determination of lead in whole blood: Comparison of LeadCare blood testing system with Zeeman longitudinal electrothermal atomic absorption spectrometry. Journal of Trace Elements in Medicine and Biology. 16:113–117.

LEAD POISONING IN WHITE-TAILED SEA EAGLES: CAUSES AND APPROACHES TO SOLUTIONS IN GERMANY

OLIVER KRONE[1], NORBERT KENNTNER[1], ANNA TRINOGGA[1], MIRJAM NADJAFZADEH[1], FRIEDERIKE SCHOLZ[1], JUSTINE SULAWA[1], KATRIN TOTSCHEK[1], PETRA SCHUCK-WERSIG[2], AND ROLAND ZIESCHANK[2]

[1]*Leibniz Institute for Zoo and Wildlife Research, P.O. Box 601103, D-10252 Berlin, Germany.* E-mail: krone@izw-berlin.de

[2]*Environmental Policy Research Centre, Free University Berlin, Ihnestrasse 22, D-14195 Berlin, Germany.*

ABSTRACT.—Our project aims to identify the causes and consequences of oral lead intoxications of the White-tailed Sea Eagle (*Haliaeetus albicilla*) as an umbrella species for other scavenging birds. A dialogue-oriented and communicative part of the project encourages involved stakeholders like hunting organizations, foresters, the ammunition industry, ammunition dealers and nature conservationists to develop potential solutions for eliminating lead risks for scavenging birds. Veterinarians, biologists, and social scientists work together to elucidate key issues of the biology of White-tailed Sea Eagle (WTSE), analyse information use and attitudes of hunters as well as conflicts between social actors, transfer knowledge quickly to stakeholders, and mediate between the different parties.

Previously, post mortem examinations of more than 390 WTSEs from Germany performed at the Leibniz Institute for Zoo and Wildlife Research, Berlin revealed that lead intoxications are the most important cause of death (23% of mortality). In this study we identified the potential sources of lead intoxications for WTSEs, being waterfowl such as geese and carcasses of game animals or their remains (gut piles) shot with lead-containing bullets. Three species of geese (n = 154) captured and x-rayed carried embedded shot pellets in 21.4% of all cases. Digital radiographs of game animals (n = 315) shot with semi-jacketed bullets revealed a large number of metallic particles. Isolated lead fragments ranged from less than 1 mm to 10 mm. Gut piles (n = 14) of animals shot with conventional bullets contained metallic particles in 100%. Preliminary results of our feeding experiments suggested that WTSE avoided large (>7.7 mm) but not small particles (<4.4 mm) which may have implications for the further design of lead-free rifle bullets.

Our data on the home-range size of WTSEs support attempts to allocate the source of lead-poisoning to the local areas frequented by eagle pairs. An adult female WTSE used an area of 8.2 km^2 (95% Minimum Convex Polygon) and 4.5 km^2 respectively (95% Fixed Kernel).

To test the toxicity of bullet metals (Pb, Zn, Cu), we conducted feeding experiments on Pekin Ducks. Highest bioavailability and organ accumulation of lead was found in liver, kidney, and brain tissue. We compared the performance of lead-free bullets and lead-containing bullets with respect to hunting/killing efficiency. Expanding bullets made of copper or its alloys offer the possibility of harvesting game that is not contaminated with bullet remains and therefore pose no risk of intoxication to humans and wildlife. Our results, together with the field tests performed by hunters using lead-free ammunition, show that the use of lead-free ammunition is possible in hunting practice. The process of reducing lead intoxications in wildlife by changing to lead-free ammunition among hunters greatly depends on the involvement of all relevant stakeholders and a broad information campaign which we tried to realize by producing a leaflet, an internet page (www.seeadlerforschung.de), and organizing several workshops. Thus far, lead-free bullets are used for hunting by two large associations, eight forestry districts in four federal states, and in one National Park in Germany. We believe our interdisciplinary approach and the early involvement of stakeholders are the keys to the success of this project and a model for problem solving in the field of biodiversity conflicts. *Received 17 June 2008, accepted 8 September 2008.*

KRONE, O., N. KENNTNER, A. TRINOGGA, M. NADJAFZADEH, F. SCHOLZ, J. SULAWA, K. TOTSCHEK, P. SCHUCK-WERSIG, AND R. ZIESCHANK. 2009. Lead poisoning in White-tailed Sea Eagles: Causes and approaches to solutions in Germany. *In* R.T. Watson, M. Fuller, M. Pokras, and W.G. Hunt (Eds.). Ingestion of Lead from Spent Ammunition: Implications for Wildlife and Humans. The Peregrine Fund, Boise, Idaho, USA. DOI 10.4080/ilsa.2009.0207

Key words: Birds of prey, conservation, game animals, heavy metal, stakeholders.

THE WHITE-TAILED SEA EAGLE (*Haliaeetus albicilla*) is the largest European eagle; females have a wingspan up to 2.60 m and a weight of more than seven kg. The population in Germany is increasing, with ~575 breeding pairs in 2007. However, Germany holds a special responsibility for the resettlement of continental Western Europe, because the western distribution ends in northeastern Germany. A slow but steady expansion of the sea eagle´s range in northern, western, and southern directions can be recognized in the last 20 years (Hauff 1998). The long-lived White-tailed Sea Eagle (WTSE) suits superbly as a bioindicator of the accumulation of environmental pollutants in fresh-water and terrestrial ecosystems which may accumulate within the food chain of this top predator. Its habitat is characterised by fresh-water lakes, large rivers, and shore lines where it uses undisturbed nesting places and trees for perching in search of prey. Susceptibility to disturbance is a result of decades of persecution by man before the species became fully protected in the 1930s; the eagle still reacts very sensitively to new or unknown disturbances, especially during nest construction and breeding. Intensive protection measures and the ban of persistent pesticides (e.g., DDT) resulted in a population increase in Germany since the 1980s. Diet depends on the availability of prey in specific habitats and varies strongly during the seasons. Fish comprise the majority of prey during spring and summer, followed by waterfowl in fall and winter, then mammals, often acquired as carcasses or gut piles, in winter (Oehme 1975, Struwe-Juhl 1998).

Studying the health status, including diseases, accumulation of pollutants, and causes of death of top predators such as the WTSE is valuable to gain information on the environmental health. The WTSE as a bioindicator is more sensitive and responds earlier to changes in ecosystem health than humans do. As a scavenger the WTSE is especially susceptible to poisoning.

Since 1996 WTSEs were routinely examined for their diseases and causes of death at the Leibniz Institute for Zoo and Wildlife Research, Berlin (IZW). At the International White-tailed Sea Eagle Conference in Björko, Sweden in 2000 the relevance of lead poisoning as an important mortality factor for WTSEs in Germany was demonstrated for the first time to an international audience of experts (Krone et al. 2003). Organ samples of eagles examined at the IZW were analysed toxicologically at the Research Institute of Wildlife Ecology in Vienna and revealed lethal concentrations of lead in liver and kidneys (Kenntner et al. 2001). The sources of lead intoxications were fragments of rifle bullets and, more rarely, lead shot ingested together with food, e.g. carcasses of game animals, gut piles, and shot waterfowl.

The results were presented at the first national workshop "Lead containing hunting bullets: Cause of death in White-tailed Sea Eagles?" in April 2005 to all relevant stakeholders such as representatives of hunting organizations, ammunition industry, ammunition suppliers, foresters, and nature conservationists in Berlin (Krone and Hofer 2005). Stakeholders at the workshop could hardly believe the results, and so all participants developed a full catalogue of open questions to be answered in a large scale investigation on the causes and potential solutions of lead poisoning in White-tailed Sea Eagles in Germany. The questions included the following:

1. What are the sources of lead intoxications in the WTSE?
2. Is there any explanation why only small bullet fragments were found in the gizzards of the WTSEs? Do the eagles avoid the ingestion of large metallic particles?
3. Do the sources of lead intoxication to adult, territorial eagles occur within specific areas?
4. Are there any options to reduce the risk of lead exposure?
5. Do lead-free bullets perform as efficiently as lead-containing bullets? How toxic are the alternative metals?

A new project containing social and natural science elements was created to answer these questions, and funded by the Federal Ministry of Education and Research. The natural science part of the project mainly covered the questions mentioned above, whereas the social component was concerned with analysing the knowledge of hunters regarding the lead problem, performing a discourse and conflict analysis, and acting as mediator in the stakeholder discussions. Natural science is important to the description and analysis of the problem of lead poisoning in birds of prey and for the evaluation of lead-free ammunition. Social science targets communication of scientific results as fundamental to societal opinion-making. Our work is focused on different groups of stakeholders, especially hunters, by analysing how they deal with the problem, which kinds of barriers and channels exist, and when changes in behavior are suggested. Our approach in social science therefore encompasses both analytical and dialogue-oriented aspects. Together with results from the ongoing research in the involved disciplines of natural sciences, the joint project is engaged in informing and communicating with stakeholders and with the public.

METHODS

Sources of Lead Intoxication.—To investigate the sources of lead intoxications in the WTSE we examined game animals including geese and game ungulates. Arctic geese such as White-fronted Geese (*Anser albifrons*) and Bean Geese (*Anser fabalis*) were caught in northern Germany during the fall migration with cannon nets or by traditional methods from the Netherlands, as well as flightless Greylag Geese (*Anser anser*) of the German population during their moult in June. Each goose was weighed, ringed, marked with collars and finally x-rayed before release back to the wild.

Game ungulates such as Roe Deer (*Capreolus capreolus*), Wild Boar (*Sus scrofa*), Fallow Deer (*Cervus dama*), Red Deer (*Cervus elaphus*), and Chamois (*Rupicapra rupicapra*) were x-rayed within 90 minutes after they were shot. The game animals were hunted in Bavaria, Berlin, Brandenburg, Mecklenburg-Western Pomerania, and Schleswig-Holstein.

We used two mobile x-ray units (Vet Ray Gamma 2000, Acona and Gamma Titan, Poscom) and imaging plates for computed radiography (Fuji CR ST-VI). The plates were scanned with a VetRay® CR35V scanner (VetRay GmbH) and subsequently processed using a laptop and the software VetRay® Vision 4.4 (VetRay GmbH). Radiographs were taken in laterolateral and ventrodorsal directions. In the game ungulates, entrance and exit wounds were marked with the help of standard medical cannulae. Hunters filled in a standardized shooting report for each animal. All radiographs were examined for wounding patterns and for the number, size, and distribution of metallic particles.

Feeding Behavior.—To understand why only small fragments were found in the gizzards of dissected WTSEs, we conducted feeding experiments on free-ranging sea eagles in Mecklenburg-Western Pomerania. Carcasses of game animals or gut piles were prepared with chamfered metallic nuts of different sizes and offered as food. The nuts were of iron (Fe), constituting no toxicological exposure for the eagles when given in controlled dosage (Fiedler and Rösler 1993, Kelly et al. 1998, Mitchell et al. 2001, Brewer et al. 2003). The feeding behavior of the experimental animals was observed by video surveillance (Scheibe et al. 2008). Baits were x-rayed after sea eagles fed on them, and we recorded the number of nuts ingested. We further examined the experimental site with a metal detector to locate remaining nuts.

Territoriality of White-tailed Sea Eagles.—To answer the question if it is possible to attribute lead intoxication in adult, territorial eagles to a specific area, we equipped adult WTSEs with satellite-reporting transmitters. Our main study area lay in northern Germany within the Lake District of the federal state of Mecklenburg-Western Pomerania. We trapped adult sea eagles with bow nets and fitted them with backpack GPS/VHF transmitters (Vectronic Aerospace, Berlin). All transmitters were programmed to receive one GPS location per day in a chronologically circulating schedule with a delay of one hour per day to insure independence of data. Data from eight tagged animals were available (five females, three males). All GPS locations were imported into a Geographical Information System (GIS, ArcView 3.3 and ArcGIS 9) using vegetation maps and high resolution digital orthophotos as base layers for habitat analysis. Resource selection patterns were analysed using methods such as the Euclidean distance approach (Conner and Plowman 2001), selection indices (Manly et al. 2002), and compositional analysis (Aebischer et al. 1993).

Reducing the Risk of Lead Exposure.—Possibilities of reducing the risk of lead exposure caused by hunting ammunition were discussed at three expert talks/workshops with the involved stakeholders. Representatives from hunting organizations, ammunition industry, ammunition suppliers, foresters, and nature conservationists participated in our workshops.

Lead-free Hunting Ammunition.—To examine if lead-free bullets performed as well as lead-containing bullets with respect to hunting/killing efficiency, x-ray images of hunted game ungulates (see above) were analysed for differences. The absorption rate of lead and metals alternatively used in lead-free bullets was tested in an avian model. We purchased zinc shot (Jagd- und Sportmunitions GmbH/Germany) and lead shot (Rottweil Tiger/Germany) in the sizes which were the recommended size to test non-toxic shot by Environment Canada (1993), similar to the US #4 shot size (~3.3 mm). Spherical copper and brass according to the U.S #4 shot size were obtained from German companies. Details of the shot sizes and weights are given in Table 1.

Six groups consisted of 40 domestic Pekin Ducks (*Anas platyrhynchos*), each held in mesh cages with ground litter, with four ducks within each cage. Six shotgun pellets were placed in the gizzard of each duck of the experimental groups by oral intubation. Application success and retention of pellets were controlled by x-raying each bird on the day after intubation, after the first week, and one day before slaughtering. During routine necropsy, the body condition and sex of each duck were also determined. Organ samples for histopathological examinations were immediately fixed in buffered formalin, whereas those for toxicological analysis were kept at -20°C prior to microwave-assisted acid digestion. We utilized graphite furnace Atomic Absorption Spectroscopy (AAS) methods for lead measurements, and flame AAS methods for analysis of copper and zinc (Analytik Jena ZEEnit 700). Gizzards and intestines of all birds were washed and sieved to screen for retained shot pellets. Retained pellets were measured and weighed.

Table 1. Sizes and weights of each test shotgun pellet (n=20 per shot) used for avian toxicity test.

Group	1	2	3	4	5
Metal	Control	Copper	Brass	Zinc	Lead
No. pellets	0	6	6	6	6
Size±SD [mm]	-	3.15±0.05	3.0±0.00	3.05±0.14	3.2±0.05
Weight±SD [mg]	-	148±0.33	120±3.2	105±12.3	191±7.1

Social Science Component.—We distinguished between two operating levels and participated accordingly in two different kinds of data collections. Levels included a macroscopic level and a microscopic level. With the former, i.e. political institutions, structures, and policy-making organizations, we implemented methods based on secondary and historical data. On the microscopic level, i.e. interactions and beliefs of individuals, we applied methods based on primary and situation-dependent data. Applied methods were conflict analysis (macroscopic level) and discourse analysis (microscopic level) and referred to all of the stakeholders involved, including hunting organizations, the ammunition industry, ammunition suppliers, forest owners, and nature conservationists, in addition to considering the affected federal agencies in Germany. Our analysis of level of information, problem acceptance, and handling of information pertained to hunters. This analysis mainly derived from a survey on knowledge of lead intoxication in raptors and acceptance of lead-free ammunition. However, we also investigated how average hunters get their information, e.g., from field experience, the internet, or other sources. We distributed the questionnaire by direct mail, by inserts in hunter's magazines, and through hand-outs. To date, more than 70,000 copies have been distributed.

RESULTS

Sources of Lead Intoxication.—Thus far, 154 geese, including 84 White-fronted Geese, 53 Bean Geese, and 17 Greylag Geese have been tested in Germany. Among the total, 21.4% had embedded shot, varying from one to seven shotgun pellets in their body tissue (Table 2). We could not differentiate between lead shot and non-lead shot in the radiographs.

Table 2. Frequency of geese with embedded shot pellets in Germany.

Species	n	shot [n]	shot [%]
Greylag Goose	17	2	12
White-fronted Goose	84	15	18
Bean Goose	53	16	30

Furthermore, we analysed radiographs of 315 game animals (110 Roe Deer, 97 Wild Boars, 103 Fallow Deer, three Red Deer, and two Chamois) shot with lead-containing bullets. Radiographs of animals shot with these semi-jacketed bullets typically revealed a large number of metallic particles located throughout the entire wound channel and its surroundings (Figure 1 and 2). Fragment size ranged from less than 1 mm to almost 10 mm, with many particles being very small. Hits of large bones were not required for the fragmentation. Table 3 gives the mean, minimum, and maximum number of fragments for some common lead-core rifle bullets.

Table 3. Number of metallic particles counted in radiographs of animals shot with lead-core bullets.

Bullet	n	Mean	Median	SD	Minimum	Maximum
Semi-jacketed	26	89	70	58	30	250
Semi-jacketed round nose	15	120	100	83	45	300
Brenneke TUG	6	230	200	157	50	480
Norma Vulkan	13	120	100	45	75	210
RWS Kegelspitz	8	104	100	55	25	200
RWS Evolution	9	279	250	130	120	500

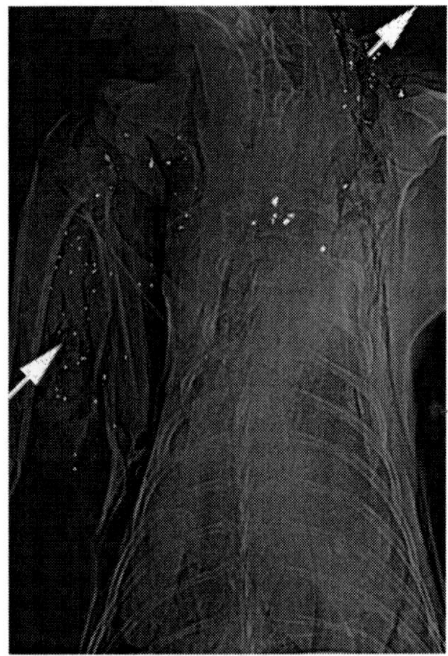

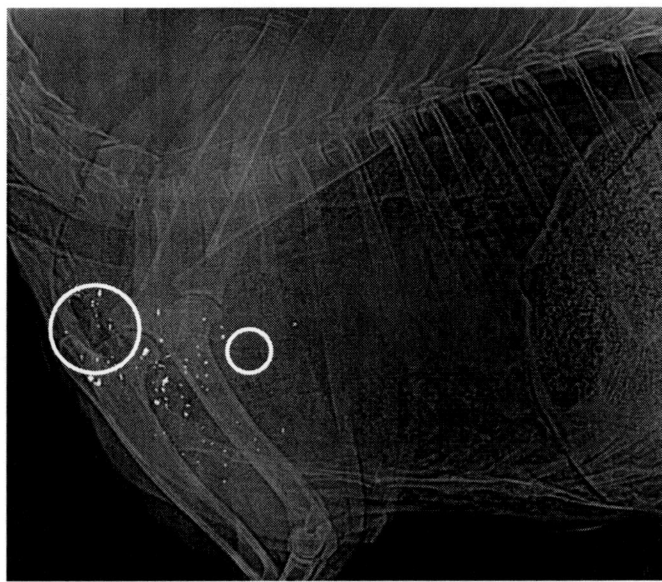

Figures 1 and 2. Ventrodorsal and laterolateral radiographs of a Roe Deer shot with a semi-jacketed bullet. Arrows in Figure 1 mark entrance and exit sites, small circle in Figure 2 indicates the entrance wound, and the larger circle indicates the exit wound.

Gut piles of animals shot with conventional bullets were interspersed with metallic particles in 100% of the cases examined (n=14). The number of fragments varied from two to 600 depending on where the animal had been hit.

None of our findings indicate a lower effectiveness of lead-free bullets. We found no difference between the frequency of retained bullets of lead-free and conventional bullets (Fisher's Exact Test, df=1, $P_{two\ sided}$=1,000, $P_{one\ sided}$=0.456).

Feeding Behavior.—Our preliminary results show that most iron nuts we inserted in carcasses or gut piles were avoided by sea eagles during feeding. In the following we present the frequency of nut avoidance during feeding by one territorial sea eagle pair as well as by roaming juveniles. In the course of three feeding experiments, the sea eagles avoided 71% of the inserted nuts (n = 28 inserted nuts). The proportions of avoided nuts of different diameters by the sea eagles varied distinctively (Figure 3). Nuts with diameters of 8.8 and 7.7 mm were completely avoided. The avoidance of smaller nuts decreased with nut size.

Territoriality of White-tailed Sea Eagles.—Here we present the data of only one adult female WTSE. The female "472" used an area of 8.2 km^2 with respect to the 95% Minimum Convex Polygon, and one of 4.5 km^2, as measured by 95% Fixed Kernel. Her activities were concentrated within a core area of 0.5 km^2 (Figure 4). Mean distances of all GPS locations (n = 476) to the habitat types "lake" and "riparian vegetation" were 416 and 359 m, compared to expected distances of 1644 and 1303 m calculated from a random point data set (Figure 5). This eagle died in January 2004 due to lead intoxication (Krone et al. 2009).

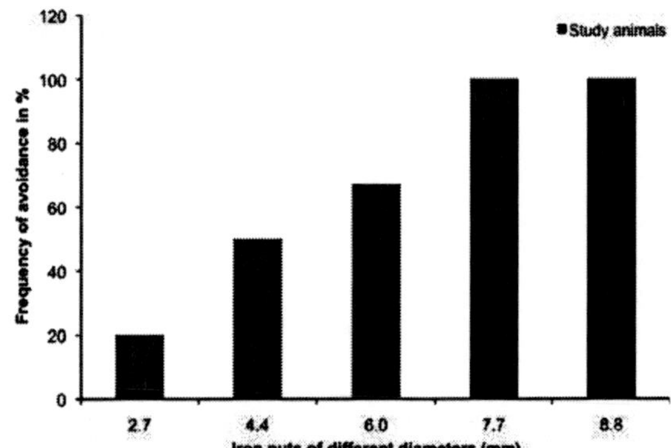

Figure 3. Proportions of avoided steel nuts of different diameters.

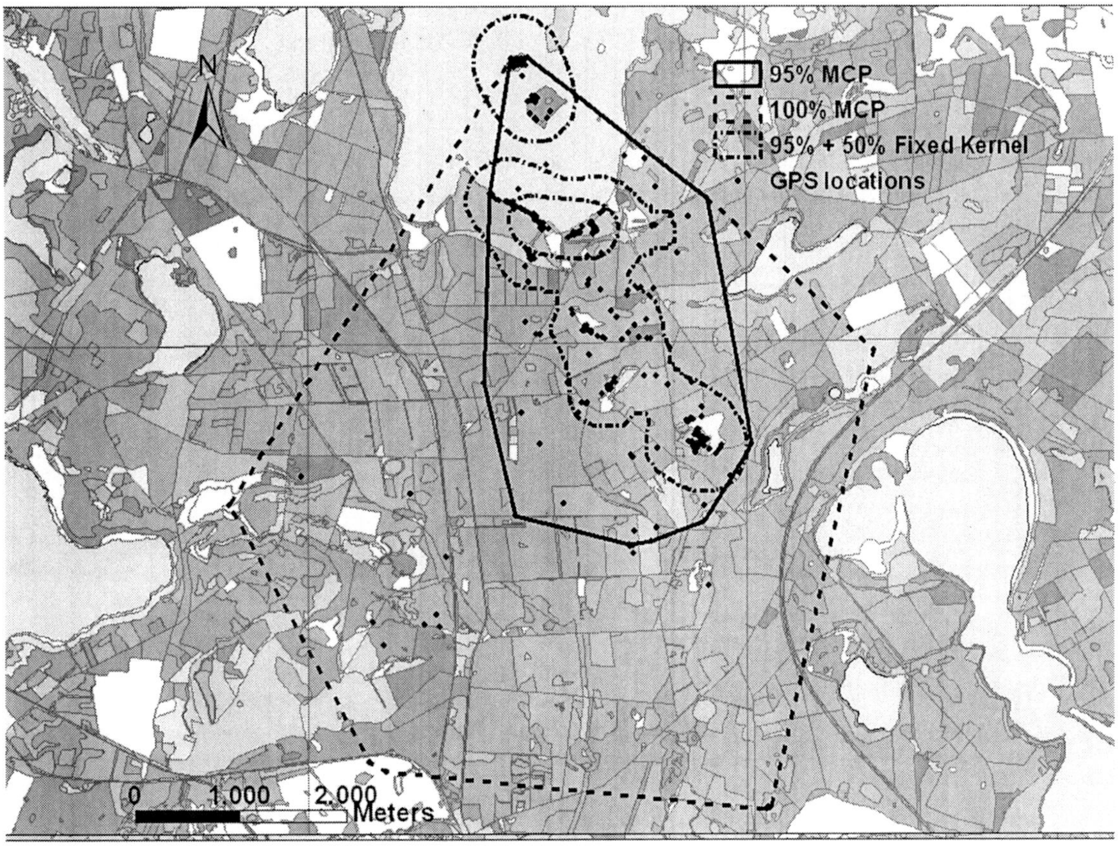

Figure 4. Home range of sea eagle "472", MCP = Minimum Convex Polygon.

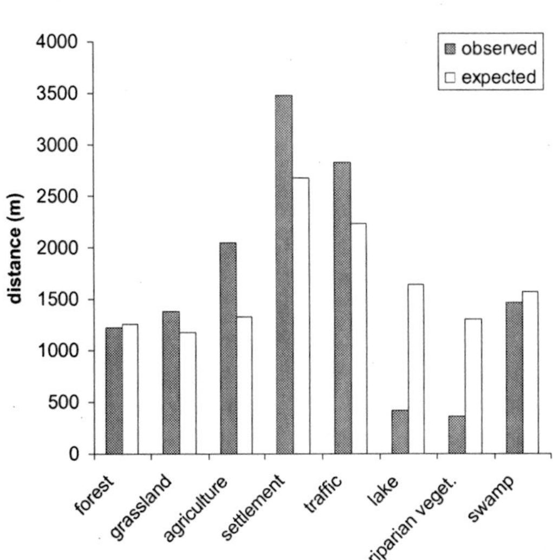

Figure 5. Mean distances to habitat types of sea eagle "472."

Reducing the Risk of Lead Exposure.—At the first workshop in April 2005, findings on the causes of mortality of WTSEs were presented and the importance of lead intoxication caused by remains of hunting ammunition was emphasized. The sources were described as bullet remains in carcasses of game animals or gut piles, and for lead shotgun pellets as embedded in small game animals such as waterfowl. Our results stimulated a heavy discussion among the stakeholders. Possibilities of reducing the risk of lead exposure caused by hunting ammunition were discussed with all involved stakeholders. Burying gut piles was advanced as a potential intermediate solution to the problem, but switching to lead-free ammunition was considered as a solution only if killing efficiency and low toxicological risk of alternative ammunition could be demonstrated.

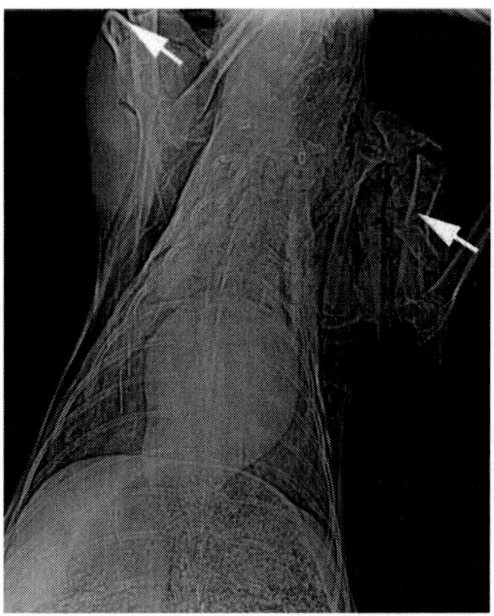

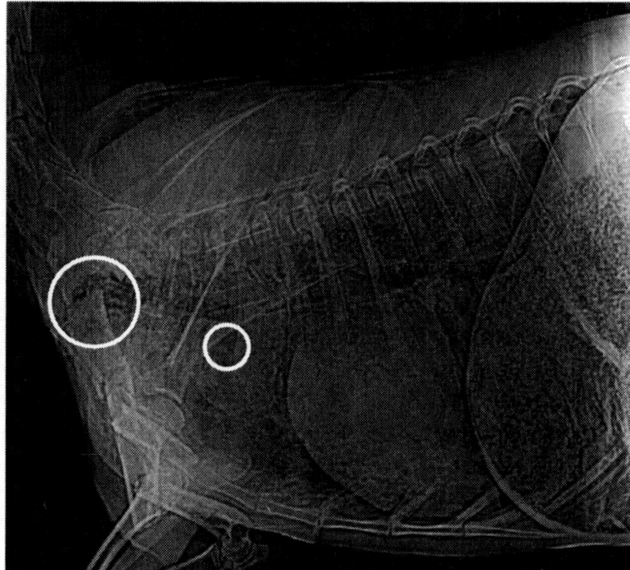

Figures 6 and 7. Ventrodorsal and laterolateral radiographs of a Roe Deer shot with an expanding copper bullet (markings as in Figures 1 and 2).

Table 4. Number of metallic particles counted in radiographs of game animals shot with lead-free bullets.

Bullet	n	Mean	Median	SD	Minimum	Maximum
Lapua Naturalis	85	0	0	0	0	0
Barnes XLC	40	0	0	0	0	0
RWS Bionic Yellow	6	23	23	12	8	40
Reichenberg HDBoH	21	6	1	22	0	100
Möller KJG	10	12	10	7	0	25

At the second workshop in March 2007, we presented a leaflet and the project homepage as information material to the stakeholders. In addition, we argued against the potential solution of burying gut piles on the basis that this method can not be realized in frozen ground in winter, the main hunting time, and because of its general impracticality. In several cases, when gut piles were buried, Wild Boar detected and unearthed them, with the result that they were again available to eagles. A major step forward was achieved during the discussion with all involved stakeholders, resulting in two milestones: first the general agreement that lead intoxication in WTSEs was caused by the ingestion of lead-containing bullet fragments, and second that action was necessary to reduce lead poisoning in WTSEs. Results of the third workshop in May 2008 are not available yet; however, the overall idea was a phase-out of lead-containing ammunition as soon as comparable lead-free ammunition became available.

Lead-free Hunting Ammunition.—Expanding copper bullets that are meant to enlarge their cross-sectional area without losing mass did not leave fragments in the wound channel even if major bones were hit (Figure 6 and 7). Other unleaded bullets which are constructed for partial fragmentation left behind some relatively large particles; the numbers we observed in our tests are given in Table 4. In three of 311 (0.97%) cases, lead-free bullets were retained in the animal, whereas one of 123 (0.81%) lead-core bullets did not produce an exit wound.

Toxicity of Alternative Lead-free Ammunition.— The preliminary results represented here show the toxicological findings of 10 ducks of each group for the element levels of lead, copper, and zinc in liver, kidney, and brain tissue. All levels of elements are given in µg/g on a wet weight (ww) basis. None of the birds died during the experimental period, and no consistent or group-specific organ alterations were detected. Weight losses were highest in the zinc and lead shot groups, whereas copper and brass did not show relevant losses after four weeks of retention in duck gizzards (Figure 8). All lead levels in organs (liver, kidney, brain) of the lead group differed significantly from the lead levels of all the other groups, but there were no such differences between the elements copper and zinc and the other groups (Figure 9-11).

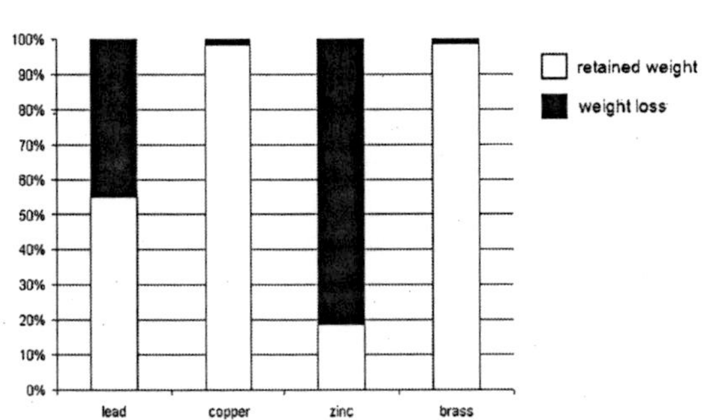

Figure 8. Percentage of weight losses of shot (lead, copper, zinc, brass) during a four-week retention period in the gizzards of Pekin Ducks.

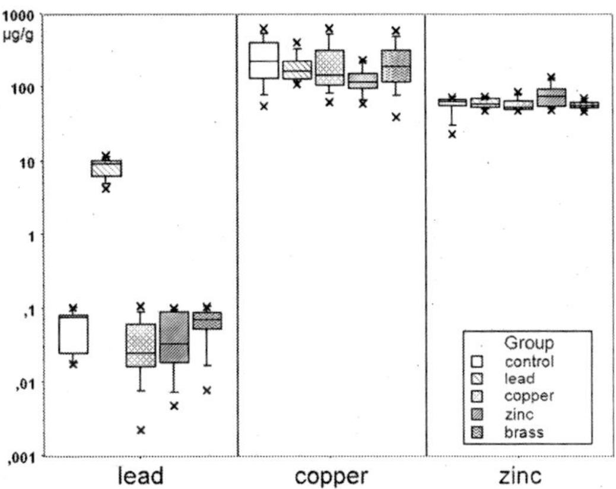

Figure 9. Levels of lead, copper and zinc in liver of Pekin Ducks fed with six shotgun pellets of either lead, copper, zinc or brass and the control group. Each group consists of 10 ducks.

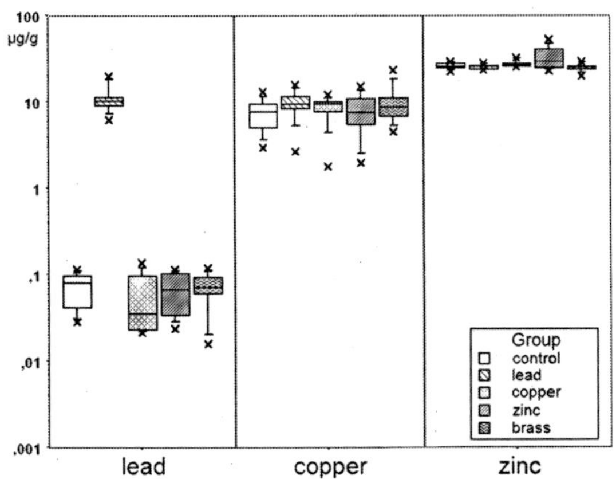

Figure 10. Levels of lead, copper and zinc in kidneys of Pekin Ducks fed with six shotgun pellets either of lead, copper, zinc or brass and the control group. Each group consists of 10 ducks.

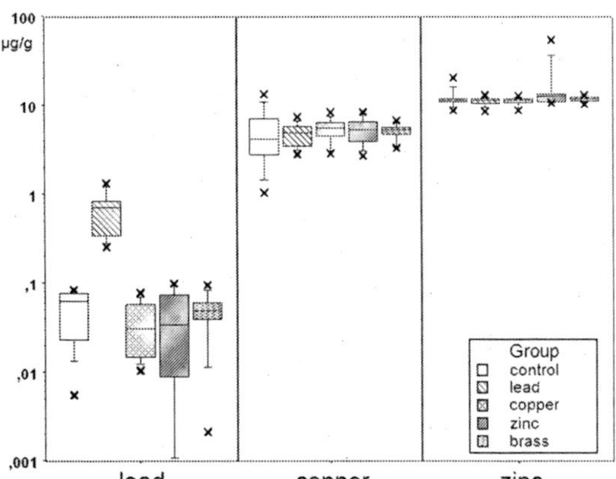

Figure 11. Levels of lead, copper and zinc in brain tissue of Pekin Ducks fed with six shotgun pellets either of lead, copper, zinc or brass and the control group. Each group consists of 10 ducks.

Social Science Component.—We have had three expert talks/workshops (see above) so far. At the first of these, the stakeholders were informed about the problem of lead intoxication in birds of prey. The discourse analysis of the second expert talk showed an agreement among all about the causes of lead intoxication of raptors and about need for action. None of the stakeholders expressed doubt about the lethal effect on sea eagles of feeding on shot game or gut piles containing particles of lead ammunition. Hunting organizations and ammunition industry/suppliers have clearly moved from their former positions.

The discourse analysis of the third expert talk is still ongoing. However, we can now state that the second part of the dialogue reconfirmed the agreement concerning need for action, although all participants do not share the same expectations about when actions can be taken. The questionnaire survey in different parts of Germany is expected to be completed in July 2008; first results will be available in October 2008 or soon thereafter.

DISCUSSION

Sources of Lead Intoxication.—In comparison with the high incidence of 30% and 18% of shot Bean Geese and White-fronted Geese, respectively, we found only 12% among Greylag Geese. However, the sample size of Greylag Geese is still small, but should soon be increased. Averbeck et al. (1990) radiographed 467 birds of 51 species found dead or moribund in northern Germany between 1985 and 1988. In total 15.8% of all birds had been shot, of which 80% had lead shot in their body tissue, and 11 birds were shot with air guns. Out of 355 Pink-footed Geese x-rayed from 1990 to 1992, one quarter (25%) of the first-year ducks and 36% of the adult ducks had embedded shot pellets (Noer and Madsen 1996). Both studies showed similar incidences of birds with embedded shot as our preliminary results.

Particle clouds, also called "lead snowstorms," a common term in forensic literature (Messmer 1998), are caused by semi-jacketed bullets (Hunt et al. 2006). Game animals harvested with conventional lead-core bullets obviously are widely contaminated with bullet material even if the bullet does not hit any bones. The fragments are found in all organs. The small fragment size and the wide area contaminated with particles make it impossible to remove all fragments by cutting off obviously crushed tissues. Complete game animal carcasses as well as gut piles containing lead fragments of semi-jacketed bullets pose a severe threat to scavenging animals and humans consuming game meat.

Feeding Behavior.—Our preliminary results suggest that sea eagles are able to feed selectively and may avoid the ingestion of large metal particles but not small ones. These findings suggest that the use of lead-free rifle bullets that deform or fragment in large pieces may present low risk for metal ingestion by sea eagles (see below). Detailed results and analyses of the selective feeding behavior by sea eagles, including data involving more experimental animals will be published soon.

Territoriality of White-tailed Sea Eagles.—The home range of sea eagle "472" was rather small, and the animal did not undertake excursions regularly (Figure 1). This first adult white-tailed sea eagle equipped with a GPS-datalogger died after 173 days due to an oral lead intoxication caused by the ingestion of fragments of a rifle bullet (Krone et al. 2009).

Lead-free Hunting Ammunition.—Expanding bullets made of copper or its alloys present a possibility of harvesting game that is not contaminated with bullet remains. In case of nose-fragmenting copper or brass bullets, the sizes of remaining fragments are larger than those of conventional bullets, so less metal surface is available for resorption by avian scavengers, and due to the larger size those particles may be avoided during ingestion.

Toxicity of Alternative Lead-free Ammunition.—Our preliminary results indicate highest bioavailability and significant organ accumulation for lead in liver, kidney, and brain tissue of ducks. Highest solubility in duck gizzards was measured for zinc, but organ levels of ducks from the zinc group were comparable with all other groups and within the physiological range of <100 µg/g in liver as summarized by Eisler (1988), and below the toxic levels of about 400 µg/g, 300 µg/g, and 2000 µg/g for liver, kidney, and pancreas, respectively, as reported by Levengood et al. (1999). Copper and brass as metals for lead-free bullets are recommended, because of their very low solubility in duck gizzards, and their non-significant accumulation effects in the investigated organs. There was no significant dissolution or organ accumulation of zinc from brass. However, considering different digestion physiology and the relatively low pH values in gizzards of raptors (Duke et al. 1975), experimental studies on the toxicity of copper and brass in raptors should certify our preliminary results.

Reducing the Risk of Lead Exposure.—Two main possibilities were discussed and tested in the field to reduce the risk of lead intoxication in WTSEs: burying gut piles and switching to lead-free ammunition. Burying gut piles was identified as being an impractical method, whereas the large scale use of lead-free ammunition has the potential to eliminate the lead in the prey and carrion that WTSE and other raptors are feeding on.

At the beginning of our project, several open questions regarding the pathways and consequences of lead poisoning in WTSEs were addressed. We also tested the wound ballistics of lead-free ammunition in game animals. Our preliminary results demonstrated the potential sources of lead as shotgun pellets in geese and lead particle clouds in gut piles and game animals shot with lead-containing bullets. Territorial WTSEs very likely acquire lethal lead concentrations when feeding on carrion or gut piles, or preying on shot waterfowl with embedded lead shot pellets, within their year-round home range. The experiments performed on WTSEs revealed a selective feeding behavior of avoiding large metallic fragments during ingestion of meat, which may be of relevance for future bullet design. Feeding experiments on ducks testing for the bioavailability of alternative metals for lead-free bullets showed that copper and brass were suitable materials for bullets because of their low toxicity in birds.

Our results, together with the field tests performed by hunters using lead-free ammunition, show that the use of lead-free ammunition is acceptable in hunting practice. When using nose-deforming lead-free bullets for hunting, no bullet fragments were found in x-rayed game animals, resulting in uncontaminated food for humans and wildlife.

So far, lead-free bullets are used for hunting in the Nature and Biodiversity Conservation Union (NABU), the Ecological Hunting Association, and eight forestry districts in four federal states and in the Mueritz National Park in Germany. In addition, hunting with lead-containing rifle bullets is forbidden in all state forest districts in the Federal country of Brandenburg by ministry decree.

The process of reducing lead intoxications in wildlife by changing to lead-free ammunition among hunters greatly depends on the involvement of all relevant stakeholders and a broad information campaign which we initiated by producing a leaflet, an internet page (www.seeadlerforschung.de), and several workshops.

Social Science Component.—The realization and documentation of the dialogues between the stakeholders played an important role within the social science part of the project. The dialogues are supposed to intensify the exchange of opinions among different protagonists concerning risks of lead ammunition and to identify options for behavioral change. Scientific results are to be communicated and interfaced with the knowledge and experience of relevant social and economic groups and responsible governmental participants in order to recognize problems and their solutions.

We were unable, in the second expert talk, to remove two conflicts. The first pertains to the demands and standards of alternative ammunition; hunters find them much too expensive in comparison with conventional lead-ammunition. Second, there is the conflict of legal regulation. Does a ban on lead ammunition make sense? Is it the best way for implementation and acceptance in the hunting community? Three alternative actions, each involving uncertainty, became evident during the course of the second expert talk: (1) voluntary application of lead-free ammunition with the option of reversal, (2) legal regulation on the basis of present scientific information, and (3) waiting for definitive results from ongoing monitoring processes.

To summarize, the joint research and communication project is characterised by four central issues:

1. Science as an actor: Research remains not in the background, looking for 'objective' results from the policy field. On the contrary, the members of the project are doing 'action research' and in this understanding they are agents of knowledge transfer.
2. This includes the exchange of information among hunters and other stakeholders about trends in ammunition development, information from state organizations about consumer risks of contaminated game, and dissemination of trends about governance in other countries.
3. Through dialogue-oriented research, the results of the questionnaire will be published in hunter magazines and publications of pressure groups.
4. Stakeholder-dialogues are proven instruments of transdisciplinary discussion of knowledge from natural sciences, social sciences, and know-how of different stakeholder communities.

The next steps of the research programme will deal with in-depth interviews of central stakeholders on topics like basic conflict attitudes or opinions, and possibilities for a self-reinforcing agreement between different organizations. One example could be the creation of a future market for lead-free ammunition together by producers and users of alternative ammunition.

ACKNOWLEDGMENTS

We are grateful to the Federal Ministry of Education and Research (BMBF) for funding and to the Project Management Juelich (PTJ) for administration of this project. We would like to thank T. Scherer, M. Duhr and A. Schulze, all staff members and former staff members of the Ministry of Rural Development, Environment and Consumer Protection of the Federal State of Brandenburg, Forest management and company head-office, who organized the monitoring of lead-free ammunition in hunting practice in the federal state of Brandenburg. We are much obliged to the staff and administration of the nature park "Nossentiner/ Schwinzer Heide" in Mecklenburg-Western Pomerania for their help conducting all the field work on White-tailed Sea Eagles, to the Ministry of Agriculture, Environment and Consumer Protection, and to the State Authority for Environment, Nature Conservation, and Geology, both in Mecklenburg-Western Pomerania, for their support and the permissions necessary for the project. Additionally we would like to thank the Reepsholt Foundation for general support and P. Hauff, Dr. W. Mewes, Dr. W. Neubauer, and C. Scharnweber for sharing their knowledge on White-tailed Sea Eagles.

LITERATURE CITED

AEBISCHER, N. J., P. A. ROBERTSON, AND R. E. KENWARD. 1993. Compositional analysis of habitat use from animal radio-tracking data. Ecology 74:1313-1325.

AVERBECK, C., E. KEMPKEN, S. PETERMANN, J. PRÜTER, G. VAUK, AND C. VISSE. 1990. Röntgenuntersuchungen zur Bleischrotbelastung tot aufgefundener Vögel in Norddeutschland. Zeitschrift für Jagdwissenschaften 36:30-42.

BREWER, L., A. FAIRBROTHER, J. CLARK, AND D. AMICK. 2003. Acute toxicity of lead, steel, and an iron-tungsten-nickel shot to Mallard Ducks (*Anas platyrhynchos*). Journal of Wildlife Diseases 39:638-648.

CONNER, L. M., AND B. W. PLOWMAN. 2001. Using Euclidean distances to assess nonrandom habitat use. Pages 275-290 *in* J. J. Millspaugh, and J. M. Marzluff (Eds.). Radio Tracking and Animal Populations. Academic Press, San Diego, California, USA.

DUKE, G. E., A. A. JEGERS, G. LOFF, AND O. A. EVANSON. 1975. Gastric digestion in some raptors. Comparative Biochemistry and Physiology 50A:649 -656.

ENVIRONMENT CANADA. 1993. Toxicity test guidelines for non-toxic shot for hunting migratory birds. Ottawa, Ontario, Canada.

EISLER, R. 1988. Copper hazards to fish, wildlife, and invertebrates: a synoptic review. US Geological Survey, Biological Resources Division, Biological Science Report USGS/BRD/BSR-1997-0002, Contaminant Hazard Reviews Report 26, Laurel, Maryland, USA.

FIEDLER, H. J., AND H. J. RÖSLER. 1993. Spurenelemente in der Umwelt. Gustav Fischer Verlag, Jena, Germany.

GANUSEVICH, S. 1996. The White-tailed Sea Eagle *Haliaeetus albicilla* in Kola Peninsula. Populationsökologie Greifvogel und Eulenarten 4:101-110.

HAUFF, P. 1998. Bestandentwicklung des Seeadlers *Haliaeetus albicilla* seit 1980 mit einem Rückblick auf die vergangenen 100 Jahre. Vogelwelt 119: 47-63.

HUNT, W. G., W. BURNHAM, C. N. PARISH, K. BURNHAM, B. MUTCH, AND J. L OAKS. 2006. Bullet fragments in deer remains: implications for lead exposure in scavengers. Wildlife Society Bulletin 34:168-171.

KELLY, M. E., S. D. FITZGERALD, R. J. AULERICH, R. J. BALANDER, D. C. POWELL, R. L. TICKLE, W. STEVENS, C. CRAY, R. J. TEMPELMAN, AND S. J. BURSIAN. 1998. Acute effects of lead, steel, tungsten-iron, and tungsten polymer shot administered to game-farm Mallards. Journal of Wildlife Diseases 34:673-687.

KENNTNER, N., F. TATARUCH, AND O. KRONE. 2001. Heavy metals in soft tissue of White-tailed Eagles found dead or moribund in Germany and Austria from 1993 to 2000. Environmental Toxicology and Chemistry 20:1831-1837.

KRONE, O., T. LANGGEMACH, P. SÖMMER, AND N. KENNTNER. 2003. Causes of mortality in White-tailed Sea Eagles from Germany. Pages 211-218 *in* B. Helander, M. Marquiss, W. Bowerman (Eds.). Sea Eagle 2000. Proceedings of the Swedish Society for Nature Conservation/SNF, Stockholm, Sweden.

KRONE, O., F. WILLE, N. KENNTNER, D. BOERTMANN, AND F. TATARUCH. 2004. Mortality factors, environmental contaminants, and parasites of White-tailed Sea Eagles from Greenland. Avian Diseases 48:417-424.

KRONE, O., AND H. HOFER (EDS.). 2005. Bleihaltige Geschosse in der Jagd – Todesursache von Seeadlern? Leibniz Institute for Zoo and Wildlife Research, Berlin, Germany.

KRONE, O., T. STJERNBERG, N. KENNTNER, F. TATARUCH, J. KOIVUSAARI, AND I. NUUJA. 2006.Mortality factors, helminth burden, and contaminant residues in White-tailed Sea Eagles (*Haliaeetus albicilla*) from Finland. Ambio 35:98-104.

KRONE, O., A. BERGER, AND R. SCHULTE. 2009. Recording movement and activity pattern of a White-tailed Sea Eagle (*Haliaeetus albicilla*) by a GPS datalogger. Journal of Ornithology 150:273-280, DOI 10.1007/s10336-008-0347-1

MANLY, B. F. J., L. MCDONALD, D. L. THOMAS, T. MCDONALD, AND W. P. ERICKSON. 2002. Resource Selection by Animals: Statistical Design and Analysis for Field Studies, 2nd ed. Kluwer Academic Publishers, Dordrecht, Germany.

MESSMER, J. M. 1998. Radiology of gunshot wounds. Pages 209-248 *in* B.G. Brogdon, (Ed.). Forensic Radiology. CRC Press, Boca Raton, Florida, USA.

MITCHELL, R. R., S. D. FITZGERALD, R. J. AULERICH, R. J. BALANDER, D. C. POWELL, R. J. TEMPELMAN, R. L. STICKLE, W. STEVENS, AND S. J. BURSIAN. 2001. Health effects following chronic dosing with tungsten-iron and tungsten-polymer shot in adult game-farm Mallards. Journal of Wildlife Diseases 37:451-458.

NOER, H., AND J. MADSEN, J. 1996. Shotgun pellet loads and infliction rates in Pink-footed Geese *Anser brachyrhynchus*. Wildlife Biology 2:65-73.

OEHME, G. 1975. Zur Ernährungsbiologie des Seeadlers (*Haliaeetus albicilla*), unter besonderer Berücksichtigung der Populationen in den drei Nordbezirken der Deutschen Demokratischen Republik. Universität Greifswald.

SCHEIBE, K. M., K. EICHHORN, M. WIESMAYR, B. SCHONERT, AND O. KRONE. 2008. Long-term automatic video recording as a tool for analysing the time patterns of utilisation of predefined locations by wild animals. European Journal of Wildlife Research 54:53–59.

STRUWE-JUHL, B. 1998. Zur Nahrungsökologie des Seeadlers in Schleswig-Hostein. Pages 51-60 *in* Projektgruppe Seeadlerschutz Dreißig Jahre Seeadlerschutz Schleswig-Holstein e.V. (Hrsg.). Ministerium für Umwelt, Natur und Forsten des Landes Schleswig-Holstein, WWF-Deutschland.

STRUWE-JUHL, B. 2000. Funkgestützte Synchronbeobachtung – eine geeignete Methode zur Bestimmung der Aktionsräume von Großvogelarten (Ciconiidae, *Haliaeetus*) in der Brutzeit. Populationsökologie Greifvogel- und Eulenarten 4:101-110.

LEAD POISONING OF STELLER'S SEA-EAGLE (*HALIAEETUS PELAGICUS*) AND WHITE-TAILED EAGLE (*HALIAEETUS ALBICILLA*) CAUSED BY THE INGESTION OF LEAD BULLETS AND SLUGS, IN HOKKAIDO, JAPAN

KEISUKE SAITO

Institute for Raptor Biomedicine Japan, Kushiro Shitsugen Wildlife Center, 2-2101 Hokuto Kushiro Hokkaido Japan, Zip 084-0922. E-mail: k_saito@r8.dion.ne.jp

ABSTRACT.—The Steller's Sea-Eagle (*Haliaeetus pelagicus*) and the White-tailed Eagle (*H.albicilla*) are among the largest eagles. The total population of the Steller's Sea-Eagle is estimated at 5,000 to 6,000 individuals, and these eagles winter in large numbers in northern Japan, on the island of Hokkaido. Lead poisoning of Steller's Sea-Eagles in Japan was first confirmed in 1996. By 2007, 129 Steller's and White-tailed Eagles had been diagnosed as lead poisoning fatalities. High lead values up to 89 ppm (wet-weight) from livers of eagles available for testing indicate that they died from lead poisoning. Necropsies and radiographs also revealed pieces of lead from rifle bullets and from shotgun slugs to be present in the digestive tracts of poisoned eagles, providing evidence that a source of lead was spent ammunition from lead-contaminated Sika Deer carcasses. Tradition and law in Japan allow hunters to remove the desirable meat from animals and abandon the rest of the carcass in the field. Deer carcasses have now become a major food source for wintering eagles.

Reacting to the eagle poisoning issue, Hokkaido authorities have regulated the use of lead rifle bullets since 2000. The Ministry of the Environment mandated use of non-toxic rifle bullets or shotgun slugs beginning the winter of 2001. However, these regulations are limited to the island of Hokkaido, and depend largely on cooperation from hunters. A nationwide ban on the use of lead ammunition for all types of hunting and cooperation from citizens would solve the problem of lead poisoning of eagles. *Received 15 July 2008, accepted 17 September 2008.*

SAITO, K. 2009. Lead poisoning of Steller's Sea-Eagle (*Haliaeetus pelagicus*) and White-tailed Eagle (*Haliaeetus albicilla*) caused by the ingestion of lead bullets and slugs, in Hokkaido Japan. *In* R. T. Watson, M. Fuller, M. Pokras, and W. G. Hunt (Eds.). Ingestion of Lead from Spent Ammunition: Implications for Wildlife and Humans. The Peregrine Fund, Boise, Idaho, USA. DOI 10.4080/ilsa.2009.0304

Key words: Hokkaido, hunting, Japan, lead poisoning, Steller's Sea-Eagle, White-tailed Eagle.

HOKKAIDO, THE MOST NORTHERN ISLAND OF JAPAN, provides habitat for four large-sized raptors. The largest species is the Steller's Sea-Eagle (*Haliaeetus pelagicus*), followed by the White-tailed Eagle (*H. albicilla*), Golden Eagle (*Aquila chrysaetos*), and Mountain Hawk Eagle (*Spizaetus nipalensis*). A high quality natural environment that has abundant food resources is necessary to sustain these species. Forests cover 71% of Hokkaido of which 67% is natural forest; those forests constitute about one quarter of the natural forest areas of Japan (Hokkaido forestry statistics in 2005 fiscal year). The natural environment of Hokkaido is comparatively well-conserved on a nationwide scale. However, currently there are human influences affecting the large birds of prey there.

SEA EAGLES

The Steller's Sea-Eagle is one of the largest eagle species in the world, and breeds almost exclusively in the coastal regions of eastern Russia. It has a high contrast white and black body color and a vivid orange beak. The body length is 89 cm (male) to 100 cm (female), wing span 210 cm (male) to 230 cm (female), and mass of 5 to 7.5 kg. The native Ainu people of Hokkaido call the Steller's Sea-Eagle "*Kapacchilikamui,*" which means "God of the Eagles."

The total population of the Steller's Sea-Eagle is estimated at 5,000 to 6,000 individuals in the limited coastal region of the Okhotsk Sea. The breeding area is the northern part of Sakhalin island, the lower Amur, Kamchatka, and the Bering Sea coast in the southern part of Coriyac. When winter comes, the eagles migrate toward south Kamchatka, the Korean peninsula, south Sakhalin, and northern Japan. Steller's Sea-Eagles winter in large numbers together with White-tailed Eagles, in northern Japan, on the Islands of Hokkaido (1,500 to 2,000 individuals) and South Kurile.

Biologists have been investigating Steller's Sea-Eagles at the breeding grounds in the north and have been radio tracking birds since the 1990s (Meyburg and Lobkov 1994, Ueta and McGrady 2000, Utekhina et al. 2000). As a result, the flyway between Japan and Russia, and also movements on the wintering grounds have gradually been clarified. The migration route of the Steller's Sea-Eagle from the breeding area to the wintering area in Hokkaido is via Sakhalin Island or the Kurile Islands. The eagles that migrate by Sakhalin come to Hokkaido through the Soya Cape (northernmost cape of Japan) in October. Afterwards, most of the eagles move along the Okhotsk Sea toward eastern Hokkaido, feeding on salmon that enter rivers. However, a portion of the population moves along the Japan Sea, and reaches the southern part of Hokkaido where it winters. Eagles that survive the winter migrate north toward the breeding ground, beginning in February until May, with the adults leaving first (Ueta and Higuchi 2002).

The Steller's Sea-Eagles select wintering areas depending on the food source. The principal diet of sea eagles, as their scientific name "*Haliaeetus*" expresses, is fish (Ueta et al. 1995, Shiraki 2001) such as salmon and pollack. Eagles also prey on ducks and scavenge stranded marine mammals. In the early 1980s, the pollack fishery was prosperous, and many sea eagles gathered around the Rausu at the Shiretoko Peninsula in Hokkaido. However, the pollack fishery has suffered recently. Around the same time, the population of Sika Deer (*Cervus nippon*) has dramatically increased. As a result, deer carcasses resulting from accidents, starvation, or hunting were easily discovered by eagles in the mountains of Hokkaido. Steller's Sea-Eagles found these deer as a new food source, and now are commonly observed inland.

The Steller's Sea-Eagle is classified as a Vulnerable Species in the IUCN (International Union for Conservation of Nature and Natural Resources) Red List of Threatened Species. The species is on the Natural Monument list (Law for the Protection of Cultural Properties), listed as a national endangered species of wild fauna and flora in Japan (Law for the Conservation of Endangered Species of Wild Fauna and Flora), and is classified as an endangered species in the Japanese Red Data Book.

The White-tailed Eagle has a 190–210 cm wingspan and 3.5–5 kg body mass. Females are significantly larger than males. The adult bird is mostly brown except for the white tail and yellow beak and legs. The species is distributed from Europe through eastern Asia. The population in Siberia migrates to winter in southern areas including Japan. Hokkaido Island is a main wintering spot for the population that migrates from Sakhalin Island and the Kamchatka Peninsula in the Far East of Russia. A small population is resident in northern Japan. The White-tailed Eagle is called by Ainu, "*Onneu,*" which means "Old and Large One."

The White-tailed Eagle mainly preys on fish, sometimes on birds and small mammals, and also scavenges carcasses or garbage left by humans. They gather in flocks at feeding sites such as fisheries, harbors, lagoons, or rivers, and also at roosting sites during the winter season. Now they also feed on Sika Deer carcasses. White-tailed Eagles are designated as a Japanese Endangered Species and also on the Natural Monument of Japan list. Moreover,

Steller's Sea-Eagles and White-tailed Eagles are included in The Conventions and Agreements for Protection of Migratory Birds between Japan and the partner countries (United States, Russia, Australia, and China).

LEAD POISONING FROM LEAD AMMUNITION

Lead poisoning in avian species has been reported in Japan previously. However, it was mainly cases of waterfowl, and was caused by the ingestion of fishing weights or lead pellets. Waterfowl swallow small stones or grit with food to assist digestion, but they also ingest lead weights used for fishing and lead pellets used in waterfowl hunting. The lead poisoned or shot waterfowl are a source of secondary poisoning to raptors that prey or scavenge on waterfowl.

However, 92 Steller's Sea-Eagles and 37 White-tailed Eagles have died since the first detection of the lead poisoned Steller's Sea-Eagle in 1996 at Abashiri, Hokkaido (Saito et al. 2000, Kurosawa 2000). In this case, a lead shot from waterfowl hunting was found in the ventriculus of the eagle. The case of eagle lead poisoning caused by lead bullets used in Sika Deer hunting was first confirmed in 1997. Since then, based on the study of preserved eagle samples, we know that the lead poisoning of the eagle was already occurring in 1986.

Twenty-one eagles (18 Steller's and three White-tailed) were confirmed to have been lead poisoned in the winter of 1997–98 and 26 more (16 Steller's and ten White-tailed) in 1998–99 (Figure 1). Afterwards, the number of eagle lead poisonings decreased slightly (ten Steller's and four White-tailed in 1999–2000), but the collection rate of the eagles varies depending on the number of persons who enter the mountain areas, and on the snowfall conditions. In 2003, two cases of lead poisoning in Mountain Hawk Eagle (*Spisaetus nipalensis orientalis*) were discovered (Figure 1). This indicates that lead poisoning caused by spent lead ammunition occurs among several raptor species.

The main cause of death of those cases was the lead rifle bullets and shotgun slugs used in Sika Deer hunting. Until 2003, tradition and law in Japan

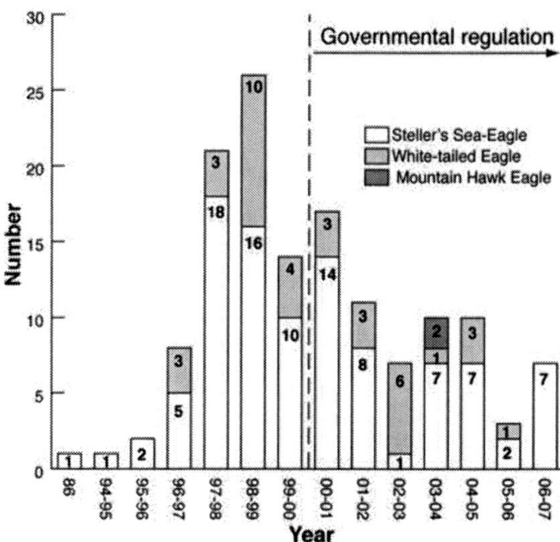

Figure 1. Number of eagles found dead from lead poisoning.

allowed hunters to remove the desirable meat from animals and abandon the rest of the carcass in the field. A number of lead fragments remain in the body of the deer, and these can be ingested by raptors when they scavenge the carcasses. Because of the reduced availability of fish or fish remains, deer carcasses have now become a major food source for these wintering eagle populations. Furthermore, in 1998, Hokkaido government authorities announced a radical deer population control program to reduce the feeding damage caused by Sika Deer. This would cull the herd in eastern Hokkaido from the current 200,000 animals to 30,000 within three years. As a result, the remains of Sika Deer left by hunters increased and Sika Deer that had been wounded and died afterwards also greatly increased.

It is understood that the Steller's and White-tailed eagles are still using the dead Sika Deer as an important winter food source. Soft tissue is exposed where the lead enters the deer, and this area becomes a convenient place for birds to begin eating. According to field observations, adult eagles are dominant at a carcass and force other birds away, and consequently, adult eagles are at high risk of consuming meat that contains a large amount of lead fragments at and near the bullet entry wound.

A large amount of deer meat and hair are frequently observed in the crop or ventriculus of eagles that have been killed by trains, probably when the eagle approached the railway track to eat from the carcass of a deer killed by collision with a train. If deer have been wounded by a lead bullet in the past, it is possible they can be a source of lead poisoning to scavengers at all times of year. Moreover, lead poisoning of eagles also is caused by ingestion of lead shot used for waterfowl hunting. This occurs when raptors prey on or scavenge waterfowl that ingested or were wounded by lead shot. The Steller's Sea-Eagle and White-tailed Eagle and waterfowl together inhabit coastal, lakeside, and riverside habitat in Hokkaido. In this situation, without a strict regulation of lead shot, eagles also are at risk when preying on lead contaminated waterfowl. Lead poisoning in sea eagles is characterized by the high mortality of the adult birds.

Steller's Sea-Eagles normally weigh 6 to 8 kg but individuals killed by lead show severe weight loss. Due to malnutrition, wasting of the pectoral muscles and reduced amounts of visceral fat were conspicuous. Green diarrhea similar to that observed among lead-poisoned waterfowl was observed in each case. Atrophy of livers and distended gallbladders filled with bile were commonly observed. Liver lead values of 2.0 to 89 ppm (wet-weight) demonstrated that these eagles were killed by lead. Necropsies and radiographs have revealed pieces of lead from rifle bullets and from shotgun slugs to be present in the digestive tracts of poisoned eagles.

Even if the amount of ingested lead was not fatal, the toxic substance can have sub-lethal effects (e.g., neurological) that could contribute to other causes of death, such as a being hit by cars. Moreover, the number of lead poisoned eagles collected is just a tip of the iceberg. It is thought that far more eagle carcasses occur in the midwinter in mountains where no one visits. It is suggested that actual casualties by lead poisoning are far more than the number found. Furthermore, the demographic impact of lead is difficult to measure because no one knows how many individuals have died during migration and in their breeding range (Russia). Lead poisoning certainly can be an important factor for the population status of these two sea eagle species (e.g., Ueta and Masterov 2000). It can be said that the population control of Sika Deer that was begun for wildlife management, has produced an unintended threat to the endangered sea eagles.

RESCUE AND TREATMENT OF LEAD POISONED SEA EAGLES

Rescued eagles usually were taken to the nearest zoo until it was learned that the common cause of eagle death was lead poisoning. For lead poisoning cases, prompt treatment becomes an important key to prevent death of the bird. This is because the toxicity of lead is very high, and the influence is remarkable in birds. Dissolution and absorption of lead are rapid in birds of prey because their gastric-acid secretion is high. Because of this, it is thought that the influence of lead is larger than in other types of birds. Lead poisoning generally causes digestive symptoms such as colic, vomiting and asitia. In addition, lead poisoning influences the hematopoietic tissue, appearing as a symptom of clinical importance.

Lead causes a decrease of the amount of hemoglobin in the red blood cell by obstruction of heme synthesis, and this causes critical anemia. Moreover, lead poisoning disrupts normal liver and kidney function and causes adverse effects on the central and the peripheral nervous systems. Therapy is provided by Ca-EDTA (injection and oral) in Japan. The side effects of this treatment of lead poisoning are strong. Therefore, it is necessary to carefully administer the amount based on the density of lead in the individual. Special equipment is required to measure the concentration level of lead, and diagnosing lead poisoning is not easily done in each case.

As a response to the frequent occurrence of raptor lead poisoning, the Japanese Ministry of the Environment together with Hokkaido local agencies decided to centralize the reception of live and dead birds at the Kushiro Shitsugen Wildlife Center, where the veterinary hospital of the Institute for Raptor Biomedicine in Japan exists. The Center utilizes the Lead Care System® (ESA Inc.), which can measure lead concentration quickly with a small blood sample (0.005 ml). The Hokkaido Institute of Public Health contributes to the diagnosis.

The criteria for assessing lead exposure in raptors have been established by using examples from waterfowl:

Hepatic level
- <0.2 ppm: normal range
- 0.2–2 ppm: high exposure
- >2 ppm: lead poisoning

Blood level
- <0.1 ppm: normal range (obstruction of the enzyme level is reported at 0.1 ppm.)
- 0.1–0.6 ppm: high exposure
- >0.6 ppm: lead poisoning

When a high density of lead is confirmed by these tests (e.g., Table 1), both routes of Ca-EDTA administration can be used under a regimen that considers the side effects and subsequent blood lead concentration.

Among the live and dead birds brought to the Wildlife Center, there are some cases that have not been obviously diagnosed as lead poisoning, but irrespective of other causes of accident or sickness, they did show a high lead level in blood. Even among individuals that had a traffic accident (collision with vehicle or train) or electrocution, many show high concentrations of lead in the blood or organs. Now that eagles are accustomed to feeding on carcasses, they are attracted to traffic accident carcasses that occur near the road or railroad and thus are vulnerable to traffic accidents. Finally, documentary research and review of preserved eagle specimens has revealed lead poisoning among cases that had been diagnosed previously as starvation.

A CIVIC GROUP "EAGLE LEAD POISONING NETWORK"

In the spring of 1997, most lead poisoning of large raptors was thought to be caused by the ingestion of lead from rifle bullets. In July 1998, the Hokkaido government formally announced that lead bullet fragments remain in the body of shot Sika Deer. Thereafter, public opinion began to move greatly for prevention. However, lead poisoning victims kept increasing. In winter 1998–1999, 26 Steller's and White-tailed Sea Eagles were found dead by lead poisoning. The total number of eagle carcasses found this winter was 33, and 78% had lead poisoning.

Table 1. Case examples of lead concentration in blood and tissues of lead poisoned raptors.

Species	Date of collection	Blood lead level (ppm)	Hepatic lead level (ppm)	Origin detected by lead isotope ratio analysis*
Steller's Sea-Eagle	2004/1/15	6.5	4.2	Rifle
Steller's Sea-Eagle	2004/1/17		10.4	Rifle
Steller's Sea-Eagle	2004/1/24		14.6	Rifle
Steller's Sea-Eagle	2004/1/31		7.8	
Steller's Sea-Eagle	2004/11/3	4.5 (tibial BM)		Rifle
Steller's Sea-Eagle	2005/3/22		10.0	
Steller's Sea-Eagle	2005/5/5		12.0	
Steller's Sea-Eagle	2007/4/20		11.2	Rifle
White-tailed Eagle	2005/2/12		9.0	Rifle
White-tailed Eagle	2006/4/16		2.0	
White-tailed Eagle	2008/2/4	1.4		Rifle

* 208/206 Pb and 207/206 Pb isotope ratio
Analyzed by Dr. Kazuo Jin, Hokkaido Institute of Public Health

Consequently, in July 1998, veterinarians from eastern Hokkaido took a leading part to establish a civic group, the "Eagle Lead Poisoning Network." Citizens, including students, teachers, company employees, hunters, and civil servants began participating in the group activities for the prevention of lead poisoning. The group's network of activity was organized as shown below.

Investigation of Large Raptors in Hokkaido.—The purpose of this field investigation was to study the behavior and ecology of wintering eagles that move inland from the historically used coastal habitats, and to understand the cause-and-effect relationships of lead poisoning. This investigation included the distribution and number of Steller's and White-tailed Sea Eagles, the occurrence of abandoned deer carcasses, and eagle dependence on the hunting remains. This information will contribute to predicting the extent of lead poisoning, and taking concrete and effective measures to reduce it.

When a debilitated or dead eagle was found during field work, we brought it to the Wildlife Center. In addition, abandoned deer carcasses were buried or transported to Deer Carcass Collection Boxes that were established by the local government. Where the boxes do not exist, deer carcasses were brought to a waste repository.

This investigation provided biological information about sea eagles that inhabit the mountain area and showed that in eastern Hokkaido, a large population of sea eagles winters inland in the mountains, feeding on deer carcasses that result from hunting activities. Also, eagles that winter mainly near the sea coast or lake shore do occasionally feed on hunting remains in addition to aquatic prey.

The inland count of eagles is small in November. An increase occurs from December to January, becoming maximum from February to March, and then greatly decreases in April. However, some young eagles have been found even in early May. This change in the occurrence of sea eagles using the inland landscape seems to reflect the deer hunting season, which usually starts in November and ends in late January or February.

Investigation to Understand Lead Poisoning Occurrence in the Environment.—It is necessary to closely examine the injured or sick individuals and to do post mortem examination to correctly understand the influence of lead poisoning in the population. Therefore, veterinarians and veterinary students use autopsy, x-ray examination, concentration measurements of lead, and they collaborate with the Hokkaido Institute of Public Health to detect the origin of lead using lead isotope ratio analysis. In addition, we have studied the extent of lead poisoning indirectly by measuring lead in fecal samples and by x-ray examination of the pellets collected in the field.

Moreover, with collaboration from the local Hokkaido agency, there is investigation of the bullet material that remains inside the deer carcass. The sample collection is done in the hunting fields of eastern Hokkaido, and when a fresh deer carcass, without evidence of eagle feeding, is found, a 30–50 cm square of soft tissue around the bullet impact area is collected. The samples are closely examined by x-ray, and when the presence of the metallic substance is detected, it is isolated, measured, and described in detail to verify the origin. This investigation continues to demonstrate the presence of the lead contaminated deer carcasses in the wintering habitat of Sea Eagles, even after the governmental ban of lead ammunition in Hokkaido.

Improvement of the Wintering Environment.—Patrol of the hunting ground and removal of abandoned dead Sika Deer is frequently done to prevent raptor lead poisoning. Historically, many remains from hunting were left on the snow. Hunters removed the desirable meat from animals and abandoned the rest of the carcass (skin, bone, intestine, and the damaged parts) in the field. Many edible parts lay under the snow, which melts in early spring. Although our understanding of lead poisoning grows, continued publicity of correct preventive measures is necessary, because deer remains contaminated with lead are still a problem.

Until the appropriate processing of game becomes widely accepted, the members of the Network take the initiative of burying or removing deer remains to reduce as much lead pollution in the environment as possible. It is not straightforward to remove more than 50 cm of snow cover, pickax the hard frozen soil, and dig a sufficient hole to bury a large deer. Despite this difficult situation, volunteers patrol the hunting field to dispose of dead deer. Along forest roads where remains left by hunters are particularly abundant, volunteers load a truck with carcasses and remains and transport them for disposal. In the maximum case, 1 ton of deer carcasses was collected at one forest road, during one day of cleaning activity.

In due course, these activities received publicity, and the government and hunting association started to express concern. The practice of dumping the carcass has gradually changed. Still, some carcasses are clandestinely dropped into the river from the bridge, or placed in culverts that pass under the forest road.

Activity to Promote the Shift to Nontoxic Bullets.—It is well known that lead is a toxic substance, but lead bullets and slugs have been long used for hunting in Japan, though most are made of imported lead. There are alternatives such as a specialized bullet that will not fragment upon impact, the copper bullet (e.g., Barnes X-bullet), and lead-core bullets of special non-fragmentive design (Winchester Fail Safe bullet). These are well known and sold as bullets that do not expose lead after impact with the target. At the time when lead poisoning of eagles became a serious problem in Hokkaido, the X-bullet was almost impossible to obtain in Japan. Therefore, if some hunters wanted to use the copper bullet, it was necessary to make them with an expensive reloading machine. To improve the situation, the Network bought a complete set for reloading and loaned it to the local hunting association.

Education Activity.—The Eagle Lead Poisoning Network used the internet to provide the latest and most accurate information about lead poisoning in raptors and prevention activities. Moreover, multiple publications including an annual report have been sent to the government, hunters, and nature conservation groups. Scientific findings have been published from academic conferences. In addition, a multitude of symposia and forums have shared findings among participants and experts (e.g., Lead Poisoning of Steller's Sea Eagle in Eastern Hokkaido, 1997).

RESPONSE OF GOVERNMENT ADMINISTRATION

Reacting to the eagle poisoning issue, the Hokkaido local government has regulated the use of lead rifle bullets for Sika Deer hunting since 2000. In addition, the government announced that hunters would be required to use non-toxic rifle bullets or shotgun slugs for Sika Deer hunting by winter of 2001. In 2003, the restriction against abandonment of shot game was announced. This regulation was put in force, not only to prevent eagle lead poisoning, but also to avoid attracting the Brown Bear. The regulation of lead rifle bullets and shotgun slugs for all big game hunting, including Brown Bear, started in the winter of 2004. Despite this, lead poisoning of eagles continues. Seven Steller's and three White-tailed Sea Eagles were victims of lead in winter 2004–2005. Moreover, one Steller's and two White-tailed Sea Eagles in winter 2005–06, and seven Steller's Sea-Eagles in winter 2006–07 were poisoned.

The lead bullet and slug restriction is limited to Hokkaido Island. Only the regulation of the use of lead shotgun pellets for waterfowl hunting is present as a legal restriction in certain limited areas elsewhere in Japan. Many hunters come from outside of Hokkaido every year. Under the condition of legality outside Hokkaido, it is difficult to verify that all "foreign" hunters change their bullets to non-toxic ones on Hokkaido Island.

When eagle lead poisoning began to appear as a serious problem, the local authority placed "garbage boxes" in the hunting fields and asked hunters to bring their game remains there as a way to decrease the amount of abandoned deer carcasses. The boxes were full on weekends or after holidays, and carcasses were occasionally stacked in piles on the roof of the garbage box. Despite this, the number of dead deer in the hunting field increased again because Sika Deer population control efforts continued after the hunting season, and the garbage boxes had been removed. Thus, disposal remained problematic. Also, after the regulation against the use of lead ammunition, the local authority discontinued the use of the garbage boxes. Because of this present situation, eagle lead poisoning still exists, and the government should try to ameliorate the situation.

The present restriction is a prohibition of "using" the lead bullets, but it does not limit sales or ownership of lead ammunition. In addition, it is difficult to catch an offender "red-handed." These factors will prolong the threat of lead poisoning. It is unrealistic to do perfect management for the resolution of the problem in the deep mountain areas during the severe winter period. However, if the police come frequently to the hunting field, collect deer tissue samples at bullet entry points, and examine for the presence of lead fragments, they will document the problem, and then the use of lead ammunition will certainly decrease.

A nationwide abolishment of all lead ammunition could solve the problem of raptor lead poisoning. It is extremely important to closely monitor the threats caused by spent lead from ammunition, not only in Hokkaido, but in all of Japan. It is expected that the elimination of this threat will be achieved most quickly by continued cooperation of government administrations, hunters, and other citizens.

LITERATURE CITED

KUROSAWA, N. 2000. Lead poisoning in Steller's Sea Eagles and White-tailed Sea Eagle. Pages 107–109 *in* First symposium on Steller's and White-tailed Sea eagles in east Asia: Proceedings of the International Workshop and Symposium in Tokyo and Hokkaido 9–15 February, 1999. Wild Bird Society of Japan, Tokyo, Japan.

MCGRADY, M. J., M. UETA, E. R. POTAPOV, I. UTEKHINA, V. MASTEROV, A. LADYGUINE, V. ZYKOV, J. CIBOR, M. FULLER, AND W. S. SEEGAR. 2003. Movements by juvenile and immature Steller's Sea-Eagles *Haliaeetus pelagicus* tracked by satellite. Ibis 145:318–328.

MEYBURG, B.-U., AND E. G. LOBKOV. 1994. Satellite tracking of a juvenile Steller's Sea-Eagle *Haliaeetus pelagicus*. Ibis 136:105–106.

SAITO, K. 1997. Lead Poisoning of Steller's Sea Eagle in Eastern Hokkaido. Proceedings, 3rd Annual Meeting of Japanese Society of Zoo and Wildlife Medicine, Gifu, Japan.

SAITO, K., N. KUROSAWA, AND R. SHIMURA. 2000. Lead Poisoning in White-tailed Eagle (*Haliaeetus albicilla*) and Steller's Sea-Eagle (*Haliaeetus pelagicus*) in Eastern Hokkaido through ingestion of shot Sika Deer, 2000. Pages 163–169 *in* J. T. Lumeij, J. D. Remple, P. T. Redig, M. Lierz, and J. E. Cooper (Eds.). Raptor Biomedicine III, Zoological Education Network, Inc, Lake Worth, Florida, USA.

SHIRAKI, S. 2001. Foraging habitats of Steller's Sea-eagles during the wintering season in Hokkaido, Japan. Journal of Raptor Research 35: 91–97.

UETA, M., E. G. LOBKOV, K. FUKUI, AND K. KATO. 1995. The food resources of Steller's Sea-Eagles in eastern Hokkaido. Pages 37–46 *in* Survey of the status and habitat conditions of threatened species, 1995. Published by Environment Agency, Tokyo, Japan.

UETA, M., AND V. MASTEROV. 2000. Estimation by a computer simulation of population trend of Steller's Sea-Eagles. Pages 111–116 *in* M. Ueta and M. J. McGrady (Eds.). First symposium on Steller's and White-tailed Sea eagles in east Asia, Proceedings of the International Workshop and Symposium in Tokyo and Hokkaido 9–15 February, 1999. Wild Bird Society of Japan, Tokyo, Japan.

UETA, M., AND M. J. MCGRADY (EDS.). 2000. First symposium on Steller's and White-tailed Sea eagles in east Asia; Proceedings of the International Workshop and Symposium in Tokyo and Hokkaido 9–15 February, 1999. Wild Bird Society of Japan, Tokyo, Japan.

UETA, M. AND H. HIGUCHI. 2002. Difference in migration pattern between adult and immature birds using satellites. Auk 119:832–835.

SUCCESS IN DEVELOPING LEAD-FREE, EXPANDING-NOSE CENTERFIRE BULLETS

VIC OLTROGGE

The Silvex Company, LLC, 11178 West 59th Place, Arvada, CO 80004, USA.
E-mail: vcoltrogge@comcast.net

ABSTRACT.—The historical practice of using lead in projectiles is declining due to its toxicity, and the search for replacements is well underway. At present the preferred replacement for shot pellets is steel and for bullets it is copper. Steel is much less dense (7.9 g/cm^3) than lead (11.3 g/cm^3), but moderate compensation is achieved with increased velocity. Copper, with a density of 8.96 g/cm^3, is considerably nearer lead, and the Barnes Bullet Company succeeded in 1985 in designing lead-free copper bullets that demonstrate good expansion without shedding copper particles. They have proper rotational moment of inertia, are made in traditional bullet weights, and despite the lower density, the over-all loaded cartridge lengths are within specification. These and other factors make them as capable as traditional lead-cored bullets. They are on the market as the X-Bullet series, in several varieties, chief of which are the Triple Shock and the MRX. The latter is shorthand for Maximum Range X-Bullet, which has an all-metal tungsten-composite core that is more dense than lead. It shoots further, with flatter trajectory, than any other lead-free bullet and surpasses many lead-containing bullets. Some of the science of achieving these lead-free, centerfire bullets is reviewed. Other companies are now making all-copper centerfire bullets, and availability is increasing. *Received 28 May 2008, accepted 20 August 2008.*

OLTROGGE, V. 2009. Success in developing lead-free, expanding nose centerfire bullets. *In* R. T. Watson, M. Fuller, M. Pokras, and W. G. Hunt (Eds.). Ingestion of Lead from Spent Ammunition: Implications for Wildlife and Humans. The Peregrine Fund, Boise, Idaho, USA. DOI 10.4080/ilsa.2009.0305

Key words: Ammunition, ballistics, bismuth, bullet, copper, lead, lead-free.

THERE ARE NOW LEAD-FREE, expanding-nose centerfire bullets that are superior to their lead-cored predecessors. Development began twenty-five years ago, long before widespread cognizance of the toxicity of lead to wildlife. The motivation was not to become lead-free, but to improve the terminal ballistics of big-game bullets of the expanding-nose type. Large "solid" bullets of the kind used in Africa on very large game have long been available, usually made of brass, but without the expanding nose; they are not legal for most hunting uses in the USA.

Lead cores have been used for a very long time, for two main reasons. First, lead is inexpensive, abundant, dense, and malleable, making it easy to buy and to form, while its high density provides excellent exterior ballistics. Second, in soft point or hollow point bullets, it provides an expanding nose. However, the core and the jacket of such bullets often separate in the terminal ballistic phase, with the lead core dispersing in both large and small particles. The jacket usually stays in one piece, but has very little inertia and does not contribute to the wound channel. Such separation of jacket and core greatly reduces bullet penetration, promotes path deviation, and reduces lethality. There followed an unsuccessful quest for a metal as dense as lead with which to make the entire bullet, sometimes imprecisely called a search for a replacement for lead.

Table 1. Selected metals by density and cost.

By density, g/cm³			By cost, in 1988 US dollars per pound			
Os	Osmium	22.57	Precious	Rh	Rhodium	5,556.00
Ir	Iridium	22.50		Os	Osmium	
Pt	Platinum	21.45		Pt	Platinum	4,809.00
Au	Gold	19.32		Lu	Lutetium	
W	Tungsten	19.30		Ir	Iridium	
Ta	Tantalum	16.60		Au	Gold	
Rh	Rhodium	12.44		Ag	Silver	
Tl	Thallium	11.85				
Th	Thorium	11.66	Refractory	Ta	Tantalum	
Pb	► Lead	11.36		W	Tungsten	12.00
Ag	Silver	10.49		(some precious metals)		
Mo	Molybdenum	10.22		Cb	Columbium	11.00
Lu	Lutetium	9.85		Mo	Molybdenum	
Bi	Bismuth	9.80				
Cu	► Copper	8.96	Other	Tl	Thallium	527.00
Ni	Nickel	8.90		Th	Thorium	305.00
Co	Cobalt	8.85		In	Indium	115.00
Cd	Cadmium	8.65		Zr	Zirconium	113.00
Cb	Columbium	8.57		Co	Cobalt	25.00
Fe	► Steel	7.87		Mn	Manganese	4.90
Mn	Manganese	7.43		Sn	Tin	3.90
In	Indium	7.31		Bi	Bismuth	3.60
Sn	Tin	7.30		Ni	Nickel	3.10
Cr	Chromium	7.19		Sb	Antimony	1.50
Zn	Zinc	7.13		Cu	► Copper	1.30
Sb	Antimony	6.62		Cd	Cadmium	1.00
Zr	Zirconium	6.49		Fe	► Steel	.60
				Zn	Zinc	.54
				Pb	► Lead	.44

The complexities in arriving at the present all-copper bullets were considerable. A few of them will now be reviewed, in appreciation of the success that was achieved and also to point out something of the remaining challenges.

The left-hand column of Table 1 shows a list of some of the metals in order of density, with flags by steel (iron), copper and lead. Lead is 1.44 times as dense as steel and 1.27 times as dense as copper. For a pellet or bullet of a given volume, the lead article is considerably heavier. In flight, it is similar to the difference between driving a golf ball versus driving a ping-pong ball. Though the golf ball is only marginally larger than the ping-pong ball, it flies dramatically further. If the ping-pong ball were the size of the golf ball, the difference in distance would be greater, but being the same size means that they would have very nearly identical aerodynamic drag. However, the heavier golf ball has far more inertial energy, so it takes much less time for aerodynamic drag to bleed away the inertia of the ping-pong ball than that of the golf ball. Examining the left-hand column, which is price-based, shows that there is no candidate for replacing lead on a density-to-price basis. Gold and platinum would work exceptionally well technically for replacing lead, but are unlikely prospects due to cost and, alas, would also experience jacket-core separation in the terminal ballistic phase. Even if gold and platinum were as inexpensive as lead, this illustrates the fallacy of "replacing lead."

Shot pellets are available in several alternative metals, including steel and bismuth, so perhaps the same change should have been made with bullets. It is steel shot that has taken over the non-toxic marketplace, which happened because its cost is very much less than the cost of other non-toxic shot, and it works fairly well in that application. Like the ping-pong ball vs. golf ball comparison, but less dramatically, a steel pellet loses velocity more rapidly than a lead pellet, but an increase in the muzzle velocity of the load has provided partial mitigation. Steel for bullets appears advantageous from the point of view of aerodynamic drag, seeming to offer an advantage not available to shot pellets. Figure 1 shows the von Kármán Vortex Street that forms behind a sphere, which indicates high drag due to the shedding of vortices. A streamlined bullet does not create such vortices, so the loss of density in going from lead to steel would appear to impose a lesser penalty on a bullet than the same change exacts from a shot pellet. Weight compensation for a bullet is discussed below.

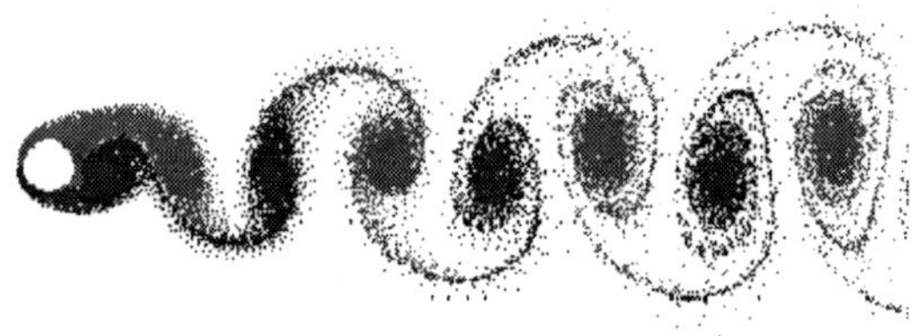

Figure 1. Von Kármán vortex street. (2007, October 3). *Wikimedia Commons,* Retrieved 17:41, February 4, 2009 from http://commons.wikimedia.org/w/index.php?title=Von_K%C3%A1rm%C3%A1n_vortex_street&oldid=7845680

These observations appear to indicate using steel for bullets. Further, during World War II, steel was used for both cores and jackets, and with the cost of steel being considerably less than the cost of copper, there is no lack of motivation to try to make it work. But it was used in war-time under the duress of material shortages, waiving its poorer performance in deference to the need for ammunition. The following discussions will show some of the reasons why steel is not adequate for the vast majority of current bullet applications, despite its continued use in some venues.

The aerodynamic drag on a bullet increases dramatically with increasing bullet diameter, but it increases very little with increasing bullet length. Therefore, a decrease in density of bullet material can be compensated with increased length of bullet, because bullet weight can be adjusted with increased length while avoiding significant increase in drag. So the new copper bullets are available in weights that match the traditional weights of lead-cored bullets by making them longer. But this led to other considerations in the development of the all-copper, expanding-nose bullets. One such item stems from the limitation on length of a loaded cartridge. A longer bullet cannot be allowed to cause a longer cartridge, so it must be seated deeper into the cartridge case. That is due to standards for the dimensions of the chamber into which the cartridge must fit, and the desirability of having the bullet be placed very close to the riflings. A bullet that protrudes too far can be pressed into the riflings and be gripped there so tightly that the bullet is pulled out of the case if the action is opened. Therefore unduly long bullets must be seated deeper than usual into the cartridge case, which can be fraught with hazards.

It is axiomatic that identical weights of materials of differing densities have proportionately differing volumes. That would require an all-copper bullet to have a greater volume than a bullet that contains considerable lead. Diameter is fixed for any given caliber, so the length of the copper bullet would necessarily be longer than its lead-bearing counterpart, in proportion to the densities. Fortunately, however, the new copper bullets were held to lengths approximately equal to traditional lengths of lead-cored bullets. That was accomplished by means such as making small changes to the ogive shape, and to boattail shape, and also by avoiding the air space that occurs in some lead-cored bullets. Though the length effect was overcome in the centerfire copper bullets, it remains a challenge in other bullets, particularly rimfire bullets and steel bullets.

The impetus to create the all-copper bullet was to eliminate jacket-core separation to develop a well-controlled expanding bullet nose. The object is for the nose of the bullet to expand during the terminal ballistic phase to a diameter significantly larger than the diameter of the as-manufactured bullet, in order to create the largest possible wound channel with a smaller diameter bullet. By treaty, the military bullets of World War II were non-expanding bullets, which aided the design of those steel bullets.

Much of the research on the all-copper bullet was directed at the design of the expanding nose, pictured in Figure 2. The nose splits into several petals upon striking the target, and they curl back but remain attached to the bullet body. They create the desired large wound channel but do not shed metal particles in the manner of lead-cored bullets. The terminology "retained bullet weight" means the weight of the bullet after it has come to a stop. The new copper bullets usually demonstrate retained bullet weights close to or equal to 100% of as-manufactured weight. If a particle does come off, it is probably too large to be ingested, unlike the very small lead particles frequently shed by lead-cored bullets. Even when a copper petal is shed, the bullet body retains a far greater percentage of initial bullet weight than a lead-cored bullet that loses its core. A fortunate attribute of the copper-petal nose design is its being a higher drag configuration than a mushroomed lead-cored bullet, so such bullets not only better endure a bone strike, but also are more likely to be brought to a stop within the target animal when traveling through tissue. Energy transfer to vital areas is increased compared to lead-cored bullets.

Another factor that was well handled in the all-copper expanding-nose bullet design was the matter of in-flight stability. For example, when a stick with a rock tied to one end is thrown into the air, it travels with the rock-end forward. The center of gravity of a single-metal bullet is aft of the mid-point of its axis, which is an unstable condition. The rock-on-a-stick illustrates that the bullet would try to fly with its base-end forward. But it is stabilized in point-forward orientation by its spin, which is imparted by the riflings. In the early days of man-made earth satellites, they were spin-stabilized, with the spin being imparted at the point of insertion into orbit. Figure 3 illustrates the satellite's spin axis remaining at a constant angle to the plane of the earth's ecliptic, because the inertial frame of reference is the solar system. It may be remembered that those satellites were rotated 360 degrees once every orbit of the earth, to "unwind the inertial guidance system," in order to avoid confusion due to its continual rotation with respect to the earth. Few modern satellites are spin-stabilized, and those that still are have on-board computers that program out that difficulty.

Likewise, a spinning bullet tries to keep its axis of rotation at a constant angle to the ecliptic, which of course means parallel to the axis of its launching tube. Figure 4 shows the launching tube, a rifle or

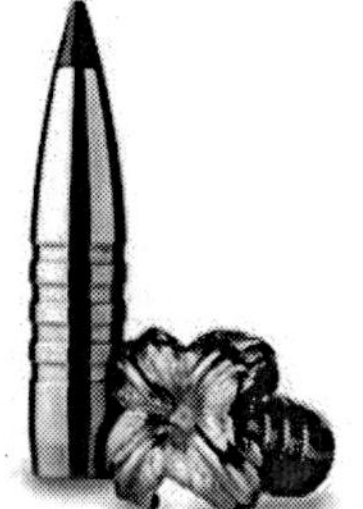

Figure 2. An all-copper expanding-nose bullet. (Reproduced with permission from Barnes Bullets, Inc.)

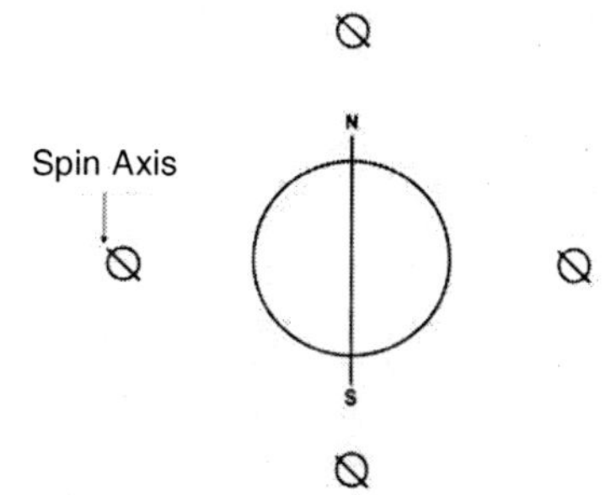

Figure 3. A spin-stabilized earth satellite illustrated.

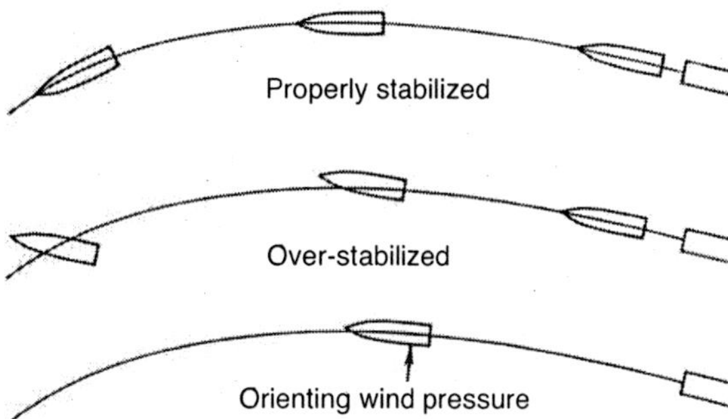

Figure 4. Bullet stability illustrated.

pistol barrel in this case, and the trajectory of the bullet. The bullet falls to earth because it lacks sufficient velocity to go into orbit, though aerodynamic drag would soon slow it to less than orbital velocity—satellites are above the atmosphere. The top diagram, marked *Properly stabilized*, shows the bullet axis remaining parallel to the trajectory. The center diagram, marked *Over-stabilized*, shows the bullet axis remaining parallel to the barrel axis, like the earth satellite, and the third diagram shows that it is air pressure that keeps the bullet properly stabilized. It is the air pressure on the bottom of the bullet that overcomes the inertial attempt to over-stabilize it, and keeps it traveling point-first through the air in properly stabilized manner. In a vacuum a bullet would always be over-stabilized. The bottom diagram shows that the angle of attack is always positive but very small—smaller than shown. Another condition, not diagramed, is an under-stabilized bullet, which tumbles and travels erratically. Over-stabilization is caused by too great a spin rate, too great a rotational moment of inertia, or too short a bullet. These matters are properly balanced in the all-copper, expanding-nose bullets, which are not marginally stable but are fully stable. The stability complexities are compatible with present barrel designs, where the rifling twist rate in extant rifles is an unyielding constraint.

Recently there has become available a somewhat surprising variation of the lead-free, expanding-nose copper bullets. They are not all copper, but have a core that is of greater density than lead. It creates the MRX bullet, for "Maximum Range X-Bullet," wherein the core is a composite material that includes tungsten. This is the ping-pong ball concept in reverse, making a bullet core of greater density than the old lead cores. The bullet exhibits exceptionally flat trajectory and long range. It is of the genre of the new, all-copper bullets in having an all-copper nose that expands with copper petals as shown earlier in Figure 2. It costs a bit more, but is indeed an exceptional performer, while also being non-toxic. The core is in the base of the bullet, so that retained bullet weight in the terminal ballistic phase is as high as the all-copper bullets. The heavy core in the bullet base (the rock at the rear of the stick) makes bullet flight stability an interesting matter, but the design achieved a very stable bullet. It has a pointed plastic tip that also contributes to its excellent ballistic coefficient.

There are some myths to be dispensed with in the bullet material arena. The metallurgical basis is first reviewed. Metals are crystalline, meaning that their atoms are arranged in an ordered, repetitive spatial pattern such as cubic, hexagonal and others. They assume such locations upon freezing from the melt due to thermodynamic factors rather than to bonding with valence electrons. Changing the temperature of the metal can cause change to a different

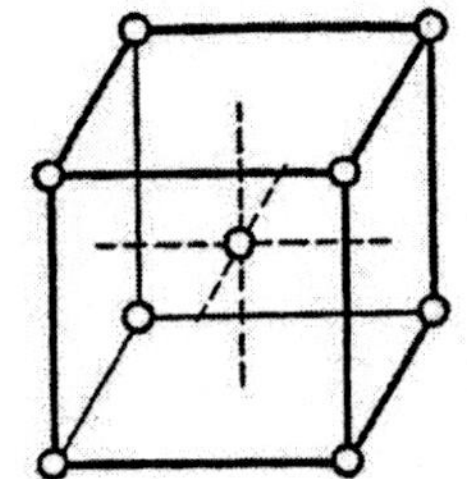

The body-centered-cubic cell cI2.

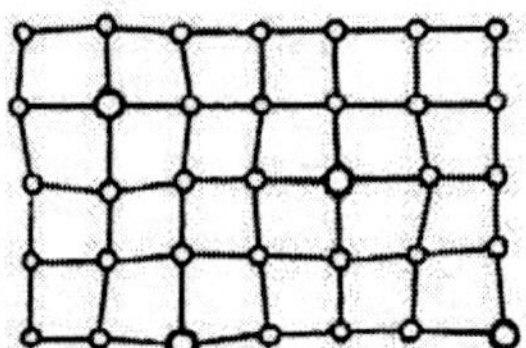

Lattice distortion in a substitional system.

Figure 5. A unit cell and lattice distortion.

crystal structure without involving melting. Figure 5 illustrates the unit cell of a cubic structure, with Pearson symbol cI2. When differing metals are mixed in the making of an alloy, the atoms of the solute metal either replace solvent metal atoms in its lattice, which is substitutional alloying, as shown in the lower portion of Figure 5, or, the solute atoms crowd between the solvent metal atoms, which is interstitial alloying (not shown.) As with the pure metal, the atomic arrangement is thermodynamically driven, and there is no valence bonding of the type seen in chemical molecules. This results in the availability of the solute atoms to outside processes, including diffusional and solubility differences, without the need to break a valence bond. For example, if lead is used as an alloying element in copper, that lead is available to the environment, whatever it may be—a swamp, a stream, a digestive tract, muscle tissue—at rather low energy cost. The bioavailability rate is a different matter and is not treated here.

The non-involvement of valence electrons is true even when an alloy phase diagram shows what appears to be a molecular structure. For example, the phase diagram of the copper-magnesium alloy system shows phases designated $CuMg_2$ and Cu_2Mg. Another illustration is the common iron-carbide structure Fe_3C present in steels and cast irons. The designator makes it appear to be a molecular structure, but it is not. It is a thermodynamically necessary atomic ratio under the circumstance.

The results are:

1. it would be futile to search for a molecule consisting of a lead atom and one or more other *metal* atoms that are combined in such a way that the lead atom is rendered inaccessible to a digestive system, because such molecules do not exist; and

2. it would be valueless to find a chemical molecule, bound with valence electrons, that included a lead atom tied up in an inaccessible manner, because the density of such a compound would be too low to be a projectile material, the lead atom notwithstanding. The conclusion is that presently known chemistry and metallurgy make pipe dreams of such concepts.

There is another pipe dream that is defeated by metallurgical realities. It consists of looking for a technique to alloy metals in such a way that the density of the alloy is greater than that of any constituent. Referring again to Figure 5, the two-dimensional representation of lattice distortion in alloying is a three dimensional phenomenon. When metals are combined in an alloy, the volume of the mixture is very close to the sum of the volumes of the separate metals. And of course the mass of the mixture is the sum of the masses of the constituents. The density of the alloy, then, is in linear proportion to the densities of the constituents, on a mass basis. For example, zinc cannot be added to copper with the result that the zinc atoms squeeze between the copper atoms in a manner that increases the density of the copper. Zinc is less dense than copper and the density of the alloy (which is brass) lies between the densities of copper and zinc. As the proportion of zinc decreases, the density of the brass approaches the density of copper, and vice-versa. Density magic of creating an alloy that is more dense than either constituent is impossible. The approach that was taken in the development of the new all-copper, expanding-nose bullets was the necessary approach.

Centerfire target bullets are not required to have an expanding nose, and are available both with and without lead cores. The word "centerfire" has persisted here because rimfire cartridges remain a dilemma. The State of California has included the .22 rimfire cartridge in its ban on lead in designated portions of the state, despite absence of a known alternative for common .22's. A copper bullet has been developed for the .22 WMR (Winchester Magnum Rimfire) but with reduced exterior ballistic performance. It represents a miniscule portion of the .22's, as the vast majority consists of the Short, Long, and Long Rifle versions, all of which have a shorter case than the WMR and do not offer the flexibilities outlined above for changing to non-lead bullets. It remains to be seen what can and will be done about lead .22 bullets, as the industry is presently non-committal[1]. Bismuth is too brittle to be used as a non-jacketed bullet, and when alloyed with tin to achieve adequate ductility, the density advantage is considerably reduced. Whether political action can catalyze a solution is questionable; perhaps research grants would be more productive.

Can lead be replaced? Not directly, because the density-cost-alloying-toxicity factors are insuperable as outlined above. Can lead be chemically tied up so strongly as to be unavailable? No. Is replacing lead desirable? Not in expanding-nose bullets, since a superior solution is in hand; for fish-line weights and other uses, yes. Doing without lead is not accomplished by direct replacement, it requires alternative product designs that come as close as possible to duplicating the function of the original product.

It has been the intent of this paper to crack open the door to some of the complexities of removing lead from ammunition. It is not exhaustive, but perhaps it somewhat acquaints the reader with the challenges involved, as well as with a real, though alternatively motivated, success.

---

[1] Editor's note: At the time of publication, February 2009, Winchester announced lead-free rimfire bullets chambered for both .22 Win. Mag. and .22 Long Rifle (LR) cartridges available beginning in 2009.

SMALL GAME HUNTER ATTITUDES TOWARD NONTOXIC SHOT, AND CRIPPLING RATES WITH NONTOXIC SHOT

JOHN H. SCHULZ[1], RONALD A. REITZ[1], STEVEN L. SHERIFF[1], JOSHUA J. MILLSPAUGH[2], AND PAUL I. PADDING[3]

[1]*Missouri Department of Conservation, Resource Science Center, 1110 South College Avenue, Columbia, MO 65201, USA.*

[2]*Department of Fisheries and Wildlife Sciences, University of Missouri, 302 Anheuser-Busch Natural Resources Building, Columbia, MO 65211, USA.*

[3]*U.S. Fish and Wildlife Service, Laurel, MD 20708-4028, USA.*

EXTENDED ABSTRACT.—In response to declining resident upland game bird populations wildlife managers are increasing managed Mourning Dove (*Zenaida macroura*) hunting opportunities (Schulz et al. 2003), and stakeholders are becoming concerned that this practice may increase avian exposure to spent lead shot. These concerns will likely progress toward nontoxic shot policy discussions that involve debates about whether nontoxic shot requirements will be acceptable to bird hunters and/or result in increased crippling loss of Mourning Doves by the use of nontoxic shot.

Our first objective was to assess the attitudes of small game hunters in Missouri, USA, toward a nontoxic shot regulation for small game hunting, specifically for Mourning Doves (Schulz et al. 2007). Most hunters (71.7–84.8%) opposed additional nontoxic shot regulations. Hunters from rural areas, hunters with a rural background, hunters who hunt doves, hunters who currently hunt waterfowl, hunters who primarily use private lands, and current upland game hunters were more likely to oppose new regulations. For Mourning Dove hunting, most small game hunters (81.1%) opposed further nontoxic shot restrictions; however, many non-dove hunters (57.1%) expressed no opinion.

We also evaluated reported waterfowl crippling rates in the United States prior to, during, and after implementation of nontoxic shot regulations for waterfowl hunting (Figure 1, Schulz et al. 2006). We used this information to make inferences about Mourning Dove crippling rates if nontoxic shot regulations are enacted. We found differences in moving average crippling rates among the three treatment periods for ducks (F = 23.232, P <0.001, n = 49). Pre-nontoxic shot period crippling rates were lower than 5-year phase-in period crippling rates (P = 0.043) but higher (P <0.001) than nontoxic shot-period crippling rates (Figure 2).

Similarly, we observed differences in moving average reported crippling rates among the three treatment periods for geese (Figure 3, F = 9.385, P <0.001, n = 49). Pre-nontoxic shot and 5-year phase-in period crippling rates were both greater than (P <0.001) nontoxic shot-period crippling rates

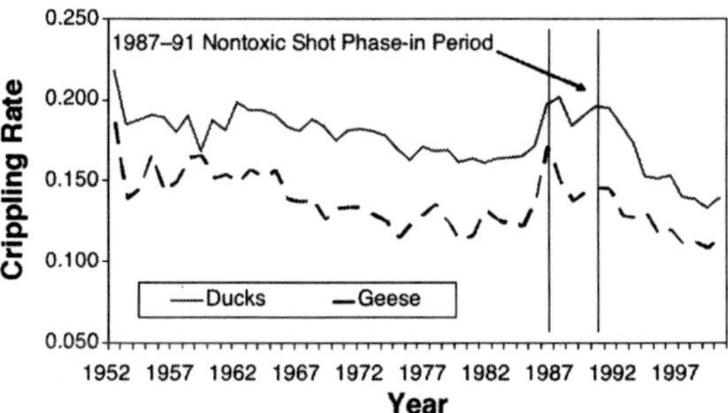

Figure 1. Duck and goose reported crippling rates from the United States Fish and Wildlife Service (USFWS) Waterfowl Harvest Survey, 1952–2001.

but did not differ from one another (P = 0.299). Regardless of why the observed increases occurred in reported waterfowl crippling rates during the phase-in period, we believe the decline that followed full implementation of the nontoxic shot regulation is of ultimate importance when inferring the impacts of lead shot restrictions for Mourning Doves.

We argue that long-term Mourning Dove crippling rates might not increase as evidenced from historical waterfowl data. Also, because our human dimensions results show most Missouri small game hunters and dove hunters are decidedly against further nontoxic shot regulations, any informational and educational programs developed to accompany future policy changes must address these concerns.

Received 15 May 2008, accepted 14 August 2008.

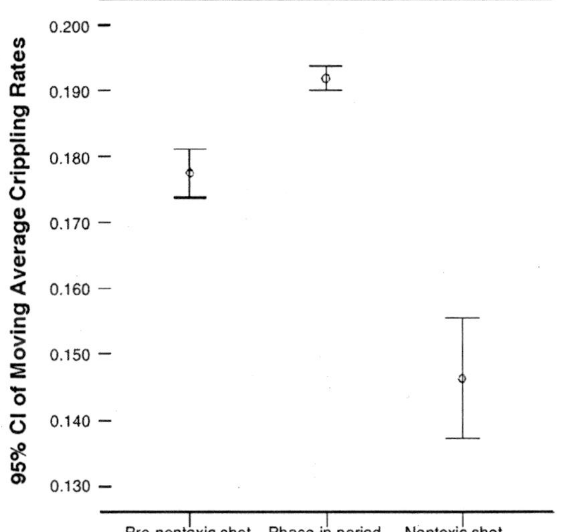

Figure 2. The 95% confidence intervals of untransformed moving average reported crippling rate values for ducks during pre-nontoxic shot (1952-1986), phase-in (1987-1991) and nontoxic shot (1992-2001) periods.

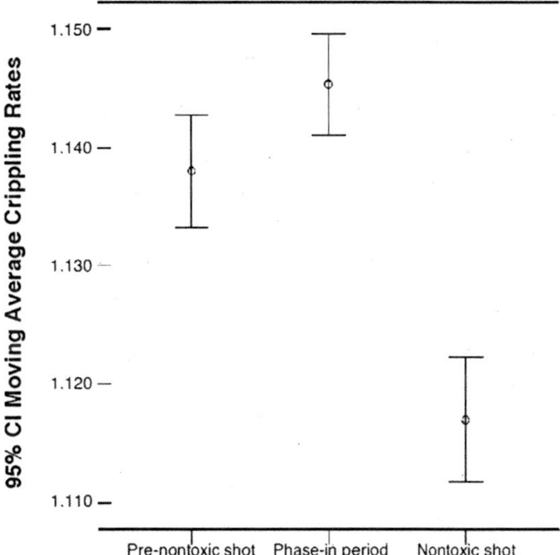

Figure 3. The 95% confidence intervals of untransformed moving average reported crippling rate values for geese during pre-nontoxic shot (1952-1986), phase-in (1987-1991) and nontoxic shot (1992-2001) periods.

SCHULZ, J. H, R. A. REITZ, S. L. SHERIFF, J. J. MILLSPAUGH, AND P. I. PADDING. 2009. Small game hunter attitudes toward nontoxic shot, and crippling rates with nontoxic shot. Extended abstract *in* R. T. Watson, M. Fuller, M. Pokras, and W. G. Hunt (Eds.). Ingestion of Lead from Spent Ammunition: Implications for Wildlife and Humans. The Peregrine Fund, Boise, Idaho, USA. DOI 10.4080/ilsa.2009.0306

Key words: Attitude survey, crippling rate, game, human dimensions, hunter, lead poisoning, nontoxic shot.

LITERATURE CITED

SCHULZ, J. H., J. J. MILLSPAUGH, D. T. ZEKOR, AND B. E. WASHBURN. 2003. Enhancing sport-hunting opportunities for urbanites. Wildlife Society Bulletin 31:565–573.

SCHULZ, J. H., P. I. PADDING, AND J. J. MILLSPAUGH. 2006. Will Mourning Dove crippling rates increase with nontoxic shot regulations? Wildlife Society Bulletin 34(3):861–865.

SCHULZ, J. H., R. A. REITZ, S. L. SHERIFF, AND J. J. MILLSPAUGH. 2007. Attitudes of Missouri small game hunters toward nontoxic shot regulations. Journal of Wildlife Management 71(2):628-633.

IMPACTS OF LEAD AMMUNITION ON WILDLIFE, THE ENVIRONMENT, AND HUMAN HEALTH – A LITERATURE REVIEW AND IMPLICATIONS FOR MINNESOTA

MOLLY A. TRANEL AND RICHARD O. KIMMEL

Minnesota Department of Natural Resources, Farmland Wildlife Population and Research Group, 35365 800th Ave, Madelia, MN 56062, USA.
E-mail: molly.tranel@dnr.state.mn.us

ABSTRACT.—The Minnesota Department of Natural Resources (MDNR) has been investigating non-toxic shot regulations for upland small game hunting because there is considerable evidence that the use of lead ammunition impacts the health of wildlife, the environment, and humans. In 2006 MDNR established a Non-toxic Shot Advisory Committee (NSAC) to provide citizen input on restricting lead shot for small game hunting. To support the NSAC discussions, we summarized available literature regarding lead ammunition and its effects on wildlife, the environment, and human health. This literature review includes more than 500 citations on lead and non-toxic ammunition related issues worldwide and summarizes studies regarding ingestion of lead shot, bullets, and fragments by wildlife species and the impacts of lead poisoning on wildlife, the environment, and humans. We found over 130 species of animals (including upland birds, raptors, waterfowl, and reptiles) have been reported in the literature as being exposed or killed by ingesting lead shot, bullets, bullet fragments, or prey contaminated with lead ammunition. The impacts of ingested lead on wildlife included decreased survival, poor body condition, behavioral changes, and impaired reproduction. We found 15 recent studies that demonstrated the impacts of lead ammunition on human health. Studies in Canada, Greenland, and Russia linked lead shot found in game animals to higher levels of lead in people who eat those game animals, and recent evidence shows that meat far from entry wounds may contain lead fragments. Effective non-toxic alternatives to lead shot are available, and at costs comparable to lead. The results of our review demonstrate the effects of lead ammunition on wildlife, the environment, and human health and support the need for the use of non-toxic alternatives to lead ammunition.
Received 30 May 2008, accepted 28 July 2008.

TRANEL, M. A., AND R. O. KIMMEL. 2009. Impacts of lead ammunition on wildlife, the environment, and human health—A literature review and implications for Minnesota. *In* R. T. Watson, M. Fuller, M. Pokras, and W. G. Hunt (Eds.). Ingestion of Lead from Spent Ammunition: Implications for Wildlife and Humans. The Peregrine Fund, Boise, Idaho, USA. DOI 10.4080/ilsa.2009.0307

Key words: Hunting, lead ammunition, lead poisoning, lead shot, non-toxic shot, Pb, Minnesota.

THE MINNESOTA DEPARTMENT OF NATURAL RESOURCES (MDNR) has been investigating non-toxic shot regulations for upland small game hunting because there is considerable evidence that lead shot impacts the health of wildlife, the environment, and humans. Currently, Minnesota's non-toxic shot regulations beyond federal waterfowl regulations are for managed dove fields, which included four wildlife management areas for 2007. In May 2006, MDNR established a Non-toxic Shot Advisory Committee (NSAC) to provide citizen input on whether to restrict lead shot for upland small game hunting. The NSAC included constituents with interests in hunting, the environment, human health, and the ammunition industry. The committee unanimously agreed restrictions on the use of lead shot were needed for hunting beyond current federal and state regulations, and that lead shot will inevitably have to be restricted for all shotgun hunting at some future time. However, opinions varied on a timetable for implementing various regulations for Minnesota (NSAC 2006). To support the NSAC discussions, we summarized available literature regarding lead ammunition and its effects on wildlife, the environment, and human health. The literature review includes more than 500 citations on lead and non-toxic ammunition-related issues worldwide (Tranel and Kimmel 2008).

IMPACTS OF LEAD AMMUNITION ON WILDLIFE

Wildlife mortality from ingestion of lead shot was first reported more than 100 years ago. In 1876, H. S. Calvert published "Pheasants Poisoned by Swallowing Shot" (Calvert 1876), and a second article about pheasants poisoned by lead shot appeared in 1882 (Holland 1882). In 1894, G. B. Grinnell published an article, entitled "Lead Poisoning," in *Forest and Stream* (Grinnell 1894). Since that time, professional journals have published literature that provides scientific evidence of lead ingestion by wildlife, lead toxicity to wildlife, and lead accumulation in wildlife and human tissues resulting from lead shot (Tranel and Kimmel 2008). The literature documents over 130 species of wildlife that have ingested lead shot, bullets, or bullet fragments (Table 1). Some wildlife species, such as raptors (e.g., hawks, eagles, and condors), are "secondarily poisoned" by consuming animals that either ate or were shot with lead ammunition.

Impacts of lead shot on wildlife include decreased survival, poor body condition, behavioral changes, and impaired reproduction. Tavecchia et al. (2001) reported decreased survival of Mallards (*Anas platyrhynchos*) from lead ingestion in France. Sileo et al. (1973) reported 25–45% reduction in body weight followed by death for Canada Geese (*Branta canadensis*) dosed with lead shot. Death as a result of poisoning from lead shot has been demonstrated for many species, including doves (Schulz et al. 2006a, Schulz et al. 2007), Mallards (Finley and Dieter 1978, Anderson and Havera 1989), and Canada Geese (Cook and Trainer 1966). Fisher et al. (2006) suggested that behavioral changes resulting from lead poisoning might influence susceptibility to predation, disease, and starvation, which increases the probability of death. Mallards experimentally dosed with lead shot had reduced immunologic cells (Rocke and Samuel 1991) and depressed antibody production (Trust et al. 1990). Experimental evidence has demonstrated impaired reproduction from lead shot ingestion in captive doves (Buerger et al. 1986) and domestic Mallards (Elder 1954).

Lead shot impacts on wildlife were most obvious in heavily hunted areas, such as wetlands that were popular waterfowl hunting areas. Because grit is essential for the digestive systems of waterfowl (and most upland game birds), and birds do not differentiate between lead shot and grit of a similar size, wildlife feeding and gathering grit in these wetlands also pick up lead shot (Osmer 1940). Wilson (1937) reported lead poisoning in ducks, geese, and swans discovered in Back Bay, Virginia, and Currituck Sound, North Carolina. He analyzed gizzards, some of which contained more than 100 full-sized No. 4 lead shot and partly ground remains. Osmer (1940) noted that "ingestion of six No. 5 shot by a duck is fatal. Even two or three shot are often fatal." Bellrose (1959) summarized historic information on duck die-offs from lead poisoning ranging from hundreds of ducks in Indiana (1922) and in Louisiana (1930) to as many as 16,000 birds in Missouri (1945–1957) and Arkansas (1953–1954).

Studies in Minnesota documented lead shot problems for Bald Eagles (*Haliaeetus leucocephalus*) and Canada Geese (Minnesota Department of Natural Resources 1981, Bengston 1984, Hennes 1985). Problems were considered severe enough at that time for a steel shot zone to be established for Canada Goose hunting at Lac Qui Parle Wildlife Management Area (Bengston 1984). Hennes (1985) noted that lead shot poisoning of Bald Eagles decreased, but wasn't eliminated. A Trumpeter Swan (*Cygnus buccinator)* die-off in 2007 at Grass Lake in Wright County, Minnesota was attributed to poisoning from lead shot (Minnesota Department of Natural Resources 2007).

Table 1. Species documented as ingesting lead shot, bullets, fragments, or contaminated prey, and species with elevated lead bone, tissue, or blood levels from lead ammunition. Due to the large amount of literature for some species, only selected references are listed.

Species	References	Location
Birds		
American Black Duck (*Anas rubripes*)	White & Stendell (1977); Zwank et al. (1985)	North America
American Coot (*Fulica americana*)	Jones (1939); Anderson (1975)	North America
[c] American Crow (*Corvus brachyrhynchos*)	NYDEC (2000) as read in Golden & Rattner (2002)	New York, USA
[b] Andean Condor (*Vultur gryphus*)	Locke et al. (1969)	Captive
[b] Bald Eagle (*Haliaeetus leucocephalus*)	Jacobson et al. (1977); Clark & Scheuhammer (2003)	North America
[b]Barn Owl (*Tyto alba*)	Mateo et al. (2003)	Spain
Black-bellied Whistling Duck (*Dendrocygna autumnalis*)	Estabrooks (1987)	Sinaloa, Mexico
Black-necked Stilt (*Himantopus mexicanus*)	Hall & Fisher (1985)	Texas, USA
Black Scoter (*Melanitta nigra*)	Lemay et al. (1989) as translated in Brown et al. (2006)	Quebec, Canada
Black Swan (*Cygnus atratus*)	Koh & Harper (1988)	Australia
Black-tailed Godwit (*Limosa limosa*)	Pain (1990)	France
Blue-headed Vireo (*Vireo solitarius*)	Lewis et al. (2001)	Georgia, USA
Blue-winged Teal (*Anas discors*)	Bellrose (1959); Zwank et al. (1985)	North America
Brant Goose (*Branta bernicla*)	National Wildlife Health Laboratory (1985)	North America
Brown Thrasher (*Toxostoma rufum*)	Lewis et al. (2001)	Georgia, USA
Brown-headed Cowbird (*Molothrus atar*)	Vyas et al. (2000)	North America
Bufflehead (*Bucephala albeola*)	Scanlon et al. (1980); Sandersen and Belrose (1986)	North America
[a] California Condor (*Gymnogyps californianus*)	Church et al. (2006); Cade (2007)	North America
California Gull (*Larus californicus*)	Quortrup & Shillinger (1941)	North America
Canada Goose (*Branta canadensis & B. hutchinsii*)	Bellrose (1959); Szymczak & Adrian (1978)	North America
Canvasback (*Aythya valisineria*)	Bellrose (1959); Havera et al. (1992)	North America
Chukar (*Alectoris chukar*)	Walter & Reese (2003); Larsen (2006)	Oregon, USA
Cinnamon Teal (*Anas cyanoptera*)	Bellrose (1959)	North America
Clapper Rail (*Rallus longirostris*)	Jones (1939)	North America
[b] Common Buzzard (*Buteo buteo*)	MacDonald et al. (1983); Battaglia et al. (2005)	France; Italy
Common Coot (*Fulica atra*)	Mateo et al. (2000)	Spain
Common Eider (*Somateria mollissima*)	Franson et al. (1995); Flint et al. (1997)	Alaska, USA
Common Goldeneye (*Bucephala clangula*)	Bellrose (1959); Anderson (1975)	North America
Common Moorhen (*Gallinula chloropus*)	Jones (1939); Locke & Friend (1992)	North America
Common Pheasant (*Phasianus colchicus*)	Hunter & Rosen (1965); Butler et al. (2005)	North America; England
Common Pochard (*Aythya ferina*)	Mateo et al. (1998); Mateo et al. (2000)	Spain

Species	References	Location
[b,a] Common Raven (Corvus corax)	Scheuhammer & Norris (1995); Craighead & Bedrosian (2008)	Canada; Wyoming, USA
Common Snipe (*Gallinago gallinago*)	Pain (1990); Olivier (2006)	France
Common Teal (*Anas crecca*)	Mateo et al. (2000)	Spain
Common Wood-pigeon (*Columba palumbus*)	Clausen & Wolstrop (1979)	Denmark
[c] Cooper's Hawk (*Accipiter cooperii*)	Martin & Barrett (2001)	Canada
Dark-eyed Junco (*Junco hyemalis*)	Vyas et al. (2000)	North America
Dunlin (*Calidris alpina*)	Kaiser et al. (1980)	British Columbia, Canada
[b] Egyptian Vulture (*Neophron percnopterus*)	Donazar et al. (2002)	Canary Islands
[b] Eurasian Eagle Owl (*Bubo bubo*)	Mateo et al. (2003)	Spain
[b] Eurasian Griffon (*Gyps fulvus*)	Mateo et al. (2003); Garcia-Fernandez et al. (2005)	Spain
[b] Eurasian sparrowhawk (*Accipiter nisus*)	MacDonald et al. (1983)	France
[c,b] European Honey-buzzard (*Pernis apivorus*)	Lumeij (1985)	The Netherlands
Gadwall (*Anas strepera*)	Bellrose (1959); Mateo et al. (2000)	North America; Spain
Glaucous-winged Gull (*Larus glaucescens*)	National Wildlife Health Laboratory (1985)	North America
[a,b] Golden Eagle (*Aquila chrysaetos*)	Craig et al. (1990); Kenntner et al. (2007)	Idaho, USA; Switzerland
[c] Gray-headed Woodpecker (Picus canus)	Mörner and Petersson (1999)	Sweden
[b] Great Horned Owl (*Bubo virginianus*)	Clark & Scheuhammer (2003)	Canada
Greater Flamingo (*Pheonicopterus ruber*)	Schmitz et al. (1990); Mateo et al. (1997)	Yucatan, Mexico; Spain
Greater Scaup (*Aythya marila*)	Bellrose (1959)	North America
Greater White-fronted Goose (*Anser albifrons*)	Zwank et al. (1985)	Louisiana, USA
Green-winged Teal (*Anas carolinensis*)	Bellrose (1959); Zwank et al. (1985)	North America
Greylag Goose (Anser anser)	Mudge (1983); DeFrancisco (2003)	England; Spain
Hardhead (*Aythya australis*)	Baxter et al. (1998)	Australia
Herring Gull (*Larus argentatus*)	National Wildlife Health Laboratory (1985)	North America
Hungarian Partridge (*Perdix perdix*)	Keymer & Stebbings (1987); Kreager et al. (2007)	England; Canada
Jack Snipe (*Lymnocryptes minimus*)	Olivier (2006)	France
Japanese Quail (*Coturnix coturnix*)	Yamamoto et al. (1993)	Japan
King Rail (*Rallus elegans*)	Jones (1939)	North America
[b] King Vulture (*Sarcorhampus papa*)	Decker et al. (1979)	Captive
[b] Laggar Falcon (*Falco jugger*)	MacDonald et al. (1983)	Captive
Lesser Scaup (*Aythya affinis*)	Bellrose (1959); Havera et al. (1992)	North America
Long-billed Dowitcher (*Limnodromus scolopaceus*)	Hall & Fisher (1985)	Texas, USA
[b] Long-eared Owl (*Asio otus*)	Brinzal (1996) as read in Fisher et al. (2006)	Spain
Long-tailed Duck (*Clangula hyemalis*)	Flint et al. (1997); Skerratt et al. (2005)	Alaska, USA; North America
Magpie Goose (*Anseranas semipalmata*)	Harper & Hindmarsh (1990); Whitehead & Tschirner (1991)	Australia

Species	References	Location
Mallard (*Anas platrhynchos*)	Bellrose (1959), Mateo et al. (2000)	North America; Spain
Maned Duck (*Chenonetta jubata*)	Kingsford et al. (1994)	Australia
Marbled Godwit (*Limosa fedoa*)	Hall & Fisher (1985); Locke et al. (1991)	Texas, USA; North America
Marbled Teal (*Marmaronetta angustirostris*)	Mateo et al. (2001); Svanberg et al. (2006)	Spain
Merganser (*Mergus spp*)	Bellrose (1959); Skerratt et al. (2005)	North America
Middendorff's Bean Goose (*Anser fabalis middendorffii*)	Chiba et al. (1999)	Japan
Mottled Duck (*Anas fulvigula*)	Merendino et al. (2005)	Texas, USA
Mourning Dove (*Zenaida macroura*)	Lewis & Legler (1968); Schulz et al. (2006a)	North America
Musk Duck (*Biziura lobata)*	Department of Sustainability and Environment (2003)	Australia
Mute Swan (*Cygnus olor*)	Bowen & Petrie (2007)	Great Lakes, Canada
Northern Bobwhite Quail (*Colinus virginianus*)	Westemeier (1966); Keel et al. (2002)	Illinois, USA
[a,b] Northern Goshawk (*Accipiter gentillis*)	Martin & Barrett (2001); Pain & Amiard-Triquet (1993)	Canada; France
Northern Pintail (*Anas acuta*)	Bellrose (1959); Mateo et al. (2000)	North America; Spain
Northern Shoveler (*Anas clypeata*)	Bellrose (1959); Mateo et al. (2000)	North America; Spain
Pacific Black Duck (*Anas superciliosa*)	Baxter et al. (1998)	Australia
Pacific Loon (*Gavia pacifica*)	Wilson et al. (2004)	Alaska, USA
[b] Peregrine Falcon (*Falco peregrinus*)	MacDonald et al. (1983); Pain et al. (1994)	Captive; England
Pink-footed Goose (*Anser brachyrhynchus*)	Mudge (1983)	England
[b] Prairie Falcon (*Falco mexicanus*)	Redig (1980); MacDonald et al. (1983)	Captive
[b] Red Kite (*Milvus milvus*)	Mateo et al. (2003); Pain et al. (2007)	England
Red-tailed Hawk (*Buteo jamaicensis*)	Sikarskie (1977); Clark & Scheuhammer (2003)	Canada
Red-crested Pochard (*Netta rufina*)	Mateo et al. (2000)	Spain
Red-legged Partridge (*Alectoris rufa*)	Butler (2005)	England
Redhead (*Aythya americana*)	Bellrose (1959); Zwank et al. (1985)	North America
Ring-necked Duck (*Aythya collaris*)	Anderson (1975); Havera et al. (1992)	North America
Rock Dove (*Columba livia*)	DeMent et al. (1987)	New York, USA
Rough-legged Hawk (*Buteo lagopus*)	Locke & Friend (1992)	North America
Ruddy Duck (*Oxyura jamaicensis*)	Perry & Artmann (1979); Sanderson & Bellrose (1986)	North America
Ruffed Grouse (*Bonasa umbellus*)	Rodrigue et al. (2005); Kendall et al. (1984)	Virginia, USA; Canada
Sandhill Crane (*Grus canadensis*)	Windingstad et al. (1984); Franson & Hereford (1994)	North America
Scaled Quail (*Callipepla squamata*)	Campbell (1950); Best et al. (1992)	New Mexico, USA
Snow Goose (*Anser caerulescens*)	Bellrose (1959); Zwank et al. (1985)	North America
[a] Snowy Owl (*Nyctea scandiaca*)	MacDonald et al. (1983)	Captive
Sora Rail (*Porzana carolina*)	Artmann & Martin (1975); Stendell et al. (1980)	Maryland, USA
Spanish Imperial Eagle (*Aquila adalberti*)	Mateo et al. (2000); Pain et al. (2005)	Spain

Species	References	Location
Spectacled Eider (*Somateria fischeri*)	Franson et al. (1995); Grand et al. (1998)	Alaska, USA
[a] Steller's Sea Eagle (*Haliaeetus pelagicus*)	Kurosawa (2000)	Japan
Trumpeter Swan (*Cygnus buccinator*)	Bellrose (1959); Blus (1994)	North America
Tufted Duck (*Aytha fuligula*)	Mudge (1983); DeFrancisco et al. (2003)	England; Spain
Tundra Swan (*Cygnus columbianus*)	Trainer & Hunt (1965); Blus (1994)	North America
[b] Turkey Vulture (*Cathartes aura*)	Clark & Scheuhammer (2003); Martin et al. (2008)	North America
Virginia Rail (*Rallus limicola*)	Jones (1939)	North America
[b] Western Marsh Harrier (*Circus aeruginosus*)	Pain & Amiard-Triquet (1993); Mateo et al. (1999)	France; Spain
[c] White-backed Woodpecker (*Dendrocopus leucotos*)	Mörner and Petersson 1999	Sweden
White-faced Ibis (*Plegadis chihi*)	Hall & Fisher (1985)	Texas, USA
White-fronted Goose (*Anser albifrons*)	Bellrose (1959); Ochiai et al. (1993)	North America; Japan
White-headed Duck (*Oxyura leucocephala*)	Mateo et al. (2001); Svanberg et al. (2006)	Spain
White Pekin (wild) (Anas platyrhychos)	Schwab & Padgett (1988)	Virginia, USA
[a] White-tailed Eagle (*Haliaeetus albicilla*)	Kurosawa (2000); Krone et al. (2004)	Japan; Greenland
White-throated Sparrow (*Zonotrichia albicollis*)	Vyas et al. (2000)	North America
Whooper Swan (*Cygnus cygnus*)	Ochiai et al. (1992); Honda et al. (2007)	Japan
Whooping Crane (*Grus americana*)	Hall & Fisher (1985)	North America
American Wigeon (*Anas americana*)	Zwank et al. (1985); Mateo et al. (2000)	Louisiana, USA; Spain
Wild Turkey (*Meleagris gallopavo*)	Stone & Butkas (1972); Kreager et al. (2007)	New York, USA; Canada
Wood Duck (*Aix sponsa*)	Bellrose (1959); Sanderson & Bellrose (1986)	North America
[b] Woodcock (*Scolopax minor*)	Scheuhammer et al. (2003)	Canada
Yellow-rumped Warbler (*Dendroica coronata*)	Lewis et al. (2001)	Georgia, USA
Mammals		
Bank vole (*Clethrionomys glareolus*)	Ma (1989)	The Netherlands
[b] Domestic cattle (*Bos taurus*)	Rice et al. (1987)	
Gray squirrel (*Sciurus carolinensis*)	Lewis et al. (2001)	Georgia, USA
Humans (*Homo sapiens*)	Engstad (1932); Larsen and Blanton (2000)	
Shrew (*Sorex araneus*)	Ma (1989)	The Netherlands
White tailed deer (*Odocoileus virginianus*)	Lewis et al. (2001)	Georgia, USA
Reptiles		
[a,b] American alligator (*Alligator mississippiensis*)	Camus et al. (1998); Lance et al. (2006)	North America; Captive
[b] Crocodile (*Crocodylus porosus*)	Hammerton et al. (2003); Orlic et al. (2003)	North America; Australia

[a] Evidence of secondary poisoning from lead bullets.
[b] Evidence of secondary poisoning from lead shot.
[c] Source of lead unknown, lead ammunition suspected.

Impacts of lead shot at a population level are variable. Butler et al. (2005) noted that 3% of pheasants on shooting estates in Great Britain had lead in their gizzards. Kreager et al. (2007) examined gizzards from upland game birds harvested in Ontario, Canada and found lead pellets ingested by 8% of the Chukars (*Alectoris chukar*) and 34% of the pheasants. They found that 13% of the livers from Chukars, Common Pheasants (*Phasianus colchicus*), Wild Turkey (*Meleagris gallopavo*), and Hungarian Partridge (*Perdix perdix*) had elevated lead concentrations. Ingestion rates may vary by species, feeding behavior, and availability of lead shot in the habitat.

Fisher et al. (2006) suggested that a lack of evidence of poisoned species does not suggest a lack of poisoning. Schulz et al. (2006a) found that Mourning Doves (*Zenaida macroura*) may expel lead shot pellets after ingesting them, indicating incidence of lead exposure in wildlife may be higher than reported. In a similar study, one in three birds that expelled all shot exhibited measurable effects of lead poisoning (Schulz et al. 2007). Die-offs and evidence of lead poisoning may not be apparent, because wildlife affected by lead poisoning may seek isolation and protective cover (Friend and Franson 1999). Furthermore, it is unknown if mortality due to non-lethal effects such as reproductive problems, lowered immunity, anemia, and weakened muscles could be higher than losses from direct lead poisoning (Michigan Department of Natural Resources 2002).

Lead shot ingestion and toxicity problems for wildlife have been documented worldwide. Table 1 documents lead ingestion or secondary lead exposure for wildlife species in more than 10 countries and several continents. Tavecchia et al. (2001) found lead pellets in the muscles and gizzards of 11% of the Mallards captured in France. Mateo et al. (1998) found lead pellets in 87.5% of Common Pochards (*Aythya ferina*) and 33% of Mallards and Northern Shovelers (*Anas clypeata*) in Spain. Mörner and Petersson (1999) found lead poisoning in two woodpecker species in forested areas in Sweden suggesting that the woodpeckers searching for food removed lead pellets shot into trees.

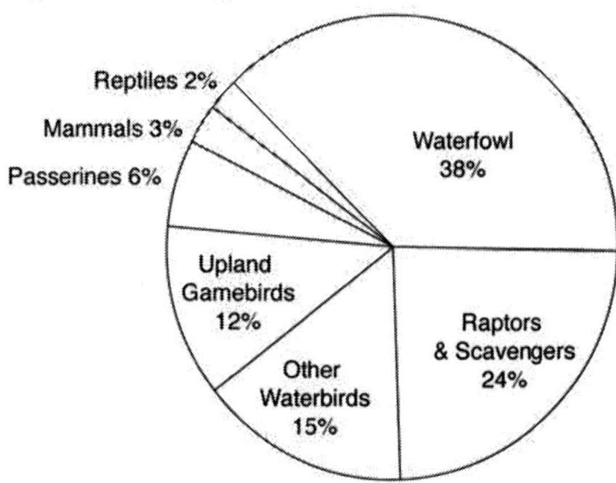

Figure 1. Categories of species reported in the literature as ingesting lead shot, bullets, fragments, or prey contaminated with lead ammunition, and species with elevated lead levels in bone, tissue, or blood from exposure to lead ammunition.

There is sufficient evidence that the problem of lead poisoning in birds extends to upland birds and raptors. Sixteen species of upland game birds, and 29 species of raptors were reported in the literature as being affected by lead ammunition (Figure 1). Butler et al. (2005) reported lead exposure over a number of years (1996–2002) for Common Pheasants in Great Britain. Fisher et al. (2006) provided a review of 59 (9 threatened) terrestrial bird species that have been documented to have ingested lead or suffered lead poisoning from ammunition sources. Exposure of lead shot on some upland game birds such as doves could rival the problem in waterfowl (Kendall et al. 1996). Ingestion of lead by wildlife, other than waterfowl and birds of prey, "appears to be extensive" (Harradine 2004). Current use of lead shot for small game hunting (such as pheasants) potentially continues to deposit lead in wetlands and expose waterfowl to lead shot.

Lead ammunition can secondarily poison wildlife that feed on hunted species. Studies have linked the likelihood of a species ingesting lead shot to feeding habits, with scavengers and predators that take game species being the most susceptible (Pain and Amiard-Triquet 1993). In Spain, Mateo et al. (2003) reported lead poisoning from exposure to lead shot in prey species in eight upland raptor species. Clark and Scheuhammer (2003) examined lead exposure in 184 dead raptors (16 species) found across Canada. They determined that, of the three most commonly encountered species, 3–4% died as a result of lead poisoning. They concluded that upland birds of prey and scavengers that eat game birds and mammals are at risk for lead poisoning from ingestion of lead shot and bullet fragments used in upland hunting. They suggested that use of non-toxic ammunition for hunting upland game would effectively remove the only serious source of high lead exposure and lead poisoning for upland-foraging raptors.

Knopper et al. (2006) reported that carcasses from squirrel populations managed by shooting had lead levels lethal to raptors and suggested either collection of carcasses shot with lead or the use of non-toxic shot. Similar to the lead shot problems described by Clark and Scheuhammer (2003), deer carcasses containing lead fragments from bullets impact California Condors (*Gymnogyps californianus*) (Cade 2007) and Bald Eagles (Franson 2007). Hunt et al. (2006) examined the remains of 38 deer killed with rifles and found that all deer killed with lead-based bullets contained bullet fragments. Mateo et al. (2003) analyzed bones from 229 birds of prey in Spain (11 species) and diagnosed lead poisoning in 8 raptor species that feed on wildlife targeted by hunters in upland habitats.

LEAD AMMUNITION IMPACTS ON THE ENVIRONMENT

The Minnesota Pollution Control Agency (1999) estimated that 1,184,202 kg of lead shot were used annually in Minnesota in hunting and shooting ranges. In their legislative report on sources and effects of lead, they state, "The fact that lead ammunition is estimated to be the single largest source of lead released to the environment qualifies it as a concern that should be examined more closely."

De Francisco et al. (2003) estimated that lead shot can take 100 to 300 years to disappear from a site, allowing for concentration of large amounts of lead in areas of heavy hunting pressure. Although the breakdown is slow, lead shot pellets accumulating in the environment are not inert and ultimately the lead will be deposited as particles in soil and water (Scheuhammer and Norris 1995). Mozafar et al. (2002) found a high degree of bioavailability in heavy metals (including lead) in soils at a shooting range in Switzerland. Uptake of this lead by terrestrial and aquatic plants and animals can occur, leading to elevated lead concentrations (Ma 1989, Manninen and Tanskanen 1993, Rooney et al. 1999).

Guitart et al. (2002) reported that a single lead shot could raise 12,000 liters of water to the European Union threshold guideline for lead in drinking water. Surface water contamination by lead shot from shooting ranges has been well documented (Stansley et al. 1992, Dames and Moore Canada 1993, Emerson 1994, United States Environmental Protection Agency 1994). Strait et al. (2007) found that shooting ranges contained areas where lead occurred at "concentrations significantly in excess of the Michigan Department of Environmental Quality criteria and therefore pose a potential risk to the human users of the land as well as to the native wildlife." While shooting ranges contain far more spent shot than typical hunting areas, these studies demonstrate the ability of lead to accumulate over time and contaminate the surrounding environment and wildlife. Areas with acidic waters or soils are at particularly high risk for contamination from lead shot, as lead is more easily mobilized at a lower pH (Stansley et al. 1992).

Contamination of human and livestock food sources due to lead shot deposition has also been documented. Guitart et al. (2002) suggested that the high lead content of rice produced in Spain was a result of hunting with lead shot near rice fields. Rice et al. (1987) reported lead poisoning of cattle from ingestion of silage contaminated with lead shot. In addition, milk production decreased and stillbirths increased in cattle (*Bos taurus*) ingesting lead contaminated hay cut from a field used for clay pigeon shooting (Frape and Pringle 1984).

IMPACTS OF LEAD AMMUNITION ON HUMANS

Lead poisoning in humans has occurred for at least 2,500 years (Eisler 1988). Today, it is widely known that lead is toxic to humans and can cause permanent learning disabilities, behavioral problems, and even death. Haldimann et al. (2002) concluded that frequent consumption of wild game meat had no significant effect on blood lead levels. However, studies in Canada, Greenland, and Russia have linked lead shot found in game animals to higher levels of lead in people who eat those game animals (Table 2). Levesque et al. (2003) stated, "Lead shots may be a major source of lead exposure to humans that consume hunted game animals." This study found that lead shot was a source of lead exposure in the Inuit population; blood lead concentrations in 7% of Inuit newborns were higher than government-recommended levels. Studies linking game meat containing lead shot and elevated blood lead levels in children (Odland et al. 1999, Smith and Rea 1995) and newborns (Dewailly et al. 2000, Hanning et al. 2003) are of particular concern.

Breurec et al. (1998) diagnosed lead poisoning in an adult patient who had frequently eaten game birds containing lead shot. Professional medical literature contains many references of humans carrying lead shot in their digestive tracts (Engstad 1932, Horton 1933, Hillman 1967, Reddy 1985, Madsen et al. 1988, Spitale and D'Olivo 1989, Moore 1994, Tsuji and Nieboer 1999, and Larsen and Blanton 2000). Lead from shot may accumulate in tissues of game animals. In upland game birds and waterfowl killed by hunters using lead shot, 40% of 123 livers (Kreager et al. 2007) and 9% of 371 gizzard tissue samples (Tsuji et al. 1999) showed lead levels greater than Health Canada's guidelines for fish.

In animals shot for human consumption, meat far from the entry wound may contain lead. Scheuhammer et al. (1998) found fragments of lead in game birds far from wounds from shotgun pellets. Hunt et al. (2006) found lead fragments in meat 15 cm away from rifle bullet wounds in game animals. Lead fragments, likely from bullets, were found in ground venison in North Dakota and Minnesota. This prompted North Dakota Health, Game and Fish, and Agriculture Departments to advise food pantries not to distribute or use donated ground venison because of the potential for lead contamination (North Dakota Department of Health 2008). Minnesota Department of Agriculture found lead fragments in 76 of 299 tested samples of ground venison donated to food shelves, prompting disposal of all remaining donated venison (Minnesota Department of Agriculture 2008).

Tsuji et al. (1999) reported that, "People who consume *any* game species harvested with lead shot risk exposure to this metal by way of ingestion of tissue-embedded lead pellets and fragments." With alternatives to lead shot readily available (Sanborn 2002), human exposure to lead through game meat is unnecessary (Rodrigue et al. 2005). Levesque et al. (2003) showed significant decreases in lead concentrations in umbilical cord blood after a public health intervention to reduce the use of lead shot by the Inuit population.

NON-TOXIC AMMUNITION

Substituting non-toxic shot for lead shot could reduce lead shot impacts on the health of wildlife, the environment, and human health. Alternatives to lead shot were not readily available in the past, especially prior to the 1991 federal ban on lead shot for waterfowl hunting in the United States. However, other types of shot, particularly steel shot, are now available at a cost comparable to lead shot ammunition (Sanborn 2002). Non-toxic shot is now also available for safe use in vintage and older shotguns (Cabela's 2008). Scheuhammer and Norris (1995) found that, while non-toxic alternatives to lead shot are more expensive than lead, they represent only a 1–2% increase in the average hunter's yearly expenses. There are currently 11 types of shot approved as non-toxic by the US Fish and Wildlife Service (US Fish and Wildlife Service 2006). Studies have demonstrated the effectiveness of steel shot. For example, Schulz et al. (2006b) evaluated crippling rates in waterfowl prior to and following implementation of non-toxic shot regulations in the US. They found that, after a five-year phase-in period, crippling rates for ducks and geese were lower after non-toxic shot restrictions were implemented.

Despite numerous reports of negative impacts of lead shot on wildlife worldwide, restrictions on the use of lead shot have been minimal (Thomas 1997). Interest in non-toxic shot regulations has resulted in discussions on restricting lead ammunition and some legislation on different continents. Denmark and the Netherlands have banned all uses

Table 2. Selected literature regarding elevated lead levels in humans consuming game meat harvested with lead shot.

Author	Country	Findings
Bjerregaard et al. (2004)	Greenland	Blood lead adjusted for age and sex was found to be associated with the reported consumption of sea birds.
Breurec et al. (1998)	Not reported	Patient diagnosed with adult lead poisoning by ingestion of game birds with small lead shots.
Dewaily et al. (2000)	Arctic Canada	Ingestion of lead shot/fragments in game meat may be responsible for higher lead levels found in Inuit newborns. Lead isotopes of shotgun cartridges were similar to those of Inuit newborns.
Dewailly et al. (2001)	Quebec, Canada	Evaluated 492 blood levels of lead and mercury in Inuit adults, revealed that smoking, age, and consumption of waterfowl were associated with lead concentrations ($r^2 = 0.30$, $p < 0.001$).
Guitart et al. (2002)	Spain	Approximately 30,000 waterfowl hunters and their families, especially children, are at risk of secondary lead poisoning from lead poisoned birds in Spain.
Hanning et al. (2003)	Canada	Traditional animal food intake, especially wild fowl, correlated significantly with umbilical cord blood lead, and reflected the legacy of using lead-containing ammunition.
Johansen et al. (2001)	Ontario, Canada	Breast meat lead values in birds killed with lead shot were 10 times higher than birds not killed with lead shot. Shot is a significant source of lead in many people in Greenland.
Johansen et al. (2004)	Greenland	Lead intake of Greenland bird eaters can largely exceed the tolerable lead intake guidelines, and the shot is a more important source of lead than previously estimated.
Johansen et al. (2006)	Greenland	Found clear relationship pointing to lead shot as the dominating lead source to people in Greenland.
Levesque et al. (2003)	Quebec, Canada	Lead from game hunting was a major source of human exposure to lead. Calls for international ban on lead shotgun ammo.
Mateo et al. (2007)	Spain	Consumption of half a pickled quail/week with embedded shot may cause the provisional tolerable weekly intake of lead by the Spanish consumer to be exceeded.
Odland et al. (1999)	Russia	Suggests lead shot as the main source of lead in population in the Kola Peninsula, Russia.
Smith and Rea (1995)	Canada	Elevated lead blood levels in children probably due to consumption of birds containing lead shot, suggest use of alternative shot.
Trebel and Thompson (2002)	Canada	Young child exhibited elevated blood lead levels after ingesting spent air rifle pellets.
Tsuji et al. (1999)	Ontario, Canada	Consumption of any game species harvested with lead shot risks exposure by way of ingestion of tissue-embedded lead pellets and fragments.

of lead shot (Thomas 1997). Broad regulatory action to restrict lead shot across Europe has been discussed by various cross-continental groups, such as the European Council, the Bonn and Bern Conservations, and by the European Union (Thomas and Owen 1996). In Australia, lead shot restrictions vary by state from a total ban, to restrictions for waterfowl hunting similar to those in the USA, or suggesting non-toxic alternatives and leaving the choice of shot up to the hunters (Green 2004).

The most significant non-toxic shot regulation in the USA was the federal ban on the use of lead shot for hunting waterfowl in 1991. This ban has been demonstrated to have a positive impact on wildlife. For example, Anderson et al. (2000) attributed an estimated 64% reduction in mortality from lead poisoning to the switch to non-toxic shot. They estimated that 1.4 million ducks in the 1997 fall continental flight of 90 million were spared from fatal lead poisoning due to the ban on lead shot. Stevenson et al. (2005) found that lead concentrations in the bones of two species of ducks decreased after the ban, but in comparison, they noted that bone lead concentrations showed no change for American Woodcock (*Scolopax minor*), a migratory upland species not impacted by the lead shot ban for waterfowl hunting.

Some small game hunters have already begun to switch to non-toxic shot. In Minnesota, a recent survey, conducted by Schroeder et al. (2008) found that 40% of pheasant hunters reported they are currently using non-toxic shot voluntarily. Case et al. (2006) surveyed USA states and Canadian provinces regarding non-toxic shot regulations and found that 45% of surveyed states and provinces have non-toxic shot regulations beyond federal waterfowl regulations. Nine states and provinces that have non-toxic shot regulations were discussing additional regulations.

ACKNOWLEDGMENTS

We thank Bill Penning for reviewing the manuscript and contributing information. We thank Jo Ann Musumeci for acquiring articles, and Bill Healy (US Forest Service, retired), Steve Hennes (Minnesota Pollution Control Agency), John Schulz (Missouri Department of Conservation), and Erik Zabel (Minnesota Department of Health) for reviews of this manuscript. Also, Lindsey Aipperspach, Tom Conroy, Kathy DonCarlos, Mike DonCarlos, Kurt Haroldson, Tonya Klinkner, Jeff Lawrence, and Dennis Simon from Minnesota Department of Natural Resources provided reviews. Roxanne Franke and Dan Smedberg (Student Interns from Minnesota State University-Mankato) developed a lead shot literature review and summaries used in this manuscript.

LITERATURE CITED

ANDERSON, W. L. 1975. Lead poisoning in waterfowl at Rice Lake, Illinois. Journal of Wildlife Management 39:264–270.

ANDERSON, W. L., AND S. P. HAVERA. 1989. Lead poisoning in Illinois waterfowl (1977–1988) and implementation of nontoxic shot regulations. Illinois Natural History Survey Biological Notes 133.

ARTMANN, J. W., AND E. M. MARTIN. 1975. Incidence of ingested lead shot in Sora Rails. Journal of Wildlife Management 39:514–519.

BATTAGLIA, A., S. GHIDINI, G. CAMPANINI, AND R. SPAGGIARI. 2005. Heavy metal contamination in Little Owl (*Athene noctua*) and Common Buzzard (*Buteo buteo*) from northern Italy. Ecotoxicology and Environmental Safety 60:61–66.

BAXTER, G. S., C. MELZER, D. BYRNE, D. FIELDER, AND R. LOUTIT. 1998. The prevalence of spent lead shot in wetland sediments and ingested by wild ducks in coastal Queensland. The Sunbird 28:21–25.

BELLROSE, F. C. 1959. Lead poisoning as a mortality factor in waterfowl populations. Illinois Natural History Survey Bulletin 27:235–288.

BENGTSON, F. L. 1984. Studies of lead toxicity in Bald Eagles at the Lac Qui Parle Wildlife Refuge. Master's thesis, University of Minnesota, Minneapolis, Minnesota.

BEST, T. L., T. E. GARRISON, AND C. G. SCHMIDT. 1992. Ingestion of lead pellets by Scaled Quail (*Callieppla squamata*) and Northern Bobwhite (*Colinus virginianus*) in southeastern New Mexico. Texas Journal of Science 44:99–107.

BJERREGAARD, P., P. JOHANSEN, G. MULVAD, H. S. PEDERSEN, AND J. C. HANSEN. 2004. Lead sources in human diet in Greenland. Environmental Health Perspectives 112(15):1496–1498.

BLUS, L. J. 1994. A review of lead poisoning in swans. Comparative Biochemistry and Physiology, Part C 108:259–267.

BOWEN, J. E., AND S. A. PETRIE. 2007. Incidence of artifact ingestion in Mute Swans and Tundra Swans on the lower Great Lakes, Canada. Ardea 95:135–142.

BREUREC, J. Y., A. BAERT, J. P. ANGER, AND J. P. CURTES. 1998. Unusual diagnosis: non occupational adult lead poisoning. Toxicology Letters 95:76.

BRINZAL. 1996. SOS venenos: Búho Chico. Quercus 124:45.

BROWN, C. S., J. LUEBBERT, D. MULCAHY, J. SCHAMBER, AND D. H. ROSENBERG. 2006. Blood lead levels of wild Steller's Eiders (*Polysticta stelleri*) and Black Scoters (*Melanitta nigra*) in Alaska using a portable blood lead analyzer. Journal of Zoo and Wildlife Medicine 37:361–365.

BUERGER, T., R. E. MIRARCHI, AND M. E. LISANO. 1986. Effects of lead shot ingestion on captive Mourning Dove survivability and reproduction. Journal of Wildlife Management 50:1–8.

BUTLER, D. A. 2005. Incidence of lead shot ingestion in Red-legged Partridges (*Alectoris rufa*) in Great Britain. Veterinary Record: Journal of the British Veterinary Association 157:661.

BUTLER, D. A., R. B. SAGE, R. A. H. DRAYCOTT, J. P. CARROLL, AND D. POTTIS. 2005. Lead exposure in Ring-necked Pheasants on shooting estates in Great Britain. Wildlife Society Bulletin 33:583–589.

CABELA'S. 2008. Cabela's Shooting and Reloading [catalogue]. Cabela's. Sidney, Nebraska, USA.

CADE, T. J. 2007. Exposure of California Condors to lead from spent ammunition. Journal of Wildlife Management 71:2125–2133.

CALVERT, H. S. 1876. Pheasants poisoned by swallowing shot. The Field. 47:189.

CAMPBELL, H. 1950. Quail picking up lead shot. Journal of Wildlife Management 14:243–244.

CAMUS, A. C., M. M. MITCHELL, J. F. WILLIAMS AND P. L. H. JOWETT. 1998. Elevated lead levels in farmed American Alligators *Alligator mississippiensis* consuming Nutria *Myocastor coypus* meat contaminated by lead bullets. Journal of the World Aquaculture Society 3:370–376.

CASE, D. J. AND ASSOCIATES. 2006. Non-toxic shot regulation inventory of the United States and Canada. D. J. Case and Associates, Mishawaka, Indiana, USA.

CHIBA, A., N. SHIBUYA, AND R. HONMA. 1999. Description of a lead-poisoned Middendorff's Bean Goose, *Anser fabalis middendorffii*, found at Fukushima-gata, Niigata Prefecture, Japan. Japanese Journal of Ornithology 47:87–96.

CHURCH, M. E., GWIAZDA, R., RISEBROUGH, R. W., SORENSON, K., CHAMBERLAIN, C. P., FARRY, S., HEINRICH, W., RIDEOUT, B. A., AND SMITH, D. R. 2006. Ammunition is the principal source of lead accumulated by California Condors reintroduced to the wild. Environmental Science and Technology 40:6143–6150.

CLARK, A. J., AND A. M. SCHEUHAMMER. 2003. Lead poisoning in upland foraging birds of prey in Canada. Ecotoxicology 12:23–30.

CLAUSEN, B., AND C. WOLSTRUP. 1979. Lead poisoning in game from Denmark. Danish Review of Game Biology 11:1–22.

COOK, R. S., AND D. O. TRAINER. 1966. Experimental lead poisoning of Canada Geese. Journal of Wildlife Management 30:1–8.

CRAIG, T. H., J. W. CONNELLY, E. H. CRAIG, AND T. L. PARKER. 1990. Lead concentrations in Golden and Bald Eagles. Wilson Bulletin 102:130–133.

CRAIGHEAD, D., AND B. BEDROSIAN. 2008. Blood lead levels of Common Ravens with access to big-game offal. Journal of Wildlife Management 72:240–245.

DAMES AND MOORE CANADA. 1993. Field investigations and environmental site assessment of outdoor military small arms ranges. Prepared for the Dept. of National Defense. Project 24903–021, Mississauga, Ontario, Canada.

DECKER, R. A., A. M. MCDERMID, AND J. W. PRIDEAUX. 1979. Lead poisoning in two captive King Vultures. Journal of the American Veterinary Medical Association 175:1009.

DE FRANCISCO, N., J. D. RUIZ TROYA, AND E. I. AGÜERA. 2003. Lead and lead toxicity in domestic and free living birds. Avian Pathology 32(1) 3–13.

DEMENT, S. H., J. J. CHISOLM, JR., M. A. ECKHAUS, AND J. D. STRANDBERG. 1987. Toxic lead exposure in the urban Rock Dove. Journal of Wildlife Diseases 23:273–278.

DEPARTMENT OF SUSTAINABILITY AND ENVIRONMENT, VICTORIA. 2003. The use of lead shot in cartridges for hunting waterfowl. Flora and Fauna Guarantee Action Statement #32. East Melbourne, Victoria, Australia.

DEWAILLY, E., B. LEVESQUE, J- F. DUCHESNES, P. DUMAS, A. SCHEUHAMMER, C. GARIEPY, M. RHAINDS, J- F. PROULX. 2000. Lead shot as a source of lead poisoning in the Canadian Arctic. Epidemiology 11:146.

DEWAILLY, E., P. AYOTTE, S. BRUNEAU, G. LEBEL, P. LEVALLOIS, AND J. P. WEBER. 2001. Exposure of the Inuit population of Nunavik (Arctic Quebec) to lead and mercury. Archives of Environmental Health 56:350–7.

DONÁZAR, J. A., C. J. PALACIOS, L. GANGOSO, O. CEBALLOS, M. J. GONZALEZ, AND F. HIRALDO. 2002. Conservation status and limiting factors in the endangered population of Egyptian Vulture (*Neophron percnopterus*) in the Canary Islands. Biological Conservation 107:89–97.

EISLER, R. 1988. Lead hazards to fish, wildlife, and invertebrates: a synoptic review. United States Fish and Wildlife Service. Biological Report 85.

ELDER, W. H. 1954. The effect of lead poisoning on the fertility and fecundity of domestic Mallard ducks. Journal of Wildlife Management 18:315–323.

EMERSON, R. 1994. Contamination of soil from gun shot: St. Thomas Gun Club (1993). Technical Memorandum, Rep. No. SDB 052–4304–94 TM, Standards Development Branch, Phytotoxicology Section, Ontario Ministry of Environment and Energy, Brampton, Ontario, Canada.

ENGSTAD, J. E. 1932. Foreign bodies in the appendix. Minnesota Medicine 15:603.

ESTABROOKS, S. R. 1987. Ingested lead shot in Northern Red-billed Whistling Ducks (*Dendrocygna autumnalis*) and Northern Pintails (*Anas acuta*) in Sinaloa, Mexico. Journal of Wildlife Diseases 23:169.

FINLEY, M. T., AND M. P. DIETER. 1978. Toxicity of experimental lead-iron shot versus commercial lead shot in Mallards. Journal of Wildlife Management 42:32–39.

FISHER, I. J., D. J. PAIN, AND V. G. THOMAS. 2006. A review of lead poisoning from ammunition sources in terrestrial birds. Biological Conservation 131:421–432.

FLINT, P. L., M. R. PETERSEN, AND J. B. GRAND. 1997. Exposure of Spectacled Eiders and other diving ducks to lead in western Alaska. Canadian Journal of Zoology 75:439–443.

FRANSON, C. 2007. Lead poisoning in wild birds: exposure, clinical signs, lesions, and diagnosis. 68th Midwest Fish and Wildlife Conference [presentation and abstract]. December 11, 2007. Madison, Wisconsin, USA .

FRANSON, J. C., AND S. G. HEREFORD. 1994. Lead poisoning in a Mississippi Sandhill Crane. Wilson Bulletin 106:766–768.

FRANSON, J. C., M. R. PETERSEN, C. U. METEYER, AND M. R. SMITH. 1995. Lead poisoning of Spectacled Eiders (*Somateria fischeri*) and of a Common Eider (*Somateria mollissima*) in Alaska. Journal of Wildlife Diseases 31:268–271.

FRAPE, D. L., AND J. D. PRINGLE. 1984. Toxic manifestations in a dairy herd consuming haylage contaminated by lead. Veterinary Records 114:615–616.

FRIEND, M. AND J. C. FRANSON (EDS.). 1999. Field Manual of Wildlife Diseases: General Field Procedures and Diseases of Birds [online]. US Geological Survey. [Online.] Available at http// www.nwhc.usgs.gov/publications/field_manual/ index.jsp.

GARCIA-FERNANDEZ, A. J., E. MARTINEZ-LOPEZ, D. ROMERO, P. MARIA-MOJICA, A. GODINO, AND P. JIMENEZ. 2005. High levels of blood lead in Griffon Vultures (*Gyps fulvus*) from Cazorla Natural Park (southern Spain). Environmental Toxicology 20:459–463.

GOLDEN, N. H., AND B. A. RATTNER. 2003. Ranking terrestrial vertebrate species for utility in biomonitoring and vulnerability to environmental contaminants. Reviews of Environmental Contamination and Toxicology 176:67–136.

GRAND, J. B., P. L. FLINT, M. R. PETERSEN, AND C. L. MORAN. 1998. Effect of lead poisoning on Spectacled Eider survival rates. Journal of Wildlife Management 62:1103–1109.

GREEN, B. 2004. The Situation in Australia. Pages 73–76 *in* Proceedings of the World Symposium on Lead Ammunition. World Forum on the Future of Sport Shooting Activities. September 9–10, 2004. Rome, Italy.

GRINNELL, G. B. 1894. Lead poisoning. Forest and Stream 42:117–118.

GUITART, R., J. SERRATOSA, AND V. G. THOMAS. 2002. Lead poisoned wildfowl in Spain: A significant threat for human consumers. International Journal of Environmental Health Research 12:301–309.

HALDIMANN, M., A. BAUMGARTNER, AND B. ZIMMERLI. 2002. Intake of lead from game meat – a risk to consumers' health. European Food Research and Technology 215:375–379.

HALL, S. L., AND F. M. FISHER. 1985. Lead concentrations in tissues of marsh birds: relationship of feeding habits and grit preference to spent shot ingestion. Bulletin of Environmental Contamination and Toxicology 35:1–8.

HAMMERTON, K. M., N. JAYASINGHE, R. A. JEFFREE, AND R. P. LIM. 2003. Experimental study of blood lead kinetics in Estuarine Crocodiles (*Crocodylus porosus*) exposed to ingested lead shot. Archives of Environmental Contamination and Toxicology 45:390–398.

HANNING, R. M., R. SANDHU, A. MACMILLAN, L. MOSS, L. J. S. TSUJI, AND E. NIEBOER, JR. 2003. Impact on blood Pb levels of maternal and early infant feeding practices of First Nation Cree in the Mushkegowuk Territory of northern Ontario, Canada. Journal of Environmental Monitoring 5:241–245.

HARPER, M. J., AND M. HINDMARSH. 1990. Lead poisoning in Magpie Geese, *Anseranas semipalmata,* from ingested lead pellets at Bool Lagoon Game Reserve (South Australia). Australia Wildlife Research 17:141–145.

HARRADINE, J. 2004. Spent lead shot and wildlife exposure and risks. Pages 119–130 *in* Proceedings of the World Symposium on Lead Ammunition. World Forum on the Future of Sport Shooting Activities. September 9–10, 2004. Rome, Italy.

HAVERA, S. P., R. M. WHITTON, AND R. T. SHEALY. 1992. Blood lead and ingested and embedded shot in diving ducks during spring. Journal of Wildlife Management 56:539–545.

HENNES, S. K. 1985. Lead shot ingestion and lead residues in migrant Bald Eagles at the Lac Qui Parle Wildlife Management Area, Minnesota. Master's thesis, University of Minnesota, Minneapolis, Minnesota, USA.

HILLMAN, F. E. 1967. A rare case of chronic lead poisoning: polyneuropathy traced to lead shot in the appendix. Industrial Medicine and Surgery 36:488–492.

HOLLAND, G. 1882. Pheasant poisoning by swallowing shot. The Field 59:232.

HONDA, K., D. P. LEE, AND R. TASUKAWA. 1990. Lead poisoning in swans in Japan. Environmental Pollution 65:209–218.

HORTON, B. T. 1933. Bird shot in verminform appendix: a cause of chronic appendicitis. Surgical Clinics of North America 13:1005–1006.

HUNT, W. G., W. BURNHAM, C. N. PARISH, K. K. BURNHAM, B. MUTCH, AND J. L. OAKS. 2006. Bullet fragments in deer remains: implications for lead exposure in avian scavengers. Wildlife Society Bulletin 34:167–170.

HUNTER, B. F., AND M. N. ROSEN. 1965. Occurrence of lead poisoning in a wild Pheasant (*Phasianus colchicus*). California Fish and Game 51:207.

JACOBSON, E., J. W. CARPENTER, AND M. NOVILLA. 1977. Suspected lead toxicosis in a Bald Eagle. Journal of American Veterinary Medical Associates 171:952–954.

JOHANSEN, P., G. ASMUND, AND F. RIGET. 2001. Lead contamination of seabirds harvested with lead shot—implications to human diet in Greenland. Environmental Pollution 112:501–504.

JOHANSEN, P., G. ASMUND, AND F. RIGET. 2004. High human exposure to lead through consumption of birds hunted with lead shot. Environmental Pollution 127:125–9.

JOHANSEN, P., H. S. PEDERSEN, G. ASMUND, AND F. RIGET. 2006. Lead shot from hunting as a source of lead in human blood. Environmental Pollution 142:93–7.

JONES, J. C. 1939. On the occurrence of lead shot in stomachs of North American Gruiformes. Journal of Wildlife Management 3:353–357.

KAISER, G. W., K. FRY, AND J. G. IRELAND. 1980. Ingestion of lead shot by Dunlin. The Murrelet 61:37.

KEEL, M. K., W. R. DAVIDSON, G. L. DOSTER, AND L. A. LEWIS. 2002. Northern Bobwhite and lead shot deposition in an upland habitat. Archives of Environmental Contamination and Toxicology 43:318–322.

KENDALL, R. J., T. E. LACHER, JR., C. BUNCK, B. DANIEL, C. DRIVER, C. E. GRUE, F. LEIGHTON, W. STANSLEY, P. G. WATANABE, AND M. WHITWORTH. 1996. An ecological risk assessment of lead shot exposure in non-waterfowl avian species: upland game birds and raptors. Environmental Toxicology and Chemistry 15:4–20.

KENDALL, R.J., G.W. NORMAN, AND P.F SCANLON. 1984. Lead concentration in Ruffed Grouse collected from Southwestern Virginia. Northwest Science 58:14–17.

KENNTNER, N., Y. CRETTENAND, H- J. FÜNFSTÜCK, M. J. JANOVSKY, AND F. TATARUCH. 2007. Lead poisoning and heavy metal exposure of Golden Eagles (*Aquila chrysaetos*) from the European Alps. Journal of Ornithology 148:173–177.

KEYMER, I. F., AND R. S. STEBBINGS. 1987. Lead poisoning in a partridge (*Perdix perdix*) after ingestion of gunshot. Veterinary Record 120:276–277.

KINGSFORD, R. T., J. L. KACPRZAK, AND J. ZIAZIARIS. 1994. Lead in livers and gizzards of waterfowl shot in New South Wales, Australia. Environmental Pollution 85(3):329–335.

KNOPPER, L. D., P. MINEAU, A. M. SCHEUHAMMER, D. E. BOND, AND D. T. MCKINNON. 2006. Carcasses of shot Richardson's Ground Squirrels may pose lead hazards to scavenging hawks. Journal of Wildlife Management 70:295–299.

KOH, T. S., AND HARPER, M. J. 1988. Lead-poisoning in Black Swans, *Cygnus atratus*, exposed to lead shot at Bool Lagoon Game Reserve, South Australia. Australian Wildlife Research 15:395–403.

KREAGER, N., B. C. WAINMAN, R. K. JAYASINGHE, AND L. J. S. TSUJI. 2007. Lead pellet ingestion and liver-lead concentrations in upland game birds from southern Ontario, Canada. Archives of Environmental Contamination and Toxicology. DOI 10.1007/s00244–007–9020–6.

KRONE, O., WILLIE, F., KENNTNER, N., BOERTMANN, D., TATARUCH, F. 2004. Mortality factors, environmental contaminants, and parasites of White-tailed Sea Eagles from Greenland. Avian Diseases 48:417–424.

KUROSAWA, N. 2000. Lead poisoning in Steller's Sea Eagles and White-tailed Sea Eagles. Pages 107–109 *in* M. Ueta and M. J. McGrady (Eds.). First Symposium on Steller's and White-tailed Sea Eagles in East Asia. Wild Bird Society of Japan, Tokyo, Japan.

LANCE, V. A., T. R. HORN, R. M. ELSEY AND A. DE PEYSTER. 2006. Chronic incidental lead ingestion in a group of captive-reared alligators (*Alligator mississippiensis*): possible contribution to reproductive failure. Toxicology and Pharmacology 142:30–35.

LARSEN, A. R., AND R. H. BLANTON. 2000. Appendicitis due to bird shot ingestion: a case study. American Surgeon 66(6):589–591.

LARSEN, R. T. 2006. Ecological investigations of Chukars (*Alectoris chukar*) in western Utah. Master's thesis. Brigham Young University, Provo, Utah, USA.

LEMAY, A., P. MCNICHOLL, AND R. OUELLET. 1989. Incidence de la grenaille de plomb dans les gesiers de canards, d'oies et de bernaches recoltes au Quebec. Direction de la gestion des especes et des habitats. Ministere du Loisir de la Chasse et de la Peche, Quebec.

LÉVESQUE, B., J. F. DUCHESNE, C. GARIÉPY, M. RHAINDS, P. DUMAS, A. M. SCHEUHAMMER, J. F. PROULX, S. DÉRY, G. MUCKLE, F. DALLAIRE, AND É. DEWAILLY. 2003. Monitoring of umbilical cord blood lead levels and sources of assessment among the Inuit. Occupational and Environmental Medicine 60:693–695.

LEWIS, J. C., AND E. LEGLER, JR. 1968. Lead shot ingestion by Mourning Doves and incidence in soil. Journal of Wildlife Management 32(3):476–482.

LEWIS, L. A., R. J. POPPENGA, W. R. DAVIDSON, J. R. FISCHER, AND K. A. MORGAN. 2001. Lead toxicosis and trace element levels in wild birds and mammals at a firearms training facility. Archives of Environmental Contamination and Toxicology 41(2):208–214.

LOCKE, L. N., G. E. BAGLEY, D. N. FRICKE, AND L. T. YOUNG. 1969. Lead poisoning and aspergillosis in an Andean Condor. Journal of American Veterinary Medical Associates 155(7):1052–1056.

LOCKE, L. N., AND M. FRIEND. 1992. Lead poisoning of avian species other than waterfowl. Pages 19–22 *in* D. J. Pain (Ed.). Lead Poisoning in Waterfowl. Proceedings of an IWRB Workshop, Brussels, Belgium, 13–15 June 1991. International Waterfowl and Wetlands Research Bureau Special Publication 16, Slimbridge, UK.

LOCKE, L. N., M. R. SMITH, R. M. WINDINGSTAD, AND S. J. MARTIN. 1991. Lead poisoning of a Marbled Godwit. Prairie Naturalist 23(1):21–24.

LUMEIJ, J. T. 1985. Clinicopathologic aspects of lead poisoning in birds: a review. Veterinary Quarterly 7:133–138.

MA, W. 1989. Effect of soil pollution with metallic lead pellets on lead bioaccumulation and organ/body weight alterations in small mammals. Archives of Environmental Contamination and Toxicology 18:617–622.

MADSEN, H. H. T., T. SKJØDT, P. J. JORGENSEN, AND P. GRANDJEAN. 1988. Blood lead levels in patients with lead shot retained in the appendix. Acta Radiologica 29:745–746.

MACDONALD, J. W., C. J. RANDALL, H. M. ROSS, G. M. MOON, AND A. D. RUTHVEN. 1983. Lead poisoning in captive birds of prey. Veterinary Records 113:65–66.

MANNINEN, S., AND N. TANSKANEN. 1993. Transfer of lead from shotgun pellets to humus and three plant species in a Finnish shooting range. Archives of Environmental Contamination and Toxicology 24:410–414.

MARTIN, P. A., D. CAMPBELL, K. HUGHES, AND T. DANIEL. 2008. Lead in the tissues of terrestrial raptors in southern Ontario, Canada, 1995–2001. Science of the Total Environment 391(1):96–103.

MARTIN, P. A., AND G. C. BARRETT. 2001. Exposure of terrestrial raptors to environmental lead - determining sources using stable isotope ratios. International Association for Great Lakes Research Conference Program and Abstracts 44. Ann Arbor, Michigan, USA.

MATEO, R., J. BELLIURE, J. C. DOLZ, J. M. AGUILAR-SERRANO, AND R. GUITART. 1998. High prevalences of lead poisoning in wintering waterfowl in Spain. Archives of Environmental Contamination and Toxicology 35(2):342–347.

MATEO, R., J. C. DOLZ, J. M. AGUILAR-SERRANO, J. BELLIURE, AND R. GUITART. 1997. An epizootic of lead poisoning in Greater Flamingos (*Pheonicopterus rubber roseus*) in Spain. Journal of Wildlife Diseases 33(1):131–134.

MATEO, R., J. ESTRADA, J- Y. PAQUET, X. RIERA, L. DOMÍNGUEZ, R. GUITART, AND A. MARTÍNEZ-VILALTA. 1999. Lead shot ingestion by Marsh Harriers *Circus aeruginosus* from the Ebro delta, Spain. Environmental Pollution 104(3):435–440.

MATEO, R., A. J. GREEN, C. W. JESKE, V. URIOS, AND C. GERIQUE. 2001. Lead poisoning in the globally threatened Marbled Teal and White-headed Duck in Spain. Environmental Toxicology and Chemistry 20(12):2860–2868.

MATEO, R., R. GUITART, AND A. J. GREEN. 2000. Determinants of lead shot, rice, and grit ingestion in ducks and coots. Journal of Wildlife Management 64(4):939–947.

MATEO, R., M. RODRÍGUEZ-DE LA CRUZ, D. VIDAL M. REGLERO, AND P. CAMARERO. 2007. Transfer of lead from shot pellets to game meat during cooking. Science of the Total Environment 372(2–3):480–485.

MATEO, R., M. TAGGART, AND A. A. MEHARG. 2003. Lead and arsenic in bones of birds of prey from Spain. Environmental Pollution 126(1): 107–114.

MERENDINO, M. T., D. S. LOBPRIES, J. E. NEAVILLE, J. D. ORTEGO, AND W. P. JOHNSON. 2005. Regional differences and long-term trends in lead exposure in Mottled Ducks. Wildlife Society Bulletin 33(3):1002–1008.

MICHIGAN DEPARTMENT OF NATURAL RESOURCES. 2002. Michigan Wildlife Disease Manual. Michigan Department of Natural Resources Wildlife Disease Laboratory, Lansing, Michigan, USA.

MINNESOTA DEPARTMENT OF AGRICULTURE. 2008. State tests confirm lead in some venison from food shelves [news release]. Minnesota Department of Agriculture, St. Paul, Minnesota, USA.

MINNESOTA DEPARTMENT OF NATURAL RESOURCES. 1981. Study of the presence and toxicity of lead shot at the Lac Qui Parle Wildlife Refuge, Watson, Minnesota from 1978 to 1979. Minnesota Department of Natural Resources, St. Paul, Minnesota, USA.

MINNESOTA DEPARTMENT OF NATURAL RESOURCES. 2007. Trumpeter Swan die-off at Grass Lake, Wright County. DNR Fact Sheet. February 28, 2007. Division of Ecological Services, St. Paul, Minnesota, USA.

MINNESOTA POLLUTION CONTROL AGENCY. 1999. Legislative report on sources and effects of lead presented to the Committees on the Environment and Natural Resources. Minnesota Pollution Control Agency, St. Paul, Minnesota, USA.

MOORE, C. S. 1994. Lead shot passed per urethrem [letter]. British Medical Journal 308:414.

MÖRNER, T., AND L. PETERSSON. 1999. Lead poisoning in woodpeckers in Sweden. Journal of Wildlife Diseases 35(4):763–765.

MOZAFAR, A., R. RUH, P. KLINGEL, H. GAMPER, S. EGLI, AND E. FROSSARD. 2002. Effect of heavy metal contaminated shooting range soils on mycorrhizal colonization of roots and metal uptake by leek. Environmental Monitoring and Assessment 79:177–191.

MUDGE, G. P. 1983. The incidence and significance of ingested lead pellet poisoning in British wildfowl. Biological Conservation 27:333–372.

NATIONAL WILDLIFE HEALTH LABORATORY. 1985. Lead poisoning in non-waterfowl avian species. Unpublished Report. US Fish and Wildlife Service, Washington, DC, USA.

NONTOXIC SHOT ADVISORY COMMITTEE. 2006. Report of the Nontoxic Shot Advisory Committee. Submitted to Minnesota Department of Natural Resources, December 12, 2006. St. Paul, Minnesota, USA.

NORTH DAKOTA DEPARTMENT OF HEALTH. 2008. Food pantries notified about lead fragments discovered in donated ground venison. News Release, March 26, 2008. North Dakota Department of Health. Bismarck, North Dakota, USA.

OCHIAI, K., K. JIN, M. GORYO, T. TSUZUKI, AND C. ITAKURA. 1993. Pathomorphologic findings of lead poisoning in White-fronted Geese (*Anser albifrons*). Veterinary Pathology 30(6):522–528.

OCHIAI, K., K. JIN, C. ITAKURA, M. GORYO, K. YAMASHITA, N. MIZUNO, T. FUJINAGA, AND T. TSUZUKI. 1992. Pathological study of lead poisoning in Whooper Swans (*Cygnus cygnus*) in Japan. Avian Diseases 36(2):313–323.

ODLAND, J. O., I. PERMINOVA, N. ROMANOVA, Y. THOMASSEN, L. J. S. TSUJI, J. BROX, AND E. NIEBOER. 1999. Elevated blood lead concentrations in children living in isolated communities of the Kola Peninsula, Russia. Ecosystem Health 5(2):75–81.

OLIVIER, G.-N. 2006. Considerations on the use of lead shot over wetlands. Pages 866–867 *in* G.C. Boere, C. A. Galbraith, and D. A. Stroud (Eds.). Waterbirds Around the World. The Stationery Office, Edinburgh, UK.

ORLIC, I., R. SIEGELE, K. HAMMERTON, R. A. JEFFREE, AND D. D. COHEN. 2003. Nuclear microprobe analysis of lead profile in crocodile bones. Nuclear Instruments and Methods in Physics Research Section B: Beam Interactions with Materials and Atoms 210:330–335.

OSMER, T. L. G. 1940. Lead shot: its danger to water-fowl. The Scientific Monthly 50(5):455–459.

PAIN, D. J. 1990. Lead shot ingestion by waterbirds in the Camargue, France: an investigation of levels and interspecific differences. Environmental Pollution 66:273–285.

PAIN, D. J. AND C. AMIARD-TRIQUET. 1993. Lead poisoning of raptors in France and elsewhere. Ecotoxicology and Environmental Safety 25:183–192.

PAIN, D. J., I. CARTER, A. W. SAINSBURY, R. F. SHORE, P. EDEN, M. A. TAGGART, S. KONSTANTINOS, L. A. WALKER, A. A. MEHARG, AND A. RAAB. 2007. Contamination and associated disease in captive and reintroduced Red Kites *Milvus milvus* in England. Science of the Total Environment 376:116–127.

PAIN, D. J., A. A. MEHARG, M. FERRER, M. TAGGART AND V. PENTERIANI. 2005. Lead concentrations in bones and feathers of the globally threatened Spanish Imperial Eagle. Biological Conservation 121(4):603–610.

PAIN, D. J., J. SEARS, AND I. NEWTON. 1994. Lead concentrations in birds of prey in Britain. Environmental Pollution 87:173–180.

PERRY, M. C., AND J. W. ARTMANN. 1979. Incidence of embedded shot and ingested shot in oiled Ruddy Ducks. Journal of Wildlife Management 43(1):266–269.

QUORTRUP, E.R., AND J. E. SHILLINGER. 1941. 3,000 wild bird autopsies on western lake areas. American Veterinary Medical Association Journal.

REDDY, E.R. 1985. Retained lead shot in the appendix. Journal of the Canadian Association of Radiologists 36:47–48.

REDIG, P. T., C. M. STOWE, D. M. BARNES, AND T. D. ARENT. 1980. Lead toxicosis in raptors. Journal of American Veterinary Medical Associates 177:941–943.

RICE, D. A., M. F. MCLOUGHLIN, W. J. BLANCHFLOWER, AND T. R. THOMPSON. 1987. Chronic lead poisoning in steers eating silage contaminated with lead shot: diagnostic criteria. Bulletin of Environmental Contamination and Toxicology 39(4):622–629.

ROCKE, T. E., AND M. D. SAMUEL. 1991. Effects of lead shot ingestion on selected cells of the Mallard immune system. Journal of Wildlife Diseases 27(1):1–9.

RODRIGUE, J., R. MCNICOLL, D. LECLAIR, AND J.-F DUCHESNE. 2005. Lead concentrations in Ruffed Grouse, Rock Ptarmigan, and Willow Ptarmigan in Quebec. Archives of Environmental Contamination and Toxicology 49(1):334–340.

ROONEY, C. P., R. G. MCLAREN, AND R. J. CRESSWELL. 1999. Distribution and phytoavailability of lead in a soil contaminated with lead shot. Water, Air, and Soil Pollution 116:535–548.

SANBORN, W. 2002. Lead poisoning of North American wildlife from lead shot and lead fishing tackle [Draft]. HawkWatch International, Inc., Salt Lake City, Utah, USA. [Online.] Available at http://www.hawkwatch.org/publications.php?id=2.

SANDERSON, G. C., AND F. C. BELLROSE. 1986. A review of the problem of lead poisoning in waterfowl. Illinois Natural History Survey, Special Publication 4. Champaign, Illinois, USA.

SCANLON, P. F., V. D. STOTTS, R. G. ODERWALD, T. J. DIETRICH, AND R. J. KENDALL. 1980. Lead concentrations in livers of Maryland waterfowl with and without ingested lead shot present in gizzards. Bulletin of Environmental Contamination and Toxicology 25(6):855–860.

SCHEUHAMMER, A. M., D. E. BOND, N. M. BURGESS, AND J. RODRIGUE. 2003. Lead and stable lead isotope ratios in soil, earthworms, and bones of American Woodcock (*Scolopax minor*) from Eastern Canada. Environmental Toxicology and Chemistry 22:2585–2591.

SCHEUHAMMER, A. M., AND S. L. NORRIS. 1995. A review of the environmental impacts of lead shotshell ammunition and lead fishing weights in Canada. Occasional Paper Number 88, Canadian Wildlife Service. National Wildlife Research Centre, Hull, Quebec, Canada.

SCHEUHAMMER, A. M., J. A. PERRAULT, E. ROUTHIER, B. M. BRAUNE, AND G. D. CAMPBELL. 1998. Elevated lead concentrations in edible portions of game birds harvested with lead shot. Environmental Pollution 102:251–257.

SCHMITZ, R. A., A. A. AGUIRRE, R. S. COOK, AND G. A. BALDASSARRE. 1990. Lead poisoning of Caribbean Flamingos in Yucatan, Mexico. Wildlife Society Bulletin 18(4):399–404.

SCHROEDER, S. A., D. C. FULTON, W. PENNING, AND K. DONCARLOS. 2008. Small game hunter lead shot study. US Geological Survey. University of Minnesota, Minnesota Cooperative Fish and Wildlife Research Unit, Department of Fisheries, Wildlife, and Conservation Biology. Draft manuscript.

SCHULZ, J. H., J. J. MILLSPAUGH, A. J. BERMUDEZ, X. GAO, T. W. BONNOT, L. G. BRITT, AND M. PAINE. 2006a. Acute lead toxicosis in Mourning Doves. Journal of Wildlife Management 70(2):413–421.

SCHULZ, J. H., X. GAO, J. J. MILLSPAUGH, AND A. J. BERMUDEZ. 2007. Experimental lead pellet ingestion in Mourning Doves (*Zenaida macroura*). American Midland Naturalist 158:177–190.

SCHULZ, J. H., P. I. PADDING, AND J. J. MILLSPAUGH. 2006b. Will Mourning Dove crippling rates increase with nontoxic-shot regulations? Wildlife Society Bulletin 34(3), 861–864.

SCHWAB, D., SR., AND T. M. PADGETT. 1988. Lead poisoning in free ranging Pekin Duck (*Anas platyrhychos*) from Chesapeake, VA. Virginia Journal of Science 39:412–413.

SKERRATT, L. F., C. FRANSON, C. U. METEYER, AND T. E. HOLLMEN. 2005. Causes and mortality in sea ducks (Mergini) necropsied at the US Geological Survey- National Wildlife Health Center. Waterbirds 28(2):193–207.

SIKARSKIE, J. 1977. The case of the Red-tailed Hawk. Intervet 8:4.

SILEO, L., R. N. JONES, AND R. C. HATCH. 1973. The effect of ingested lead shot on the electrocardiogram of Canada Geese. Avian Diseases 17(2):308–313.

SMITH, L. F., AND E. REA. 1995. Low blood levels in northern Ontario-what now? Canadian Journal of Public Health 86:373–376.

SPITALE, L. S., AND M. A. D'OLIVO. 1989. Cecal appendix with pellets. Revista de la Facultad de Ciencias Médicas de Córdoba 47(1–2):23–25.

STANSLEY, W., L. WIDJESKOG, AND D. E. ROSCOE. 1992. Lead contamination and mobility in surface water at trap and skeet ranges. Bulletin of Environmental Contamination and Toxicology 49:640–647.

STENDELL, R. C., J. W. ARTMANN, AND E. MARTIN. 1980. Lead residues in Sora Rails from Maryland. Journal of Wildlife Management 44(2):525–527.

STEVENSON, A. L., A. M. SCHEUHAMMER, AND H. M. CHAN. 2005. Effects of nontoxic shot regulations on lead accumulation in ducks and American Woodcock in Canada. Archives of Environmental Contamination and Toxicology 48(3):405–413.

STONE, W. B., AND S. A. BUTKAS. 1972. Lead poisoning in a Wild Turkey. New York Fish and Game Journal 25:169.

STRAIT, M. M., J. E. NAILE, AND J. M. L. HIX. 2007. Lead analysis in soils and sediments at the Saginaw Field and Stream Club. Spectroscopy Letters 40(3):525–536.

SVANBERG, F., R. MATEO, L. HILLSTRÖM, A. J. GREEN, M. A. TAGGART, A. RAAB, A. A. MEHARG. 2006. Lead isotopes and lead shot ingestion in the globally threatened Marbled Teal (*Marmaronetta angustirostris*) and White-headed Duck (*Oxyura leucocephala*). Science of the Total Environment 370(2–3):416–24.

SZYMCZAK, M. R., AND W. J. ADRIAN. 1978. Lead poisoning in Canada Geese in southeast Colorado. Journal of Wildlife Management 42:299–306.

TAVECCHIA, G., R. PRADEL, J.-D. LEBRETON, A. R. JOHNSON, AND J.-Y. MONDAIN-MONVAL. 2001. The effect of lead exposure on survival of adult Mallards in the Camargue, southern France. Journal of Applied Ecology 38(6):1197–1207.

THOMAS, V. G. 1997. The environmental and ethical implications of lead shot contamination of rural lands in North America. Journal of Agricultural and Environmental Ethics 10(1):41–54.

THOMAS, V. G., AND M. OWEN. 1996. Preventing lead toxicosis of European waterfowl by regulatory and non-regulatory means. Environmental Conservation 23(4):358–364.

TRAINER, D. O., AND R. A. HUNT. 1965. Lead poisoning of Whistling Swans in Wisconsin. Avian Diseases 9(2):252–264.

TRANEL, M. A., AND R. O. KIMMEL. 2007. Nontoxic and lead shot literature review. Pages 116–155 *in* Summaries of Wildlife Research Findings 2007. Minnesota Department of Natural Resources. Wildlife Populations and Research Unit. St. Paul, Minnesota, USA.

TREBEL, R. G., AND T. S. THOMPSON. 2002. Case Report: elevated blood lead levels resulting from the ingestion of air rifle pellets. Journal of Analytical Toxicology 26(6):370–373.

TRUST, K. A., M. W. MILLER, J. K. RINGELMAN, AND I. M. ORME. 1990. Effects of ingested lead on antibody production in Mallards (*Anas platyrhynchos*). Journal of Wildlife Diseases 26(3):316–322.

TSUJI, L. J. S., AND E. NIEBOER. 1997. Lead pellet ingestion in First Nation Cree of western James Bay region of Northern Ontario, Canada: implications for nontoxic shot alternative. Ecosystem Health 3:54–61.

TSUJI, L. J. S., E. NIEBOER, J. D. KARAGATZIDES, R. M. HANNING, AND B. KATAPATUK. 1999. Lead shot contamination in edible portions of game birds and its dietary implications. Ecosystem Health 5 (3):183–192.

UNITED STATES ENVIRONMENTAL PROTECTION AGENCY. 1994. Proceeding Under Section 7003 of the Solid Waste Disposal Act. Westchester Sportmen's Center. Administrative Order of Consent. Docket No. II RCPA–94–7003–0204.

UNITED STATES FISH AND WILDLIFE SERVICE. 2006. Nontoxic shot regulations for hunting waterfowl and coots in the US. USFWS, Division of Migratory Bird Management, Arlington, Virginia, USA.

VYAS, N. B., J. W. SPANN, G. H. HEINZ, W. N. BEYER, J. A. JAQUETTE, AND J. M. MENGELKOCH. 2000. Lead poisoning of passerines at a trap and skeet range. Environmental Pollution 107 (1):159–166.

WALTER, H., AND K. P. REESE. 2003. Fall diet of Chukars (*Alectoris chukar*) in eastern Oregon and discovery of ingested lead pellets. Western North American Naturalist 63:402–405.

WESTEMEIER, R. L. 1966. Apparent lead poisoning in a wild Bobwhite. Wilson Bulletin 78(4):471–472.

WHITE, D. H., AND R. C. STENDELL. 1977. Waterfowl exposure to lead and steel shot on selected hunting areas. Journal of Wildlife Management 41(3):469–475.

WHITEHEAD, P. J., AND K. TSCHIRNER. 1991. Lead shot ingestion and lead poisoning of Magpie Geese, *Anseranas semipalmata,* foraging in a northern Australian hunting reserve. Biological Conservation 58:99–118.

WINDINGSTAD, R. M., S. M. KERR, AND L. N. LOCKE. 1984. Lead poisoning in Sandhill Cranes (*Grus canadensis*). Prairie Naturalist 16:21–24.

WILSON, I. D. 1937. An early report of lead poisoning in waterfowl. Science, New Series 86(2236):423.

WILSON, H. M., J. L. OYEN, AND L. SILEO. 2004. Lead shot poisoning of a Pacific Loon in Alaska. Journal of Wildlife Diseases 40(3):600–602.

YAMAMOTO, K., M. HAYASHI, M. YOSHIMURA, H. HAYASHI, A. HIRATSUKA, AND Y. ISII. 1993. The prevalence and retention of lead pellets in Japanese Quail. Archives of Environmental Contamination and Toxicology 24:478–482.

ZWANK, P. J., V. L. WRIGHT, P. M. SHEALY, AND J. D. NEWSOM. 1985. Lead toxicosis in waterfowl in two major wintering areas in Louisiana. Wildlife Society Bulletin 13(1):17–26.

POLICY CONSIDERATIONS FOR A MOURNING DOVE NONTOXIC SHOT REGULATION

JOHN H. SCHULZ[1], JOSHUA J. MILLSPAUGH[2], AND LARRY D. VANGILDER[1]

[1]*Missouri Department of Conservation, Resource Science Center, 1110 South College Avenue, Columbia, MO 65201, USA.* E-mail: John.H.Schulz@mdc.mo.gov

[2]*Department of Fisheries and Wildlife Sciences, University of Missouri, 302 Anheuser-Busch Natural Resources Building, Columbia, MO 65211, USA.*

EXTENDED ABSTRACT.—The use of lead in sport hunting is quickly becoming a priority conservation policy issue as demonstrated by this symposium. Within the context of making policy, decision makers must balance the relative importance of multiple data streams, and simultaneously assign relative certainty to the available knowledge (Reitz et al. 2007). As previously demonstrated, we know with certainty (1) that lead is a well established broad-spectrum ecological poison (Sanderson and Bellrose 1986, Eisler 1988, Kendall et al. 1996), (2) hunters can deposit relatively large amounts of lead shot on areas that are popular Mourning Dove (*Zenaida macroura*) feeding sites (Lewis and Legler 1968, Best et al. 1992, Schulz et al. 2002), (3) a certain proportion of the dove population feed on these sites and ingest lead pellets (Otis et al. 2008, Franson et al. 2009, this volume), and (4) virtually all doves that ingest pellets succumb to the direct or indirect effects of lead poisoning (Schulz et al. 2006, Schulz et al. 2007). Conversely, there is considerable uncertainty about the (1) relationship between lead pellet availability and pellet ingestion on areas with different levels of hunter and bird use, (2) uncertainty about the actual proportion of the population impacted by lead poisoning and whether that proportion is significant, (3) uncertainty about the impacts of other surface-feeding seed-eating songbirds and upland game-birds, (4) uncertainty about the eventual fate of doves dying of lead poisoning (e.g., consumption by scavengers), and (5) uncertainty about how the increased cost of nontoxic shot ammunition may negatively influence future small-game hunter participation rates. Given this information, natural resource policy makers face two types of risk; the risk of taking an unnecessary action (Type I Error), and the risk of failing to enact a needed action (Type II Error; Lee 1993). Our professional culture has a strong tradition of avoiding Type I policy errors as demonstrated by the length of time required to enact nontoxic shot regulations for waterfowl hunting (i.e., more research was needed to ensure the action being taken as necessary). However, the risks associated with Type II policy errors cannot be overlooked while potential harm to the resource may be occurring; this concept is often viewed as the precautionary principle.

Economic considerations, especially short-term economic goals, often conflict with effort to achieve ecological sustainability, and these differences become a source of social strife and conflict in the policy-making process. In other words, policy decisions are affected by application of the risk paradigm (i.e., risk analysis) and/or the ecological paradigm (i.e., the precautionary principle). The risk paradigm views environmental hazards as manageable and uses risk management as its primary scientific and policy-making tool. In comparison, the ecological paradigm is based upon the precautionary principle and begins from the view that scientific knowledge of complex systems is incomplete and imprecise, and precautionary measures are warranted (Raffensperger and Tickner 1999, Tickner 2003). This suggests that substances that can be reasonably and scientifically judged to have the potential to cause widespread long-term, and severe forms of environmental damage (e.g., extensive use of lead-based ammunition for sport hunting) should be replaced, whenever feasible, with safer alternatives rather than continued use with acceptable amounts of environmental risk.

Given these challenges, suggested actions for policy makers include (1) an explicit recognition of all stakeholders, (2) a long-term vision among the stakeholders that identifies an ultimate desired future condition that transcends immediate concerns and issues (e.g., spent lead-based ammunition is an environmental poison with several available alternatives), (3) a continued emphasis on research to reduce key uncertainties related to *a priori* policy decisions (compared to research aimed at determining the legitimacy of the problem), and (4) an explicit recognition that sufficient reliable information currently exists to suggest some preliminary and/or incremental policy decisions can be immediately contemplated. *Received 30 April 2008, accepted 14 August 2008.*

SCHULZ, J. H., J. J. MILLSPAUGH, AND L. D. VANGILDER. 2009. Policy considerations for a Mourning Dove non-toxic shot regulation. Extended abstract *in* R. T. Watson, M. Fuller, M. Pokras, and W. G. Hunt (Eds.). Ingestion of Lead from Spent Ammunition: Implications for Wildlife and Humans. The Peregrine Fund, Boise, Idaho, USA. DOI 10.4080/ilsa.2009.0308

Key words: Decisions, error, hunting, lead, Mourning Dove, nontoxic, shot, policy, regulation, risk.

LITERATURE CITED

BEST, T. L., T. E. GARRISON, AND C. G. SCHMITT. 1992. Availability and ingestion of lead shot by Mourning Doves (*Zenaida macroura*) in southeastern New Mexico. The Southwestern Naturalist 37:287–292.

EISLER, R. 1988. Lead hazards to fish, wildlife, and invertebrates: a synoptic review. United States Fish and Wildlife Service Biological Report 85. Patuxent Wildlife Research Center, Laurel, Maryland, USA.

FRANSON, J. C., S. P. HANSEN, AND J. H. SCHULZ. 2009. Ingested shot and tissue lead concentrations in Mourning Doves. *In* R. T. Watson, M. Fuller, M. Pokras, and W. G. Hunt (Eds.). Ingestion of Lead from Spent Ammunition: Implications for Wildlife and Humans. The Peregrine Fund, Boise, Idaho, USA. DOI 10.4080/ilsa.2009.0202

KENDALL, R. J., T. E. LACHER JR., C. BUNCK, B. DANIEL, C. DRIVER, C. E. GRUE, F. LEIGHTON, W. STANSLEY, P. G. WATANABE, AND M. WHITWORTH. 1996. An ecological risk assessment of lead shot exposure in non-waterfowl avian species: upland game birds and raptors. Environmental Toxicology and Chemistry 15:4–20.

LEE, K. N. 1993. Compass and Gyroscope: Integrating Science and Politics for the Environment. Island Press, Washington, D.C., USA.

LEWIS, J. C., AND E. LEGLER, JR. 1968. Lead shot ingestion by Mourning Doves and incidence in soil. Journal of Wildlife Management 32:476–482.

OTIS, D. L., J. H. SCHULZ, D. A. MILLER, R. E. MIRARCHI, AND T. S. BASKETT. 2008. Mourning Dove (*Zenaida macroura*). *In* A. Poole (Ed.). The Birds of North America Online. Ithaca: Cornell Lab of Ornithology. Retrieved from the Birds of North America Online Database: http://bna.birds.cornell.edu/bna/species/117.

RAFFENSPERGER, C., AND J. A. TICKNER (EDS.). 1999. Protecting Public Health and the Environment: Implementing the Precautionary Principle. Island Press, Washington, D.C., USA.

REITZ, R. A., H. J. SCROGGINS, AND J. H. SCHULZ. 2007. Survey questionnaire wording and interpretation: implications for policy makers. Proceedings of the Annual Conference of Southeastern Association of Fish and Wildlife Agencies 61:1-5

SANDERSON, G. C., AND F. C. BELLROSE. 1986. A review of the problem of lead poisoning in waterfowl. Special Publication 4. Natural History Survey, Champaign, Illinois, USA.

SCHULZ, J. H., J. J. MILLSPAUGH, B. E. WASHBURN, G. R. WESTER, J. T. LANIGAN III, AND J. C. FRANSON. 2002. Spent-shot availability and ingestion on areas managed for Mourning Doves. Wildlife Society Bulletin 30:112–120.

SCHULZ, J. H., X. GAO, J. J. MILLSPAUGH, AND A. J. BERMUDEZ. 2007. Experimental lead pellet ingestion in Mourning Doves (*Zenaida macroura*). American Midland Naturalist 158:177–190.

SCHULZ, J. H., J. J. MILLSPAUGH, A. J. BERMUDEZ, X. GAO, T. W. BONNOT, L. G. BRITT, AND M. PAINE. 2006. Experimental Acute Pb Toxicosis in Mourning Doves. Journal of Wildlife Management 70:413–421.

TICKNER, J. A. (ED.). 2003. Precaution, Environmental Science, and Preventive Public Policy. Island Press, Washington, D.C., USA.

VOLUNTARY LEAD REDUCTION EFFORTS WITHIN THE NORTHERN ARIZONA RANGE OF THE CALIFORNIA CONDOR

RON SIEG[1], KATHY A. SULLIVAN[1], AND CHRIS N. PARISH[2]

[1]*Arizona Game and Fish Department, 3500 S. Lake Mary Road, Flagstaff, AZ 86001, USA.* E-mail: rsieg@azgfd.gov

[2]*The Peregrine Fund, 5668 West Flying Hawk Lane, Boise, ID 83709, USA.*

ABSTRACT.—Lead exposure is a significant factor affecting the success of the California Condor (*Gymnogyps californianus)* reintroduction program in northern Arizona and southern Utah. Lead toxicity is the leading cause of mortality, with 12 confirmed cases, and the primary obstacle to a self-sustaining condor population. Research has identified incidental ingestion of spent lead ammunition found in animal carcasses and gut piles as the major lead exposure pathway. Peaks in condor lead exposure rates have corresponded with big game hunting seasons on the Kaibab Plateau in northern Arizona.

In response, the Arizona Game and Fish Department (AGFD) initiated a public education campaign in 2003 promoting voluntary lead reduction actions within condor range, including the use of non-lead ammunition by hunters. In addition, the AGFD implemented a free non-lead ammunition program for the 2005 and 2006 fall big game hunting seasons. This program resulted in 50–60% voluntary participation from Kaibab deer hunters. Although this represented an unprecedented voluntary effort, condor lead exposure data suggested that a 50–60% reduction in lead-laden carrion was not sufficient to maintain a self-sustaining population of free-foraging condors. Consequently, the Arizona Game and Fish Department intensified lead reduction efforts in 2007. Modifications included improved hunter outreach in the form of articles in sportsman's publications; distribution of an educational DVD and brochure; increased field communication; and added incentives for gut pile retrieval. Despite non-lead ammunition supply problems, 2007 voluntary efforts were successful and yielded over an 80% compliance rate from hunters. No lead toxicity fatalities occurred during the 2007 hunting season and preliminary data revealed that condor lead exposure rates declined. Voluntary lead reduction efforts must be further augmented to achieve a self-sustaining condor population. Future lead reduction efforts should also include southern Utah. *Received 16 May 2008, accepted 18 June 2008.*

SIEG, R., K. A. SULLIVAN, AND C. N. PARISH. 2009. Voluntary lead reduction efforts within the northern Arizona range of the California Condor. *In* R. T. Watson, M. Fuller, M. Pokras, and W. G. Hunt (Eds.). Ingestion of Lead from Spent Ammunition: Implications for Wildlife and Humans. The Peregrine Fund, Boise, Idaho, USA. DOI 10.4080/ilsa.2009.0309

Key words: ammunition, Arizona, condor, hunting, lead, non-lead, voluntary.

FOR NEARLY TWO DECADES, biologists have linked lead poisoning in wild California Condors (*Gymnogyps californianus*) to the ingestion of spent lead ammunition in animal carcasses (Janssen et al. 1986, Weimeyer et al. 1988, Snyder and Snyder 1989, 2000, Pattee et al. 1990). More recently lead from spent ammunition has been linked to lead exposure and lead toxicity in recently reintroduced condors in both California and Arizona (Meretsky et al. 2000, Snyder and Snyder 2000, Fry and Maurer 2003, Cade et al. 2004). In Arizona, significant efforts to verify the association between spent lead ammunition and condor lead exposure, as well as to educate the public and engage hunters in voluntary lead reduction efforts, began in 2003.

The first release of California Condors in Arizona occurred on 12 December 1996. As of 15 March 2008, 102 condors have been released in northern Arizona. Sixty-three condors, including six wild-hatched chicks, inhabit northern Arizona and southern Utah. Although the project is making progress towards its goal of 150 free-flying birds, 40 condors have died since the initial release. The leading cause of death is lead toxicity with 12 confirmed cases. The first major lead exposure event in Arizona occurred in June 2000, resulting in the death of three condors (Woods et al. 2007). Since that time extensive trapping and testing of condors for lead exposure has occurred in Arizona. Condor blood tests have identified over 300 cases of lead levels indicative of lead exposure, while in 124 cases condors have been treated with chelation therapy to reduce dangerously high lead levels. Further, ingested lead pellets or more frequently bullet fragments have been recovered from 14 individual condors (Parish et al. 2007). Without the intervention of chelation therapy and other measures, additional condors would have succumbed to lead poisoning.

As elsewhere in their current range, the condors are supplied with a clean lead-free supplemental food source of calf carcasses at the release site in Arizona. As condors disperse from the release site, they forage on carcasses of wild animals such as Mule Deer (*Odocoileus hemionus*), Elk (*Cervus elaphus*) and Coyotes (*Canus lantrans*). Since 2000, the highest frequency of lead exposure in condors has been associated with increased condor movements away from the release site, and the consumption of non-proffered carcasses potentially containing lead (Hunt et al. 2007). Although field biologists have managed to reduce the number of condor deaths due to lead toxicity by pursuing a rigorous monitoring and treatment protocol (Parish et al. 2007), these efforts are highly invasive, labor intensive and costly. Moreover, the long-term sublethal effects of lead exposure and chelation therapy in condors are unknown (Snyder 2007). It is unlikely that condors in Arizona will achieve a self-sustaining population at the current lead exposure rates.

While California has implemented a ban on the use of lead ammunition within the condor range starting in July 2008, efforts in Arizona have focused on voluntary measures to reduce the amount of lead from spent ammunition available to condors in the wild. This is due to a consensus among the main project cooperators that voluntary measures are the best course of action to take in Arizona. Also, unlike releases in California, condors in Arizona are released under Section 10(j) of the Endangered Species Act, which provided assurances to people in the release area that no changes would occur in land management practices, including hunting (US Fish and Wildlife Service, 1996).

COLLECTING BACKGROUND INFORMATION

In May 2003, the lead mitigation subcommittee of the California Condor Recovery Team produced a report on condor-lead issues (Redig et al. 2003). As one of several recommendations, the U.S. Fish and Wildlife Service (USFWS) contracted with Wildlife Management Institute (WMI) to conduct surveys of hunters' knowledge and attitudes on the condor-lead issue in California, Arizona and Utah. WMI contracted with Responsive Management for the phone survey and D. J. Case and Associates (D. J. Case) for the focus group work in Arizona.

In late fall 2003, Responsive Management conducted phone surveys of 205 hunters who held tags that year in the core condor range (Responsive Management 2003). There were three key questions for the hunters in these phone surveys, 1) were they aware that lead poisoning was a problem faced by condors; 2) were they aware of any educational efforts to try and raise awareness of this issue; and 3) would they be willing to take action to help reduce lead exposure in condors (Responsive Management 2003). Key findings from this survey were that only 23% of surveyed hunters were aware that lead poisoning was a problem faced by condors and only nine percent were aware of any educational efforts to reduce condor deaths from lead poisoning (Responsive Management 2003). At this time, information had been published in the 2003 Arizona Hunting Regulations, each hunter surveyed had been mailed a letter regarding this issue and any successful hunter had been asked questions about use of lead ammunition when they completed the mandatory check in of their harvested deer. The survey did reveal that between 83 and 97% of surveyed hunters would be somewhat to very willing, depending on the requested action, to take some action to help condors (Responsive Management 2003). The actions requested included removing all carcasses from the field, burying or hiding all gut piles, removing bullets and surrounding impacted flesh, and using non-lead ammunition (Responsive Management 2003).

Once the survey results were in, D. J. Case interviewed condor professionals and reviewed the literature to develop some conservation and lead reduction test messages. In December 2003, they conducted three focus group meetings in Arizona where the test messages were discussed and rated on a five-point scale (D. J. Case 2005). The best scoring (1.89) communication message based on these focus groups was "Hunters and ranchers have a long history of caring for the land and conserving all kinds of wildlife. They can continue this tradition and help prevent lead poisoning in California Condors by taking one or more of the following actions in condor range: remove all carcasses from the field; hide or bury carcasses and gut piles; remove bullet and surrounding affected flesh or use non-lead ammunition" (D. J. Case 2005).

Focus groups also revealed that hunters and ranchers were not yet convinced that lead from spent ammunition was a problem for condors and requested credible data linking lead from spent ammunition and condor lead poisoning (D. J. Case 2005). They expressed a willingness to help condors if shown the data link and if asked by a credible source, such as the Arizona Game and Fish Department or sportsman's groups (D. J. Case 2005). Based on this information, D. J. Case proposed a communications strategy that included increased education, communication and cooperation between program partners and the hunting community, continued research on the condor-lead link, and consider implementation of a non-lead ammunition program (D. J. Case 2005).

LEAD RESEARCH

Based on the phone survey and focus group information, it was apparent that more information on the link between lead from spent ammunition and condor lead poisoning needed to be provided to hunters (D. J. Case 2005). AGFD and The Peregrine Fund (TPF) responded by funding and conducting research projects related to the issue. First, TPF biologists detailed lead exposure and lead ammunition ingestion by condors starting in 1999 (summarized to 2005 in Parish et al. 2007). Second, TPF condor biologists summarized lead mortality rates (Woods et al. 2007). Data from these two studies verified that lead exposure was a critical management issue for the Arizona condor program. Third, starting in 2003, AGFD purchased 21 GPS satellite transmitters to more precisely track condor movements and relate movements to lead exposure rates (Hunt et al. 2007). This comparison showed that the highest lead exposure period coincided with the hunts on the Kaibab Plateau (Game Management Unit 12A Figure 1). Fourth, TPF conducted research from 2002 to 2004 to determine the extent of lead bullet fragmentation in rifle-killed deer (Hunt et al. 2006). This study demonstrated that standard lead bullets fragment into hundreds of pieces before exiting the deer and that these fragments remain in the deer carcasses as well as the gut pile. The study also confirmed that pure copper bullet fragmentation is minimal (Hunt et al. 2006). The final study is an ongoing lead isotope study funded by AGFD and conducted by the University of Arizona in Tucson using TPF provided biological samples as well as lead fragments removed from condors. Lead isotope ratios of condor blood and the removed fragments are being compared to lead isotope ratios from ammunition and other environmental sources (Chesley et al. 2006). Preliminary results have established a direct match between lead ammunition and lead found in condor blood samples and digestive tracts (Chesley, pers. comm.).

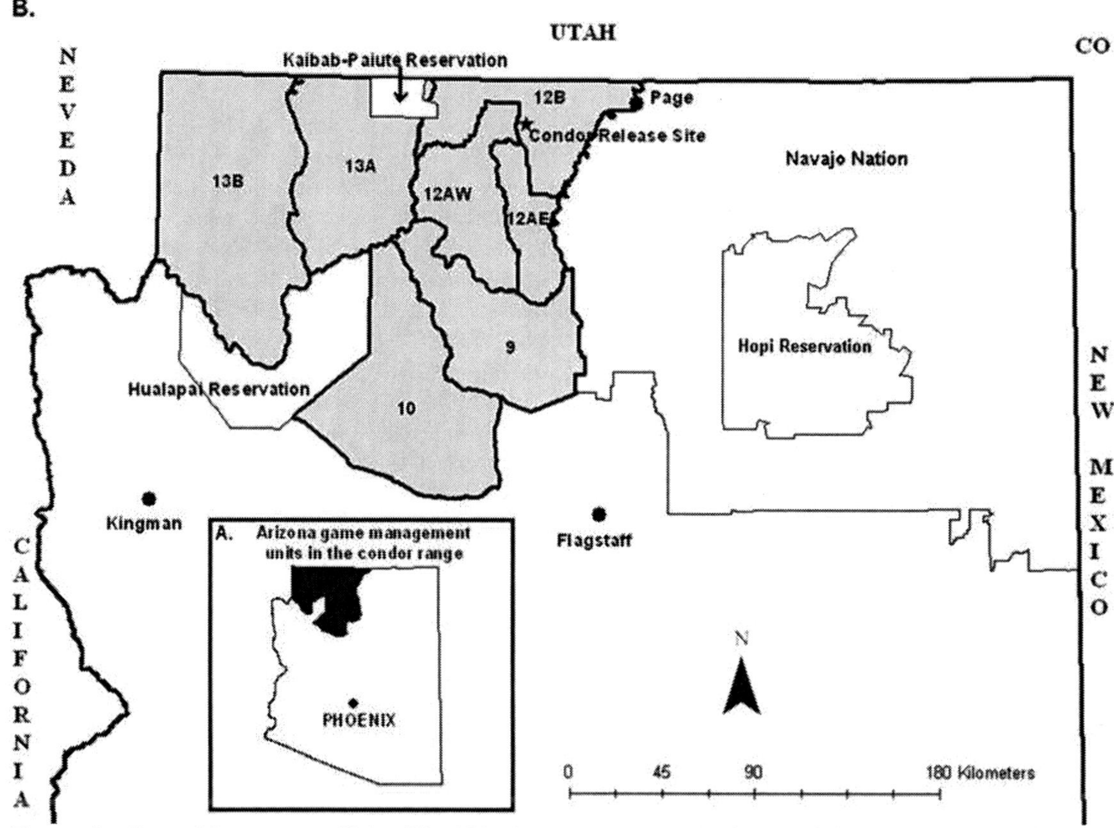

Figure 1. Game Management Units (B) within the condor range in Arizona (A). Hunters drawn for deer, pronghorn, buffalo, and bighorn sheep hunts in Units 12AE, 12AW, and 12B qualified for the free non-lead ammunition program. Hunters drawn for big game hunts in Units 9, 10, 13A, and 13B were mailed letters asking them to take voluntarily lead reduction actions.

COMMUNICATION WITH HUNTERS

Using the information from both the phone survey and focus groups AGFD set out to create an education and communication strategy to encourage hunters to support voluntary lead reduction efforts in Arizona's condor range. In 2003 and 2004, these efforts included a full page information piece in the annual hunting regulations booklet as well as mailings to between 2,000 and 7,000 hunters drawn for a big-game tag in the condor range (Figure 1). During this same period, AGFD made presentations to all the major sportsmen's organizations in the state asking them to join the "condor coalition" and lend their name and support for voluntary lead reduction efforts in the condor range. The current members of the Arizona coalition are the Arizona Antelope Foundation, the Arizona Desert Bighorn Sheep Society, the Arizona Deer Association, the Arizona Elk Society and the Arizona Chapter of the National Wild Turkey Foundation. Also in August 2005, WMI and D. J. Case presented two sessions of "one-voice" communication training for program partners and hunter group representatives to encourage uniform, consistent and accurate information dissemination in all outreach efforts regardless of who initiates the outreach. In addition to these efforts, the general public started to receive the condor conservation and lead reduction message in all outreach forums such as educational presentations, wildlife fair displays, legislative contacts, the AGFD web page and through general media outlets, including the AGFD Wildlife Views magazine and television program.

In the fall 2005, using money allocated to AGFD through the Arizona State Lottery, AGFD implemented a voluntary free non-lead ammunition program for hunters in the core condor range. AGFD partnered with Sportsman's Warehouse® for in store purchases and Cabela's® for mail order sales and sent each hunter drawn for a deer, pronghorn, sheep or bison tag in Game Management Units 12A and 12B (Figure 1) a coupon they could redeem for two free boxes of non-lead ammunition. These coupons came with a letter outlining condor lead poisoning issues and asking hunters to voluntarily help in reducing the amount of lead available to condors from spent ammunition. In 2005, 65% (n = 1,551) of eligible hunters redeemed their coupons with 50% of those harvesting deer using non-lead ammunition.

To evaluate the first year of this program, AGFD worked with D. J. Case to develop two post-hunt surveys, one for non-lead ammunition program participants and one for non-participants. Surveys were mailed to all 2,393 eligible hunters with 46% (1,105) surveys returned, including 943 participants (61%) and 162 non-participants (19%). For the participants, 85% tested the ammunition before their hunt, 60% rated the ammunition accuracy as excellent or above average, 70.5% said it performed as well as lead with 22.6% saying it performed better than lead. Most would use it again if provided free, with 55.8% saying they would use it again even if not provided free. The majority (72%) said they would recommend the ammunition to other hunters and 81 percent used it on their hunt with 41.6% using it to harvest their deer. When asked why they participated, the majority said because AGFD asked them to, followed by it helped condors, because it was free and because they had heard or read that non-lead ammunition had good ballistics (D. J. Case 2006).

The primary reason for those not participating in the program was that the non-lead ammunition was not available in their caliber. The next most important reasons were that the non-lead ammunition was not available in their preferred bullet weight and that it takes too long to sight in new ammunition. The next most important reasons were that redeeming the coupon was too complicated or too much hassle, that they were not convinced that lead from spent ammunition is a problem for condors and that they think the program is an effort by anti-hunters to ban the use of lead. Other reasons were that they hand load their own ammunition, non-lead bullets were not covered by the program, they had heard that non-lead didn't perform as well as lead and they had tried non-lead and it didn't meet their expectations (D. J. Case 2006).

When non-participants were asked what could be done to encourage more participation they offered that the ammunition should be offered in more calibers and bullet weights, that more information should be provided on how lead from spent ammunition is a problem for condors, that bullets for reloading should be offered and that sports groups should endorse the program (D. J. Case 2006). Fortunately, at least Federal Ammunition, using Barnes bullets was increasing the variety of calibers and bullet weights each year and AGFD started offering reloading components as part of the program.

In 2006, the voluntary free non-lead ammunition program was continued in nearly the same manner as 2005. The primary difference was an effort to provide significantly more information about the link between lead from spent ammunition and condor lead poisoning. Along with the free ammunition program, individual mailings to hunters in non-core areas were sent information which also requested their voluntary help. Although we were responding to what we thought hunters wanted on providing a link between lead and condors (D. J. Case 2006) we received negative responses to providing too much information and found that most hunters did not read it. Participation in the voluntary free non-lead ammunition program was similar to the previous year, but due to increased field outreach during hunts, 60% of successful Kaibab deer hunters took lead reduction actions during their hunt, an increase of 10% from 2005. Even with this level of participation, 95% of the birds were exposed to lead (Parish et al. 2009, this volume). One factor that is likely contributing to this continued high exposure is that the condors are increasing their use of southern Utah habitats (Figure 2). To date Utah has not implemented any extensive outreach or programs for raising awareness on this issue, but are working on plans to do this in 2009.

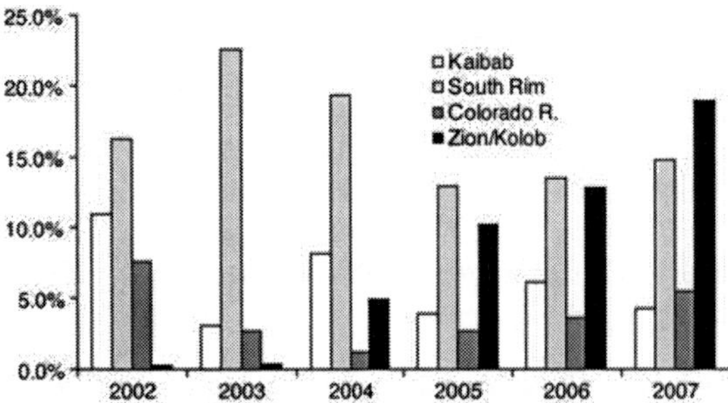

Figure 2. Condor roost locations by region. Condor foraging in southern Utah (Zion/Kolob region) has increased steadily since 2004 and several lead exposure incidents have been directly linked to this area. To be effective, future lead reduction efforts should therefore include southern Utah.

With this high lead exposure rate and only slightly increasing participation, we gathered a group of people involved in the program and brainstormed ideas for trying to achieve at least 80% participation for 2007. This effort resulted in eight new actions to improve outreach efforts. First, we asked our sports group supporters to publish articles in their magazines about the program because they were viewed as credible sources of information. Second, we increased general media stories about how hunters were helping to recover condors through their voluntary efforts and this was another example of how hunters were aiding recovery of a species. Third, we decided to simplify our outreach message to "use non-lead ammunition." Previous outreach had included options such as hiding or burying gut piles or removing impacted flesh but these messages seemed to confuse people and were not clear enough direction. Fourth, we developed an 11-minute DVD, with Nolan Ryan as host, entitled "How to be successful in your upcoming deer hunt." The DVD provided about 6 minutes of successful hunt information followed by 5 minutes of information on lead exposure, and asked hunters for their help. Based on field visits with hunters, the majority of people said they had viewed the video before their hunt. Fifth, we combined the outreach material and the DVD in the mailing with their tag. In previous years the information had been mailed separately. Sixth, any hunter not redeeming their coupon within two months of their hunt dates was sent follow up information encouraging them to participate in the program. Seventh, we dramatically increased our field staff to directly contact hunters in the field during all hunt weekends between October and December. One staff member for each 200 permitted hunters allowed us to achieve between 60 and 70% direct field contact with hunters in the field. And finally, we implemented a gut pile raffle. This came from the realization that once a hunter was in the field with lead ammunition, options for asking for their help were more limited. Trash bags were provided, along with a flyer, during field contacts and hunters were asked to bring their gut pile, if shot with lead ammunition, to the mandatory check station when they checked in their deer. The Peregrine Fund provided $1,000 to purchase gift certificates to a sporting goods store as an incentive for hunters assisting with this effort. In 2006, without the incentive, only a handful of hunters brought in gut piles. In 2007, the number rose to 170, resulting in 54% of hunters who used lead ammunition to kill their deer to carry their gut pile out of field. Overall, with these changes, participation in the voluntary program increased to 83%, with 62% of successful hunters using non-lead ammunition and 21% participating in the gut pile raffle. One of the biggest obstacles to increasing participation was the lack of available non-lead ammunition from our vendors for anyone who waited until close to their hunt and then looked for the free ammunition. This was in spite of the fact that this was the third year of the program, and the number of eligible hunters was provided early to the vendors. Plans are underway to continue the program in 2008, retaining all of the 2007 outreach changes while at the same time working with vendors to increase supply and make it easier to find non-lead ammunition in the stores with displays located in one area of the store.

DISCUSSION

There are many factors to consider when designing an outreach program that asks people to do something different. Using social psychology and marketing principles can aid in outreach design. We used six principles of influence identified from the field of social psychology (Cialdini 1993) to design our outreach program:

1. Reciprocity—give someone something in exchange for their action—achieved through the free ammunition program.
2. Commitment and consistency—insure dedication to what you are asking for—achieved by a multi-year dedication to a voluntary program and a consistent message.
3. Social proof—show that others are also participating—achieved by use of sports group publications for outreach.
4. Liking—show that others like them are also participating—achieved by the use of hunter quotes in outreach materials and also by consistently thanking hunters for their help.
5. Authority—exert influence on the decision—achieved through the use of AGFD, the regulatory agency for hunting and fishing, doing primary outreach.
6. Scarcity—indicate that not participating might limit future actions—achieved by stating that voluntary efforts could reduce calls for mandates or regulations.

We also incorporated lessons learned from similar experiences in the past. In the early 1990s, a ban was put in place on the use of lead ammunition for waterfowl hunting throughout the United States. A survey among people involved in that ban revealed useful ideas on what they would have done differently in hindsight (Association of Fish and Wildlife Agencies 2007). Among their ideas were:

1. More thought, study, and action should have been invested in obtaining input from hunters before any decision was made.
2. There should have been more analysis of supply issues.
3. Moving too fast on the issue didn't allow groups to be informed, educated, and convinced, and this included agencies, nongovernmental organizations, manufacturers, dealers, and the media.
4. Education is the key to a smooth transition.
5. One negative media article can nullify all the factual information.
6. Training sales people, especially in large stores, is important because they may be the main sources of information for buyers.
7. Unlike the 1990s, there is an increased emphasis today on hunter recruitment and retention throughout the nation, and mandates could be an obstacle to this objective.
8. Any program needs to establish sources for reliable, accurate information, and a common understanding of the goal.
9. Ideas should be advanced through leaders of change in hunting and sports groups as well as outdoor retailers.
10. Having a consistent, united voice by all parties is important.
11. Hunter education instructors can play an important role in getting the message to new hunters.
12. The use of focus groups to develop and refine messages can aid the process.
13. Technical articles can hinder, rather than help, the process so using marketing professionals to tailor messages is important.

In Arizona, we have found that manufacturers respond slowly to demand, so a significant transition time is needed to reach appropriate production and distribution levels. We have found that, like us, people respond better to requests so we should ask for their help and bring them along, rather than taking the short term fix of a mandate. We realize that the cost of non-lead ammunition is going to be a continuing issue. While non-lead ammunition is comparable in price to premium lead ammunition, and moving those using premium lead over to non-lead ammunition may be relatively easy, many hunters buy the cheapest lead ammunition available and non-lead ammunition can be up to three times more expensive. Focus groups can help refine outreach messages, but more importantly they can also aid in determining who should do the outreach. A continuing challenge is working with those groups and organizations that the focus groups view to be non-credible to keep them engaged in the program while limiting their outreach efforts. We are proud of the response of our hunters and partners to the call for a voluntary effort to reduce the amount of lead from spent ammunition available to condors, and think our program can serve as an example to others.

LITERATURE CITED

ASSOCIATION OF FISH AND WILDLIFE AGENCIES. 2007. Final Report and Recommendations to Association of Fish and Wildlife Agencies from Ad Hoc Mourning Dove and Lead Toxicosis Working Group. Unpublished report, 19 September 2007.

CADE, T. J., S. A. H. OSBORN, W. G. HUNT, AND C. P. WOODS. 2004. Commentary on released California Condors *Gymnogyps californianus* in Arizona. Pages 11–25 *in* R.D. Chancellor and B.-U Meyburg (Eds.), Raptors Worldwide: Proceedings of VI World Conference on Birds of Prey and Owls. World Working Group on Birds of Prey and Owls/MME-Birdlife, Hungary.

CHESLEY, J., P. N. REINTHAL, T. CORLEY, C. PARISH, AND J. RUIZ. 2006. Radioisotopic analysis of potential sources of lead contamination in California Condors. Invited Symposium on Applications of Stable and Radiogenic Isotopes in Wildlife and Fisheries. Joint Annual Meeting of the Arizona/New Mexico chapters of the Wildlife Society and the American Fisheries Society, 2–4 February 2006.

CIALDINI, R. B. 1993. Influence: The Psychology of Persuasion, Rev. Ed. William Morrow and Company, New York, USA.

D. J. CASE AND ASSOCIATES. 2005. Communicating with hunters and ranchers to reduce lead available to California Condors. Unpublished report, Wildlife Management Institute, Washington, DC, and US Fish and Wildlife Service, Sacramento, California.

D. J. CASE AND ASSOCIATES. 2006. Nonlead Ammunition Program Hunter Survey. Unpublished report, Arizona Game and Fish Department, Flagstaff, Arizona, USA.

FRY, D. M. AND J. R. MAURER. 2003. Assessment of lead contamination sources exposing California Condors. Final report to the California Fish and Game, Sacramento, California, USA.

HUNT, W. G., W. BURNHAM, C. N. PARISH, K. BURNHAM, B. MUTCH, AND J. L. OAKS. 2006. Bullet fragments in deer remains: implications for lead exposure in avian species. Wildlife Society Bulletin 34:168–171.

HUNT, W. G., C. N. PARISH, S. G. FARRY, T. G. LORD, AND R. SIEG. 2007. Movements of introduced California Condors in Arizona in relation to lead exposure. Pages 79–96 *in* A. Mee and L. S. Hall (Eds.). California Condors in the 21st Century. Series in Ornithology, no. 2. The Nuttall Ornithological Club, Cambridge, Massachusetts, and The American Ornithologists' Union, Washington, DC, USA.

JANSSEN, D. L., J. E. OOSTERHUIS, J. L. ALLEN, M. P. ANDERSON, D. G. KELTS, AND S. N. WIEMEYER. 1986. Lead poisoning in free ranging California Condors. Journal of the American Veterinary Medical Association 189:1115–1117.

MERETSKY, V. J., F. R. SNYDER, S. R. BEISSINGER, D. A. CLENDENEN, AND J. W. WILEY. 2000. Demography of the California Condor: Implications for reestablishment. Conservation Biology 14:957–967.

PARISH, C. N., W. G. Hunt, E. Feltes, R. Sieg, and K. Orr. 2009. Lead exposure among a reintroduced population of California Condors in northern Arizona and southern Utah. *In* R. T. Watson, M. Fuller, M. Pokras, and W. G. Hunt (Eds.). Ingestion of Lead from Spent Ammunition: Implications for Wildlife and Humans. The Peregrine Fund, Boise, Idaho, USA. DOI 10.4080/ilsa.2009.0217

PARISH, C. N., W. R. HEINRICH, AND W. G. HUNT. 2007. Lead exposure, diagnosis, and treatment in California Condors released in Arizona. Pages 97–108 *in* A. Mee and L. S. Hall (Eds.), California Condors in the 21st Century. Series in Ornithology, no. 2. The Nuttall Ornithological Club, Cambridge, Massachusetts, and The American Ornithologists' Union, Washington, D.C., USA.

PATTEE, O. H., P. H. BLOOM, J. M. SCOTT, AND M. R. SMITH. 1990. Lead hazards within the range of the California Condor. Condor 92:931–937.

REDIG, P., N. ARTZ, R. BYRNE, B. HEINRICH, F. GILL, J. GRANTHAM, R. JUREK, S. LAMSON, B. PALMER, R. PATTERSON, W. SANBORN, S. SEYMOUR, R. SIEG AND M. WALLACE. 2003. A report from the California Condor lead exposure reduction steering committee. Unpublished report to the US Fish and Wildlife Service, California Condor Recovery Team.

RESPONSIVE MANAGEMENT. 2003. Hunters' knowledge of and attitudes towards threats to California Condors. Unpublished report. D. J. Case and Associates, Mishawaka, Indiana, USA.

SNYDER, N. F. R. 2007. Limiting factors for wild California Condors. Pages 9–33 *in* A. Mee and L. S. Hall (Eds.), California Condors in the 21st Century. Series in Ornithology, no. 2. The Nuttall Ornithological Club, Cambridge, Massachusetts, and The American Ornithologists' Union, Washington, DC, USA.

SNYDER, N. F. R., AND H. F. SNYDER. 1989. Biology and conservation of the California Condor. Current Ornithology 6:175–267.

SNYDER, N. F. R. AND H. A. SNYDER. 2000. The California Condor: A Saga of Natural History and Conservation. Academic Press, San Diego, California, USA.

US FISH AND WILDLIFE SERVICE. 1996. Final Rule: Endangered and threatened wildlife and plants: establishment of a nonessential experimental population of California Condors in northern Arizona. Federal Register 61:54044–54060.

WIEMEYER, S. N., J. M. SCOTT, M. P. ANDERSON, P. H. BLOOM AND C. J. STAFFORD. 1988. Environmental contaminants in California Condors. Journal of Wildlife Management 52:238–247.

WOODS, C. P., W. R. HEINRICH, S. C. FARRY, C. N. PARISH, S. A. H. OSBORN, AND T. J. CADE. 2007. Survival and reproduction of California Condors released in Arizona. Pages 57–78 *in* A. Mee and L. S. Hall (Eds.). California Condors in the 21st Century. Series in Ornithology, no. 2. The Nuttall Ornithological Club, Cambridge, Massachusetts, and The American Ornithologists' Union, Washington, DC, USA.

TAKING THE LEAD ON LEAD: TEJON RANCH'S EXPERIENCE SWITCHING TO NON-LEAD AMMUNITION

HOLLY J. HILL

Tejon Ranch, P.O. Box 1000, Lebec, CA 93243

ABSTRACT.—In February 2007, Tejon Ranch Company, which operates the largest State licensed private hunting operation in California, became the first to voluntarily discontinue and ban the use of all lead ammunition in its hunting and Ranchwide operations. Implemented in January 2008, this decision has been generally met with wide-spread support from resource agencies and other large-scale hunting operators such as Fort Hunter-Ligget who have since considered or have implemented their own lead ban policies. Hunters on the Ranch must all agree to comply with the ban; however, the ban presents a number of problems for hunters ranging from the availability of non-lead ammunition to the effectiveness of these products. Likewise, philosophical issues regarding gun rights and increased government oversight on hunting activities, not to mention the protection of endangered species, have also been necessary to address with hunters on the Ranch. Health-related concerns to wildlife and hunters regarding the effects of spent lead ammunition are generally not considered major concerns, but Tejon Ranch will continue to forward scientific information to hunters on the benefits of switching to non-lead ammunition.

HILL, H. J. 2009. Taking the lead on lead: Tejon Ranch's experience switching to non-lead ammunition. Abstract *in* R. T. Watson, M. Fuller, M. Pokras, and W. G. Hunt (Eds.). Ingestion of Lead from Spent Ammunition: Implications for Wildlife and Humans. The Peregrine Fund, Boise, Idaho, USA. DOI 10.4080/ilsa.2009.0310

Key words: Ammunition, hunting, lead, non-lead, ranch.

THE POLICY AND LEGISLATIVE DIMENSIONS OF NONTOXIC SHOT AND BULLET USE IN NORTH AMERICA

VERNON G. THOMAS

Department of Integrative Biology, College of Biological Science, University of Guelph, 50 Stone Road East, Guelph, Ontario N1G 2W1, Canada. E-mail: vthomas@uoguelph.ca

ABSTRACT.—This paper addresses the policy and legislative considerations for moving North American society towards the use of nontoxic shot and bullets for all types of hunting and shooting. Progress in one or more areas of lead use reduction by society has not facilitated transitions in other areas of lead use, and the two solitudes of conservationists (anti-lead) and hunters (pro-lead) is real. Regulators must emphasize the gains in wildlife to both constituencies that will attend adoption of nontoxic products. Sixteen years of nontoxic shot use in waterfowl hunting is the most cost-effective conservation tool to date in conserving waterfowl populations. Similar savings could be expected from the use of lead-free shot such as for hunting migratory doves and upland birds. New ballistic materials are available for use on upland species, and in all gauges of modern and old guns. Industry has adapted materials for use in rifle cartridges of varying calibers. Although industry has responded well to the quest for nontoxic ballistic materials, industry requires enforceable regulations to create and assure the market demand for their products. Different policy and legislative options are presented. Regulatory progress would best be based on precedents under the Migratory Bird Treaty Act, entailing its application to species that fall under federal jurisdiction. The use of this Act would constitute the rationale for Canada to implement similar provisions for the same species under its Migratory Birds Convention Act. Individual states and provinces could then be petitioned to adopt complementary measures for hunting upland bird and mammalian species that fall under their jurisdiction. The development of nontoxic bullets for big game hunting could also be applied to the smaller caliber lead bullets used for small mammals, because they constitute a source of secondary lead poisoning of carrion feeders. Any legislation developed to phase out all lead use must be harmonized between the USA and Canada, and among the states and provinces to ensure consistency of regulation and its application. Progress in this task has to be based on the premise that use of nontoxic materials benefits all wildlife, the sport of proactive hunters, and society that experiences less lead in the environment. *Received 16 May 2008, accepted 6 August 2008.*

THOMAS, V. G. 2009. The policy and legislative dimensions of nontoxic shot and bullet use in North America. *In* R. T. Watson, M. Fuller, M. Pokras, and W. G. Hunt (Eds.). Ingestion of Lead from Spent Ammunition: Implications for Wildlife and Humans. The Peregrine Fund, Boise, Idaho, USA. DOI 10.4080/ilsa.2009.0311

Key words: Canada, bullets, law, lead, nontoxic shot, poisoning, policy, USA, weights.

THE EVOLUTION OF HUMAN CULTURE has been accompanied by an extensive use of lead products that have, inevitably, become released to the general environment. Awareness of the toxic properties of lead and lead compounds has existed for over two millennia (Nriagu 1983, 1998). However, such awareness has resulted in only a slow rate of amelioration of the problem of toxicosis in the human environment, and especially the environment of aquatic and terrestrial wildlife. Instances where the use of a toxic lead compound has been banned, based on extensive scientific evidence, should, in theory, facilitate the removal of lead products from other human uses, especially where equally compelling scientific evidence exists (Lanphear 1998).

Such facilitation has not been widespread in the human environment[1], and even less so for the environment of wildlife subject to lead poisoning from discharged lead shot bullets, and fishing weights (Thomas 1997, Thomas and Guitart 2005). Thus a complete ban on the use of lead shot for hunting has not been reciprocated by a ban on the use of lead fishing gear for angling (and vice versa), as shown in the USA, Canada, and a number of states in the European Union (Thomas 1997, Beintema 2001). This is despite scientific evidence of the need to adopt consistent policy on lead reduction across different user groups in these countries (Thomas and Guitart 2003, 2005). Thus each situation of primary or secondary lead poisoning (i.e., lead shot toxicosis in waterbirds and upland game birds, sinkers and piscivorous birds, and bullet fragments and carrion feeders) has been dealt with separately, each with its own peculiar user constituencies, jurisdictions, biases, and scientific researchers (USFWS 1986, Twiss and Thomas 1998, Scheuhammer et al. 2002, Sidor et al. 2003, Fisher et al. 2006, Hunt et al. 2006, Cade 2007).

The short history of transitions to nontoxic shot, bullets, and fishing weights has been exceedingly contentious, and resisted by all the principal user groups who viewed requirements for nontoxic materials to be unwarranted and infringements on their rights to practice their sport (Anderson 1992, Williams 1994, Center for Biological Diversity 2006, Schultz et al. 2007). Thus any proposal to extend the use of nontoxic materials to all forms of hunting and angling needs to address the sustainability of these sports, and the benefits of using these new materials to hunters and anglers, the wildlife, and the general environment.

[1]As in the cases of prohibiting the use of leaded paints in interior use, banning the use of leaded gasoline, lead in glass and glazes, and rehabilitating urban soils contaminated with lead from smelters.

The abundant scientific literature documenting the extent and impact of lead poisoning on wildlife does not, by itself, make decisions about its use. Such decisions are rooted in social value systems, popular beliefs, economics, policy, court decisions and laws. As such, they may support, confound, or refute current scientific thought about the issues. This paper accepts the enormous legacy of published research that links primary and secondary lead toxicosis of birds and mammals to discharged lead shot and bullets and lost fishing weights (Church et al. 2006, Fisher et al. 2006, Cade 2007, Rattner et al. 2008, Papers in this Conference Proceedings). The paper focuses on spanning the gulf between science and policy, and how the emerging scientific "message" could be translated into policy options that may result in the creation of progressive law.

STARTING THE POLICY PROCESS AND THE TRANSITION TO NONTOXIC MATERIALS

The adversarial debate around the replacement of lead products has often polarized the positions of the hunting and angling communities from that of the largely non-hunting "conservationists". Given that hunting and angling are socially and politically legitimized pursuits, and will continue, the real quest is to improve the apparent sustainability of both sports by finding and approving lead substitutes that leave no toxic legacy in the environment (Cade 2007). This should be in the interest of both hunting-angling communities and those with a non-consumptive approach to wildlife's conservation.

Selecting the Appropriate Policy Options.—For those advocating further use of lead substitutes, knowing precisely what one wishes to achieve of the policy process is paramount. Thus the following graded options can be identified *a priori* for consideration:

- Requiring use of nontoxic shot for an additional particular species (e.g., Mourning Doves, *Zenaidura macroura*) nation-wide.
- Requiring use of nontoxic shot for hunting all species of migratory birds across all habitats in the USA.

- Requiring use of nontoxic shot for hunting all migratory and non-migratory game birds nation-wide.
- Requiring nontoxic shot for all bird hunting and nontoxic rifle bullets for hunting both large and small mammals.
- Requiring nontoxic shot and bullets for all hunting and all types of target shooting.
- Requiring use of nontoxic shot and bullets for all hunting and shooting in addition to use of non-toxic sinkers, weights, and jigs for sport fishing nation-wide.

These options cut across federal and state/provincial jurisdictions, and apply, progressively, to more recreational constituencies in society. While the above are presented as discrete policy options, it is recognized that further subdivision, or re-combinations, of some options is possible. For example, should requirements for using nontoxic shot apply across all public lands and privately-operated and owned shooting preserves dedicated to the hunting of game farm, non-migratory birds? Or, would nontoxic fishing gear be required in those states beyond the natural range of the piscivorous species most commonly afflicted?

The choice of option has to reflect the perceived scientific "message," in that the array of scientific evidence has to be able to withstand considerable challenge. The option of a voluntary use of non-toxic products has not been considered in this paper. The disadvantages of this approach have been identified in Thomas and Owen (1996), and this author is of the view that only a regulated approach is capable of resolving the issue of lead poisoning of wildlife in North America. Moreover, given the present availability of lead-free products, concerned sportsmen could have already made the transition to lead-free products, were they so inclined. The option chosen may also reflect a desire of regulators and politicians to proceed successfully, in an incremental manner across time, rather than to attempt an unsuccessful simultaneous ban on all lead products. Whichever approach is taken it is vital to consider the legislative vehicle(s) that might be used to support transitions to nontoxic materials and important legal precedents that would support a given policy option. Where appropriate legislation does not exist, new legislation has to be created, or existing legislation has to be amended. It is critical that any legislation selected has to be sufficiently robust to withstand repeated challenges by groups opposed to the removal of lead products (see Anderson 1992). For the purpose of this paper, the fourth policy option (requiring nontoxic shot use for the hunting of all migratory and non-migratory birds, and nontoxic rifle bullets for the hunting of big and small game mammals) will be used to address the legislative considerations that apply. The strength of the available research on lead toxicosis supports this policy option very well, as do the available legislative tools in the USA. Since lead discharged by hunting is the principal focus of the Conference Proceedings, this paper will not deal, overtly, with lead from fishing weights, especially since it involves a different public constituency.

Continuation of Applied Research Supporting Policy Process.—The continuation of applied research into the extent and distribution of primary and secondary lead toxicosis of birds and mammals is vital. It is important to have up-to-date, peer-reviewed, journal research that documents further the case for using lead substitutes, especially when proposing extension of nontoxic product use to other categories of hunting and shooting (e.g., Church et al. 2006, Cade 2007). Where gaps in the scientific coverage exist, it is advisable to address them. Similarly, where evidence of lead build-up applies to one part of a species range (as for American Woodcock, *Scolopax minor,* Scheuhammer et al. 1999), it is advisable to extend such studies throughout the species annual range. There will certainly be those from the angling and shooting communities whose role is to undermine or negate the science supporting change (Williams 1994, Center for Biological Diversity 2006), and their assertions must be countered in the policy process. Science, by its nature, cannot provide absolute certainty to society, but scientists can indicate that the state of understanding of the problem of lead poisoning of wildlife has become asymptotic, and that the issue does transcend political and geographic boundaries. A huge body of independently-replicated research reveals consistently that there is a single syndrome of ingested lead toxicosis, whose collective scientific credibility exceeds the burden of proof used in other forms of environmental chemical regulation (e.g., cigarette smoking and public health).

Awareness of Nontoxic Substitutes and their Applications.—The arms and the fishing tackle industries have done much in recent years to develop lead substitutes, and to provide an array of government-approved products for use. It is necessary for the proponents of change to be aware of what these new materials are and how they can be adapted to other shooting, hunting, and angling applications beyond their current use[2]. Thus some materials (for example sintered tungsten-tin, sintered tungsten-bronze, and tungsten matrix shot) could be used in upland and wetland shooting situations and serve as excellent dense materials for fishing weights. Some of the federally-approved materials (for example, sintered tungsten-tin, and tungsten-matrix) have a physical softness that allows their safe use as shot in small gauge guns[3] that are sometimes favored for use in upland game bird shooting. Such awareness reduces the impact of critics' remarks that no effective lead substitutes are available for shooting and angling. American manufacturers of rifle ammunition have already made effective substitutes for lead rifle bullets and shotgun slugs. Here, pure copper has been the metal of choice in a range of rifle calibers and bullet weights, including partition bullets in which an approved, nontoxic, tungsten formulation provides a dense lower core[4].

It is also important to bring into the policy process constructive precedents and information gained in other jurisdictions that have already begun a broader adoption of lead substitutes, whether in other nations (e.g., Denmark and the shooting of upland game birds) or states that have introduced requirements for nontoxic shot when hunting state regulated upland game. Such information can often assuage concerns of skeptics, and guide the policy process by setting legal precedents.

Emphasizing the Acknowledged Success of Existing Nontoxic Shot Regulations.—The evidence that nontoxic shot use has been an extremely successful management approach is vital in the policy process. The USA began the national adoption of nontoxic shot for waterfowl hunting in 1991, and the intervening 16 years have provided opportunities for agencies to assess the efficacy of nontoxic shot use. The evidence, to date, has favored the transition, is conducive to a broader use of lead substitutes, and should form the basis of any policy proposition. Samuel and Bowers (2000) analyzed the impact of a ban on the use of lead shot on elevated blood lead levels of American Black Ducks (*Anas rubripes*) and reported a decline of 44% in the prevalence of high blood lead levels. Stevenson et al. (2005) reported declines in wing bone lead levels in American Black Ducks and Mallards (*Anas platyrhynchos*) in Canada of 11.0 to 4.8 µg/g, and in Ring-necked Ducks (*Aythya collaris*) of 28.0 µg/g to 10 µg/g, over the period of 1989–90 to 2000. These results attest to the speed with which declines in body lead can become achieved.

Perhaps, the most compelling evidence supporting the transition to nontoxic shot use comes from the research of Anderson et al. (2000), who observed that the use of nontoxic shot reduced the mortality of Mallard from lead toxicosis by 64%, and generated a national saving of approximately 1.4 million ducks a year from ingested lead shot mortality. These figures were generated from research undertaken only 5–6 years following the 1991 US national ban on lead use, and do not include estimates for Canada. The previous three studies illustrate that adoption of nontoxic shot for waterfowl hunting has been the most effective tool used by the individual hunter in the conservation of waterfowl in North America. Its contribution to the survivorship of birds exceeds the contributions to waterfowl numbers made by continental habitat manipulations and improvements (Anderson et al. 2000, Thomas

---

[2] In the USA, non-toxic materials are presently required for the hunting of waterfowl and coots, for sport fishing in several locations, and upland game hunting in some states. In Canada, in addition to the hunting of waterfowl, non-toxic fishing weights are required for use in all national parks and national wildlife areas.

[3] As the gauge of shotguns increases, the pressure in the barrel chamber also increases. Thus shot that is harder than lead contributes to higher chamber pressures. Accordingly, some of the approved lead shot substitutes can not be used safely in cartridges smaller than 20 gauge because they might exceed safety standards.

[4] As made by the companies Barnes, Lapua, Nosler, and Remington, and sold either as complete rifle or shotgun cartridges, or bulk bullets/slugs for hand loaders and muzzle loading guns.

and Guitart 2005). This is not to imply that habitat improvement and preservation is unimportant, but that hunters' purchase and use of nontoxic shot is an activity that directly complements and enhances the benefits of all types of habitat improvement and expansion. For maximum improvement to wildlife populations, do both: create new habitats and ensure that they remain uncontaminated by spent lead shot and bullet fragments. The resulting situation is then a "win-win" for hunters and wildlife, in that there are now more surviving birds, their populations are afflicted less by lead shot toxicosis, and there is less secondary lead toxicosis affecting predatory and scavenger bird populations (Kramer and Redig 1997, Wayland and Bollinger 1999).

Primary and secondary lead toxicosis are not confined to waterfowl and their predators/scavengers. Upland game species that have been hunted heavily during the past two centuries also display characteristic lead poisoning from ingested spent lead shot (Kendall et al. 1996, Butler et al. 2005, Fisher et al. 2006), as do their predators and scavengers (Kramer and Redig 1997, Wayland and Bollinger 1999, Mateo et al. 2001, 2007). Some of these hunted species are migratory and fall under federal jurisdiction in the USA and Canada (migratory doves and American Woodcock), while non-migratory species of pheasant, quail, grouse, and partridge are under state or provincial jurisdiction. The various forms of nontoxic shot that have been developed for shooting waterfowl could be used to great effect on upland species. Upland birds are usually shot at closer ranges than waterfowl. They are also less heavily feathered and have thinner skins than waterfowl, so promoting shot passage deeper into the body. The same advantages that accrue to waterfowl hunters using nontoxic shot (Anderson et al. 2000) should also extend to upland game bird species. Furthermore, the rapid reduction in secondary lead poisoning of scavengers would also justify an end to the use of lead shot for such hunting. Light steel shot shotgun loads (24–28g) for 12 and 20 gauge guns are already available to hunters[5], and would, by virtue of their relatively greater pellet count, be effective for the hunting of migratory species of small-bodied doves, Common Snipe (*Capella gallinago*), and American Woodcock. This is an important consideration in light of concerns about increased wounding losses of game birds by the use of nontoxic shot (Schultz et al. 2006a).

The use of lead-free rifle bullets can be expected to bring about a rapid decline in the prevalence of lead poisoning of scavengers, especially those that acquire the lead from bullet fragments in the discarded viscera of big game (Hunt et al. 2006, Cade 2007, Craighead and Bedrosian 2008). Where fragmenting, small-caliber, lead bullets are used to kill nuisance rodents in agricultural areas, discarded carcasses could also be a source of lead fragments to aerial and ground scavengers (Pauli and Buskirk 2007). The use of nontoxic fragmenting bullets, available in all of the common calibers[6] would reduce this problem of secondary toxicosis.

CHOICE OF LEGISLATIVE VEHICLES AND CONSIDERATION

Federally-regulated Migratory Birds.—The Migratory Bird Treaty was signed initially between the USA and Great Britain for Canada (Lyster 1985) and is administered in the USA and Canada by The Migratory Birds Treaty Act (MBTA), and The Migratory Birds Convention Act (MBCA), respectively, the two articles of law that regulate all management of the two nations' migratory birds. The regulations of the MBTA were used in 1991 to regulate the use of nontoxic shot for the hunting of waterfowl and American Coots (*Fulica americana*) throughout the USA, as well as the composition of lead shot substitutes. The MBCA was used in Canada in 1999 to effect the same transition throughout Canada. At the time of passage, the regulations were applied only to the hunting of waterbirds because this was the area of greatest primary and secondary lead poisoning that federal authorities wished to address (USFWS 1986). Hunted migratory species, such as Mourning Doves and American Woodcocks were excluded from the regulations requiring nontoxic shot use in both Canada and the USA, a situation that exists to the present. The hunting of migratory doves (*Columba fasciata, Zenaidura macroura, Zenaida asiatica*) occurs

[5] As made by Kent Cartridge Co., and Remington Arms, in the USA.

[6] As made by the Barnes Bullet Co., in the USA.

across most states (40 states) of the USA, where present, but only in the Province of British Columbia in Canada. Thus federal legislation is preferable for reasons of jurisdiction as well as geographical coverage to provide consistent regulation.

The case that Mourning Doves ingest lead shot and succumb readily to lead poisoning has been made (Lewis and Legler 1968, Schultz et al. 2002, 2006b), including the suggestion that the impacts from recreational hunters may be greater than once believed (Schultz et al. 2002). In a later paper Schultz et al. (2006b) advocated that a national nontoxic shot regulation be implemented for Mourning Doves, based on the numbers of doves suspected to be afflicted by lead poisoning and their susceptibility to ingested lead shot.

Because the MBTA already has the authority to manage all migratory birds in the USA, it is the appropriate legislative vehicle to use in extending requirements for nontoxic shot when hunting these species. This Act applies throughout the entire USA. The Act has withstood successfully numerous legal challenges when used to enforce nontoxic shot requirements for hunting waterfowl (Anderson 1992), and it should act as the legal precedent for the greater protection of all hunted migratory birds. It is the opinion of this author that the use of federal legislation to manage the hunting of federally-regulated species is preferable, because of regulative efficiency, to situations in which individual states or provinces pass laws requiring use of nontoxic shot for hunting the same migratory species. Then a consistent approach to management is achieved throughout the species range.

An important consideration is that the regulations of the MBTA and the MBCA which determine the forms of lead shot substitutes that are acceptable for hunting waterfowl in the USA and Canada would also apply to the shot types used for hunting "upland" species of migratory birds. Given that migratory doves are hunted legally in only one Canadian province, there is greater need for the USA to extend nontoxic shot requirements to these species than Canada. There are no studies of the levels of lead in the bodies of migratory doves in British Columbia, but there is no reason to suppose that the prevalence of lead shot ingestion by doves and its impacts on birds in the province differs from that in the USA. If Canada were to advocate the use of nontoxic shot for the hunting of doves, it could do so on the basis of protecting a species throughout its entire migratory range (Thomas and Owen 1996). In Canada, the MBCA could be used to regulate the hunting of doves, snipe and American Woodcock with nontoxic shot as it has done for regulating the shot required for hunting waterfowl. Then, Canada would be acting in concert with the USA, both nations using the legislation under the Migratory Bird Treaty as the basis for a common, consistent approach to management at the continental level. The USA and Canada are bound to manage their common migratory birds in a complementary manner, and the harmonization of legislative approaches on extending a ban on use of lead shot would be highly desirable (Thomas 2003).

State-regulated, Non-migratory, Game Species.— The requiring of nontoxic shot for the hunting of non-migratory small game has already received much attention within the USA, but a large range of policy decisions exists. There are states that have already regulated the use of nontoxic shot for all such hunting, states that still favor the use of lead, and positions between these two extremes. This situation across the entire USA and Canada has been investigated by the Nontoxic Shot Advisory Committee (NSAC 2006) for the State of Minnesota. The analysis reveals that, as of 2006, 26 jurisdictions had nontoxic shot regulations that extended beyond those set federally for waterfowl and American Coots. However, there was considerable variation in the use of the regulations, depending on the target species, and whether hunting occurred on public or private lands (NSAC 2006, Tables 2–7). As an example, South Dakota requires nontoxic shot be used when hunting grouse, quail, and pheasants, except when hunting on private lands, walk-in areas, state school lands, and on US Forest Service National grasslands. There was no consistency in the application of nontoxic shot requirements across the states, and the exceptions to legislated use varied greatly among states (NSAC 2006). These findings argue in support of a consistent approach to regulation, because that facilitates enforcement, supports habitat improvement across a species range, and makes compliance easier for the public. The response to banning lead shot use by individual states is encouraging. More than 1.3 million acres of habitat across 23 states have been secured from further lead deposition, including more than 400,000 acres in Nebraska and South Dakota, combined[7].

Acting as a precedent in this matter is the existing federal requirement that nontoxic shot be used for hunting state-regulated species when hunting occurs on federally-regulated lands. Thus, the public has been introduced to this requirement and the need to comply. What emerges from this analysis of nontoxic shot requirements for upland game hunting is a patchwork of regulatory application, which may cross jurisdictional lines. As an example of this, 15 of the 40 states that allow dove hunting require use of nontoxic shot over some specified lands (NSAC 2006, Table 4). Thus, in these instances the approved type of shot for hunting this migratory species has been governed by the states, and not the federal government. The fact that 26 states are already engaged in regulating the use of nontoxic shot bodes well for extending this regulation further, both within compliant states and to states that still have to embark on this initiative. Individual states and provinces value their particular autonomy and right to manage wildlife within their jurisdiction. So the most successful approach would be to encourage more states to regulate nontoxic shot use. Then, the successes and experiences of other states become valuable tools in the transition, especially if there is an existing federal requirement for nontoxic shot use for all migratory birds in that state.

The passage of California Assembly Bill 821, the Ridley-Tree Condor Preservation Act in 2007 has been a major landmark in requiring use of lead-free bullets for big game hunting in the range of the California Condor (*Gymnogyps californianus*). This state Act is complemented by a state-assisted voluntary use of nontoxic bullets/slugs in the adjacent state of Arizona (Cade 2007). The deposition of lead bullet fragments in the gut piles and unretrieved carcasses of large and small game annually across the USA presents an enormous toxic risk to all species of scavengers, especially mobile Bald Eagles (*Haliaeetus leucocephalus*) and Golden Eagles (*Aquila chrysaetos*). Although the California and Arizona initiatives were predicated on the preservation of the California Condor, adoption of available nontoxic rifle bullets by other states and provinces would be highly appropriate to reduce further risk of lead ingestion.

DISCUSSION

Resolving lead toxicosis of wildlife requires the application of regulation across all forms of hunting and angling, and across all federal and state/provincial jurisdictions. Regulatory efficiency would suppose that suitable federal legislation existed in both the USA and Canada that would enable the phase-out of lead in sporting uses, as for other forms of environmental pollution. This assumption is problematic. The segregation of lead pollution into human environmental pollution and wildlife pollution compounds the issue because different agencies and different laws have been applied.

In 2005, the Canadian Wildlife Service of Environment Canada was considering application of the Canadian Environmental Protection Act (CEPA) to regulate a national ban on the importation, manufacture and sale of lead fishing weights. Application of the Act's provisions would have over-ridden any objection of a province to the proposed ban on lead weights[8]. The same Act's provisions could also have extended to a ban on all forms and uses of lead shot in Canada (Caccia 1995). There is no direct US equivalent of the Canadian CEPA, and historically, two separate US federal agencies have been involved in the regulation of lead sporting products. The US Fish and Wildlife Service still regulates the hunting of migratory birds under the MBTA. However, the US Environmental Protection Agency, a different agency in the Department of the Interior, had intended[9] using the Toxic Substances Control Act to regulate the use of lead fishing weights and their consequent poisoning of piscivo-

[7] Minnesota, Department of Natural Resources. The case for nontoxic shot. http://www.dnr.state.mn.us/outdoor_activities/hunting/nts/index.html

[8] This intent was not translated into a Parliamentary Bill, however.

[9] This intent has not been pursued into law.

rous birds (Thomas 2003). Here is a clear case of two federal agencies needing to agree on how best to regulate a lead product to obviate secondary lead poisoning of species that are, for the most part, protected under the MBTA.

Regulatory efficiency has to be tempered with practicality. Use of a strong article of federal law might not be politically expedient if it were perceived to usurp the roles and rights of states and provinces. Because, historically, the MBTA and the MBCA have been used successfully in the USA and Canada to regulate hunting of waterfowl and American Coots, it might be more expedient to retain these same legal tools for regulating other migratory bird hunting. Then the quest is to have states and provinces complement the federal initiative with respect to game animals under their jurisdiction.

The final policy option presented earlier (a regulated national ban on all uses of lead shot, bullets and sinkers) is certainly supported by a wealth of scientific evidence and precedents from other nations, and is favored by many (e.g., Cade 2007). The US Environmental Protection Agency (USEPA 1979) has already identified the need to control the use of lead in the environment, and the Toxic Substances Control Act would likely be the legislation used to implement a total ban on use of lead sporting products. The legislation would have to specify what was being banned, whether use, sale, manufacture or importation of certain lead products. Use would be difficult to enforce, but controlling the commercial availability would have a greater impact on public compliance. The legislation would also have to consider whether a national ban would apply to hunters and target shooters. The Toxic Substances Control Act would require amendment to include provisions for regulating the approval of all lead substitutes (shot, bullets, and fishing weights), as in the present regulations of the MBTA for shot (Thomas 2003). This policy option would pit federal powers against state powers, and would affect a broad range of sporting constituencies for whom hunting, shooting and angling form the basis of their heritage. Most important, the legislation proposed from this policy option would have to withstand the inevitable massive opposition from such constituencies, and would require broad support among both Houses and the Presidency to avoid rejection.

The US Toxic Substances Control Act is an appropriate legislative tool to regulate the deposition of a known toxin in the nation's environment. However, given the reluctance of the USEPA to pursue a ban on lead sinkers (Thomas 2003), it is questionable as to whether the same agency is suited for pursuing a national ban on all uses of shot, bullets, and sinkers. The US Fish and Wildlife Service is already effective in regulating nontoxic shot use for hunting over wetlands, and could extend this requirement further. Over two dozen individual states are already proceeding with nontoxic shot use for upland game hunting (NSAC 2006). Only the political process will determine whether an incremental, cooperative, approach versus a blanket, federal ban will be attempted.

Complementing any federal initiatives assumes that requirements for nontoxic shot and bullets are supported by the state agencies administering wildlife and the public, so a period of analysis, consultation and education is required (NSAC 2006, Schultz et al. 2007). The public, if compliant, demands a period of phase-in of lead substitutes. The duration of such a period is difficult to determine. For the ammunition producers, a 2–3 year period is probably adequate, given that new nontoxic loads have to be developed, made, and distributed widely. Most of the large ammunition producers have already been making suitable nontoxic shot and bullets, so the technology is already in place. The length of a phase-in for hunters is more problematic. If a period of five years (for example) were allowed, very little transition would be expected until the last year. Moreover, because nontoxic shot and bullets would not be required, legally, until the fifth year, there would be no incentive for manufacturers to distribute before that time because the market demand might be low. It is fallacious to assume that knowledge about lead poisoning of waterfowl and use of nontoxic shot will translate into a realization of the same problem and its resolution in upland game hunting and angling (Thomas 1997). The sporting public behaves as very different communities (i.e., upland bird hunters, big game hunters and anglers) and each may have valid concerns that must be addressed (Schultz et al. 2007). Even where a state agency has diligently prepared its case for adopting nontoxic shot for hunting upland game, met with the user groups, engaged in education, and appeared to act progressively and environmentally responsibly (NCAS 2006), there is no assurance of political support for the case.[10] Thus it is important to consider the procedures used by states such as Nebraska and South Dakota, which have successfully implemented nontoxic shot requirements, and use them to good effect.

The widespread availability of nontoxic shot and bullets rests on the reality of an assured market demand for those products that is provided only by regulation and enforcement. Failing that, people will continue to use lead products. This "Catch-22" can be resolved only by regulations being created first. For nontoxic shot and bullet regulations to work at the Continental level, especially to reduce the secondary lead poisoning of highly mobile species, there has to be a "buy-in" from a majority, or more, of the states and provinces. The 2006 figure of 26 states is a promising start (NSAC 2006). A greater participation by states and provinces would reduce the price of ammunition by an economy of scale effect, and would encourage competition among manufacturers eager to increase their sales in an expanding nontoxic product market.

The extension of nontoxic requirements to shot and bullets on a wider scale would complement other conservation initiatives in the USA. The Sonoran Desert population of the Bald Eagle was re-listed as "Threatened" under the US Endangered Species Act in May, 2008 (USFWS 2008). If lead poisoning were to afflict this population, as it had afflicted other populations (Anderson 1992; USFWS 1986), then relief from local lead poisoning would assist recovery. In California, passage of Assembly Bill 821, The Ridley-Tree Condor Preservation Act in 2007, requires the use of nontoxic bullets and shot when hunting large game and small game in central and southern California, the range of the endangered California Condor. The action of the state of California is to be applauded for action that deliberately complements the intent of the Endangered Species Act, and protects California Condors and other scavengers from secondary lead poisoning. In this regard, it is advisable to expand the requirement of nontoxic shot for the hunting of all migratory birds so that the MBTA is seen to complement and potentiate the provisions of the Endangered Species Act.

Convincing the different hunting and angling organizations that use of lead-free products is in their interest is the key to effecting policy and legislative change. These same groups have done much, historically, to enhance conservation of all wildlife and their habitats across North America. Perhaps it has been easier for these organizations to promote conservation when threats to wildlife and their habitats have originated from agricultural, urban, and industrial development. However, using lead-free products requires that hunters and anglers change their *own individual behavior* to contribute to species conservation, and then often not their preferred target species. Industry in North America has provided an array of effective substitutes for all lead products. Thus, the policy process has to focus on communicating the message that using these products promotes the sustainability of hunting and angling, bolsters all wildlife populations, and leaves no long-term toxic legacy in the environment. This is the "win-win-win" situation. Part of that message should be the research of Anderson et al. (2000), which shows that investment in nontoxic shot yields an enormous dividend in wildlife that compounds the gains arising from private investments in habitat. The successful articulation of this message in the policy process of the federal and state/provincial agencies will be the precursor to passage of appropriate legislation.

Preventing further lead toxicosis in birds requires regulating the use of nontoxic shot, bullets and fishing weights for all types of hunting and angling. Nontoxic substitutes exist, and their use in waterfowl hunting has constituted an enormous saving of birds each year in North America. Reduction in the mortality of birds from primary and secondary lead toxicosis would be expected following use of nontoxic products for all hunting and angling.

Resolution of the problem at the policy and legislative level demands careful choice of which policy option(s) to pursue in the short and long terms.

[10] Minnesota was required to abandon its proposal to use nontoxic shot for small game hunting in 2008.

Continuing research on the prevalence of lead toxicosis and how it is reduced by use of lead substitutes is vital to support the policy process and inform the sporting community. It is argued that the greatest regulative efficiency is achieved by using the Migratory Bird Treaty Act to regulate, nationally, nontoxic shot use for hunting all species of migratory birds. That also provides incentives for individual states (many already embarking on this initiative) to regulate nontoxic shot and bullet use for all hunting of species under their jurisdiction.

The arms industries are already able to make effective nontoxic shot and bullets widely available, but need the assured market demand provided by law to make it happen. Central to any policy at both levels of government is communication with public user groups, who should perceive nontoxic shot, bullets and fishing sinkers as an investment in the sustainability of their sport, the complementing of habitat conservation, a direct generator of wildlife, and a less polluted environment.

LITERATURE CITED

ANDERSON, W. L. 1992. Legislation and lawsuits in the United States and their effects on nontoxic shot regulations. Pages 56–60 *in* D. J. Pain (Ed.). Lead Poisoning in Waterfowl. Proceedings of an IWRB Workshop, Brussels, Belgium, 13–15 June 1991. International Waterfowl and Wetlands Research Bureau Special Publication 16, Slimbridge, UK.

ANDERSON, W. L., S. P. HAVERA, AND B. W. ZERCHER. 2000. Ingestion of lead and nontoxic shotgun pellets by ducks in the Mississippi flyway. Journal of Wildlife Management 64:848–857.

BEINTEMA, N. 2001. Third international update report on lead poisoning in waterbirds. Wetlands International. Wageningen, The Netherlands.

BUTLER, D. A., R. B. SAGE, R. A. H. DRAYCOTT, J. P. CARROLL, AND D. POTTS. 2005. Lead exposure in Ring-necked Pheasants on shooting estates in Great Britain. Wildlife Society Bulletin 33:583–589.

CACCIA, C. L. 1995. It's about our health. Towards pollution prevention. Report of the House of Commons Standing Committee on Environment and Sustainable Development. Public Works and Government Services Canada, Ottawa, Canada.

CADE, T. J. 2007. Exposure of California Condors to lead from spent ammunition. Journal of Wildlife Management 71:2125–2133.

CENTER FOR BIOLOGICAL DIVERSITY. 2006. Governor's response to request for immediate action. [Online.] Available at: www.biologicaldiversity.org/swcbd/species/condor/DFG-Governor-response.pdf. Accessed 1 May 2008.

CHURCH, M. E., R. GWIAZDA, R. W. RISEBROUGH, K. SORENSON, C. P. CHAMBERLAIN, S. FARRY, W. HEINRICH, B. A. RIDEOUT, AND D. R. SMITH. 2006. Ammunition is the principal source of lead accumulated by California Condors reintroduced to the wild. Environmental Science and Technology 40: 6143–6150.

CRAIGHEAD, D., AND B. BEDROSIAN. 2008. Blood lead levels of Common Ravens with access to big game offal. Journal of Wildlife Management 72:240–245.

FISHER, I. J., D. J. PAIN, AND V. G. THOMAS. 2006. A review of lead poisoning from ammunition sources in terrestrial birds. Biological Conservation 131:421–432.

HUNT, W. G., W. BURNHAM, C. N. PARISH, K. K. BURNHAM, B. MUTCH, AND J. L. OAKS. 2006. Bullet fragments in deer remains: implications for lead exposure in avian scavengers. Wildlife Society Bulletin 34:167–170.

KENDALL, R. J., T. E. LACHER, C. BUNCK, B. DANIEL, C. DRIVER, C. E. GRUE, F. LEIGHTON, W. STANSLEY, P. G. WATANABE, AND M. WHITWORTH. 1996. An ecological risk assessment of lead shot exposure in non-waterfowl avian species: upland game birds and raptors. Environmental Toxicology and Chemistry 15:4–20.

KRAMER, J. L., AND P. T. REDIG. 1997. Sixteen years of lead poisoning in eagles, 1980–1995: an epizootiologic view. Journal of Raptor Research 31:327–332.

LANPHEAR, B. 1998. The paradox of lead poisoning prevention. Science 281:1617–1618.

LEWIS J. C., AND E. LEGLER. 1968. Lead shot ingestion by Mourning Doves and incidence in soil. Journal of Wildlife Management 32:476–482.

LYSTER, S. 1985. International Wildlife Law. Grotius Publications, Cambridge, UK.

MATEO, R., R. CADENAS, M. MANEZ, AND R. GUITART. 2001. Lead shot ingestion in two raptor species from Donana, Spain. Ecotoxicology and Environmental Safety 48:6–10.

MATEO, R., A. J. GREEN, H. LEFRANC, R. BAOS, AND J. FIGUEROLA. 2007. Lead poisoning in wild birds from southern Spain: A comparative study of wetland areas and species affected, and trends over time. Ecotoxicology and Environmental Safety 66:119–126.

NONTOXIC SHOT ADVISORY COMMITTEE (NSAC). 2006. Report of the Nontoxic Shot Advisory Committee. Minnesota Department of Natural Resources, Fish, and Wildlife Division. St. Paul, Minnesota, USA.

NRIAGU, J. O. 1983. Lead and lead poisoning in antiquity. Wiley, New York, New York, USA.

NRIAGU, J. O. 1998. Tales told in lead. Science 281 (5383): 1622–1623.

PAULI, J. N., AND S. W. BUSKIRK. 2007. Recreational shooting of prairie dogs: A portal for lead entering wildlife food chains. Journal of Wildlife Management 71:103–108.

RATTNER B. A., J. C. FRANSON, S. R. SHEFFIELD, C. I. GODDARD, N. J. LEONARD, D. STANG, AND P. J. WINGATE. 2008. Sources and implications of lead-based ammunition and fishing tackle to natural resources. Technical Review 08–01. The Wildlife Society, Bethesda, Maryland, USA.

SAMUEL M. D., AND E. F. BOWERS. 2000. Lead exposure in American Black Ducks after implementation of non-toxic shot. Journal of Wildlife Management 64:947–953.

SCHEUHAMMER, A. M., C. A. ROGERS, AND D. BOND. 1999. Elevated lead exposure in American Woodcock (*Scolopax minor*) in eastern Canada. Archives of Environmental Contamination and Toxicology 36:334–340.

SCHEUHAMMER, A. M., S. L. MONEY, D. A. KIRK, AND G. DONALDSON. 2002. Lead fishing sinkers and jigs in Canada: review of their use patterns and toxic impacts on wildlife. Canadian Wildlife Service Occasional Paper No. 108. Environment Canada. Ottawa, Canada.

SCHULZ, J. H., J. J. MILLSPAUGH, B. E. WASHBURN, G. R. WESTER, J. T. LANIGAN III, AND J. C. FRANSON. 2002. Spent-shot availability and ingestion on areas managed for Mourning Doves. Wildlife Society Bulletin 30: 112–120.

SCHULZ, J. H., P. I. PADDING, AND J. J. MILLSPAUGH. 2006a. Will Mourning Dove crippling rates increase with nontoxic-shot regulations? Wildlife Society Bulletin 34:861–865.

SCHULZ, J. H., J. J. MILLSPAUGH, A. J. BERMUDEZ, X. GAO, T. W. BONNOT, L. G. BRITT, AND M. PAINE. 2006b. Acute lead toxicosis in Mourning Doves. Journal of Wildlife Management 70:413–421.

SCHULZ, J. H., R. A. REITZ, S. L. SHERIFF, AND J. J. MILLSPAUGH. 2007. Attitudes of Missouri small game hunters toward non-toxic shot regulations. Journal of Wildlife Management 71:628–633.

SIDOR , I. F., M. A. POKRAS, A. R. MAJOR, K. M. TAYLOR, AND R. M. MICONI. 2003. Mortality of Common Loons in New England, 1987–2000. Journal of Wildlife Diseases 39:306–315.

STEVENSON, A. L., A. M. SCHEUHAMMER, AND H. M. CHAN. 2005. Effects of nontoxic shot regulations on lead accumulation in ducks and American Woodcock in Canada. Archives of Environmental Contamination and Toxicology 48:405–413.

THOMAS, V. G. 1997. Attitudes and issues preventing lead bans on toxic lead shot and sinkers in North America and Europe. Environmental Values 6:185–199.

THOMAS, V. G. 2003. Harmonizing approval of nontoxic shot and sinkers in North America. Wildlife Society Bulletin 31:292–295.

THOMAS, V. G., AND M. OWEN. 1996. Preventing lead toxicosis of European waterfowl by regulatory and non-regulatory means. Environmental Conservation 23:358–364.

THOMAS, V. G., AND R. GUITART. 2003. Lead pollution from shooting and angling, and a common regulative approach. Environmental Policy and Law 33:150–154.

THOMAS, V. G., AND R. GUITART. 2005. Role of international conventions in promoting avian conservation through reduced lead toxicosis: progression towards a non-toxic agenda. Bird Conservation International 15:147–160.

TWISS, M. P., AND V.G. THOMAS. 1998. Preventing fishing-sinker-induced lead poisoning of Common Loons through Canadian policy and regulative reform. Journal of Environmental Management 53:49–59.

US ENVIRONMENTAL PROTECTION AGENCY (USEPA). 1979. The health and environmental impacts of lead and an assessment of a need for limitations. US Environmental Protection Agency Report 560/2-79-001, Government Printing Office, Washington, D.C., USA.

US FISH AND WILDLIFE SERVICE (USFWS). 1986. Use of Lead Shot for Hunting Migratory Birds in the United States: Final Supplemental Environmental Impact Statement. US Department of the Interior, Arlington, Virginia, USA.

US FISH AND WILDLIFE SERVICE (USFWS). 2008. Endangered and threatened wildlife and plants; Listing the potential Sonoran Desert Bald Eagle Distinct population segment as threatened under the Endangered Species Act. Federal Register 73 (85): 23966-23970.

WAYLAND, M., AND T. BOLLINGER. 1999. Lead exposure and poisoning in Bald Eagles and Golden Eagles in the Canadian prairie provinces. Environmental Pollution 104:341–350.

WILLIAMS, W. 1994. The Lead Industries Association. Part 2, pages 31–61 *in* United States Environmental Protection Agency, Lead Fishing Sinker Rule, Docket 62134, Public Hearing. USEPA, Washington, DC, USA.

COMMENTARY

JOHN FREEMUTH

Andrus Center for Public Policy, Boise State University, 1910 University Drive, Boise, ID 83725, USA.
E-mail: JFREEMU@boisestate.edu

Transcribed from Conference Expert Panel 15 May 2008.

FREEMUTH, J. 2009. Commentary. *In* R. T. Watson, M. Fuller, M. Pokras, and W. G. Hunt (Eds.). Ingestion of Lead from Spent Ammunition: Implications for Wildlife and Humans. The Peregrine Fund, Boise, Idaho, USA. DOI 10.4080/ilsa.2009.0313

Key words: Ammunition, environment, EPA, health, lead, policy, public.

MY COMMENTS are concerned particularly with the development of public policy on environmental matters. Here are thoughts based on some of what I've heard, stemming in particular from Professor Thomas's remarks earlier today.

First idea, if you read the article in the Idaho Statesman by Rocky Barker, who is a fine reporter that has much on his plate, notice how important it is to have a strategy to be able to justify or validate your science versus it immediately being portrayed as "that's from The Peregrine Fund, therefore, they are an advocacy group, therefore it's all nonsense." This happens all the time. It's part of the "politics" of science. That is a classic move in American politics; just discount all the science because it is from a certain group's perspective. Of course those that do it take the opposite perspective. There needs to be somebody that is neutral, that is respected, to say that what came from the effort is good science, even though it was "sponsored", in part, by the Peregrine Fund.

Somebody asked a question earlier today about "why do we need all of this science? Why can't we just move ahead?" Of course, the obvious point there with many of our environmental laws is that science is demanded before we move ahead. So we need good science. I also like to say that science is a necessary, but insufficient condition, for public policy making. Though, on the other hand, in a democratic society we *could* certainly empower scientists to make decisions for us. That might be a way around the dilemma. I don't think I would advise it necessarily because, as an example: after all, I have a Ph.D. in Political Science, I know more than you do about politics and policy, shouldn't my vote count 100 times more than yours? Of course not, because I am going to vote based on my values. That is also hugely important when we use science. We have to make sure that we state up front what our values are that may have influenced our choice of research topic and our hypotheses. We also need to be as clear as we can about when it's our science speaking, and when it's our values speaking. Back to my Ph.D., my "hidden" value in studying public policy is that I value democracy, even with all its flaws. But there isn't a way I know of to assert scientifically that democratic decision-making is "best."

Federalism is an important topic, and it's important to Canada, too, but I'll refer to the United States right now. Some pretty interesting things come out of federalism. If one is going to use the EPA in a strategy for solving problems that are due to lead contamination, it's going to be very important who runs the next EPA. Number one, we will clearly have a different philosophy whether it is President McCain or President Obama, but probably not President Clinton (I'm a political scientist remember, that's not a hard and fast scientific prediction). They will have a different philosophy about EPA and these people probably will be consumed at first by climate issues and figuring out if there are regulatory ways to attack those issues. Right, wrong, or indifferent, they will be consumed at first. That is something to be aware of. They have many science advisory boards that struggle through this. The strategy there, obviously, is regulatory change. This is what President Bush has done, and this is what other Presidents of both persuasions do when Congress can't do anything. One way to change policy in America is to rewrite regulations. This is why we are spending a lot of time idiotically right now in this part of the West trying to open National Parks to people so that they can carry their weapons openly into the park. Now there is a burning issue folks, but they are trying to change the regulations to allow that. But that is the way you would get EPA involved in the spent lead issue, though some might think it is a reach. That doesn't stop people from trying. There is nothing wrong with that strategy.

One can also try to use laws to attack problems best left to other laws. To be blunt, the Endangered Species Act is not the vehicle to attack climate change, but it is being used in the case of the polar bear. Those kinds of strategies will continue. Probably it is a difficult strategy unless you are able to find some hook in a law that courts are likely to give you some room to move on an issue.

Another danger is, quite frankly, in the United States we are developing a hollow state. We want our agencies to do all sorts of things, but we continue to cut their environmental budgets and reduce their staff, rendering them functionally incompetent, then we blame them for not being able to get things done quickly while making it impossible to solve problems. The reality is that the EPA is a great example of this. They are overworked. You can't imagine the many kinds of regulation and policy they have to develop, that they are tasked with under all our environmental laws. So, maybe using EPA is not a bad strategy, you just need to be prepared for that.

Under federalism in the United States, states remain important, so if you are going to think through trying to federalize this issue, and by that I mean making it a national issue, involve the states. The states on climate change issues have been more proactive than the national government has. We have an inherent tension in the United States between states and national government. Sometimes we like what the states do, sometimes we don't. Sometimes we like what the federal government does, sometimes we don't. You have a careful with a national strategy. Also remember, what you nationalize under a more "green" administration can be changed under a non-green presidential administration. Nothing is really set in stone.

Finally, consider time. This is an important issue, because as I said in my opening comments, there will be a lot of issues fighting to get on the policy agenda. Work hard, but be patient. It took, after all, 10 years to pass the Wilderness Act in the United States. It's just the nature of our political system. You just have to push ahead, but understand we don't act quickly most of the time, unless there is a big crisis.

Biography.—**John Freemuth, Ph.D.**, is the Senior Fellow of the Andrus Center for Public Policy, Director of the Energy Policy Institute, and Professor of Public Policy and Administration at Boise State University. Dr. Freemuth teaches in the Master of Public Administration Program and Political Science Department. His MPA specialty is Natural Resources and Public Land Policy and administration. He also teaches Organizational Theory, Introduction to Public Administration and Public Service and Democracy. In addition Dr. Freemuth teaches American Government, Environmental Policy, and Public Land Policy in the Political Science Department. Dr. Freemuth has published numerous articles and one book on public land policy, and has worked on several projects with federal and state natural resources agencies, including the Forest Service, Bureau of Land Management, and National Park Service at the federal level and the Departments of Fish and Game, Parks and Recreation, and Environmental Quality of the State of Idaho. He was the chair of the Bureau of Land Management's National Science Advisory Board.

COMMENTARY

MILTON FRIEND

US Geological Survey, National Wildlife Health Center, 6006 Schroeder Road, Madison, WI 53711, USA.

Transcribed from Conference Expert Panel 15 May 2008.

FRIEND, M. 2009. Commentary. *In* R. T. Watson, M. Fuller, M. Pokras, and W. G. Hunt (Eds.). Ingestion of Lead from Spent Ammunition: Implications for Wildlife and Humans. The Peregrine Fund, Boise, Idaho, USA. DOI 10.4080/ilsa.2009.0314

Key words: Ammunition, competitive, industry, lead, non-lead, science, shooting, stakeholders.

I WOULD LIKE TO THANK The Peregrine Fund for putting this program together. It's nice to be re-immersed in an issue that I spent so much of my professional career working on. I have learned a lot from the conference, particularly of the potential association of lead from ammunition that I had not considered previously, such as lead in the food that I am eating. A couple of important thoughts come to mind.

When I left the Wildlife Health Center after directing it for 23 years, I guessed I had moved away from the lead issue. The Secretary of the Interior asked me to go out to California and take on the issue of putting the scientific underpinning to the Salton Sea Restoration Project so that political and management decisions could be made in terms of that particular issue. The Salton Sea is near the California-Mexican border and supports 418 species of birds, so there are a lot of conservation consequences to what takes place there. As the Executive Director of the Salton Sea Science Committee I was handed a large group of political appointees representing a wide diversity of backgrounds, and I thought it was going to be impossible to come to any kind of consensus. When the process started, for example, there was absolutely no consideration of the 418 species of birds that were there. This was a water issue and a development issue in terms of major cities like Los Angeles and San Diego. Yet, I was absolutely flabbergasted at the ability of such diverse stakeholders to come together, because we were able to put together a good underpinning of science. We walked away from that experience with one of the common goals being the preservation of the bird life. Now that was a major advancement in an otherwise totally hostile environment.

I mention that experience because it reflects one of the challenges in front of us here. We've not had participation, even though there are a few people in the audience, of the major stakeholders in the lead vs. non-lead ammunition debate. So, it becomes our task to visit and engage them, take our ideas to them and try to find common ground to move forward. If we do not do that, we will probably be unsuccessful in getting to where we need to be.

There were a lot of very good presentations in the conference. I want to focus on one. I commend Dr. Oliver Krone for doing it right in Germany, but some of you may not have picked up on an important issue, and I'm using this as an example of the nuances of dealing with lead and dealing with hunter-killed venison in Germany, compared to dealing with it in this country. There is an economic incentive associated with hunting game in Germany because game is marketed in Europe, whereas it is not in the U.S. The thought I want to convey, from my perspective, is that to change human behavior in the direction that we would like it to go involves two powerful motivating factors. One is that you are experiencing personal, unwanted impacts so it becomes a personal situation, not an abstraction in dealing with the conservation of something "out there." But if you are impacted personally, you are inclined to pursue solutions. So, for example, we have shifted from eating red meat to depleting the oceans because we were concerned that red meat was bad for our diet. Likewise, I see the human health aspect here as weighing more heavily, potentially, than it did in the waterfowl wars.

Another stakeholder and driving force is the competitive shooters. As a former competitor I understand that if you think you have an advantage, or if there is something that will give you an advantage, you are going to pursue it. I think here is an opportunity to work with industry in terms of product enhancement that would cause people to want to go out and use it. Look at the quality of the steel shot today compared with early offerings – they were awful. That is because of the competitive marketplace. I strongly encourage the involvement of industry to work together to develop a product that is in the best interest of competitive shooters and the best interest of wildlife conservation.

Biography.—**Milton Friend, Ph.D.**, is Director Emeritus at the US Geograpical Survey's National Wildlife Health Center in Madison, Wisconsin. He has worked as a university researcher and with state and federal conservation organizations. Dr. Friend entered the wildlife conservation field in 1956 with the Vermont Department of Fish and Game. In 1975 he was assigned to Madison, Wisconsin, to develop the National Wildlife Health Center. He served as Center Director until 1998, when he was asked by the Secretary of Interior to accept a special assignment as the Executive Director of the Salton Sea Science Committee to develop and oversee the science program for the Salton Sea Restoration Project. He returned to the Center in 2002 to complete a book on emerging diseases. His many honors and awards include the Department of the Interior´s Meritorious Service and Distinguished Service Awards.

COMMENTARY

MICHAEL KOSNETT

Division of Clinical Pharmacology and Toxicology, Department of Medicine, University of Colorado Denver, Aurora, CO 80045, USA.

Transcribed from Conference Expert Panel 15 May 2008.

KOSNETT, M. J. 2009. Commentary. *In* R.T. Watson, M. Fuller, M. Pokras, and W.G. Hunt (Eds.). Ingestion of Lead from Spent Ammunition: Implications for Wildlife and Humans. The Peregrine Fund, Boise, Idaho, USA. DOI 10.4080/ilsa.2009.0315

Key words: Ammunition, child, childhood, child bearing, food, game, hunters, lead, toxicosis, women.

I HAVE ENJOYED THIS CONFERENCE; there has been a lot of good information presented. I hope it doesn't come as a surprise to you that this issue is not on the radar screen of my colleagues in the medical and public health community. There is an enormous amount of effort to combat lead poisoning in humans in the United States. Virtually every state in the country has a childhood lead poisoning prevention program; 35 states monitor blood lead concentrations and follow up on them through lead registries; and many states have full time occupational health lead poisoning prevention programs. There are hundreds of people doing research on the effects of lead intoxication; and the scientific literature on lead intoxication is vast, numbering tens of thousands of articles. Despite all this, I don't think there's a single study been done on blood lead levels and lead exposure in sportsmen who consume game animals in the United States. There have been some studies on native communities in Canada and people in Greenland, and they have raised some concern. The key take home message that I have learned from this Conference is to connect this concern about the human health aspects of consuming lead in game meat with the need for human health studies. I am a member of the Advisory Committee on Childhood Lead Poisoning Prevention for the Centers for Disease Control. I intend to bring this up at the CDC to promote funding of research and population surveys to look into this issue. I think it will be very helpful to focus studies on children and women of reproductive age in hunters' families rather than only on hunters because there may be more impetus to do something about this source of lead exposure if we find that this is a public health problem and document the magnitude of the problem.

Another issue that I was not really aware of, that really opened my eyes, was the vast amount of lead that is being discharged onto public and private lands through hunting. We heard from one of the talks that about 72,600 tons per year of lead are being emitted at 9,000 outdoor non-military target ranges in the United States. We heard about Mourning Dove hunting fields where tens of thousands of rounds of lead ammunition are discharged each season, year after year. There's a need to involve the Environmental Protection Agency and other environmental groups to investigate the impact of loading the environment with lead. This is an issue that has just not been looked at.

Finally, I want to say something about the issue of perception. It floored me to hear papers about wildlife, particularly Bald Eagles and condors, that had what I would consider to be lead poisoning with median lead levels of 40 µg/dL during the hunting season. The Bald Eagle is a very symbolic bird; it is the symbol of our country. A lot of powerful publicity could be made by pointing out to the media that our national symbol is being poisoned by lead. This message would concern the general public in a way that poisoning of other wildlife, which is not insignificant, maybe would not.

In closing, there is a real opportunity now to make progress with this issue. There is a lot of attention given to being "green." If you bought coffee downstairs, you saw they are selling it in "green, earth-friendly" mugs. Everywhere you go now, people and businesses want to be "green." Well, lead is not a "green" material—this is a toxic material that has a deleterious effect on the environment and human health. If we, as a group, can let people know that lead ammunition is a hazardous material for which there are substitutes, then we can benefit from the environmental shift to prefer safer products.

Some of you are concerned about the difficulties involved and resistance to mandatory bans. Bear in mind that there are now statewide bans on second-hand smoke; restrictions on smoking in restaurants and public places. People thought that would never happen; that people would be incredibly resistant. Twenty-five percent of the American public smoke cigarettes, and in California they went and banned smoking in restaurants and public places entirely. Colorado now has a complete ban. There was a lot of resistance at first, but it went through, and what we are seeing now is that it is really having an effect, not only decreasing second-hand smoke, but also decreasing the numbers of young people taking up smoking. If something as entrenched as smoking can be changed, then it puts the issue of lead ammunition in a different light. I look forward to working with many of you and my colleagues in the medical community to join arms to work on this issue.

Biography.—**Michael J. Kosnett, M.D., M.P.H.,** is Associate Clinical Professor in the Division of Clinical Pharmacology and Toxicology, Department of Medicine, University of Colorado Denver. He is a medical toxicologist specializing in occupational and environmental toxicology with clinical and research interest in heavy metals. Dr. Kosnett is a current member of the Advisory Committee on Childhood Lead Poisoning Prevention of the US Centers for Disease Control. He recently served as President of the American College of Medical Toxicology, and past assignments include the Committee on Toxicology, National Research Council, World Health Organization, US State Department, US Environmental Protection Agency, and others.

COMMENTARY

DEBORAH PAIN

Wildfowl and Wetlands Trust, Slimbridge, Gloucestershire, GL2 7BT, UK
E-mail: debbie.pain@wwt.org.uk

Transcribed from Conference Expert Panel 15 May 2008.

PAIN, D. 2009. Commentary. *In* R. T. Watson, M. Fuller, M. Pokras, and W. G. Hunt (Eds.). Ingestion of Lead from Spent Ammunition: Implications for Wildlife and Humans. The Peregrine Fund, Boise, Idaho, USA. DOI 10.4080/ilsa.2009.0316

Key words: Ammunition, birds, game, global, health, human, hunting, lead, poisoning, wildlife.

THE INGESTION OF LEAD FROM AMMUNITION affects large numbers of birds annually resulting in sub-lethal effects and mortality. In some species, especially raptors (California Condor, Steller's Sea-eagle), population effects occur; in others lower levels of mortality occur along with sub-lethal effects that compromise welfare and may reduce survival. The problem is global in nature—where lead ammunition is used and birds are exposed they will be affected.

Lead is a non-essential highly toxic persistent heavy metal and lead poisoning in birds is likely to occur wherever birds feed in areas where lead has been deposited, or where predators or scavengers feed on game species. Whilst the majority of research over the last 125 years has been conducted in North America and Europe, this is a global problem, and wherever we look we find new species to add to the list of lead poisoning casualties. Scientific evidence identifying lead from ammunition as a major cause of lead poisoning and mortality in birds has existed for decades—evidence for this goes beyond all reasonable doubt. People that are not convinced by the mountain of existing evidence are unlikely to be convinced by additional research into the causes of poisoning and mortality.

On day one of this meeting, Milton Friend indicated that in any such situation three things are needed: (1) identification of the problem, (2) finding acceptable alternatives, and (3) ensuring that there is the authority to act.

There were several components to identifying the problem, i.e. What, Where, When, Who and Why. The answers to these have been comprehensively covered over the last three days:

What—birds eat lead from ammunition or ammunition fragments, suffer sub-lethal effects and mortality from lead poisoning—in their millions.

Where—globally wherever lead ammunition is used for any purpose and where birds feed in areas of lead deposition, or prey upon or scavenge game or other hunted species.

When—every year—constantly in some areas, temporally and spatially correlated to hunting season in others.

Who—caused by anyone using lead ammunition for hunting or target shooting.

Why—It appears that lead ammunition continues to be used, possibly because hunters (a) do not believe or are not fully aware that there is a problem—additional research into the problem is unlikely to help here although better communication may, (b) are resistant to change—and they are not alone in this, and (c) perceive this to be anti-hunting, which it is not.

This last point seems to be one of the main stumbling blocks—where much of the problem lies, and one of the areas that we need to look to for a solution.

The second point described by Milton Friend was the identification of acceptable alternatives. Legislative change can happen in the absence of these when a problem is deemed sufficiently serious: however, it is always far better if acceptable alternatives can be found, as solutions require compliance, and the social and economic impacts of change must be managed. Alternatives to lead shot and expanding nose bullets are available, as we heard from Barnett Rattner and Vic Oltrogge. Indeed, John Harradine from BASC commented that alternative gunshot in the UK is very effective. Slightly different shooting techniques may be required, some alternatives may cost a bit more (others may cost less), and the work of John Schulz indicated that alternative shot are unlikely to increase crippling rates. Alternatives exist, they work well, and their effectiveness, cost and choice will all improve with market forces. However, as Vernon Thomas indicated, market forces require that a guaranteed market exists, and this requires legislation. I believe that the 'acceptability' of alternatives goes back to the 'Why' part of the equation above. Many hunters may not find it acceptable to use alternatives to lead, and there is work to do here; motivating a change in behavior requires excellent communication and clear messages.

So what are the options for seriously reducing this problem?

On the first day, Barnett Rattner suggested that options included:

(1) Restricting the use of lead ammunition in localities where it poses an unacceptable hazard, or

(2) Phasing out the use of lead ammunition with a goal of complete elimination.

I believe that phasing out lead ammunition where it poses an unacceptable hazard is not a practical option. First, what is 'unacceptable'? Is unacceptable different in different circumstances, cultures, and for different species? After all, lead poisoning affects birds wherever lead ammunition is used and they are exposed, i.e. across much of the globe. Even within individual countries or states, do we really want to do detailed research to define the level of lead poisoning everywhere, and then argue about what is acceptable and what is not? This would certainly take a 'totally unacceptable' amount of time and resources, and the delays caused would result in considerable additional wildlife mortality.

To guarantee a significant reduction in the risk to birds a phase out of the use of lead with the goal of complete elimination is needed. This would also have the advantage of solving the majority of other environmental and wildlife problems associated with the use of lead ammunition—and—importantly—would tackle the human health issues. It is in this that I believe there may be a way forward that will help us to deal with the apparently widespread belief by hunters that anti-lead is anti-hunting. The human health risks from lead ammunition provide an argument against lead ammunition that has been demonstrated throughout this meeting, and one that should help change hunters' views of the acceptability of using alternatives to lead.

One of the excellent things about this conference is that it has bridged an important gap. To my knowledge, this is the first time that experts on the impact of lead on human health and lead in wildlife have been brought together. It is also very timely, as there is now good evidence to show that lead ammunition can affect human health. We heard from Lori Verbrugge of risks from lead ammunition to subsistence hunting communities, and to people using indoor shooting ranges. Exposure was through ingestion of lead particles in ingested meat and inhalation of dust, and some exposure levels were of concern for human health, including children with elevated blood lead levels. Grainger Hunt and other authors showed the extent to which lead ammunition fragments on hitting a target, even without hitting bone. There is consequently the real and concerning potential for anyone cooking and eating game to be exposed to unacceptably high levels of a very toxic metal. And finally, and most importantly, we heard from Michael Kosnett about the health effects of low dose lead exposure in children and adults. We heard of evidence that the risk of abortion in pregnant women increases as blood lead level increases above 5 μg/dL and that increased pre-natal exposure to lead is associated with a reduction in post-natal IQ, with the steepest declines in IQ at maternal blood leads of <10 μg/dL. We heard that gastrointestinal absorption of lead by children was higher than by adults, and that eating just one game bird, even cooked after the removal of obvious shot pellets, is likely to result in increased exposure. Much of this information is from studies published over the last few years.

The impacts of lead from ammunition on wildlife alone have long been sufficient to justify the phase out of all lead ammunition. However, there is now an additional concern, that of human health. Together they make a more than compelling case. The potential risks to human health may be key in helping to persuade hunters that being anti-lead is not anti-hunting—it is simply common sense. Chris Parish made me laugh when he said yesterday that we are reticent to quit things even when they are bad for us—this is certainly true of myself where chocolate is concerned. However, on a rather more serious note, we are rather less reticent to change things that have been shown to be bad for our children and for pregnant women.

To remove the threat of lead from ammunition to wildlife, and to humans, as rapidly as possible:

(1) We must work towards the phase out and eventual elimination of all lead ammunition. Identifying the most appropriate legislation through which to work in different geopolitical regions will be key to this.
(2) In the short term, in cases where lead from ammunition poses a serious conservation problem, it is important to continue to work with hunters at a local level to gain acceptance for and voluntary use of alternatives. Several studies described during this meeting have shown that with a great deal of effort, at some cost and with excellent communication and participation programs, voluntary use of alternatives can reduce the problem for wildlife. Examples include parts of the range of the California Condor, and the White-tailed Eagle in Germany described by Oliver Krone. Better knowledge, including social science studies, of the best ways of influencing stakeholders will help.
(3) We must make sure that we have sufficient information on the toxicity of alternative ammunition types. As Barnett Rattner told us, a non-toxic approval protocol exists for shotgun ammunition and many alternatives have been approved through this process. We need to ensure that the materials used in all ammunition are suitably non-toxic, and to have the available information at our fingertips to prevent unnecessary delays in eliminating lead.
(4) Further work is needed to investigate the impacts and potential impacts of lead from ammunition on children and adults, both in food that is eaten, and in the people that eat it.
(5) We must create public awareness and develop good education programs on the public health and wildlife effects of lead from ammunition. We must continue to find creative ways of engaging hunters and other stakeholders in this process, and make it clear that this is not an anti-hunting agenda. It is a sustainable and wise-use of wildlife resources agenda, and increasingly also a human health agenda.

Biography.—**Deborah Pain, Ph.D.** is Conservation Director at the Wildfowl and Wetlands Trust and was, until recently, Head of International Research at the Royal Society for the Protection of Birds, UK. She has studied the effects of lead shot on waterfowl and published extensively about lead in wildlife.

COMMENTARY

MARK POKRAS

Center for Conservation Medicine, Tufts University, Cummings School of Veterinary Medicine, 200 Westboro Rd., North Grafton, MA 01536, USA.
E-mail: mark.pokras@tufts.edu

Transcribed from Conference Expert Panel 15 May 2008.

POKRAS, M. 2009. Commentary. *In* R.T. Watson, M. Fuller, M. Pokras, and W.G. Hunt (Eds.). Ingestion of Lead from Spent Ammunition: Implications for Wildlife and Humans. The Peregrine Fund, Boise, Idaho, USA. DOI 10.4080/ilsa.2009.0317

Key words: Ammunition, bullets, fishing, game, hunting, shot, sinker.

I AM A LONG-TIME ANGLER and was involved in both firearm and bow hunting in my younger years, but when it comes to problem solving, I always like to keep a lot of arrows in my quiver. Most of the work I have been doing in the eastern US is on the fishing end of lead, not the hunting end, so I'm coming at this from a little bit different point of view.

New England is a collection of 6 tiny states that, if you add them all together, is about the size of one western state. It's not an area where hunting and fishing are as important to the economies of the states, unlike many in the west, so it's a little harder to get the attention of some agencies. But in terms of fishing, I think we have some perspective that can be helpful in the present discussion.

First, let's talk about big box stores and how to get their attention for the sales of non-toxic ammunition, fishing gear, and other materials. I know from working with Walmart in New England that they have a "green" program. They think of themselves as a "green" organization. Each state has a Walmart representative. If you can have a meeting with that person, I've found that they are very congenial to this sort of thing and, if they are interested, they will put up a little exhibit and maybe have some state agency brochures, information brochures, and that sort of thing. So, go talk with them. They are looking for a business edge, and we can encourage them to find it through an improved marketing of non-toxic gear. I think the same is true for manufacturers. I know from talking with many sinker manufacturers, such as Water Gremlin. They were one of the first US companies to put out a line of nontoxic sinkers in the early '90s; but Gremlin Green® never sold well, and I know that this frustrated the company. Perhaps there's an opportunity here for many of our organizations to help. One reason that Gremlin Green® may not have sold well is that it was not effectively marketed. None of the organizations that I know of—state fish and game agencies, conservation NGOs—nobody helped. None of us stepped forward to put articles in our newsletters about using Gremlin Green® (or other nontoxic alternatives) when we fish. I think that all of our organizations together need to jump on the bandwagon to help create market demand for these non-toxic products. We all have our own means of communication with newsletters and web sites. We can help increase demand and help market these products among our memberships, readers and the groups we influence.

I have a question for people here today. As we consider trying to develop and implement policy: is it important to unify the lead issue so that we are talking about all sorts of lead objects? This would include such things as fishing gear, wheel weights, bullets, sheathing for roofs, and things of that nature. Conversely, it might be important to separate out each class of products and seek different solutions for each one by having separate interactions with each separate constituency? I'm not sure of the right answer... and it may vary from state to state. I know from talking with the Massachusetts Wildlife Federation a couple of weeks ago, that about 80% of the people there are both hunters and anglers. Thus with that constituency I'd like to discuss both non-lead hunting and fishing gear together. But I think we have to know our audience, as many people have said, and develop arguments that meet the needs of each situation.

On the human health side of things, I think there is a lot we can do, but in many cases this will mean stepping outside of our comfort zones. Every state public health agency has a lead poisoning prevention program. I am sure, to a 100% certainty, that more money is spent in this country on lead poisoning prevention on children and in occupational settings than is spent on all our sporting activities put together. It is amazing the dollars spent trying to do lead poisoning detection and remediation. Every state public health agency has human data. When I go to Maine, New Hampshire, Vermont, or Massachusetts, they can tell me of homes in which people were making bullets or making sinkers or making dive weights and where the children got lead poisoning. We have to get those data together with the sportsmen's data to figure out the magnitude of these problems and the educational programs that we need to go after.

Coming back to one of the things I said at the beginning, this is a marvelous meeting, but this should only be the beginning. I would challenge us all, myself included, to do similar presentations at other meetings. These might include The Wildlife Society meeting, the Wildlife Disease Association meeting, various state and regional meetings such as the 68th Midwest Fish & Wildlife Conference which many of us attended earlier this year, as well as medical, public health and veterinary meetings. Many times organizations are looking for program topics for a small symposium or session. Propose one. I think there are a great many people interested in expanding discussions of nontoxic gear. For example, the group we've assembled here in Boise is very generous with their expertise. Everybody will share data. We have the same agenda. We are all interested in protecting both wildlife and human health. We're not anti-hunting; we're not anti-fishing. We DO want to protect the things that are precious to us on this planet. We have enough threats as it is.

One meeting coming up in Mexico the first week in December 2008, is the Eco-health II meeting. Eco-Health I was held in Madison, Wisconsin a couple of years ago. It was an attempt to bring together people who were looking at environmental health and human health issues. A session on lead and health has been proposed for the meeting in Mexico. There is great concern about lead (Pb) in Latin America including issues of water quality, wildlife, and human health. We can continue to build bridges and expand the discussions we've had here in Boise to include the whole hemisphere and, in fact, the world.

Let's get this on the agenda for many other meetings. Let's keep talking about it. Let's bring everybody in under the tent because it is extremely important to do so.

Going back to my experience with state agencies in New England, I know how strapped state fish and wildlife agencies are because I work with these people every day. I've been sitting on a state nongame advisory council for 22 years. Everybody is worried about sales of licenses, everybody is worried about income, and funding for many of the state nongame check-offs is going down the tubes. We've had our state budget line item zeroed out this year and we're fighting that battle (NOTE: funding has since been restored to the program). So, what can we do to help the state agencies? What can we do to increase their funding for education, research and monitoring?

Many states have developed programs to encourage new people to become involved in outdoor shooting and fishing activities and many such activities are directed at women and children. We have National Fishing Day activities and we have a program in Massachusetts called Project WoodsWoman which is trying to get urban and suburban women out in the woods learning basic hunting and fishing skills. I think this is wonderful. Everyone needs to get outdoors and get involved in recreational and conservation activities. But from a health perspective, we are taking the two most sensitive components of our population, women of childbearing age and children, and potentially making them susceptible to lead exposure. We have an issue there that needs to be addressed. Any programs aimed at women and children need to be lead-free!

Last, we need to work together with many of the state agencies to try and find nontraditional sources of revenue for our state wildlife agencies. We've relied on license sales for revenue in the past and that has been very successful. We have not been so successful in getting nonconsumptive users of our environment, the campers, birdwatchers, and others, to pay their fair share of the costs of conservation. People have proposed taxes on camping equipment, binoculars, birdseed, and other items but none of those proposals have proven to be politically viable. We need to work harder on such efforts. We need to get this other component of people who love wildlife and the environment to pay their fair share and support a wide variety of conservation efforts that benefit everyone.

Biography.—**Mark Pokras, D.V.M.**, is Associate Professor and former head of the Tufts University's Cummings School of Veterinary Medicine's Wildlife Clinic. He has been recognized for his work in education, wildlife rehabilitation, and wildlife health, and has been published extensively in these areas. As a cofounder of the Tufts Center for Conservation Medicine and member of many conservation and veterinary organizations, Dr. Pokras is strongly committed to building cross-disciplinary research and educational bridges to address health and conservation issues.

COMMENTARY

ANTON M. SCHEUHAMMER

Environment Canada, National Wildlife Research Centre, Carleton University, Ottawa, ON, Canada, K1A 0H3. E-mail: Tony.Scheuhammer@ec.gc.ca

Transcribed from Conference Expert Panel 15 May 2008.

SCHEUHAMMER, A. M. 2009. Commentary. *In* R. T. Watson, M. Fuller, M. Pokras, and W. G. Hunt (Eds.). Ingestion of Lead from Spent Ammunition: Implications for Wildlife and Humans. The Peregrine Fund, Boise, Idaho, USA. DOI 10.4080/ilsa.2009.0318

Key words: Ammunition, lead, health, hunting, non-lead, science, stakeholders.

I WOULD LIKE TO THANK The Peregrine Fund and other sponsors for organizing an extremely engaging conference. I don't remember a conference where I was so eager to hear each and every talk that was presented! I recommend continuing this conference in future years, if possible – not annually, but perhaps every few years, to summarize science and policy progress.

As a toxicologist, I tend to look at lead in a rather simplistic way. Nutritionally, lead is completely nonessential to humans and other living organisms. On the other hand, lead is intrinsically very toxic. So, isn't it just basic good sense to get rid of lead wherever we can and replace it with other less toxic materials, so that environmental and human exposure to lead is reduced? Many countries have, in fact, recognized the need to phase out the use of lead, and over the past 40 years or so, lead has gradually been removed from numerous items and uses for which it was historically quite common. In North America, it has largely been removed from paints, pottery glazes, gasoline products, solder, and plumbing pipes. And isn't that ironic—we don't use lead in plumbing anymore, even though the word "plumbum" means "lead" in Latin.

So, we've made progress, but it's been slow at times, slower than most of us would like to see. We have made progress, also, in the area of lead ammunition. In Canada and the USA, and in some other countries, the use of lead shot for hunting waterfowl is now generally prohibited. But new issues involving lead ammunition have arisen, and I think that one of the great successes of this Conference is that it has succeeded in bringing together the wildlife researchers and issues, with the human health researchers and issues. This is very important because dietary exposure to fragments of metallic lead from ammunition has potentially important consequences for both wildlife and people. And the recognition of a human health component to this issue may help to accelerate progress in "getting the lead out".

I'll tell you one anecdote that illustrates why it can be important to include considerations of both environmental and human health in issues such as this. In Canada, in 2003, the Department of the Environment released a report on the environmental hazards of lead in recreational angling. It caused quite a stir in some quarters. Initially, one of the major opponents to removing lead from small fishing sinkers and jigs was the Canadian Sport Fishing Industry Association (CSIA), a group of manufacturers, distributors, and retailers of tackle equipment. Our dialogue with the CSIA was quite contentious at first, as some of their members believed that restricting the use of lead would have a major and unnecessary negative impact on the tackle industry, and on recreational angling in general. Needless to say, we didn't get along very well at the beginning of this endeavor. However, what we were trying to do was to bring them on board as partners, to help us move forward to determine the best strategies for removing lead from terminal fishing tackle, especially small sinkers and jigs. CSIA was very much opposed to that concept, initially. But eventually, once they became aware that people who cast their own lead sinkers might be at risk from lead exposure, and that children might accidentally swallow split-shot sinkers and be at risk for lead exposure, their attitude changed dramatically. Our discussions became much more cordial, and CSIA is now no longer opposed to controlling the use of lead for manufacturing sinkers and jig; rather, they want to be part of the process for determining the best strategies for reducing the use of lead.

That's a great lesson and I hope that the group that has come together in Boise for this Conference will continue to work together, and perhaps will also expand to include other stakeholder groups. It's important in the crafting of policies to reduce the use of lead that all significant stakeholders be actively involved in a consultative approach. For the lead-from-ammunition issue, this includes federal and state/provincial environmental and human health agencies; the ammunition industry; non-governmental environmental and wildlife organizations; and the hunting community.

Biography.—**Tony Scheuhammer, Ph.D.**, is a research scientist at the National Wildlife Research Centre at Carleton University, Canada. He conducts research on the ecotoxicology of metals, especially lead and mercury, and their hazards to wildlife. His research influenced the Canadian ban on lead shot and recommendations to replace lead sinkers with other non-toxic materials in fishing equipment. Some of his current research includes investigations of the environmental impacts of lead from various sources, especially as it relates to toxicity in avian wildlife.

COMMENTARY

VERNON G. THOMAS

Department of Integrative Biology, College of Biological Science, University of Guelph, 50 Stone Road East, Guelph, Ontario N1G 2W1, Canada. E-mail: vthomas@uoguelph.ca

Transcribed from Conference Expert Panel 15 May 2008.

THOMAS, V. G. 2009. Commentary. *In* R. T. Watson, M. Fuller, M. Pokras, and W. G. Hunt (Eds.). Ingestion of Lead from Spent Ammunition: Implications for Wildlife and Humans. The Peregrine Fund, Boise, Idaho, USA. DOI 10.4080/ilsa.2009.0319

Key words: Ammunition, bullets, game, health, hunting, human, lead, medicine, non-lead, nontoxic, scavengers, science, subsistence, wildlife.

THE CONFERENCE, what I've learned, and what it has achieved. To me, a major accomplishment has been shortening this enormous disconnect that has existed between the medical human health issue around lead and the wildlife issue around lead. This is a major accomplishment, I can assure you. So, shortening this disconnect is tremendous.

Also, the array of papers dealing with upland game bird species and birds of prey, have meant that in those different categories of wildlife lead toxicosis, we can now see that there is really a single toxic lead syndrome that manifests itself in different ways in those different categories of birds. So the different constituencies that dealt with waterfowl and upland game birds really are one in addressing this single toxic lead syndrome. That is another success as far as I am concerned.

From the human health side, I think we also need to recognize that there are within Canada and the United States, and perhaps also in parts of Mexico, native people whose consumption of wild game far, far exceeds that of the non-native segment of the population. I have spent months living with the Cree and Inuit in Canada and I can attest to the enormous importance of shot birds and shot mammals to their daily food existence.

The paper that we saw last in the conference, the paper dealing with consumption of shot game in Alaska, I think needs to be taken a step further so that we say, "What is the lead burden presented to these people. What are the consequences of potential lead ingestion to people who have very little option in terms of their red meat consumption across the year." Also remember, these are people who are often economically disadvantaged. That is another dimension that needs to be considered because they don't have the freedom to go elsewhere and import the sort of food luxuries that we can. So, if we have this chance to develop a theme around native people, potential lead intake and use of non-leaded ammunition, that would be a tremendous future section in a conference.

I've been impressed by the amount of science that has been given over to scavengers, birds of prey and secondary lead toxicosis. This is a tremendous step forward. And also, when we look across the papers, we have seen the reliance upon isotope ratio analysis. Not that it, of itself, proves a source of lead was the contributor, but that it enables us to explain a potential use of lead that may not be, or is, contributing to the lead problem in birds. So that sophistication in the science is good, simply because it gives us a chance to remove a level of criticism from some naysayers.

I knew The Peregrine Fund invited members of the hunting and ammunition community, but they to a large extent chose not to show so that, to your credit, they were invited. I hope that in future meetings they do attend because they are part of the equation in the resolution process and they need to be here. We need to understand their production concerns and their economic business concerns, so that we can seek compliance, and go forth, where it is possible.

We also need disciples. Despite the fact that Chris Parish is a fantastic guy, he is just one guy. And we can easily overuse him, I'm sure. I'm not a Christian religious sort, but I was told that Jesus Christ had a ratio of 12 to 1, 11 to 1 if you exclude Judas. So I think that we need to have a segment of our society that basically is there communicating our ideas to the various segments that need to be informed, educated, and convinced.

An important point is that this Conference is going to produce a book. I would urge all who have contributed to actually contribute their papers to this book because having a tangible product is so important. I can't tell you the number of times I've consulted the 1992 proceedings of the Conference in Europe convened and authored by Debbie Pain. It's almost a bible to me and I look at how useful that has been in bringing forth change in Europe. I think this book will have a similar profile in North America because there we have the repository of information. We can use it, we can wack people over the head with it. It's that important.

I think that even though we are scientists, and we like to be "pure," let's be prepared to talk to the media whenever it is possible. We might be misquoted, but then we can go back to the media and correct it. And as Oscar Wilde said, "The worst thing about being quoted, is not being quoted."

I would urge people, also, to consult with politicians. Let your Senators and Congresspersons know your address. It's a good investment in personal relationships. It's a good investment in the political process. And I hope that, in future years, we can see a greater involvement of the scientists with the policy makers; and I'm not saying that to try and get subsequent invitations. John Schulz made a fantastic contribution in terms of describing the complexities of the policy process. We need to have more done on that front. What is going to be the nature of the policy-option horse that we ride? We need discussion around it. And let's face it, saying so does not make it so, we need analysis in that area.

Biography.—**Vernon Thomas, Ph.D.** is an associate professor at The University of Guelph, Canada. His principal interest is the application of science to situations where human activities impact wildlife and the revision of policies related to these management problems. Dr. Thomas is currently leading research on the problem of lead toxicity in wildlife, specifically how it affects waterfowl and loons.

COMMENTARY

LORI VERBRUGGE

Alaska Department of Health and Social Services, Division of Public Health, Section of Epidemiology, 3601 C Street Suite 540, Anchorage, AK 99503, USA.
E-mail: lori.verbrugge@alaska.gov

Transcribed from Conference Expert Panel 15 May 2008.

VERBRUGGE, L. 2009. Commentary. *In* R.T. Watson, M. Fuller, M. Pokras, and W.G. Hunt (Eds.). Ingestion of Lead from Spent Ammunition: Implications for Wildlife and Humans. The Peregrine Fund, Boise, Idaho, USA. DOI 10.4080/ilsa.2009.0320

Key words: Alaska, ammunition, bullets, fish, food, game, hunting, shot, subsistence.

IT IS HUMBLING to be the only public health person here. I'll be talking specifically about the Alaska public, but I think many of my comments and ideas will apply to other rural regions in the lower 48 states as well.

Dr. Titus set the stage well when he talked about how much Alaska Natives rely on fish and game. I often talk to audiences about that situation too, but my talks have a bit of a different focus because my emphasis is on the nutritional importance of these foods. The nutritional value of fish and game in Alaska is so much greater than any alternative foods that Alaska Natives may have. Even though they eat a considerable amount of red meat, it is very lean in comparison to red meat from cows, and much healthier. Our fish, too, are highly nutritious with the omega-3 fatty acids that are so important to healthy hearts and to brain development in children.

These traditional foods are very important to the culture and even the identity of Alaska Native people. Only a small part of Alaska has a road system, which makes it difficult to acquire alternative commercial foods. For example, small planes deliver food at great cost to a store that may be open a few hours a week. The types of food that are there are shelf-stable and non-perishable things like spam and hot dogs that are not nearly as nutritionally sound as traditional foods.

Therefore, one of the main things we often do when we go out and talk to people in Alaska is to actually promote traditional food consumption, and promote hunting and fishing. Essentially, we advocate the maintenance of a traditional way of life. Unfortunately, what we have seen is that, over time, due to a variety of different forces, reliance on traditional foods has slowly declined. As people are switching to market foods, we are seeing an epidemic of heart disease, obesity, and diabetes in our Alaska Native communities that we didn't see previously. So to us, this is a major health concern, and that is why we promote traditional food use.

When we talk to people about this issue of lead in bullets, we need to make sure we are not talking people out of hunting. Also, we need to be very, very sensitive to the economic issues. These villages have few local jobs, except for local government or the local school system, and many people are unemployed. You saw that gasoline is very expensive—market foods are incredibly expensive and limited—and people simply do not have money. The situation is getting better, but there are many villages without running water or sewage systems; people manually dump their wastes into ponds called sewage lagoons. Conditions can be primitive, and money is often scarce. Consequently, asking rural villagers to buy more expensive ammunition could be a big problem. It's not that they are frugal, rather that they simply do not have the money. We need to think of ways to help them acquire better ammunition. I cannot stress enough that we need to be very sensitive about the culture and the economic situation.

I'll tell you what I've learned here. I saw the abstracts of the venison studies come out a few weeks ago. That was the first time I heard about lead fragments contaminating meat. And now, from what I have learned from this conference, I am very concerned. One thing that I am going to do is add this as a potential source when I am doing lead follow-up. I made a decision that I will never hunt with anything but a copper bullet. And I think that if other Alaskans knew, many of them would make the same decision.

Alaskans have a special relationship to the environment. They would not want to contaminate it or their children. I think that if they knew what we know, they would want to change. The challenge is, how do we help them to know what we know? It is difficult because it is hard to reach into remote villages throughout Alaska. We do not have much access to them, and, when we do, we cannot simply tell them what we know as scientists. Instead, we have to work within the context of their way of learning and knowing; namely, they must have the experience themselves. It is very intensive work and our resources are limited. We will have to collaborate with partners to get it done. One thing I'm going to do is to intensify our efforts to screen people for blood lead levels in Alaska, and consider lead ammunition as a potential exposure source. There is interest within the health department to do screening of children, and so we will be testing children for lead. If you do have another conference in three to five years, I'll have results and be able to tell you all about them.

This has been a great conference, but I would like to challenge each of you. Choose one thing to do, put it in your planner, and do it. My one thing is this additional bio-monitoring, and I am also going to add consumption of game meat shot with lead to my list of things I check for.

Biography.—**Lori Verbrugge, Ph.D.**, is the Environmental Public Health Program manager for the Alaska Division of Public Health. She has been working to assess the human health implications of contaminants, including lead, in Alaska's environment since 1997. Dr. Verbrugge has coordinated the development of analytical chemistry capacity and programs for the Alaska Public Health Laboratory, and currently works in the Section of Epidemiology to provide expert toxicological support and policy advice to the Division. Dr. Verbrugge oversees various environmental health programs, including human biomonitoring, blood lead surveillance, subsistence food safety, environmental health research, and an ATSDR cooperative agreement to assess the public health implications of contaminated sites in Alaska.

SUMMARY OF THE MAIN FINDINGS AND CONCLUSIONS OF THE CONFERENCE "INGESTION OF SPENT LEAD AMMUNITION: IMPLICATIONS FOR WILDLIFE AND HUMANS"

IAN NEWTON

Centre for Ecology and Hydrology, Monks Wood, Abbots Ripton, Huntington PE28 2LS, England. E-mail: ine@ceh.ac.uk

Transcribed from Conference Expert Panel 15 May 2008.

NEWTON, I. 2009. Summary of the main findings and conclusions of the conference "Ingestion of spent lead ammunition: Implications for wildlife and humans." *In* R. T. Watson, M. Fuller, M. Pokras, and W. G. Hunt (Eds.). Ingestion of Lead from Spent Ammunition: Implications for Wildlife and Humans. The Peregrine Fund, Boise, Idaho, USA. DOI 10.4080/ilsa.2009.0321

Key words: Ammunition, bullet, food, game, humans, hunting, lead, lead poisoning, policy, regulation, shot, wildlife.

I WOULD LIKE TO THANK our expert panel for their helpful comments, and would now like to make a few general points arising from the conference, on which we might hang part of our discussion.

First, the science: what have we learnt?

1. Many problems of lead in people and wildlife have been long recognized, and some have been addressed. The banning of the use of lead in ammunition over wetlands has greatly reduced the huge mortalities in waterfowl and others which were formerly so apparent. We can assume that this measure has reduced lead consumption by people, and also by some scavengers such as Bald Eagles. However, other uses of lead in ammunition have continued unabated, as a result of which incidental mortalities in many birds and mammals are still staggeringly and unnecessarily high.

2. A second finding concerns the behavior of bullets: the way that lead from lead-based bullets scatters on impact, distributing fragments widely within carcasses, and making it impossible for people or scavenging animals to avoid ingesting lead along with meat. No normal butchery can remove it, so if you eat lead-killed meat, inevitably you eat lead. While this fact may have been known to some for years, new studies have re-emphasized it in a most dramatic way. This is clearly a problem of huge geographical extent, potentially affecting large numbers of people in North America and elsewhere, especially hunters and their families.

3. Lead has been shown to affect adults and children at far lower concentrations in body tissues than formerly thought, and at lower concentrations than current regulations acknowledge. Lead obtained from wild meat shot with lead-based ammunition has been linked with elevated blood lead levels in people.

4. It follows from these findings that we have on our hands a bigger human health problem caused by lead from ammunition than previously recognized, potentially affecting people over most of the continent, but particularly in the many areas where wild game forms a significant part of the human diet.

5. Lead is also causing huge incidental mortality in wildlife. Some species ingest spent gunshot along with grit, while others ingest lead fragments from the carcasses and gut piles of shot animals on which they feed. More than 130 species are regularly affected in this way, and in some species thousands or tens of thousands of individuals die from lead ingestion every year in North America. For most of them, we have no assessment of the effect of this lead-caused mortality on population levels. However, it is clear that in current conditions one species, the California Condor, can no longer maintain a self-sustaining population in its historic range: the mortality from lead-based ammunition obtained from game carcasses and gut-piles well exceeds its natural reproductive rate. While ever lead-based bullets of current design are used as now in game hunting, the condor is unlikely to survive anywhere in North America. It is being kept from extinction in the wild only by a program of intensive conservation management-cum-veterinary care, involving frequent capture and chelation therapy to reduce the blood levels of individuals. This spectacular and charismatic species is one of the largest flying birds in the world, which has inhabited North America for millions of years, long before humans evolved and arrived on the continent. What a pity if it disappeared, lost to all future generations from a problem that could so easily be solved, with benefits for all.

So what needs to be done?

6. On many aspects more research is desirable. We can always benefit from more targeted research. Some specific gaps in knowledge have been identified, and will need to be filled, apart from continually updating our information base. Everyone loves 'recent information,' even though much of it may tell those of us in the field no more than we already know. A major requirement is for a study of the blood levels and impacts of lead in hunters and their families, especially those living in the lower States, outside the northern and native communities already known to be affected. But I believe firmly that we already have sufficient scientifically-robust information to go public with some of the new findings. Indeed, some would argue that it may be irresponsible not to make our findings more widely known, especially those concerning the distribution of lead fragments in meat. We need to spread our information as widely and assiduously as possible.

7. Some of us came to this meeting as wildlife biologists, others from the medical/public health arenas. But if something is to be done quickly about the problems we have discussed here, I believe – as others have already stated – that we will have to act primarily on the basis of human health rather than on wildlife impacts. Some hunters are unaware or unconcerned about secondary poisoning of non-target animals, but they do care about their own health and their own families.

8. Those sectors of society most in need of this new information are the hunters themselves, along with other consumers of lead bullet-killed and lead shot-killed meat. It is to these people that I believe our efforts at disseminating information should be directed with greatest urgency. Public health departments and community food centres could also respond responsibly to new scientific findings. Some public authorities are already working on reducing lead contamination from other sources. There may be no need for advocacy here: just the targeted distribution of unequivocal scientific findings by appropriate messengers.

9. During the course of this conference, two approaches for reducing the use of lead ammunition (in favor of less toxic kinds) have been suggested. One is the 'top down' mandatory approach, in which case the job would be to convince the authorities (state or federal) to introduce appropriate legislation or regulation. This process would inevitably take time, may not be considered as a priority by the authority, and in the end may be unsuccessful. And even if legislation were introduced, the twin problems of compliance and enforcement would remain (never easy in the hunting community). This approach could also be seen as confrontational, perhaps bringing resentment and other undesirable consequences. Nevertheless this approach is currently on trial in condor range in California and, with a different hunting culture, in Hokkaido, Japan.

Legislation was previously used successfully in the banning of DDT and other organo-chlorine pesticides. However, in that case the public were well prepared beforehand, following (among other things) the publication of Rachel Carson's book *Silent Spring*, which produced a ground-swell of public opinion in support of a ban, despite vigorous and dirty attempts by the agrochemical industry to quash it. Similarly with smoking, the public were prepared beforehand before any attempt was made to curb the excesses of the tobacco industry. With lead ammunition, we do not yet have the benefit of an educated and supportive public.

The second approach is 'bottom up,' through which hunters are informed of the human health problems, and are asked for help on the wildlife issues, in the hope that they appreciate the advantages of switching to non-toxic ammunition, and voluntarily change their own behavior. This method seems to have worked with measurable success, at least in the short term, in Arizona, as well as in Germany where hunters are switching to non-lead ammunition to ensure that their saleable product is considered fit for human consumption. However, to solve the wildlife problems, high compliance would be crucial. I can imagine that hunters would change ammunition when hunting to feed themselves and their families, but may be less inclined to do so merely for recreational varmint hunting.

Neither approach (top down or bottom up) is ideal, but nor are they mutually exclusive. My own view is that we need both, beginning now with the bottom up contact with hunters and the wider public, and in the process paving the way for possible legislation at some later date. Whatever approaches are taken, we can expect that obstacles outside our control will get in the way: the availability of non-toxic ammunition, the price of non-toxic metal, negative and ill-informed press comment, including attempts to portray an anti-lead argument as an anti-hunting argument. At the moment we lack any formal system for rapidly countering the ill-informed criticism and opposition that will surely arise in any publicized attempt to replace lead by less toxic alternatives. None of this should deter us from making a start: we have the pioneering experience from Alaska, Arizona, and North Dakota as encouragement.

Biography.—**Professor Ian Newton, D. Phil., D.Sc., FRS, OBE,** Senior Ornithologist (Ret.) Natural Environment Research Council, UK, Chairman of the Board, The Peregrine Fund, Chairman of the Council, the Royal Society for Protection of Birds in the United Kingdom. Ian received his D.Phil. and D.Sc. degrees from Oxford University. He has studied a wide range of bird species, but may be best known for his work on raptors, and his landmark book *Population Ecology of Raptors* first published in 1979. His 27-year study of a Sparrowhawk population nesting in southern Scotland resulted in what many consider to be the most detailed and longest-running study of any population of birds of prey. He is author of more than 300 papers and several books, including *The Sparrowhawk* (1986), *Population Limitation in Birds* (1998), *The Speciation and Biogeography of Birds* (2003), and *The Ecology of Bird Migration* (2007).